Healthy Ageing and Longevity

Volume 22

Rapidly changing demographics worldwide towards increased proportion of the elderly in the population and increased life-expectancy have brought the issues, such as "why we grow old", "how we grow old", "how long can we live", "how to maintain health", "how to prevent and treat diseases in old age", "what are the future perspectives for healthy ageing and longevity" and so on, in the centre stage of scientific, social, political, and economic arena. Although the descriptive aspects of ageing are now well established at the level of species, populations, individuals, and within an individual at the tissue, cell and molecular levels, the implications of such detailed understanding with respect to the aim of achieving healthy ageing and longevity are ever-changing and challenging issues. This continuing success of gerontology, and especially of biogerontology, is attracting the attention of both the well established academicians and the younger generation of students and researchers in biology, medicine, bioinformatics, bioeconomy, sports science, and nutritional sciences, along with sociologists, psychologists, politicians, public health experts, and health-care industry including cosmeceutical-, food-, and lifestyle-industry. Books in this series will cover the topics related to the issues of healthy ageing and longevity. This series will provide not only the exhaustive reviews of the established body of knowledge, but also will give a critical evaluation of the ongoing research and development with respect to theoretical and evidence-based practical and ethical aspects of interventions towards maintaining, recovering and enhancing health and longevity.

Suresh I. S. Rattan • Ceren Gezer
Editors

Supplements and Drugs for Healthy Longevity

Springer

Editors
Suresh I. S. Rattan
Department of Molecular Biology
and Genetics
Aarhus University
Aarhus, Denmark

Ceren Gezer
Department of Nutrition and Dietetic
Eastern Mediterranean University
Famagusta, Cyprus

ISSN 2199-9007 ISSN 2199-9015 (electronic)
Healthy Ageing and Longevity
ISBN 978-3-032-11607-9 ISBN 978-3-032-11608-6 (eBook)
https://doi.org/10.1007/978-3-032-11608-6

This Springer imprint is published by the registered company Springer Nature Switzerland AG
The registered company address is: Gewerbestrasse 11, 6330 Cham, Switzerland

Preface

Ageing research and interventions-landscape is undergoing a transformational change from "anti-ageing" to health maintenance, recovery, and enhancement for healthy longevity. While no single "longevity pill" currently exists, and may never exist, a number of dietary supplements and potential drugs are claimed to promote longevity based on emerging scientific evidence coming mainly from the animal model systems.

The two categories of compounds, supplements and drugs, represent two distinct approaches toward ageing interventions. While supplements are mainly the same natural compounds and conditions, which are a regular part of the lifestyle, drugs are usually the target-oriented formulations against a problem situation. For example, nutrients and micronutrients including vitamins, which are normally obtained from the food and diet, can be taken in extra (supplementary) quantities both to compensate for any deficiency and for attaining any additional benefits. Drugs, on the other hand, are generally defined as natural or synthetic substances used for the diagnosis, treatment, or prevention of diseases or other maladies, and are primarily used in clinical practice. Therefore, the critical question is whether the use of supplements and drugs should be recommended as a strategy to promote healthy ageing and longevity without specifically targeting any diseases.

The aim of this book is to identify and discuss the underlying rationale and the scientific evidence for the use of specific supplements and drugs in terms of their dosage, frequency, and timing of administration. The book is a compilation of 24 chapters contributed by internationally acknowledged and active researchers and practitioners, and is arranged in three parts. The first part includes specific micronutrients, hormones, and pro-, post-, and prebiotics for health and longevity as dietary supplements. The second part includes various plant-based compounds and extracts, including dietary practices and formulations in different cultural folk-medicinal traditions, such as Traditional Chinese Medicine, Ayurveda, and African medicinal systems used for health and longevity. The third part includes synthetic and natural compounds such as metformin, dasatinib, quercetin, curcumin, resveratrol, carotenoids, and calorie restriction mimetics as drugs for health with specific or multiple biochemical targets, including hormesis and hormetins. Each chapter critically examines various supplements and pharmacological agents, evaluates the robustness of scientific evidence underpinning their efficacy, outlines evidence-based dosage recommendations through dietary intake or supplementation, and highlights key research gaps in the context of healthy ageing and longevity.

We hope that graduate and postgraduate students, ageing researchers, medical doctors, and other healthcare personnel, and longevity-medicine-oriented organizations will find this book a valuable source of information and knowledge.

Aarhus, Denmark Suresh I. S. Rattan
Famagusta, Cyprus Ceren Gezer

Contents

Part I
Dietary Supplements

Chapter 1
Vita-Amines, 'Molecules for Life': Do They Live Up to Their Name?

Marios Kyriazis

Abstract Since the discovery of vitamins over 100 years ago, there have been debates and controversies regarding their true efficacy on health. With regards to preventing age-related degenerative conditions, we encounter a considerable degree of exaggerated claims, and at the same time, hidden benefits, which are unveiled by continuing research. In this chapter there is a balanced discussion based on a synthesis of the latest findings regarding all existing vitamins, with an emphasis on improving overall health outcomes. Apart from the well-accepted actions of vitamins, other emerging and less well-known benefits of supplementation will be examined, with emphasis on the latest research. Dosage, specific active configurations of each vitamin (vitamers), the presence or not of deficiency, the aim of the treatment, combinations of vitamins and of other active nutrients, all are parameters which define the final effectiveness of vitamins. It will be shown that most vitamins do have a positive effect for maintaining health and preventing age-related degenerative conditions, particularly in cases of deficiency, when used in combination, in a focused manner, and in accordance with hormetic principles.

Keywords Vitamins · Disease prevention · Age-related degeneration · Nutrition · Hormesis

1.1 Introduction

In 1912, Polish biochemist Casimir Funk published a paper describing four compounds against beriberi, scurvy, pellagra, and rickets (Funk 1912) calling these 'vital-amines' or vitamins. Since then, several more vitamins have been described and divided into water-soluble and fat-soluble classes. Water-soluble vitamins are absorbed quickly but are not stored effectively, needing regular consumption, whereas fat-soluble vitamins are more easily stored in adipose tissue and their consumption is required less frequently. Subsequently, it was clarified that vitamins

M. Kyriazis (✉)
National Gerontology Centre, Larnaca, Cyprus

S. I. S. Rattan, C. Gezer (eds.), *Supplements and Drugs for Healthy Longevity*, Healthy Ageing and Longevity 22, https://doi.org/10.1007/978-3-032-11608-6_1

may occur in several active forms called vitamers, in other words in related forms which have similar biological activities. Examples include Vitamin A (with the vitamers: retinol, retinoic acid and retinyl palmitate), vitamin E (with the vitamers: tocopherols α, β, γ, and δ, and tocotrienols α, β, γ, and δ) and so on. The following discussion will concentrate on the main vitamins without particular reference to each individual vitamer.

1.2 Individual Vitamins

1. Vitamin A (all-trans-retinols, all-trans-retinyl-esters, as well as all-trans-β-carotene and other provitamin A carotenoids)

Since humans cannot synthesize carotenoids, it is necessary to obtain these through food and/or supplementation. It is therefore important to ensure an adequate dietary intake, as these have antioxidant actions, among others. Most fruits and vegetables contain at least some form of carotenoids, but depending on the needs for health prevention, oral supplementation may be required, particularly for eye health. Vitamin A is proven to prevent and treat night blindness and xerophthalmia. In addition, strong clinical evidence supports its role in preventing keratomalacia.

Furthermore, evidence points at other benefits of carotenoids, such as improvements in cardiovascular health and cognitive function, although other studies were not so supportive, and may even increase mortality in high doses (Eggersdorfer and Wyss 2018). There is no strong evidence showing that vitamin A in non-deficient populations is useful as a general 'health-booster', or as an anti-cancer remedy (Gutiérrez-Torres et al. 2024). It is also well accepted that there is a real danger of toxicity with higher intakes of vitamin A, leading to liver problems and an increased fracture risk due to osteoporosis. Overall, the risks of toxicity are higher than potential benefits in non-deficient individuals.

2. Vitamin B (general) and Vitamin B1 (thiamine)

Vitamins of the B group are water-soluble and are mainly derived from food sources although vitamin B12 is synthesized by gut bacteria and isolated from animal foods (Kundra et al. 2022). In this group there are: thiamine (B1), riboflavin (B2), niacin (B3), pantothenic acid (B5), pyridoxine (B6), biotin (B7), folate (B9), and cobalamin (B12). It is worth noting that the following are no longer considered vitamins (they are not believed to be essential in humans): Vitamin B4, B8, B10, B11, B13, B14, B15, B16, and B17.

The risk of vitamin B deficiency increases with advancing age (Ali et al. 2022). This is particularly true with regards to the vitamins B6, B9 and B12, where deficiency has been linked to an increased risk of cognitive decline. One reason for this, is that deficiency of vitamin B leads to increased homocysteine levels and this has been linked not only to age-related cognitive decline but also to an increased cardiovascular risk. High intake of these vitamins reduces this risk. Older people who may have early signs of cognitive decline may benefit from vitamin B supplementation, as these are involved in neurotransmission and repair of myelin, and protect against neurodegeneration.

With regards to vitamin B1 (thiamine), there are certain diseases related to its deficiency and it is important to distinguish these from the diagnosis of dementia. For instance, deficiency of thiamine can manifest as Wernicke's encephalopathy and Korsakoff syndrome, with confusion, memory impairment and ataxia, which may mimic symptoms of dementia. In addition, thiamine deficiency has been linked to reduced α-ketoglutarate and pyruvate dehydrogenase activity, which is found in Alzheimer's disease (Wang et al. 2018, Sakimura et al. 2024). Other studies however did not find evidence of thiamine deficiency in older patients in hospital (Pourhassan et al. 2018) something that reinforces the fact that older patients are not a homogeneous group, but have personalised needs and demands.

3. Vitamin B2 (riboflavin)

Riboflavin has been found to protect against several medical problems such as sepsis, circulatory conditions, certain cancers, and also in reducing unwanted effects of medication (Suwannasom et al. 2020). It has antioxidant and anti-inflammatory benefits.

Oxidation has been implicated as one of the multiple factors causing type 2 diabetes via a nitric oxide dysregulation (Bahadoran et al. 2023). Riboflavin has antioxidant effects and studies show that supplementation with 1.8 mg/day riboflavin may reduce the risk of type 2 diabetes by 44% over 5 years, at least in women (Eshak et al. 2019).

Hormesis is also a factor here. The belief that 'more is always beneficial' is not valid in most physiological situations, where the benefit follows an inverse U relationship with the dose. It was shown that, where low or moderate levels of riboflavin are associated with health benefits, high levels have the opposite effect. In a hormetic way, depletion of riboflavin has been associated with increased lifespan and health span. Riboflavin-depleted animals exhibit improved expression of AMP-activated protein kinase (AMPK), the mitochondrial unfolded protein response, and FOXO/daf-16 (Yerevanian et al. 2022). Low riboflavin levels may be perceived by the organism as a positive stress, and activate the hormetic response, resulting in clear and advantageous effects.

It is therefore recommended to avoid excessive intake of riboflavin supplements, and follow a balanced diet with liver, beef, mushrooms, clams, and other nutritional sources of riboflavin, providing an average intake of 1.6 or 1.8 mg daily, rather than taking additional oral nutritional supplements.

4. Vitamin B3 (niacin)

Niacin has been positively associated with a reduced LDL concentrations and a reduction of oxidative stress (Kim and Park 2022). An interesting study has also found a negative correlation between levels of niacin and hearing loss in older people, suggesting that it could protect against age-related hearing decline (Nakagawa-Nagahama et al. 2023).

One of the most important physical actions in later life is the ability to get up from a chair. This may indicate that the person is likely to retain their independence without help from others. Studies show that supplementation with niacin is associated with an improved ability to get up unaided from a low chair, and is also associated with an improved hand-grip strength (Grootswagers et al. 2021). This may be relevant to healthy older people who do not exercise frequently, and it may be an important health supplement for these people. However, there is still no strong evidence supporting this, as most studies focus on niacin's role in cholesterol metabolism, cardiovascular health, and skin function rather than muscle performance. Nevertheless, some studies highlight the role of niacin supplementation to preventing the risk of sarcopenia (Xiang et al. 2023).

Frequent unwanted effects of niacin supplementation include flushing (redness, warmth, and itching of the face, neck, and chest), nausea, vomiting or diarrhea, and liver toxicity with long-term high doses, leading to elevated liver enzymes or hepatitis. Overall, a targeted use of niacin seems to have more benefits than drawbacks.

5. Vitamin B5 (pantothenic acid)

In a relevant study, pantothenic acid decreased the levels of C-reactive protein, and therefore reduced inflammation (Jung et al. 2017). On the other hand, it was also shown that having high levels of pantothenic acid is associated with increased amyloid-β and consequent deterioration of symptoms of Alzheimer's dementia (AD) (Lee et al. 2018). Studies have also confirmed this, showing that excessive supplementation is associated with increased amyloid beta expression, worsening AD via increased CoA activity and consequent increase in the concentration of pro-inflammatory cytokines and overall inflammation (Lee et al. 2018). Clearly more research is needed in this area, in order to clarify the issue.

An opposite effect was described in another study, where patients with AD were found to have severe and widespread deficiency of pantothenic acid in the brain, particularly the hippocampus and other areas which are normally affected in AD (Xu et al. 2020). These authors conclude that supplementation with vitamin B5 may prevent or even reverse early-stage neurodegeneration. We know that pantothenic acid is involved in the metabolism of acetyl-coenzyme A, which, in turn, is the precursor of acetylcholine.

Pantothenic acid deficiency may also be observed in Parkinson's disease, Huntington's disease, and Lewy body dementia (Scholefield et al. 2024) suggesting that deficiency is closely related to neurodegeneration. There is limited evidence suggesting that supplementation with pantothenic acid may be useful in preventing these neurodegenerative conditions in healthy people, but the studies are rather old (Ousset et al. 2008; Kamphuis and Scheltens 2010; Gu and Scarmeas 2011). Newer research is still needed in this area.

Therefore, the concept of hormesis comes to mind once again. A dose-response situation where a mild intake promotes positive health outcomes, when a higher dose has the opposite effect.

6. Vitamin B6 (pyridoxine, pyridoxal, and pyridoxamine)

Deficiency of pyridoxine may follow poor nutritional habits in older people, although it is also encountered in people who are not malnourished (Kjeldby et al. 2013). Even a mild deficiency of pyridoxine may affect long-chain fatty acid concentration with an increased risk of cerebrovascular disease (Das 2024). Men with an increased triglyceride profile may experience an improvement following supplementation with pyridoxine (Hlais et al. 2012).

It is not surprizing to note that both vitamin B6 deficiency and an excessive B6 intake have been associated, as risk factors, with peripheral neuropathy. While excessive intake of B6 may lead to neurotoxic effects with resulting peripheral neuropathy, the role of B6 deficiency in neuropathy is not completely understood and further studies are needed in this respect. For instance, low vitamin B6 has been seen in patients with peripheral neuropathy, although there is no firm evidence that monotherapy with vitamin B6 has benefits (Muhamad et al. 2023).

The problem with assessing vitamins is that there is overall limited research about each individual vitamin. The bulk of the research concerns combinations of vitamins and their impact on health. For instance, research shows that vitamin B complex, including vitamin B6, influence important aspects of biology such as energy metabolism, DNA repair and immunity, with effects on diverse areas such as neurodegeneration, immune imbalance, inflammation, and ageing in general (Mikkelsen et al. 2024). To mention an example, combination of vitamin B6 and other vitamins, omega-3-fatty acids, and lycopene was found to improve the health of post-menopausal women with deficiency of such factors (Wylenzek et al. 2024). Therefore, a balanced intake, not by supplementation, but from good nutritional sources, is important. Good sources are poultry, fish, potatoes, chickpeas, bananas, and fortified cereals.

7. Vitamin B7 (biotin)

Low levels of biotin have been associated with immune system disorders, and conditions such as diabetes, elevated low-density lipoprotein (LDL), liver disease, skin issues and neuropathies. In addition, chronic alcoholism in older people causes a reduction of biotin in the blood (Watanabe et al. 1998). Older people who are at risk may benefit from biotin supplementation for prevention of these conditions.

Biotin assists in the conversion of glucose into energy and is thus sometimes used in the control of blood glucose, with effects on synthesis of fatty acids, and breakdown of amino acids, although larger clinical trials did not support this. It must be also noted, that high intake of biotin may interfere with laboratory tests and cause false positive or false negative results. Specifically, Ylli et al. (2021) showed that a daily dose of 10 mg can interfere with specific assays for thyroid function such as thyroid stimulating hormone, free thyroid hormone 4, triiodothyronin, and thyroglobulin. There are cases in the literature where this interference was misdiagnosed as actual hyperthyroidism, which was unsuccessfully treated with medication, and which improved after discontinuation of biotin (James et al. 2023).

There is no strong evidence that biotin is useful in improving hair growth or thickness in healthy people, for enhancing cognitive function, or as a general 'energy-booster' (Perry and Butterick 2024). In healthy, well-nourished people, biotin supplementation should be avoided.

8. Vitamin B9 (folic acid and folates)

Folates play an important role in several processes such as DNA synthesis, repair, and methylation. It is known that cigarette smoking may influence folate metabolism, and this, in turn, negatively affects homocysteine concentrations, a fact connected to an increased risk of coronary diseases (Zhou et al. 2018). This, however, does not necessarily mean that increased dietary supplementation with folic acid will reduce the risk. In smokers who had high levels of homocysteine, folic acid supplementation failed to reduce the overall risk of heart disease (Okumura and Tsukamoto 2011). This relationship therefore remains controversial. In addition, a low concertation of folic acid has been linked to inflammation, and this, in turn, is associated with Alzheimer's disease. Supplementation with folic acid may play a beneficial role in this respect (Chen et al. 2016). There is a link between low folate and the risk of age-related degenerative diseases, and this risk may be reduced with an adequate intake of folate (Porter et al. 2016).

Regarding combinations of folate with other nutrients, it was shown that folate with other vitamins of the B group (such as vitamin B12), which are involved in DNA and RNA function, may improve cognitive function (Ma et al. 2019). In this study it was shown that this combination has a significant effect in reducing inflammatory cytokines and thus improves mild cognitive impairment.

On the other hand, excessive intake of folic acid has caused concerns particularly with regards to possible carcinogenesis via alteration of methylation and other mechanisms. It is important to highlight that there are differences between naturally-occurring folate and folic acid in relation to mechanisms of action (Fardous and Heydari 2023). In conclusion, it appears that excessive intake of folic acid has to be avoided, whereas moderate but targeted dietary intake is beneficial.

9. Vitamin B12 (cobalamins)

Deficiency in vitamin B12 has several drawbacks, some of which are subtle and difficult to diagnose. For instance, age-related impairment of the function of vitamin B12 has been associated with cognitive decline (Stover 2010). In addition, there is a link between B12 (and folate) deficiency and cardiovascular disease, neuropathy, bone health and depression (Obeid et al. 2024). Although, obviously, vitamin B 12 supplementation is essential in people who are deficient, prophylactic use of B12 must be considered in high-risk groups such as those with a family history of deficiency, history of gastrectomy, autoimmune diseases, type 1 diabetes, Crohn's disease, strict vegetarian diets etc.

With regards to bowel health, B12 has a beneficial effect of gut microbiota, which has a knock-on effect on inflammatory bowel conditions, immunity, oxidation and other physiological aspects of health (Feng et al. 2024). In a meta-analysis of 95 studies with a total of 46,175 participants it was shown that B vitamins improve the scores of the Mini Mental State examination and slow cognitive decline (Wang et al. 2022). While assessing the effects of vitamin B12 on health, it is necessary to consider the variation that exists between the vitamers of B12, the route of administration, combination with other vitamins or compounds, and even genetic polymorphism and gene methylation patters. Undoubtedly, this is a complex subject which has no immediate answer (Batista et al. 2022).

10. Vitamin C (ascorbic acid and ascorbates)

The role of vitamin C as an antioxidant is well known. Deficiency in vitamin C has been linked to an increased risk of heart disease, diabetes, cancer (Mason et al. 2023), depression, as well as cognitive problems such as those affecting executive function, memory, and concentration. Studies also show the benefits of vitamin C against age-related macular degeneration, cataract, wound healing, and general immunity. Therefore, avoidance of vitamin C deficiency may help protect against age-related degeneration (Sato et al. 2024).

With regards to viral infections (such as the common cold) there is no solid evidence of the effectiveness of vitamin C supplementation, although this area remains debateable (Hemilä and Chalker 2023). The same is true in the case of heart disease, where large-scale studies do not show strong evidence for the effectiveness of vitamin C supplements in non-deficient people.

Therefore, individuals who are not deficient in vitamin C may not experience these protective benefits of additional supplementation. Although older people may have an increased oxidative stress, high doses of vitamin C supplementation have not been found to be linked to any additional benefits, and may interfere with endogenous antioxidant responses. High doses may also increase the risk of kidney stones (especially in those predisposed to oxalate stones), and interfere with certain medications, including chemotherapy and anticoagulants. Having said that, vitamin C supplementation may be relevant in smokers, those with a high risk of cognitive decline, and those who do not consume enough fruits and vegetables even if they have no clinical deficiency.

11. Vitamin D (calciferols)

The role of vitamin D in preventing osteoporosis, and in bone health in general, is well accepted. However, the actions of vitamin D are far wider. Appropriate vitamin D concentrations are necessary for preventing cognitive decline in later life, perhaps through an improvement of blood flow to the brain, acting also as a neuroprotective agent and as a modulator of brain plasticity (Shea et al. 2023). Vitamin D is involved in the production of dopamine, serotonin, and other neurotransmitters, and has thus an effect on mood, depression, and emotional health. Although vitamin D supplementation has shown promise in reducing the risk of age-related cognitive problems (via for example, improving neuroprotection and neurotransmission, as well as brain plasticity), more studies are needed in this respect.

It is relevant to note here that vitamin D deficiency is also found even in countries with abundant sunlight (India, for instance), and it may be worth adding vitamin D into the everyday diet, even if sunlight is not restricted. In this respect, vitamin D was also found to play a positive role in preventing sarcopenia in older people. On the other hand, vitamin D supplementation in healthy people was not found to reduce cardiovascular risks and no significant reduction in overall mortality in healthy individuals. Vitamin D has been used during the covid 19 pandemic in order to reduce the risk of infection, and there was much controversy about this (Contreras-Bolívar et al. 2023).

It is important to highlight that supplementation with vitamin D in older individuals who are deficient in this vitamin, is most likely to have the benefits outlined above (bone, muscle, immunity, cognition). On the contrary, people who are not deficient, are unlikely to obtain further benefits following oral supplementation. For example, a randomized controlled trial found no significant reduction in overall mortality with vitamin D supplementation in healthy individuals and supplementation has not consistently shown benefits in preventing heart disease (Hahn et al. 2022). In addition, excessive intake of vitamin D in non-deficient individuals, carries a risk of toxicity.

12. Vitamin E (tocopherols and tocotrienols)

As with the case of most other vitamins, vitamin E works and cooperates with other factors and its effects cannot clearly be separated from this cooperation. We know that vitamin E has strong antioxidant properties, and that it interferes with the antioxidant action of other factors such as selenium, vitamin C, and glutathione. This antioxidant benefit can be translated into clinical protection against cognitive decline (e.g. it protects against lipid peroxidation in cell membranes, among others).

Age-related inflammation may also be affected by vitamin E status. Increased levels of vitamin E were associated with lower body fat, lower risk of inflammation, and better metabolic function (Szewczyk et al. 2025). In older people, supplementation above commonly-accepted dietary recommendations has been linked to an overall improved immune functions and a consequent reduced risk of infection, possibly through modulation of T cell function (Lewis et al. 2019).

Another, less well-known effect of vitamin E supplementation, is a possible improvement of the health of the human microbiome (Gothandapani and Makpol 2023), and consequently, improvement of some of the effects of ageing. Several other studies confirmed the modulating effects of vitamin E on balancing gut bacteria such as Actinobacteria, Bifidobacterium, Faecalibacterium, Akkermansia and Verrucomicrobia (Pham et al. 2021).

Realistically, it is necessary to examine the combined effects of vitamin E with other nutritional agents in order to get a situation more relevant to everyday life. For example, in a double-blind, placebo-controlled, randomised clinical trial, it was shown that, a combination of vitamin E together with carotenoids (lutein and zeaxanthin) and omega-3 fatty improves memory even in cognitively healthy older people (Power et al. 2022). This is an example of how different nutritional factors work synergistically to achieve a biologically beneficial effects, and it may be irrelevant in clinical practice to examine the benefits of each such factor individually.

13. Vitamin K (phylloquinones, menaquinones, and menadiones)

Vitamin K modulates blood clotting via synthesis of coagulation factors, and is also involved in apoptosis, cell growth and proliferation (Simes et al. 2020). Older people with deficiency in vitamin K are at higher risk of cognitive decline, osteoarthritis, and other low-grade inflammatory conditions. Supplementing the diet with green leafy vegetables such as broccoli, cabbage and kale, as well as foodstuffs from marine sources and fermented food, provides adequate vitamin K, preventing deficiency.

There is evidence that vitamin K has definite health benefits, not only for coagulatory health, but also for cognitive decline and bone health. The effectiveness depends on the form, with vitamin K2 (menaquinone) having more noticeable effects on bone health than the K1 (phylloquinone) variety, with some studies suggesting it may slow bone loss and reduce fracture risk, particularly in postmenopausal women (Liu et al. 2022). With regards to bone health, some scientific evidence suggests that vitamin K has a strong synergistic effect with vitamin D, although overuse of vitamin D promotes hypercalcemia (Moore et al. 2023). On the contrary, no significant side effects are reported with vitamin K, even in higher doses, except for those who are on anticoagulants such as warfarin, where there is a risk of interference.

Some studies suggest that supplementation with vitamin K2 may help reduce arterial stiffness and improve vascular function, but large-scale clinical trials in non-deficient people are missing (Hariri et al. 2021). The same is true in the case of diabetes, where small studies suggest that vitamin K may improve insulin sensitivity. However, the effects appear small and further research is required (Lacombe and Ferron 2024).

Postmenopausal women who are at risk of osteoporosis may benefit from supplementation. Otherwise, if dietary intake is already adequate, extra supplementation may not provide significant extra benefits.

1.3 Discussion and Conclusion

Perhaps one of the most feared conditions in later life is cognitive impairment. Members of the public may decide to take significant steps to prevent or lessen the impact of age-related memory problems and the risk of dementia. Vitamins, minerals, and other nutritional factors play an important role in this, and it appears from available current research that, in cases of deficiency, it is best to take oral supplements of certain vitamins than not take them (Fekete et al. 2023). However, it is important to highlight that there is no single vitamin that has a worthwhile clinical effect on its own. There is a complex interplay between the different nutritional factors, and matters such as bioavailability, interactions, dosage, personal situations, age and sex, physical activity status, side effects and many more, all of which may affect the final clinical result (Table 1.1).

Table 1.1 Summary of the overall evidence of effectiveness of vitamins

Vitamin	Conclusion
Vitamin A	Appropriate for eye health, perhaps also for cardiovascular and cognitive health, but only in deficient individuals. Danger of toxicity
Vitamin B1	Contributes to cognitive, and other aspects of health, if indicated
Vitamin B2	An antioxidant, may help in combination with other B vitamins
Vitamin B3	General effects on cardiovascular and lipid health, if there is a deficiency
Vitamin B5	General effects on metabolism, useful mainly if there is a deficiency
Vitamin B6	Worth sourcing it from natural foodstuffs for multiple benefits, including brain health
Vitamin B7	Best used in addition to other B group vitamins but only if there is deficiency
Vitamin B9	Best used in addition to other B group vitamins if there is deficiency
Vitamin B12	Targeted supplementation is necessary for several aspects of disease prevention
Vitamin C	Generally useful as an integrated plan for protection against age-related diseases
Vitamin D	Useful mostly when there is a proven deficiency
Vitamin E	Antioxidant actions, suitable as a general health supplement (from food sources and perhaps oral form)
Vitamin K	Has several newly-described benefits, best taken from food sources rather than oral supplementation

The combined actions of vitamins B, C, D, and E may reduce the risk of cognitive decline, and of other age-related degenerative conditions. Association with certain other nutrients (minerals, antioxidants, etc) mentioned in other chapters of this book, also plays a part but mainly during a wholistic approach with a generally healthy and balanced nutrition, and a mentally and physically active lifestyle.

The final verdict regarding prevention of age-related degeneration in general is that supplementation with vitamins is clearly worthwhile in cases of vitamin deficiency or in poor nutrition. In cases where there is no deficiency of vitamins, it is generally not advisable to use vitamin supplements, although it may be worth balancing the pros and the cons of additional (oral) supplementation, keeping in mind the hormesis effect: The dose should not be excessive (risk of toxicity) and should not be too low (ineffective and worthless result), but somewhere in the middle. A generally-balanced diet containing foodstuffs high in vitamins (according to the condition that needs preventing) is also worth following.

It is a scientific 'urban myth', that a question mark in the title of a scholarly paper means that the answer to the question being asked is usually 'No'. However, in the case of the title of this chapter, the answer is not a clear 'No', but an equivocal one. Perhaps the best answer would be 'Yes in deficiency, No otherwise'.

References

Ali MA, Hafez HA, Kamel MA, Ghamry HI, Shukry M, Farag MA (2022) Dietary Vitamin B complex: orchestration in human nutrition throughout life with sex differences. Nutrients 14(19):3940. https://doi.org/10.3390/nu14193940

Bahadoran Z, Mirmiran P, Kashfi K, Ghasemi A (2023) Vascular nitric oxide resistance in type 2 diabetes. Cell Death Dis 14(7):410. https://doi.org/10.1038/s41419-023-05935-5

Batista KS, Cintra VM, Lucena PAF, Manhães-de-Castro R, Toscano AE, Costa LP, Queiroz MEBS, de Andrade SM, Guzman-Quevedo O, Aquino JS (2022) The role of vitamin B12 in viral infections: a comprehensive review of its relationship with the muscle-gut-brain axis and implications for SARS-CoV-2 infection. Nutr Rev 80(3):561–578. https://doi.org/10.1093/nutrit/nuab092

Chen H, Liu S, Ji L, Wu T, Ji Y, Zhou Y, Zheng M, Zhang M, Xu W, Huang G (2016) Folic acid supplementation mitigates Alzheimer’s disease by reducing inflammation: a randomized controlled trial. Mediators Inflamm 2016:5912146. https://doi.org/10.1155/2016/5912146

Contreras-Bolívar V, García-Fontana B, García-Fontana C, Muñoz-Torres M (2023) Vitamin D and COVID-19: where are we now? Postgrad Med 135(3):195–207. https://doi.org/10.1080/00325481.2021.2017647

Das UN (2024) Pyridoxine, essential fatty acids, and protection against doxorubicin-induced cardiotoxicity. J Biochem Mol Toxicol 38(1):e23639. https://doi.org/10.1002/jbt.23639

Eggersdorfer M, Wyss A (2018) Carotenoids in human nutrition and health. Arch Biochem Biophys 652:18–26. https://doi.org/10.1016/j.abb.2018.06.001

Eshak ES, Iso H, Muraki I, Tamakoshi A (2019) Among the water-soluble vitamins, dietary intakes of vitamins C, B2 and folate are associated with the reduced risk of diabetes in Japanese women but not men. Br J Nutr 121:1357–1364. https://doi.org/10.1017/S000711451900062X

Fardous AM, Heydari AR (2023) Uncovering the hidden dangers and molecular mechanisms of excess folate: a narrative review. Nutrients 15(21):4699. https://doi.org/10.3390/nu15214699

Fekete M, Lehoczki A, Tarantini S, Fazekas-Pongor V, Csípő T, Csizmadia Z, Varga JT (2023) Improving cognitive function with nutritional supplements in aging: a comprehensive narrative review of clinical studies investigating the effects of vitamins, minerals, antioxidants, and other dietary supplements. Nutrients 15(24):5116. https://doi.org/10.3390/nu15245116

Feng C, Yan J, Luo T, Zhang H, Zhang H, Yuan Y, Chen Y, Chen H (2024) Vitamin B12 ameliorates gut epithelial injury via modulating the HIF-1 pathway and gut microbiota. Cell Mol Life Sci 81(1):397. https://doi.org/10.1007/s00018-024-05435-5

Funk C (1912) The etiology of the deficiency diseases. Beri-beri, polyneuritis in birds, epidemic dropsy, scurvy, experimental scurvy in animals, infantile scurvy, ship beri-beri, pellagra. J State Med 20:341–368

Gothandapani D, Makpol S (2023) Effects of vitamin E on the gut microbiome in ageing and its relationship with age-related diseases: a review of the current literature. Int J Mol Sci 24(19):14667. https://doi.org/10.3390/ijms241914667

Grootswagers P, Mensink M, Berendsen AAM, Deen CPJ, Kema IP, Bakker SJL, Santoro A, Franceschi C, Meunier N, Malpuech-Brugère C, Bialecka-Debek A, Rolf K, Fairweather-Tait S, Jennings A, Feskens EJM, de Groot LCPGM (2021) Vitamin B-6 intake is related to physical performance in European older adults: results of the new dietary strategies addressing the specific needs of the elderly population for healthy aging in Europe (NU-AGE) study. Am J Clin Nutr 113(4):781–789. https://doi.org/10.1093/ajcn/nqaa368

Gu Y, Scarmeas N (2011) Dietary patterns in Alzheimer’s disease and cognitive aging. Curr Alzheimer Res 8:510–519. https://doi.org/10.2174/156720511796391836

Gutiérrez-Torres DS, Kim S, Albanes D, Weinstein SJ, Inoue-Choi M, Albert PS, Freedman ND (2024) Changes in smoking use and subsequent lung cancer risk in the alpha-tocopherol, beta-carotene cancer prevention study. J Natl Cancer Inst 116(6):895–901. https://doi.org/10.1093/jnci/djae012

Hahn J, Cook NR, Alexander EK, Friedman S, Walter J, Bubes V, Kotler G, Lee IM, Manson JE, Costenbader KH (2022) Vitamin D and marine omega 3 fatty acid supplementation and incident autoimmune disease: VITAL randomized controlled trial. BMJ 376:e066452. https://doi.org/10.1136/bmj-2021-066452

Hariri E, Kassis N, Iskandar JP, Schurgers LJ, Saad A, Abdelfattah O, Bansal A, Isogai T, Harb SC, Kapadia S (2021) Vitamin K2-a neglected player in cardiovascular health: a narrative review. Open Heart 8(2):e001715. https://doi.org/10.1136/openhrt-2021-001715

Hemilä H, Chalker E (2023) Vitamin C reduces the severity of common colds: a meta-analysis. BMC Public Health 23(1):2468. https://doi.org/10.1186/s12889-023-17229-8

Hlais S, Abou Reslan DR, Sarieddine HK, Nasreddine L, Taan G, Azar S, Obeid OA (2012) Effect of lysine, vitamin B(6), and carnitine supplementation on the lipid profile of male patients with hypertriglyceridemia: a 12-week, open-label, randomized, placebo-controlled trial. Clin Ther 34:1674–1682. https://doi.org/10.1016/j.clinthera.2012.06.019

James A, Stalan J, Kuzhively J (2023) Biotin induced biochemical hyperthyroidism: a case report and review of the literature. J Med Case Rep 17(1):266. https://doi.org/10.1186/s13256-023-04002-z

Jung S, Kim MK, Choi BY (2017) The long-term relationship between dietary pantothenic acid (vitamin B5) intake and C-reactive protein concentration in adults aged 40 years and older. Nutr Metab Cardiovasc Dis 27:806–816. https://doi.org/10.1016/j.numecd.2017.05.008

Kamphuis PJ, Scheltens P (2010) Can nutrients prevent or delay onset of Alzheimer's disease? J Alzheimers Dis 20:765–775. https://doi.org/10.3233/JAD-2010-091558

Kim C, Park K (2022) Dietary niacin intake and risk of dyslipidemia: a pooled analysis of three prospective cohort studies. Clin Nutr 41(12):2749–2758. https://doi.org/10.1016/j.clnu.2022.10.018

Kjeldby IK, Fosnes GS, Ligaarden SC, Farup PG (2013) Vitamin B6 deficiency and diseases in elderly people-a study in nursing homes. BMC Geriatr 13:13. https://doi.org/10.1186/1471-2318-13-13

Kundra P, Geirnaert A, Pugin B, Morales Martinez P, Lacroix C, Greppi A (2022) Healthy adult gut microbiota sustains its own vitamin B12 requirement in an in vitro batch fermentation model. Front Nutr 9:1070155. https://doi.org/10.3389/fnut.2022.1070155

Lacombe J, Ferron M (2024) Vitamin K-dependent carboxylation in β-cells and diabetes. Trends Endocrinol Metab 35(7):661–673. https://doi.org/10.1016/j.tem.2024.02.006

Lee JH, Ahn SY, Lee HA, Won KS, Chang HW, Oh JS, Kim HW (2018) Dietary intake of pantothenic acid is associated with cerebral amyloid burden in patients with cognitive impairment. Food Nutr Res 62. https://doi.org/10.29219/fnr.v62.1415

Lewis ED, Meydani SN, Wu D (2019) Regulatory role of vitamin E in the immune system and inflammation. IUBMB Life 71(4):487–494. https://doi.org/10.1002/iub.1976

Liu Y, He Y, He B, Kong L (2022) The anti-osteoporosis effects of vitamin K in postmenopausal women. Curr Stem Cell Res Ther 17(2):186–192. https://doi.org/10.2174/1574888X16666210512020103

Ma F, Zhou X, Li Q, Zhao J, Song A, An P, Du Y, Xu W, Huang G (2019) Effects of folic acid and vitamin B12, alone and in combination on cognitive function and inflammatory factors in the elderly with mild cognitive impairment: a single-blind experimental design. Curr Alzheimer Res 16(7):622–632. https://doi.org/10.2174/1567205016666190725144629

Mason SA, Parker L, van der Pligt P, Wadley GD (2023) Vitamin C supplementation for diabetes management: a comprehensive narrative review. Free Radic Biol Med 194:255–283. https://doi.org/10.1016/j.freeradbiomed.2022.12.003

Mikkelsen K, Trapali M, Apostolopoulos V (2024) Role of vitamin B in healthy ageing and disease. In: Subcell Biochem, vol 107, pp 245–268. https://doi.org/10.1007/978-3-031-66768-8_12

Moore AE, Dulnoan D, Voong K, Ayis S, Mangelis A, Gorska R, Harrington DJ, Tang JCY, Fraser WD, Hampson G (2023) The additive effect of vitamin K supplementation and bisphosphonate on fracture risk in post-menopausal osteoporosis: a randomised placebo-controlled trial. Arch Osteoporos 18(1):83. https://doi.org/10.1007/s11657-023-01288-w

Muhamad R, Akrivaki A, Papagiannopoulou G, Zavridis P, Zis P (2023) The role of vitamin B6 in peripheral neuropathy: a systematic review. Nutrients 15(13):2823. https://doi.org/10.3390/nu15132823

Nakagawa-Nagahama Y, Igarashi M, Miura M, Kashiwabara K, Yaku K, Fukamizu Y, Sato T, Sakurai T, Nakagawa T, Kadowaki T, Yamauchi T (2023) Blood levels of nicotinic acid negatively correlate with hearing ability in healthy older men. BMC Geriatr 23(1):97. https://doi.org/10.1186/s12877-023-03796-3

Obeid R, Andrès E, Češka R, Hooshmand B, Guéant-Rodriguez RM, Prada GI, Sławek J, Traykov L, Ta Van B, Várkonyi T, Reiners K, The Vitamin B Consensus Panelists Group (2024) Diagnosis, treatment and long-term management of vitamin B12 deficiency in adults: a Delphi expert consensus. J Clin Med 13(8):2176. https://doi.org/10.3390/jcm13082176

Okumura K, Tsukamoto H (2011) Folate in smokers. Clin Chim Acta 412(7–8):521–526. https://doi.org/10.1016/j.cca.2011.01.003

Ousset PJ, Nourhashemi F, Reynish E, Vellas B (2008) Nutritional status is associated with disease progression in very mild Alzheimer disease. Alzheimer Dis Assoc Disord 22:66–71. https://doi.org/10.1097/WAD.0b013e31815a9dbb

Perry CA, Butterick TA (2024) Biotin. Adv Nutr 15(7):100251. https://doi.org/10.1016/j.advnut.2024.100251

Pham VT, Fehlbaum S, Seifert N, Richard N, Bruins MJ, Sybesma W, Rehman A, Steinert RE (2021) Effects of colon-targeted vitamins on the composition and metabolic activity of the human gut microbiome- a pilot study. Gut Microbes 13:1875774. https://doi.org/10.1080/19490976.2021.1875774

Porter K, Hoey L, Hughes CF, Ward M, McNulty H (2016) Causes, consequences and public health implications of low B-vitamin status in ageing. Nutrients 8(11):725. https://doi.org/10.3390/nu8110725

Pourhassan M, Biesalski HK, Angersbach B, Lueg G, Klimek C, Wirth R (2018) Prevalence of thiamine deficiency in older hospitalized patients. Clin Interv Aging 13:2247–2250. https://doi.org/10.2147/CIA.S183102

Power R, Nolan JM, Prado-Cabrero A, Roche W, Coen R, Power T, Mulcahy R (2022) Omega-3 fatty acid, carotenoid and vitamin E supplementation improves working memory in older adults: a randomised clinical trial. Clin Nutr 41(2):405–414. https://doi.org/10.1016/j.clnu.2021.12.004

Sakimura K, Kawai T, Nashida R, Ishida Y, Harada K, Suzuki T, Okuma C, Cole GM (2024) A novel PDHK inhibitor restored cognitive dysfunction and limited neurodegeneration without affecting amyloid pathology in 5xFAD mouse, a model of Alzheimer's disease. Alzheimers Res Ther 16(1):197. https://doi.org/10.1186/s13195-024-01552-2

Sato A, Kondo Y, Ishigami A (2024) The evidence to date: implications of l-ascorbic acid in the pathophysiology of aging. J Physiol Sci 74(1):29. https://doi.org/10.1186/s12576-024-00922-7

Scholefield M, Church SJ, Xu J, Patassini S, Cooper GJS (2024) Localized pantothenic acid (vitamin B5) reductions present throughout the dementia with Lewy bodies brain. J Parkinsons Dis 14(5):965–976. https://doi.org/10.3233/JPD-240075

Shea MK, Barger K, Dawson-Hughes B, Leurgans SE, Fu X, James BD, Holland TM, Agarwal P, Wang J, Matuszek G, Heger NE, Schneider JA, Booth SL (2023) Brain vitamin D forms, cognitive decline, and neuropathology in community-dwelling older adults. Alzheimers Dement 19(6):2389–2396. https://doi.org/10.1002/alz.12836

Simes DC, Viegas CSB, Araújo N, Marreiros C (2020) Vitamin K as a diet supplement with impact in human health: current evidence in age-related diseases. Nutrients 12(1):138. https://doi.org/10.3390/nu12010138

Stover PJ (2010) Vitamin B12 and older adults. Curr Opin Clin Nutr Metab Care 13:24–27. https://doi.org/10.1097/MCO.0b013e328333d157

Suwannasom N, Kao I, Pruß A, Georgieva R, Bäumler H (2020) Riboflavin: the health benefits of a forgotten natural vitamin. Int J Mol Sci 21(3):950. https://doi.org/10.3390/ijms21030950

Szewczyk K, Bryś J, Brzezińska R, Górnicka M (2025) Nutritional status of vitamin E and its association with metabolic health in adults. Nutrients 17(3):408. https://doi.org/10.3390/nu17030408

Wang C, Fei G, Pan X, Sang S, Wang L, Zhong C, Jin L (2018) High thiamine diphosphate level as a protective factor for Alzheimer's disease. Neurol Res 40:658–665. https://doi.org/10.1080/01616412.2018.1460704

Wang Z, Zhu W, Xing Y, Jia J, Tang Y (2022) B vitamins and prevention of cognitive decline and incident dementia: a systematic review and meta-analysis. Nutr Rev 80(4):931–949. https://doi.org/10.1093/nutrit/nuab057

Watanabe T, Yasumura S, Shibata H, Fukui T (1998) Biotin status and its correlation with other biochemical parameters in the elderly people of Japan. J Am Coll Nutr 17:48–53. https://doi.org/10.1080/07315724.1998.10720454

Wylenzek F, Bühling KJ, Laakmann E (2024) A systematic review on the impact of nutrition and possible supplementation on the deficiency of vitamin complexes, iron, omega-3-fatty acids, and lycopene in relation to increased morbidity in women after menopause. Arch Gynecol Obstet 310(4):2235–2245. https://doi.org/10.1007/s00404-024-07555-6

Xiang S, Li Y, Li Y, Zhang J, Pan W, Lu Y, Liu S (2023) Increased dietary niacin intake improves muscle strength, quality, and glucose homeostasis in adults over 40 years of age. J Nutr Health Aging 27(9):709–718. https://doi.org/10.1007/s12603-023-1967-0

Xu J, Patassini S, Begley P, Church S, Waldvogel HJ, Faull RLM, Unwin RD, Cooper GJS (2020) Cerebral deficiency of vitamin B5 (d-pantothenic acid; pantothenate) as a potentially-reversible cause of neurodegeneration and dementia in sporadic Alzheimer's disease. Biochem Biophys Res Commun 527(3):676–681. https://doi.org/10.1016/j.bbrc.2020.05.015

Yerevanian A, Murphy LM, Emans S, Zhou Y, Ahsan FM, Baker D, Li S, Adedoja A, Cedillo L, Stuhr NL, Gnanatheepam E, Dao K, Jain M, Curran SP, Georgakoudi I, Soukas AA (2022) Riboflavin depletion promotes longevity and metabolic hormesis in Caenorhabditis elegans. Aging Cell 21(11):e13718. https://doi.org/10.1111/acel.13718

Ylli D, Soldin SJ, Stolze B, Wei B, Nigussie G, Nguyen H, Mendu DR, Mete M, Wu D, Gomes-Lima CJ, Klubo-Gwiezdzinska J, Burman KD, Wartofsky L (2021) Biotin interference in assays for thyroid hormones, Thyrotropin and Thyroglobulin. Thyroid 31(8):1160–1170. https://doi.org/10.1089/thy.2020.0866

Zhou Z, Li J, Yu Y, Li Y, Zhang Y, Liu L, Song Y, Zhao M, Wang Y, Tang G, He M, Xu X, Cai Y, Dong Q, Yin D, Huang X, Cheng X, Wang B, Hou FF, Wang X, Qin X, Huo Y (2018) Effect of smoking and folate levels on the efficacy of folic acid therapy in prevention of stroke in hypertensive men. Stroke 49(1):114–120. https://doi.org/10.1161/STROKEAHA.117.018273

Chapter 2
Micronutrients and Nutraceuticals for Health and Longevity

Claudio Santelli

Abstract Micronutrients consist of vitamins and minerals, which are essential for basic metabolic reactions in the human body, while nutraceuticals are dietary supplements used for health purposes, other than just nutrition. Recent studies have evaluated the favorable relationship between micronutrients, nutraceuticals, health and longevity. In this chapter, some of the micronutrients and nutraceuticals that are associated with greater longevity in health have been discussed. Among macro-minerals, magnesium (Mg), among micro-minerals (or trace elements), zinc (Zn) and selenium (Se) have been discussed. In addition, ergothioneine (ERGO), a sulfur-containing amino acid derived from food, which has been considered as a presumed longevity vitamin, and pyrroloquinoline-quinone (PQQ), a biofactor with vitamin-like action associated with biological processes such as growth, aging, reproduction, and mitochondrial genesis, has been discussed. Among nutraceuticals derived from foods that offer health benefits, beyond simple nutrition, omega-3 (ω-3) fatty acids are among those most frequently associated with benefits for healthy longevity.

Keywords Aging · Health · Biofactors · Macrominerals · Micronutrients

2.1 Introduction

The longevity of populations around the world has increased, but there is still little evidence that this increase is accompanied by an equally long period of good health (Beard et al. 2016). While extended life expectancy is a remarkable achievement, morbidity rates are expected to increase further as aging is the predominant risk factor for age-related diseases. The commitment of scientific research is to promote an extension of healthy longevity (Grootswagers et al. 2024) by implementing interventions that slow down molecular and cellular decline due to the aging process

C. Santelli (✉)
Santelli Medical Office, Viale Europa, Rapagnano, FM, Italy
e-mail: claudio.santelli.kqn4@mc.omceo.it

S. I. S. Rattan, C. Gezer (eds.), *Supplements and Drugs for Healthy Longevity*, Healthy Ageing and Longevity 22, https://doi.org/10.1007/978-3-032-11608-6_2

(Farrelly 2024). As we age, senescent cells accumulate in numerous tissues and organs where they secrete a range of inflammatory cytokines, chemokines, proteases, and tissue factors, collectively known as senescence-associated secretory phenotype (SASP) (Islam et al. 2023) that causes chronic inflammation and tissue dysfunction (Zhang et al. 2023).

Aging is a gradual process, accompanied by a general decline in cell function (Izadi et al. 2024) and is the most important risk factor for chronic diseases and frailties related to limitation or loss of function. It is estimated that by 2050 there will be over two billion people aged >60 (Farrelly 2024). In general, delaying aging is a more effective method than treating the diseases associated with it (Izadi et al. 2024). The relationship between micronutrients and health represents an area of growing interest and research in the medical and nutritional literature. Nutrition, by improving the bioavailability of micronutrients and nutraceuticals, plays a fundamental role in promoting optimal well-being (Townsend et al. 2023). The use of micronutrients and nutraceuticals with senolytic action, reducing the increase in age-related β-galactosidase associated with senescence (Islam et al. 2023) aims to counteract cellular aging by extending the duration of health in organisms (Zhang et al. 2023). Oral frailty, due in part to periodontal disease, which tends to increase with age, precedes malnutrition and physical frailty. For most non-communicable diseases, there is a direct relationship with periodontal diseases. Micronutrients can support oral gum health with vitamins A, D, and C, dentin with vitamins D and C, and mucosal health with vitamins A, B_2, B_3, B_6, and B_{12} (Inui et al. 2021). In addition to vitamins, essential minerals such as calcium, zinc, iron and magnesium are important for periodontal health. Calcium for the maintenance of bone density and structure, zinc for its role in the host's immune response to periodontal pathogens, iron for supporting the immune system and inflammation control (Berg et al. 2024), and magnesium for its role as a mineral improving bone structure (Rajanna et al. 2024). Further help can be obtained from nutrigenetics, which studies the interaction between genes and nutrients, allowing us to go beyond generic recommendations to find out how the body, with its unique genetic makeup, responds to and metabolizes different foods and understand which nutrients can enhance health and optimize the physical and mental performance of the individual. Lately, thanks to the significant decrease in genotyping and sequencing costs, nutrigenetics tests are increasingly used to obtain precise and personalized nutritional indications (Floris et al. 2020). The study of aging and human diseases can also be understood in a precise and personalized way through metabolomics, which allows an in-depth characterization of small molecule metabolites in biological systems at particular times and conditions, with these data it is then possible to evaluate the final response of the body to genomic, transcriptomic, proteomic and epigenetic alterations (Abdullah et al. 2024).

In this chapter, the micronutrients and nutraceuticals that are associated with increased longevity and optimal health have been discussed (Inui et al. 2021):

1. Magnesium (Mg)
2. Zinc (Zn)
3. Selenium (Se)

4. Ergothioneine (ERGO)
5. Pyrroloquinoline-quinone (PQQ)
6. Omega-3 (ω-3)

2.2 Definition of Micronutrients

Micronutrients consist of vitamins and minerals, which are essential for basic metabolic reactions in the human body. The main function of micronutrients is their catalytic effect in enzyme systems, both as cofactors and as components of metalloenzymes. They perform antioxidant activity, regulate cellular immunity and wound healing, influence enzymatic activities in intermediate metabolism, and regulate the genomic expression of structural proteins. Micronutrients, with their antioxidant activity, can neutralize reactive oxygen species (ROS) and reactive nitrogen species (RNS) of free radicals (Lahoda Brodska et al. 2023). Micronutrients are also necessary as immune system support, especially vitamins C and D and zinc (Gombart et al. 2020).

2.2.1 Vitamins

Vitamins are organic compounds that the body cannot synthesize in sufficient amounts, apart from a small amount (vitamin D), so they must be taken through the diet. They fall into two main groups (Lahoda Brodska et al. 2023):

1. *Water-soluble vitamins:*

   These vitamins, such as vitamin C and B vitamins, dissolve in water and are easily eliminated from the body. It is important to get them regularly through food. B vitamins include B_1 (thiamine), B_2 (riboflavin), B_3 (niacin), B_5 (pantothenic acid), B_6 (pyridoxine), B_7 (biotin), B_9 (folic acid), and B_{12} (cyanocobalamin). These vitamins are important for energy metabolism, nervous system health, and red blood cell production (Lahoda Brodska et al. 2023).
2. *Fat-soluble vitamins:*

   These are vitamins A, D, E, K, which dissolve in fats and can be stored in the body. They are important for various functions, such as skin health, vision, and blood clotting (Lahoda Brodska et al. 2023).

2.2.2 Minerals

Minerals, being inorganic elements, cannot be synthesized endogenously and must be taken through diet or supplementation (Fekete et al. 2023). Minerals are also divided into two categories:

1. *Macro-minerals:*

   These minerals, such as calcium, potassium, magnesium, sodium (Makevic et al. 2023) and phosphorus (Wagner 2024) are needed in larger amounts (Makevic et al. 2023).
2. *Micro-minerals* or *trace elements*:

   These minerals, such as iron, zinc, copper, manganese, cobalt, molybdenum, selenium and iodine (Makevic et al. 2023), chromium, chloride, sulfur (Ames 2018), are needed in smaller quantities as, although essential for health and physiological functions, trace elements in high concentrations can exert toxicity (Makevic et al. 2023).

2.3 Definition of Nutraceuticals

The term nutraceutical was coined by Dr. Stephen De Felice from "Nutrition"and "Pharmaceutical" in 1989 (Kalra 2003). Nutraceuticals are dietary supplements (Sachdeva et al. 2020) that, in addition to nutrition, have a physiological benefit to improve health, delay the aging process, prevent chronic diseases, increase life expectancy (Nasri et al. 2014) based on scientific evidence (Golubnitschaja et al. 2024).

Difference between micronutrients and nutraceuticals: Micronutrients, i.e., vitamins and minerals, are essential components of the diet (de Man et al. 2024), can be present in foods and are also extracted or synthetically produced to form supplements, while nutraceuticals are dietary supplements that are used for health purposes, other than nutrition (Nasri et al. 2014).

2.3.1 Relationship with Aging and Longevity

The relationship between micronutrients, nutraceuticals, aging and healthy longevity is a topic of current interest as many countries struggle with malnutrition while others have age-related diseases due to excessive consumption of foods rich in calories but lacking in micronutrients and other nutrients essential for optimal human health. Improper nutrition is a widespread problem all over the world and affects people from all walks of life and ages. Nutritional deficiencies affect health and longevity as one in five deaths worldwide can be attributed to a suboptimal diet (Townsend et al. 2023). Adequate micronutrient intake is crucial not only for overall well-being, but also to support healthy aging and potentially increase longevity, as they act in a coordinated manner to maintain the cellular and tissue physiological functions of organs (Razzaque 2018). Recent studies have evaluated the favorable relationship between micronutrients, nutraceuticals, healthy aging and longevity, many of them play a dual role for both survival and longevity. Among the most studied, as alleged longevity vitamins, are the fungal antioxidant ergothioneine, the bacterial metabolites pyrroloquinoline quinone (PQQ) and queuin, the plant antioxidant carotenoids lutein, zeaxanthin, lycopene, alpha and beta carotene, beta cryptoxanthin and the marine carotenoid astaxanthin (Ames 2018).

2.4 Micronutrients and Nutraceuticals

2.4.1 Magnesium (Mg)

Magnesium (Mg) plays a role in numerous biological processes, including oxidative phosphorylation, energy production, glycolysis, and the synthesis of proteins, nucleic acids, and adenosine triphosphate (ATP) to form Mg ATP. Mg regulates cellular homeostasis, organ physiology, and metabolic pathways and is a cofactor in over 600 enzymatic reactions (Barbagallo et al. 2021). About 20% of people consistently take lower than recommended amounts of Mg while supplementing it contributes to the maintenance of robust bones with regular mineral density (Rondanelli et al. 2021). Mg is essential for maintaining overall cardiovascular health. Its deficiency is associated with several disorders such as metabolic syndrome and skeletal deformities. In the human body, the Mg content is about 24–29 g (Barbagallo et al. 2021) and is localized for 40% in the intracellular space and 60% in the bones and teeth (Razzaque 2018), only less than 1% of the total Mg is extracellular (Barbagallo et al. 2021). In bone physiology, magnesium plays a role in the formation, turnover, and maintenance of skeletal structure, via its effects on parathyroid hormone and vitamin D metabolism (Adomako and Yu 2024). Food sources rich in Mg include almonds, bananas, black beans, broccoli, leafy greens (spinach), whole grains, brown rice, oatmeal, dried fruits, cashews, seeds (flax, pumpkin, sesame, sunflower), soybeans, sweet corn, tofu. Mg taken orally is mainly absorbed in the small intestine and stored in the bone as a mineral. Excess Mg is excreted via the kidneys (Razzaque 2018).

Vitamin D can positively influence the absorption of Mg and support its metabolism (Razzaque 2018). The optimal requirement of Mg with food is 420 mg/day for men and 320 mg/day for women but, in some physiological conditions such as pregnancy, aging, sports physical activity and in some pathological conditions, greater needs may be necessary (Barbagallo et al. 2021). Mg plays an important role in many physiological functions, but its absorption is often insufficient due to its low content in the modern Western diet (Fiorentini et al. 2021), due to Mg-poor farmland and food processing techniques that further reduce its content in the most common foods (Adomako and Yu 2024). With aging, several changes occur in Mg metabolism due to reduced Mg intake, kidney loss and altered intestinal absorption. Symptoms are usually absent or non-specific such as asthenia, sleep disturbances, cognitive impairments and hyperemotionality which are common in the elderly and can often be confused with age-related symptoms (Barbagallo et al. 2021). Given the pathophysiological and clinical importance of Mg and the possible effects of its deficiency on several human pathologies and on the increase in free radicals that lead to various chronic age-related disorders, the intake of Mg should be taken into account in the correction of functional deficits, as optimal levels of Mg throughout life can help prevent oxidative stress and chronic diseases associated with aging (Barbagallo et al. 2021). Mg supplementation is associated with improved cognitive function and a reduced risk of dementia in older adults, as it protects nerve cells (Fekete et al. 2023).

2.4.2 Zinc (Zn)

Zinc (Zn) is an essential micronutrient for human metabolism, catalyzing enzymes, facilitating protein folding and modulating gene expression (Saper and Rash 2009). In biology, Zn plays essential roles in all metabolic pathways. Some examples in humans include alkaline phosphatase, carbonic anhydrase, RNA and DNA polymerases, and alcohol dehydrogenase. Zinc-finger proteins control the transcription of DNA into RNA. Zn plays an important role in the synthesis of proteins and nucleic acids to promote body growth and wound healing, and participates in the antioxidant defense system (Berger et al. 2024). Zn is the second most abundant naturally occurring micronutrient after iron, the main dietary sources of Zn include meat, seafood, whole grains, and oilseeds. The recommended daily allowance (RDA) of Zn, to maintain overall health, is 11 mg/day for men and 8 mg/day for women. It is an important trace element for regulating blood glucose levels and reducing the risks of developing diabetic complications (Ahmad et al. 2024). Zn deficiencies can lead to growth retardation, delayed sexual development and bone maturation, alopecia, skin rash, difficult wound healing, reduced immune function, diarrhea, reduced taste and smell (Berger et al. 2024). Aging is an inevitable biological process with gradual biochemical and physiological changes and increased susceptibility to disease. Some nutritional factors such as Zn, niacin, selenium can modulate these changes by promoting healthy aging, improving immune functions, metabolic homeostasis and antioxidant defense (Mocchegiani et al. 2008). Zn is essential for the functionality of the enzyme superoxide dismutase, which promotes the elimination of free radicals, also supports cognitive functions and memory (Fekete et al. 2023).

2.4.3 Selenium (Se)

Selenium (Se) is an essential trace element with antioxidant and anti-inflammatory capacity that has been associated, in experimental studies, with beneficial effects on appetite control, regulation of the gut microbiota and control of anabolic-catabolic balance, the results suggest that adequate dietary intake of Se is needed to prevent undernutrition in the elderly (García-Esquinas et al. 2023). Selenonein has a strong radical removal ability (Cordell and Lamahewage 2022). The Se is mainly obtained through food sources such as seafood, dairy, meat, eggs, and Brazil nuts. The Se content in cultivated foods is highly dependent on soil quality and the availability of Se during cultivation, just as the nutritional quality of animal-based foods depends to a large extent on the quality of the forage provided during growth (Demircan et al. 2024). The recommended daily dose is 55 mcg/day (Berg et al. 2024). Se has a wide range of pleiotropic effects, influences redox homeostasis, thyroid hormone metabolism, and protects against oxidative stress and inflammation. Serum Se levels are reduced in the elderly and are associated with reduced survival (Giovannini et al. 2018).

Se is a key component of antioxidant enzymes such as glutathione peroxidase. These enzymes protect the brain from oxidative stress, which leads to both aging and neurodegenerative diseases. Oxidative stress, in fact, damages neurons and impairs cognitive function. Higher levels of Se in the blood are associated with a reduced risk of cognitive decline and certain types of cancer. The self is also crucial for the proper functioning of the thyroid and plays a critical role in the immune system (Fekete et al. 2023). Se plays a role in modulating inflammatory bowel disease by promoting the expression of tight junction proteins, such as ZO-1 and occludin, which are critical for strengthening gut barrier function (Liu et al. 2024). A cohort study, aimed at evaluating the association between plasma minerals and health duration in older adults aged 90 years and older, showed that higher plasma levels of Se, Mg, and Fe play an active role in maintaining health and longevity in older adults (Ye et al. 2024).

2.4.4 Ergothioneine (ERGO)

Ergothioneine (ERGO), ovothiol and selenonein are natural products derived from histidine containing sulfur and selenium that are widely distributed in different organisms. They exhibit significant antioxidant properties, and growing evidence suggests that ERGO is positively correlated with healthy and long-lived aging (Chen et al. 2024). ERGO is a natural amino acid and thiol antioxidant, capable of capturing many reactive oxygen and nitrogen species, with cytoprotective and metal chelation properties. After age 60, blood levels of ERGO decrease, and lower levels of ERGO are associated with faster cognitive decline, while elevated levels of ERGO in plasma have been associated with cardiovascular protection and longevity (Tian et al. 2023). The amino acid ERGO can be obtained with the diet through some species of mushrooms and foods.

Edible mushrooms, which are one of the main sources, synthesize ERGO at concentrations ranging from a very high level (>100 mg/kg wet weight) in *Pleurotus ostreatus* and porcini mushrooms, to a much lower level (0.5 mg/kg) in commercial button mushrooms. Other foods that contain moderate amounts of ERGO include beef, pork, chicken, lamb, oat bran, black and red kidney bean (Ames 2018). Studies on this thiol/thiothione derivative of histidine date back to the early 1900s, when it was first identified in the mushroom ergot, from which it derived its name (Halliwell et al. 2018). With the increase in the geriatric population worldwide, healthy aging is the main focus of scientific research. ERGO, a sulfur-containing amino acid derived from food, has recently been considered as a purported vitamin for longevity and protection from age-related diseases. Reduced plasma levels of ERGO are found in the elderly and individuals with cognitive impairment, Parkinson's disease and other pathologies while high plasma levels of ERGO reduce the risks of coronary, cardiovascular and cerebral diseases. In conclusion, since ERGO mitigates the fragilities of aging and various age-related disorders, it may be an important vitamin-like compound that promotes healthy longevity (Katsube et al. 2024).

2.4.5 Pyrroloquinoline Quinone (PQQ)

Methoxatine or pyrroloquinoline quinone (PQQ) is a novel cofactor that has recently attracted attention from nutritionists and healthcare professionals. PQQ occurs naturally in various food sources, including fermented foods, vegetables, and human breast milk (Yan et al. 2024) and is available, as a dietary supplement, in its disodium crystalline form (Charrier et al. 2024). PQQ is a redox-active quinone molecule, an essential cofactor for oxidative enzymes, vital for mitochondrial function and ATP synthesis with significant implications for human health. PQQ has remarkable antioxidant properties that protect against ROS-mediated oxidative stress, aging, metabolic disorders, and age-related diseases. It also improves cognitive abilities (Yan et al. 2024). It is a biofactor with an action similar to vitamins and has a strong potential as a powerful therapeutic nutraceutical, whose absence in diets causes symptoms similar to a vitamin deficiency with dose-dependent recovery to PQQ supplementation. It is associated with biological processes such as growth, aging, reproduction and mitochondrial genesis. A neonatal or perinatal deficiency of PQQ reduces mitochondria-associated gene expression, resulting in a reduction in their cell content. It modulates mitochondrial functions by enhancing NAD+ dependent sirtuin activity, via modulation of lactic acid by PQQ. It also attenuates clinical dysfunctions such as those related to ischemia, lipotoxicity and inflammation. PQQ has a role as an enzyme cofactor whose integration is able to bring health benefits through immuno-neuroprotection and the improvement of metabolic functions (Jonscher et al. 2021).

The ergogenic effect generated by PQQ supplementation on aerobic performance is not direct but consequent to mitochondrial biogenesis through important increases in PGC-1α protein content (Hwang et al. 2019). PQQ improves lipid metabolism and is useful for reducing visceral and liver fat (Mohamad Ishak and Ikemoto 2023). *In vivo* studies have demonstrated the ability of PQQ to reduce the accumulation of body fat and, recently, further studies have shown that PQQ is able to improve muscle health and counteract the dysfunctional metabolism of adipose tissue (Charrier et al. 2024). By improving cellular energy metabolism, reducing oxidative stress, and promoting neuronal survival, PQQ exerts a neuroprotective action by enhancing neuroplasticity, protecting neurons from damage induced by oxidative stress, mitochondrial dysfunction, and inflammation (Xie et al. 2025). Several studies have shown that PQQ is considered a safe dietary supplement, with no toxicity or genotoxicity. Future research will focus on determining optimal dosages of PQQ to achieve specific health outcomes and evaluating its long-term efficacy and safety. The results of this research applied in clinical practice could offer new strategies for managing metabolic disorders, improving cognitive health, and potentially extending lifespan. In summary, PQQ is a promising molecule with broad potential benefits for healthy longevity (Yan et al. 2024) and for delaying aging (Ames 2018) positioning it as a novel candidate for the prevention of age-related debilitating conditions (Xie et al. 2025), having been approved by the Food and Drug Administration (FDA) as a dietary supplement (Xue et al. 2024).

2.4.6 Omega-3 (ω-3)

There is already a dedicated chapter on omega-3 (ω-3) fatty acids in this book, so I will only mention their importance in relation to healthy aging for their multiple health promotion benefits (Shahidi and Ambigaipalan 2018). In addition to the main dietary sources such as oily fish, flaxseed, chia seeds and walnuts (Shahidi and Ambigaipalan 2018), a high concentration of ω-3 is contained in the plant Sacha inchi (*Plukenetia volubilis L.*) native to the tropical Amazon rainforest of Peru and northwestern Brazil (Goyal et al. 2022) with recognized anti-aging properties (Lourith et al. 2024). High levels of ω-3 are present in the central nervous system where they support the structure and function of the brain and retina (Ames 2018). The intake of ω-3 is associated with healthy brain aging and, if supplemented in a high way as early as middle age, can help maintain the integrity of gray and white brain matter even in old age until the last decades of life (Tsiknia et al. 2023). ω-3 polyunsaturated fatty acids, being critical components of cell membranes, in the brain have the potential to improve memory, mood, attention, and executive functions while reducing the risk of chronic degenerative diseases, either alone or in combination with carotenoids and vitamin E (Fekete et al. 2023).

2.5 Discussion

For healthy longevity, the importance of micronutrients and nutraceuticals to improve metabolic and cognitive functions is emphasized, as, with advancing age, nutritional deficiencies become more frequent, due to an inadequate diet or reduced absorption of nutrients. Minerals, since they cannot be synthesized endogenously, must be taken through diet or supplementation to play a fundamental role in supporting health, particularly in the elderly population. When considering dietary supplementation with micronutrients and nutraceuticals, it is crucial to take into account several individual factors such as age, gender, nutritional status, lifestyle, physical activity, stress levels, season, eating habits, and specific diets. As we age, the body's ability to absorb nutrients decreases and nutritional needs change. Men and women have different nutritional needs, just as active individuals with higher metabolic demands may need more of certain micronutrients and antioxidants to counteract increased oxidative stress. People who follow particular diets may have to compensate for some deficiencies with supplements. It is suggested to take high-quality supplements from reputable sources to ensure safety and efficacy. By taking into account the specific needs and circumstances of each individual, this personalized approach offers optimal benefits for healthy longevity and cognitive function. A one-size-fits-all approach is not enough, you need personalized plans based on individual needs and conditions to achieve optimal benefits with a holistic approach, including diet, lifestyle and supplementation. Supplementation alone is not a panacea but should be seen as a useful component in managing a comprehensive approach to healthy aging. Regular physical activity is another pillar of this holistic plan, to improve physical health, mood and reduce stress (Fekete et al. 2023). I call for new research on the role of food engineering in food and markers for assessing effects. Understanding examples of physical and cognitive health could lead to the important development, for healthy longevity, of new supplements to be combined with a balanced and varied diet, rich in fruits, vegetables, whole grains, lean proteins and healthy fats.

2.6 Conclusion

The compounds examined are already on the market as supplements for human use. There are definitive and reliable human studies supporting the use of these compounds as supplements for healthy longevity but it is necessary to comply with the recommended doses as micro-minerals or trace elements, although essential for health and physiological functions, are needed in small quantities, otherwise, in high concentrations, they can exert toxicity (Makevic et al. 2023).

Since nutritional deficiencies are highly prevalent, it is advisable to take an adequate intake of micronutrients via an improved diet and/or personalized supplementation (Ames 2018) not only for overall well-being, but also to support healthy aging and, potentially, increase longevity by maintaining the cellular and tissue physiological functions of organs (Razzaque 2018). PQQ can be considered as a novel candidate for the prevention of age-related debilitating conditions (Xie et al. 2025).

The ideal way to obtain a sufficient dose of these compounds is to take them, through the diet, from the natural foods from which they are isolated but incorrect nutrition is a widespread problem all over the world and affects people of all walks of life and ages (Townsend et al. 2023). Micronutrient uptake is often insufficient as today's soil is depleted and, as a result, the crops growing there are no longer as rich in minerals as they once were (Razzaque 2018), and food processing techniques further reduce their content in food (Adomako and Yu 2024). To support health and delay aging, targeted and personalized supplementation is recommended, avoiding overdoses, to be combined with a balanced diet (Fekete et al. 2023).

Future research will be focused on determining optimal dosages of PQQ to achieve specific health outcomes and evaluating its long-term efficacy and safety (Yan et al. 2024). Furthermore, further and broader analyses are needed to confirm observations on the neuroprotective effects of minerals and antioxidants, thus interpreting the currently available observational evidence with the necessary caution (Wang et al. 2024).

References

Abdullah G, Akpan A, Phelan MM, Wright HL (2024) New insights into healthy ageing, inflammageing and frailty using metabolomics. Front Aging 5:1426436. https://doi.org/10.3389/fragi.2024.1426436

Adomako EA, Yu ASL (2024) Magnesium disorders: core curriculum 2024. Am J Kidney Dis 83(6):803–815. https://doi.org/10.1053/j.ajkd.2023.10.017

Ahmad R, Shaju R, Atfi A, Razzaque MS (2024) Zinc and diabetes: a connection between micronutrient and metabolism. Cells 13(16):1359. https://doi.org/10.3390/cells13161359

Ames BN (2018) Prolonging healthy aging: longevity vitamins and proteins. Proc Natl Acad Sci USA 115(43):10836–10844. https://doi.org/10.1073/pnas.1809045115

Barbagallo M, Veronese N, Dominguez LJ (2021) Magnesium in aging, health and diseases. Nutrients 13(2):463. https://doi.org/10.3390/nu13020463

Beard JR, Officer A, de Carvalho IA, Sadana R, Pot AM, Michel JP, Lloyd-Sherlock P, Epping-Jordan JE, Peeters GMEEG, Mahanani WR, Thiyagarajan JA, Chatterji S (2016) The World report on ageing and health: a policy framework for healthy ageing. Lancet (London, England) 387(10033):2145–2154. https://doi.org/10.1016/S0140-6736(15)00516-4

Berg Y, Gabay E, Božić D, Shibli JA, Ginesin O, Asbi T, Takakura L, Mayer Y (2024) The impact of nutritional components on periodontal health: a literature review. Nutrients 16(22):3901. https://doi.org/10.3390/nu16223901

Berger MM, Shenkin A, Dizdar OS, Amrein K, Augsburger M, Biesalski HK, Bischoff SC, Casaer MP, Gundogan K, Lepp HL, de Man AME, Muscogiuri G, Pietka M, Pironi L, Rezzi S, Schweinlin A, Cuerda C (2024) ESPEN practical short micronutrient guideline. Clin Nutr (Edinburgh, Scotland) 43(3):825–857. https://doi.org/10.1016/j.clnu.2024.01.030

Charrier D, Cerullo G, Carpenito R, Vindigni V, Bassetto F, Simoni L, Moro T, Paoli A (2024) Metabolic and biochemical effects of Pyrroloquinoline Quinone (PQQ) on inflammation and mitochondrial dysfunction: potential health benefits in obesity and future perspectives. Antioxidants (Basel, Switzerland) 13(9):1027. https://doi.org/10.3390/antiox13091027

Chen L, Zhang L, Ye X, Deng Z, Zhao C (2024) Ergothioneine and its congeners: anti-ageing mechanisms and pharmacophore biosynthesis. Protein Cell 15(3):191–206. https://doi.org/10.1093/procel/pwad048

Cordell GA, Lamahewage SNS (2022) Ergothioneine, Ovothiol A, and Selenoneine-histidine-derived, biologically significant, trace global alkaloids. Molecules (Basel, Switzerland) 27(9):2673. https://doi.org/10.3390/molecules27092673

de Man AME, Amrein K, Casaer MP, Dizdar OS, van Zanten ARH, Gundogan K, Lepp L, Rezzi S, Shenkin A, Berger MM, ESPEN MN Special Interest Group (2024) LLL 44-4: micronutrients in acute disease and critical illness. Clinical nutriti ESPEN 61:437–446. https://doi.org/10.1016/j.clnesp.2024.04.011

Demircan K, Chillon TS, Bang J, Gladyshev VN, Schomburg L (2024) Selenium, diabetes, and their intricate sex-specific relationship. Trends Endocrinol Metab 35(9):781–792. https://doi.org/10.1016/j.tem.2024.03.004

Farrelly C (2024) Aging, equality and the human Healthspan. HEC Forum 36(2):187–205. https://doi.org/10.1007/s10730-022-09499-3

Fekete M, Lehoczki A, Tarantini S, Fazekas-Pongor V, Csípő T, Csizmadia Z, Varga JT (2023) Improving cognitive function with nutritional supplements in aging: a comprehensive narrative review of clinical studies investigating the effects of vitamins, minerals, antioxidants, and other dietary supplements. Nutrients 15(24):5116. https://doi.org/10.3390/nu15245116

Fiorentini D, Cappadone C, Farruggia G, Prata C (2021) Magnesium: biochemistry, nutrition, detection, and social impact of diseases linked to its deficiency. Nutrients 13(4):1136. https://doi.org/10.3390/nu13041136

Floris M, Cano A, Porru L, Addis R, Cambedda A, Idda ML, Steri M, Ventura C, Maioli M (2020) Direct-to-consumer nutrigenetics testing: an overview. Nutrients 12(2):566. https://doi.org/10.3390/nu12020566

García-Esquinas E, Carballo-Casla A, Ortolá R, Sotos-Prieto M, Olmedo P, Gil F, Plans-Beriso E, Fernández-Navarro P, Pastor-Barriuso R, Rodríguez-Artalejo F (2023) Blood selenium concentrations are inversely associated with the risk of undernutrition in older adults. Nutrients 15(22): 4750. https://doi.org/10.3390/nu15224750

Giovannini S, Onder G, Lattanzio F, Bustacchini S, Di Stefano G, Moresi R, Russo A, Bernabei R, Landi F (2018) Selenium concentrations and mortality among community-dwelling older adults: results from IlSIRENTE Study. J Nutr Health Aging 22(5):608–612. https://doi.org/10.1007/s12603-018-1021-9. https://www.sciencedirect.com/science/article/pii/S1279770723020924?via%3Dihub

Golubnitschaja O, Kapinova A, Sargheini N, Bojkova B, Kapalla M, Heinrich L, Gkika E, Kubatka P (2024) Mini-encyclopedia of mitochondria-relevant nutraceuticals protecting health in primary and secondary care-clinically relevant 3PM innovation. EPMA J 15(2):163–205. https://doi.org/10.1007/s13167-024-00358-4

Gombart AF, Pierre A, Maggini S (2020) A review of micronutrients and the immune system-working in harmony to reduce the risk of infection. Nutrients 12(1):236. https://doi.org/10.3390/nu12010236

Goyal A, Tanwar B, Kumar Sihag M, Sharma V (2022) Sacha inchi (Plukenetia volubilis L.): an emerging source of nutrients, omega-3 fatty acid and phytochemicals. Food Chem 373 (Pt B):131459. https://doi.org/10.1016/j.foodchem.2021.131459

Grootswagers P, Bach D, Biemans Y, Behrouzi P, Horvath S, Kramer CS, Liu S, Manson JE, Shadyab AH, Stewart JD, Whitsel E, Yang B, de Groot L (2024) Discovering the direct relations between nutrients and epigenetic ageing. J Nutr Health Aging 28(9):100324. https://doi.org/10.1016/j.jnha.2024.100324

Halliwell B, Cheah IK, Tang RMY (2018) Ergothioneine – a diet-derived antioxidant with therapeutic potential. FEBS Lett 592(20):3357–3366. https://doi.org/10.1002/1873-3468.13123

Hwang PS, Machek SB, Cardaci TD, Wilburn DT, Kim CS, Suezaki ES, Willoughby DS (2019) Effects of Pyrroloquinoline Quinone (PQQ) supplementation on aerobic exercise performance and indices of mitochondrial biogenesis in untrained men. J Am Coll Nutr 39(6):547–556. https://doi.org/10.1080/07315724.2019.1705203

Inui T, Hanley B, Tee ES, Nishihira J, Tontisirin K, Van Dael P, Eggersdorfer M (2021) The role of micronutrients in aging in Asia: what can be implemented with existing knowledge. Nutrients 13(7):2222. https://doi.org/10.3390/nu13072222

Islam MT, Tuday E, Allen S, Kim J, Trott DW, Holland WL, Donato AJ, Lesniewski LA (2023) Senolytic drugs, dasatinib and quercetin, attenuate adipose tissue inflammation, and ameliorate metabolic function in old age. Aging Cell 22(2):e13767. https://doi.org/10.1111/acel.13767

Izadi M, Sadri N, Abdi A, Zadeh MMR, Jalaei D, Ghazimoradi MM, Shouri S, Tahmasebi S (2024) Longevity and anti-aging effects of curcumin supplementation. GeroScience 46(3):2933–2950. https://doi.org/10.1007/s11357-024-01092-5

Jonscher KR, Chowanadisai W, Rucker RB (2021) Pyrroloquinoline-Quinone is more than an antioxidant: a vitamin-like accessory factor important in health and disease prevention. Biomolecules 11(10):1441. https://doi.org/10.3390/biom11101441

Kalra EK (2003) Nutraceutical--definition and introduction. AAPS PharmSci 5(3):E25. https://doi.org/10.1208/ps050325

Katsube M, Ishimoto T, Fukushima Y, Kagami A, Shuto T, Kato Y (2024) Ergothioneine promotes longevity and healthy aging in male mice. GeroScience 46(4):3889–3909. https://doi.org/10.1007/s11357-024-01111-5

Lahoda Brodska H, Klempir J, Zavora J, Kohout P (2023) The role of micronutrients in neurological disorders. Nutrients 15(19):4129. https://doi.org/10.3390/nu15194129

Liu S, Lin T, Wang W, Jing F, Sheng J (2024) Selenium deficiency in inflammatory bowel disease: a comprehensive meta-analysis. Heliyon 10(22):e40139. https://doi.org/10.1016/j.heliyon.2024.e40139

Lourith N, Kanlayavattanakul M, Chaikul P (2024) Sacha Inchi: the promising source of functional oil for anti-aging product. J Oleo Sci 73(4):429–435. https://doi.org/10.5650/jos.ess23147

Makevic V, Milovanovich ID, Popovac N, Janković S, Janković V, Stefanović S, Bukumiric Z, de Luka SR (2023) Oligoelements in serum and intestinal tissue of pediatric IBD patients. J Trace Elem Med Biol 79:127239. https://doi.org/10.1016/j.jtemb.2023.127239

Mocchegiani E, Malavolta M, Muti E, Costarelli L, Cipriano C, Piacenza F, Tesei S, Giacconi R, Lattanzio F (2008) Zinc, metallothioneins and longevity: interrelationships with niacin and selenium. Curr Pharm Des 14(26):2719–2732. https://doi.org/10.2174/138161208786264188

Mohamad Ishak NS, Ikemoto K (2023) Pyrroloquinoline-quinone to reduce fat accumulation and ameliorate obesity progression. Front Mol Biosci 10:1200025. https://doi.org/10.3389/fmolb.2023.1200025

Nasri H, Baradaran A, Shirzad H, Rafieian-Kopaei M (2014) New concepts in nutraceuticals as alternative for pharmaceuticals. Int J Prev Med 5(12):1487–1499. https://pmc.ncbi.nlm.nih.gov/articles/PMC4336979/

Rajanna H, Chaudhry K, Pandey AK, Somvanshi S, Singh S, Kumar P, Chugh A (2024) Efficacy of Oral magnesium on bone regeneration and implant stability in immediate dental implants: a randomized controlled trial. Int J Oral Maxillofac Implants:1–31. https://doi.org/10.11607/jomi.11017. Advance online publication

Razzaque MS (2018) Magnesium: are we consuming enough? Nutrients 10(12):1863. https://doi.org/10.3390/nu10121863

Rondanelli M, Faliva MA, Tartara A, Gasparri C, Perna S, Infantino V, Riva A, Petrangolini G, Peroni G (2021) An update on magnesium and bone health. Biometals 34(4):715–736. https://doi.org/10.1007/s10534-021-00305-0

Sachdeva V, Roy A, Bharadvaja N (2020) Current prospects of nutraceuticals: a review. Curr Pharm Biotechnol 21(10):884–896. https://doi.org/10.2174/1389201021666200130113441

Saper RB, Rash R (2009) Zinc: an essential micronutrient. Am Fam Physician 79(9):768–772

Shahidi F, Ambigaipalan P (2018) Omega-3 polyunsaturated fatty acids and their health benefits. Annu Rev Food Sci Technol 9:345–381. https://doi.org/10.1146/annurev-food-111317-095850

Tian X, Thorne JL, Moore JB (2023) Ergothioneine: an underrecognised dietary micronutrient required for healthy ageing? Br J Nutr 129(1):104–114. https://doi.org/10.1017/S0007114522003592

Townsend JR, Kirby TO, Marshall TM, Church DD, Jajtner AR, Esposito R (2023) Foundational nutrition: implications for human health. Nutrients 15(13):2837. https://doi.org/10.3390/nu15132837

Tsiknia AA, Bergstrom J, Reas ET (2023) Midlife omega-3 fatty acid intake predicts later life white matter microstructure in an age- and APOE-dependent manner. Cereb Cortex (New York, NY: 1991) 33(5):2143–2151. https://doi.org/10.1093/cercor/bhac196

Wagner CA (2024) The basics of phosphate metabolism. Nephrol Dial Transplant 39(2):190–201. https://doi.org/10.1093/ndt/gfad188

Wang J, Huang Y, Bei C, Yang H, Lin Z, Xu L (2024) Causal associations of antioxidants with Alzheimer's disease and cognitive function: a Mendelian randomisation study. J Epidemiol Community Health 78(7):424–430. https://doi.org/10.1136/jech-2023-221184

Xie T, Zhang Z, Feng M, Kong L (2025) Current study on Pyrroloquinoline quinone (PQQ) therapeutic role in neurodegenerative diseases. Mol Biol Rep 52(1):397. https://doi.org/10.1007/s11033-025-10491-6

Xue Q, Li J, Qin R, Li M, Li Y, Zhang J, Wang R, Goltzman D, Miao D, Yang R (2024) Nrf2 activation by pyrroloquinoline quinone inhibits natural aging-related intervertebral disk degeneration in mice. Aging Cell 23(8):e14202. https://doi.org/10.1111/acel.14202

Yan T, Nisar MF, Hu X, Chang J, Wang Y, Wu Y, Liu Z, Cai Y, Jia J, Xiao Y, Wan C (2024) Pyrroloquinoline Quinone (PQQ): its impact on human health and potential benefits: PQQ: human health impacts and benefits. Current Res Food Sci 9:100889. https://doi.org/10.1016/j.crfs.2024.100889

Ye X, Xu T, Yang L, Hu X, Xie X, Lan G, Lu X, Huang Z, Wang T, Wu J, Lan J, Zhang Q, Zhan Z, Guo Y, Xie X (2024) Association between plasma metal exposure and health span in very elderly adults: a prospective cohort study with mixture statistical approach. BMC Geriatr 24(1):388. https://doi.org/10.1186/s12877-024-05001-5

Zhang L, Pitcher LE, Prahalad V, Niedernhofer LJ, Robbins PD (2023) Targeting cellular senescence with senotherapeutics: senolytics and senomorphics. FEBS J 290(5):1362–1383. https://doi.org/10.1111/febs.16350

Chapter 3
Pro-, Post- and Pre-biotics for Health and Longevity

Murat Baş and Meryem Kahrıman

Abstract Aging is a natural process characterized by the gradual deterioration of cellular and physiological processes. Although there are many factors underlying this process, changes in the microbiota also play an important role. It is thought that microbial elements that encourage the development of beneficial bacteria in the gut microbiome may be advantageous in this respect. Probiotic strains and their derived post/parabiotics that can beneficially attenuate basic aging mechanisms and reduce physiological aging processes have been defined as gerobiotics. In addition, prebiotics, which are selectively used by host microorganisms, have been accepted as alternatives because they positively modulate the gut microbiota. These components can modulate the immune response, reduce inflammation, exhibit antioxidant effects through various cellular, metabolic and physiological pathways, and improve many aspects of health such as metabolic health, skin health and mental health. They can reduce physiological aging processes through these mechanisms. In this chapter, we describe these underlying mechanisms and summarize their possible functions.

Keywords Microbiota · Healthy aging · Microorganisms · Gerobiotics · Immunobiotics · Parabiotics

Abbreviations

AMPK	5′ AMP-activated protein kinase
CA3	Caspase-3
COX-2	Cyclooxygenase-2
GSH	Glutathione
IIS	Insulin/IGF-1 Signaling
IL-1β	Interleukin 1 beta

M. Baş (✉) · M. Kahrıman
Department of Nutrition and Dietetics, Faculty of Health Sciences, Acibadem Mehmet Ali Aydinlar University, İstanbul, Türkiye
e-mail: Murat.Bas@acibadem.edu.tr; Meryem.Kahriman@acibadem.edu.tr

S. I. S. Rattan, C. Gezer (eds.), *Supplements and Drugs for Healthy Longevity*, Healthy Ageing and Longevity 22, https://doi.org/10.1007/978-3-032-11608-6_3

IL-6	Interleukin-6
IL-8	Interleukin-8
iNOS	Inducible Nitric Oxide Synthase
ISAPP	International Scientific Association for Probiotics and Prebiotics
MAPK	Mitogen-activated Protein Kinase
mTOR	Mammalian Target of Rapamycin
MUC 2	Mucin-2
NF-κB	Nuclear Factor kappa B
NO	Nitric Oxide
Nrf2	Nuclear factor erythroid 2-related factor 2
PI3K	Phosphoinositide 3-kinase
ROS	Reactive Oxygen Species
SCFA	Short Chain Fatty Acid
SIRT-1	Sirtuin-1
SKN-1	Skinhead-1
SOD	Superoxide Dismutase
TLR	Toll-like Receptor
TNF-α	Tumor Necrosis Factor alpha
WHO	World Health Organization

3.1 Introduction

The world is aging and life expectancy is increasing. According to the World Health Organization (WHO), life expectancy rose from 66.8 years in 2000 to 73.1 years in 2019 (WHO 2024a), and between 2015 and 2050, the percentage of individuals over 60 would rise from 12% to 22% (WHO 2024b). This trend, referred to as the 'Silver Tsunami', is a potential burden for both health systems and socio-economically. This is because the National Council on Aging (2024) reports that almost 95% of older people have at least one chronic disease and about 80% have two or more chronic diseases. Therefore, it is important to elucidate the processes related to aging.

Cellular and physiological processes continuously deteriorate as a result of aging, a complicated and multidimensional biological process (Mahamud et al. 2025). Multiple factors underlie this process, including genomic instability, telomere attrition, loss of proteostasis, epigenetic alterations, mitochondrial dysfunction, disabled macroautophagy, dysregulated nutrient sensing, cellular senescence, altered intercellular communication, stem cell exhaustion, chronic inflammation and dysbiosis (López-Otín et al. 2023). Ineffective coordination between various systems through integrated circuits, desynchronization of circadian rhythms, loss of integrity of internal and external barriers, incapacity to maintain long-term homeostasis and regeneration, and incapacity to react appropriately to stress are all consequences of disruptions in these processes. This can affect the entire organization of the organism across different classes of molecules, organelles, and cell types (López-Otín et al. 2023).

Research into the factors underlying health and disease during the aging process has revealed that the intestinal microbiota also plays an important role (Claesson et al. 2011; Biagi et al. 2016). In this regard, it has been reported that there are age-related changes in the phyla ratios and diversity in the microbiota of individuals (Claesson et al. 2011). Therefore, in recent years, emphasis has been placed on probiotics, prebiotics, postbiotics and parabiotics based on their effects in promoting healthy aging, extending health span and reducing the risk of age-related chronic diseases by modulating the microbiota (Mahamud et al. 2025). The microbiota as a marker of aging will be covered first in this part, followed by the impact of probiotics, postbiotics, and prebiotics on the aging process.

3.2 Changes in Gut Microbiota as a Hallmark of Aging

The gastrointestinal tract is home to a very large and complex ecosystem called the gut microbiota. Dynamic homeostasis of the gut microbiota, which hosts a large number of microbial species, plays an active role in maintaining health. The enzymatic pathways that support the digestion of nutrients, the metabolism of xenobiotics, and the synthesis of different bioactive compounds are provided by these microbial communities (Belizário et al. 2018). The intestinal microbiota is dynamic throughout life, and the formation of a stable microbiota is driven by various internal and external factors. The microbial population begins to develop rapidly from birth, changes rapidly, especially during weaning, followed by a stable phase during adulthood, and changes occur with aging (Ling et al. 2022). With aging, the composition of intestinal commensals changes and microbial diversity decreases. Beneficial microbes such as members of *Verrucomicrobia* decrease while proinflammatory commensals accumulate (Ragonnaud and Biragyn 2021). A mice study on the subject reported a decrease in anti-inflammatory bacterial species, including *Faecalibacterium prauznitzii* and *Bifidobacteria* spp. (Thevaranjan et al. 2017). According to another study that looked at the age-related microbial dynamics in healthy people, relative abundance of the phyla *Actinobacteria* and *Firmicutes* rose while *Bacteroidetes* declined, and this led to an increase in the *Firmicutes/Bacteroidetes* ratio (Vaiserman et al. 2020). Systemic health and lifespan may suffer as a result of this age-related dysbiosis, which can hasten aging, inflammation, and frailty processes (Boyajian et al. 2021).

Different dietary habits have a significant impact on the balance, functioning, and diversity of the microbial community in our guts, and nutrition plays a crucial role in determining the composition, function, and diversity of the gut microbiome (Ross et al. 2024). In the literature, emphasis has been placed on intermittent fasting, Mediterranean diet, probiotics, prebiotics and postbiotics, but it has also been emphasized that nutrition should be personalized (Xiao et al. 2025). From a biotics perspective, the concept of 'gerobiotics' has been defined from geroscience principles. This concept represents a new class of probiotics, postbiotics and parabiotics designed to regulate systemic biological pathways and address mechanisms related to aging (Tsai et al. 2021). It has also become a focus, particularly due to its functions in aging-related processes.

3.3 Gerobiotics

International Scientific Association for Probiotics and Prebiotics (ISAPP) defines probiotics as 'live microorganisms that, when administered in adequate amounts, confer a health benefit on the host' (Salminen et al. 2021). In addition, ISAPP reported in its consensus that the mechanisms mediating the health effects of probiotics, such as acid and short chain fatty acid (SCFA) production, colonization resistance, regulation of intestinal transit, modulation of microbiota, increased enterocyte turnover and exclusion of pathogens, are common in most probiotic strains (Hill et al. 2014). However, some mechanisms are strain specific. For example, bacteria that support health by activating the mucosal immune apparatus are 'immunobiotics' (Clancy 2003), while those that affect mental and neurological health are 'psychobiotics' (Del Toro-Barbosa et al. 2020).

Gerobiotics are defined as 'probiotic strains and their derived postbiotics and paraprobiotics that can beneficially attenuate the basic mechanisms of aging, reduce physiological aging processes and thereby prolong the health span of the host' (Tsai et al. 2021).

3.3.1 Probiotics and Longevity: Possible Mechanisms

It has been reported that probiotics can reduce age-related diseases and contribute to longevity by regulating various biological pathways such as inflammatory pathways nuclear factor kappa B (NF-κB), antioxidant pathways (nuclear factor erythroid 2-related factor 2 (Nrf2) and nitric oxide (NO)), insulin/IGF-1 signaling (IIS) pathway, stress response pathways (such as (mitogen-activated protein kinase) MAPK and 5′ AMP-activated protein kinase (AMPK)), mitochondrial maintenance and stress resistance pathways (PI3K/Akt/mTOR) in various models such as humans, rodents, *Drosophila melanogaster* and *Caenorhabditis elegans* (Mahamud et al. 2025).

An important pathway affected by gerobiotics is the IIS pathway, which plays a central role in aging. It has been reported that probiotics down-regulate the IIS pathway and prolong life span by increasing dFOXO levels, a transcription factor that promotes stress resistance and autophagy, in *D. melanogoster* (Lee et al. 2023) and *C. elegans* (Zhang et al. 2022). In line with these findings, probiotics promote cellular repair mechanisms by down-regulating the IIS pathway and delay age-related cellular damage by reducing oxidative stress. Through the activation of genes that guard against cellular damage, the MAPK pathway—a crucial regulator of oxidative stress—supports mitochondrial maintenance (Mahamud et al. 2025). In this regard, a study has shown that probiotic prolongs life span by activating p38 MAPK/SKN-1 pathways (Zhang et al. 2022). Another important pathway targeted by probiotics is AMPK enzyme, which affects metabolism, cellular homeostasis, proteostasis and mitochondrial functions (Mahamud et al. 2025). In particular, probiotics such as *Lactobacillus* contribute to mitochondrial survival by promoting energy balance and reducing inflammation through AMPK activation (Lew et al. 2019).

Besides metabolic pathways, probiotics activate the Nrf2 pathway and show antioxidant effects by increasing the expression of antioxidants such as GSH-Px, SOD and GSH, which protect cells from damage by reactive oxygen species (Lin et al. 2018). In addition, aging is defined as chronic inflammation (inflammaging). Probiotics also show anti-inflammatory effects by down-regulating inflammatory markers such as COX-2 and iNOS (Woo et al. 2014). Another important pathway associated with chronic inflammation is inflammatory pathways such as TLR/NF-κB (Mahamud et al. 2025). According to a study conducted on mice, *Bifidobacterium longum* and *Bifidobacterium animalis* together protected age-related cognitive loss by inhibiting neuroinflammation mediated by the NF-κB/TLR4 pathway (Xia et al. 2020).

The PI3K/Akt/mTOR pathway, which directly targets key features of aging and regulates cellular maintenance, autophagy, stress resistance, nutrient sensing and loss of proteostasis, is another important pathway affected by probiotics. Probiotics have the ability to promote autophagy and shift cellular activities to maintenance and repair by reducing mTORC1 activity (Mahamud et al. 2025). Probiotics may also inhibit Akt activation, which is connected to improved DNA repair and tolerance to oxidative stress. This plays an important role in the regulation of metabolic homeostasis, redox stress and inflammation (Kim et al. 2021).

Another protein family involved in DNA repair, metabolic regulation and cellular stress protection and activated by probiotics are sirtuins (especially sirtuin-1 (SIRT-1) (Mahamud et al. 2025). For example, in animal studies with probiotic interventions, SIRT-1 has been reported to be upregulated and effective in reducing age-related cellular senescence (Cheng et al. 2024; Chen et al. 2019). In summary, probiotics may have significant effects on lifespan by targeting cellular repair, stress resistance, oxidative stress, reduction of inflammation and metabolic regulation along conserved pathways (Fig. 3.1).

3.3.2 Functions of Probiotics in Aging Related Processes

3.3.2.1 Probiotics and Antioxidant Defence

Probiotics show antioxidant effects in three different ways such as chelating metal ions, producing metabolites with antioxidant activity such as glutathione and

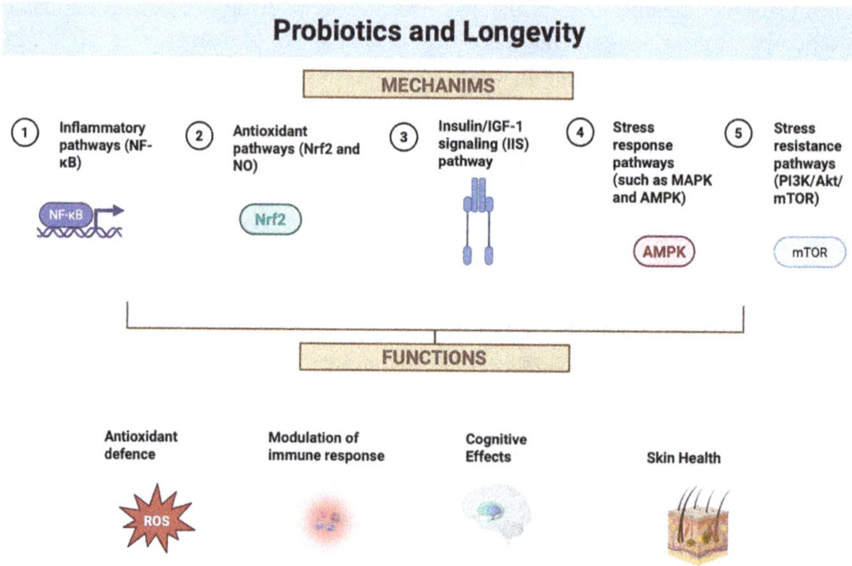

Fig. 3.1 Probiotics and their effects on the aging process

producing their own antioxidant enzyme system. For example, it has been reported that *Lactobacillus casei* and *Streptococcus thermophilus* can show antioxidant activity by chelating Fe2+ and Cu2+ ions (Choudhary et al. 2023). In addition, probiotics may also show antioxidant activity by activating the Nrf2 pathway (Lin et al. 2018), increasing the expression of antioxidants such as SOD, GSH-Px and GSH (Lin et al. 2018) and suppressing Akt activity (Kim et al. 2021). Probiotics may help prevent age-related diseases in this way, given that oxidative stress is a factor in the pathophysiology of age-related conditions like cancer, neurodegenerative diseases, chronic kidney disease, cardiovascular disease, and chronic obstructive pulmonary disease (Liguori et al. 2018).

3.3.2.2 Probiotics and Immune Response

The immune system plays an important role in the origin of many diseases. Immunosenescence accelerates with age and is responsible for the emergence of age-related rheumatoid arthritis, neurodegenerative diseases, cardiovascular, cancer, and metabolic diseases (Barbé-Tuana et al. 2020). Probiotics and their metabolites prevent immunosenescence by stimulating the maturation and functionality of immune cells (Landete et al. 2017). Additionally, probiotics stimulate the productivity of macrophages and dendritic cells, the activity of natural killer cells, and increase the number of IgA+ cells in intestinal mucosal areas (Choudhary et al. 2023). Research on this topic using *Lactobacillus* strains has highlighted the fact that they help reduce inflammatory responses and increase the number of macrophages,

IgA-producing cells, and specific IgA antibody responses to antigens (Jeong et al. 2018; Sureshkumar et al. 2019).

3.3.2.3 Probiotics and Cognition

Brain aging is a natural process that occurs with oxidative stress and a decrease in the structure and function of neurons, which can cause cognitive decline. This aging process can alter the levels of neural monoamines such as serotonin and dopamine, leading to anxiety and neurodegenerative diseases such as Parkinson's (Choudhary et al. 2023). Probiotics can positively affect the structure and functionality of the nervous system by affecting the gut-brain axis. In this regard, it has been observed that probiotic interventions with *Lactobacillus* and *Bifidobacterium* strains improve intestinal dysbiosis, and cognitive functions improve as a result of a decrease in Aβ levels in the hippocampus, trimethylamine-N-oxide (TMAO) synthesis, and neuroinflammation (Yang et al. 2020; Wang et al. 2020).

3.3.2.4 Probiotics and Skin Aging

Skin aging occurs as a result of the interaction of genetic and hormonal factors and environmental factors such as trauma, ultraviolet light, infections, pollution, and smoking (Choudhary et al. 2023). These factors cause the skin's pH to rise and the epidermal and dermal layers to be disrupted (Man et al. 2009). Probiotics and their metabolites produce acidic molecules and support skin pH, maintaining prosthetic activity. In a study on this subject, *Lactobacillus plantarum* CJLP55 intervention was reported to lower skin pH (Han et al. 2019). Ultraviolet lights are also an important environmental factor in skin aging. Previous findings emphasized that *Lactobacillus plantarum* HY7714 (Lee et al. 2021) and *Bifidobacterium breve* (Satoh et al. 2015) may have protective effects against aging caused by UV irradiation.

3.4 Postbiotics in Aging and Longevity

Postbiotics are defined by ISAPP as 'preparation of inanimate microorganisms and/or their components that confers a health benefit on the host' (Salminen et al. 2021). This concept encompasses various derivatives, metabolites and dead cells derived from the gut microbiota (Da et al. 2024). Due to their anti-inflammatory, immune and lipid metabolism regulation and intestinal barrier strengthening properties, they are thought to play a role in the aging process (Da et al. 2024) and are classified under the concept of 'gerobiotic' (Tsai et al. 2021).

Microbial processes such the fermentation of carbohydrates and the production of enzymes, different peptides, and vitamins result in the formation of postbiotics.

Postbiotics also include some of the building blocks of bacteria (Hernández-Granados and Franco-Robles 2020). Particularly studied are exopolysaccharides, short-chain fatty acids, cell wall fragments and bacterial lysates (Da et al. 2024).

3.4.1 *Possible Mechanisms for the Role of Postbiotics in Aging*

Several mechanisms have been identified for the role of postbiotics in health and the aging process (Fig. 3.2). One is the regulation of the host's gut health, gut microbiota and immune response (Kumar et al. 2024). Accordingly, postbiotics have been reported to be important mediators in metabolic pathways that play an important role in maintaining gut microbiota and homeostasis by regulating natural probiotic strains (Ozma et al. 2022). It may also provide therapeutic benefits by preserving the barrier function of the intestinal mucosa and preventing pathogen formation (Kumar et al. 2024). In addition, postbiotics can control the immune and nervous systems through various cellular and molecular mechanisms (Hernández-Granados and Franco-Robles 2020). In this regard, it has been reported to be effective in maintaining the balance between Th-1 and Th-2 (Kumar et al. 2024), reduce interleukin 6 (IL-6), interleukin 8 (IL-8), interleukin 1 beta (IL-1β) and tumor necrosis factor alpha (TNF-α), improve claudin and occlusion expression and mucin 2 (MUC 2) production, and prevent apoptosis of enterocytes by reducing caspase-3 (Ca3) (Hernández-Granados and Franco-Robles 2020).

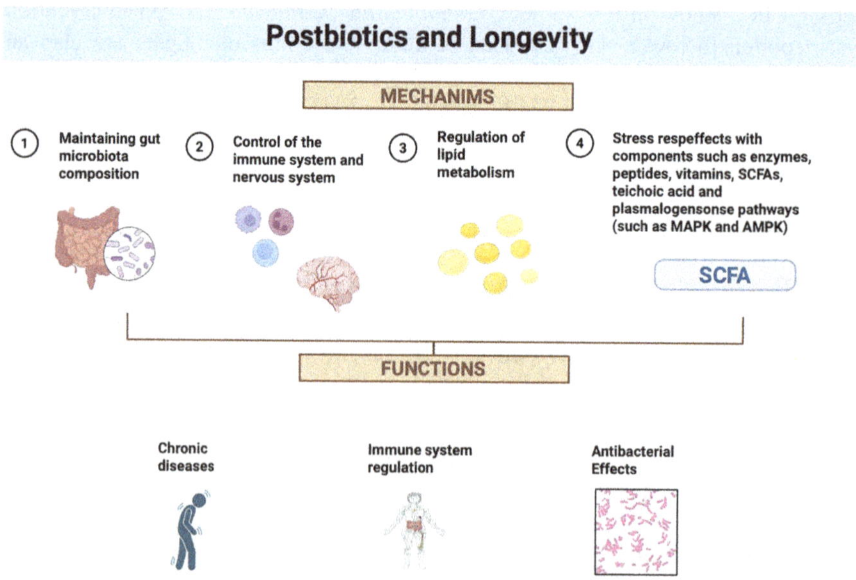

Fig. 3.2 Postbiotics and their effects on the aging process

Another important effect of postbiotics is on the regulation of lipid metabolism and inhibition of lipid absorption. Postbiotics are thought to regulate the cardiovascular system and exert antiobesity effects through these pathways (Park et al. 2023). Another mechanism by which postbiotics may mediate the development of age-related diseases is their antitumor properties. It may increase apoptosis in cancer cells by reducing oxidative stress and acting cytotoxic against these cells (Kumar et al. 2024).

Postbiotics also exert their effects through components such as enzymes, peptides, vitamins, SCFAs, teichoic acid and plasmalogens (Hernández-Granados and Franco-Robles 2020). Peptides inhibit cell wall synthesis and exert antimicrobial effects, especially for the brain and intestinal tract. Enzymes can break down substances of the extracellular polymer and reduce reactive oxygen species (ROS), thus affecting the skin, pancreas and intestine. SCFAs are the energy source for intestinal epithelial cells and regulate glucose and lipid metabolism. Vitamins are important for the absorption of thiamine and free riboflavin. Teichoic acid has antitumor and antioxidant activity, which is particularly important for bone marrow and dendritic cells. Finally, plasmalogen improves membrane dynamics and affects the lung, brain and adipose tissue (Aggarwal et al. 2022).

3.4.2 Functions of Postbiotics in Aging Related Processes

3.4.2.1 Postbiotics and Chronic Diseases

Aging can affect the entire system at the organelle, molecule and cellular level and is associated with a variety of chronic diseases such as loss of skeletal muscle and function, elevated cholesterol levels and related diseases, and impaired memory and learning (López-Otín et al. 2023). Previous studies have emphasized the positive effects of *Lactobacillus plantarum* postbiotics on age-related processes such as lowering cholesterol levels, alleviating decline in memory abilities and muscle mass.

Oxidative damage is a characteristic feature of aging and is caused by ROS associated with tumors, diabetes, atherosclerosis and many chronic diseases (Yang et al. 2024a, b). Since many metabolic intermediates, including lactate, glutamate and pyruvate, can be used in energy production, postbiotics with antioxidant activity are seen as therapeutics for these age-related diseases (Da et al. 2024). It has been reported that *Lactobacillus plantarum* postbiotics provide antioxidant activity and may be effective in reducing oxidative stress (Bagher Hashemi et al. 2022).

3.4.2.2 Postbiotics and Immune System Regulation

Postbiotics are of vital importance in regulating the immune response. Cellular and molecular mechanisms mediate this effect of postbiotics. However, it has been reported that stabilization of the host-microbiome balance, especially through preservation and strengthening of the integrity of the intestinal mucosal barrier, is effective in inhibiting pathogens and protecting against infections (Szydłowska and Sionek 2022). Regarding this subject, previous studies have confirmed the immunomodulatory effects of *Lactobacillus plantarum* (Yılmaz and Şimşek 2020), *Bifidobacterium breve* (Ruiz et al. 2017) and *Bifidobacterium longum* exopolysaccharides (Bottacini et al. 2017).

The most prominent immunomodulatory metabolites are thought to be SCFAs. SCFAs are the main energy source for colon and ileum cells and may thus affect the intestinal epithelial barrier. Additionally, they control the activity of innate immune cells such dendritic cells, neutrophils, and macrophages. They might also control how T and B cells differentiate and how they mediate antigen-specific adaptive immunity (Yao et al. 2022). Due to these properties, they are considered as potential modulators in immunomodulation.

Postbiotics are also believed to have immunomodulatory effects through their actions on nucleotide-binding oligomerization domain-containing protein-like receptors (NLRs) and toll-like receptors (TLRs). A previous study investigated the effects of postbiotics derived from *Streptococcus thermophilus* and *Bifidobacterium breve* on colon cancer cells and confirmed this by reporting a decrease in TNF-α release as a result of interaction with a TLRs (Menard et al. 2004).

3.4.2.3 Postbiotics and Antibacterial Activity

Bacteria have the ability to form biofilms, which can lead to negative outcomes such as chronic infections (Da et al. 2024). It is reported that postbiotics have an inhibitory effect on the formation of pathogen biofilms. In a study using Man-Rogosa-Sharpe fermented by six probiotic strains belonging to the *Bifidobacterium* and *Lactobacillus* genera, it was found that these strains showed strong antibacterial activity against *E. coli* isolates and had antibiofilm ability by inhibiting biofilm formation. The highest inhibition rate was determined in *Bifidobacterium longum* (Abdelhamid et al. 2018). However, there are some factors such as preparation process and lack of target that may limit the antibiofilm effect of postbiotics. Therefore, which metabolites show this effect should be well defined and the effectiveness of preparation technologies should be emphasized (Che et al. 2024).

3.5 Prebiotics in Aging

According to ISAAP, prebiotics are 'substrates that are selectively utilized by host microorganisms conferring a health benefits' (Salminen et al. 2021). A substance must be resistant to hydrolysis, gastric acidity, and upper gastrointestinal tract absorption in order to be categorized as a prebiotic. It is also crucial that it offers possible health advantages and specifically promotes the proliferation and/or activity of bacteria through fermentation by the gut microbiota (Gibson and Roberfroid

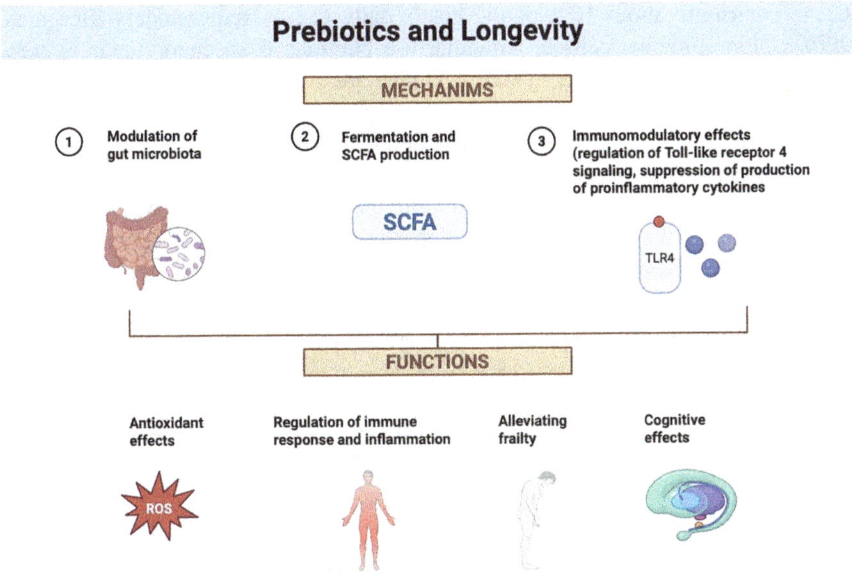

Fig. 3.3 Prebiotics and their effects on the aging process

1995). Most prebiotics are carbohydrates of plant origin (fructooligosaccharides, galactooligosaccharides, mannanoligosaccharides, xylooligosaccharides, etc.). But other compounds such as phenols and other phytochemicals, conjugated linolenic acid, polyunsaturated fatty acids and human milk oligosaccharides also fit this definition. The health effects of prebiotics are not only limited to the gut microbiota but also affect many systems of the organism through immunomodulatory, anti-inflammatory and antioxidant effects (Bevilacqua et al. 2024). These effects of prebiotics are seen as potential components in mitigating the effects of aging (Fig. 3.3).

3.5.1 Prebiotics and Healthy Aging: Possible Mechanisms

Generally, enzymes capable of hydrolyzing the polymer bonds of prebiotics are not present in the human gastrointestinal tract. Prebiotics are therefore resistant to hydrolysis and can reach the colon where they can be fermented by bacteria such as *Lactobacilli* and *Bifidobacteria* (Guarino et al. 2020; van der Beek et al. 2017). Accordingly, prebiotics may modulate the gut microbiota and work effectively in maintaining a healthy microbiota (Yoo et al. 2024).

An important mechanism in both colonic and systemic effects of prebiotics is fermentation and SCFA production. SCFAs (propionate, acetate and butyrate) are formed by bacterial fermentation of these carbohydrates (Guarino et al. 2020).

SCFAs contribute about 10% of the host's daily energy requirements (Bergman 1990). Colon epithelial cells in particular use butyrate as an energy source, even when substrates such as glucose and glutamine are available (Slavin 2013). Therefore, butyrate may control the metabolic activity and development of the colon and may have a protective function in maintaining colon health (Hodgkinson et al. 2023). Fermentation and SCFAs also lower luminal and colony pH. Acidic pH lowers bacterial enzyme activity and inhibits peptide breakdown, which leads to the production of harmful substances such ammonia, amines, and phenolic compounds. This mechanism also prevents the growth of pathogenic bacteria.

Another potential mechanism of prebiotics is their immunomodulatory effects. These components can regulate TLR 4 signaling, suppress the production of proinflammatory cytokines such as IL-6, IL-12, TNF-α, reduce the infiltration of colonic mucosa by leukocytes, and induce the differentiation of mucosal Treg cells. It is thought that through these effects, prebiotics may have positive effects on the immunosenescence process (van der Beek et al. 2017).

3.5.2 *Functions of Prebiotics in Aging Related Processes*

3.5.2.1 Antioxidant Effects of Prebiotics

Antioxidants can be defined as various anti-aging compounds that can provide longevity because aging is a process characterized by the increase of reactive oxygen species and oxidative stress. Some microbial active ingredients such as prebiotics are also known for their antioxidant properties (Mounir et al. 2022). Regarding this issue, it is reported that gastric acid secretion can promote the oxidation of lipids and other nutrients, and prebiotics can be effective in preventing lipid peroxidation in the stomach (Kanner and Lapidot 2001; Guarino et al. 2020). In general, it is emphasized that prebiotics are effective in protecting against oxidative stress (Busserolles et al. 2003). Among the prebiotic components, emphasis is placed on xylooligosaccharides, inulin, oligofructose, arabinoxylan, mannan, arabinogalactan, glucomannan and anthocyanins (Mounir et al. 2022).

3.5.2.2 Effects of Prebiotics on the Immune System and Inflammation

The aging process is characterized by immunosenescence, which refers to the destruction of the immune organ structure and the innate and adaptive immune dysfunction that occurs with aging. This is associated with an increase in infections, age-related diseases, and malignancies. In addition, during the aging process, organisms tend to develop elevated pro-inflammatory markers and chronic inflammation. This is also associated with immunosenescence (Liu et al. 2023). Pro-inflammatory cytokines like IL-1ß have been shown to decrease when prebiotics are taken, while anti-inflammatory cytokines like IL-10 have been shown to rise (Vulevic et al.

2015). In addition, prebiotics can act on local and systemic immune cells and gut-associated epithelial cells via G-protein-coupled receptor-mediated pathways, and can have immunomodulatory effects via histone deacetylase inhibition and inflammasomes. As a result of these effects, modulation of innate immunity can occur via switching in macrophage polarization and function, neutrophil recruitment and migration, and regulatory T cell differentiation (Pujari and Banerjee 2021).

3.5.2.3 Prebiotics and Frailty in Aging

It is characterized by frailty, chronic malnutrition, reduced physical activity and mobility impairments. This situation is considered the most challenging aspect of population aging as it significantly increases the risk of falls, disability, long-term care and even mortality (Yang et al. 2024a, b). It is reported that frailty syndrome is associated with age-related changes in microbiota. The basis of this relationship is the loss of beneficial taxa, the increase of harmful taxa and changes in the general microbiota composition. Additionally, microb-aging may play a role in this pathophysiology by negatively affecting the ability to metabolize basic metabolic factors such as carbohydrates, proteins, and bile acids (Le Cosquer et al. 2024). At this point, it is thought that prebiotics may be a promising alternative for both microb-aging and frailty syndrome. In a study in which fructooligosaccharide and inulin intervention was given to elderly individuals, although there was no change in the overall frailty rate, significant improvements were found in fatigue and hand grip strength, which are criteria for frailty (Buigues et al. 2016). In another study in which fructooligosaccharide and inulin intervention was performed on elderly individuals, it was emphasized that it could reduce the frailty levels of nursing home residents, especially those with high frailty levels (Theou et al. 2019).

3.5.2.4 Prebiotics and Cognition

Previous evidence has established the role of gut microbiota in cognitive function, highlighting that cognitive functions may be compromised with dysbiosis (Fröhlich et al. 2016). Interventions targeting the growth of beneficial gut bacteria may support cognitive function (Baldi et al. 2021). In this regard, prebiotics are particularly emphasized because these components are found in many food groups such as fruits, vegetables, legumes and grains (Dalile et al. 2024). In addition, the role of these components in cognitive functions such as learning ability, memory, anxiety and social behavior has been revealed in previous studies (de Paiva et al. 2023; Waworuntu et al. 2014). Prebiotics show these effects through the production of SCFAs which can affect the brain through vagal, endocrine, immune and other humoral pathways (La Torre et al. 2021). However, there is limited data on which prebiotics, in which dosages and for which time periods can be effective on cognitive function in the elderly. This should not be ignored, and emphasis should be placed on including dietary sources of prebiotics in the diets of elderly individuals.

3.6 Conclusions

Microbiota is one of the important factors that can affect the physiological aging process. Pre/pro/postbiotics are components that can promote the growth of healthy bacteria in the microbiota and thus alleviate the physiological effects of aging. In this regard, the concept of gerobiotics has been defined, which includes probiotic strains and post/probiotics derived from them, which can beneficially weaken basic aging mechanisms and thus prolong the health span of the host. These components can have antioxidant, anti-inflammatory effects and modulate the immune response. With these aspects, they can positively affect many aspects of health such as metabolic health, mental health and skin health. Therefore, it is considered a promising approach that can be used in the aging process.

However, data on which dose and strain can be used in which age group is limited. Although these components are available on the market as commercial formulations, the uncertainty about which dose and which strain can be used for anti-aging effects should not be ignored. The intake of these components from natural foods (Fig. 3.4) depends on the effectiveness of probiotics, postbiotics and prebiotics, the type of food, the frequency of consumption, the amount of microorganisms and prebiotic components contained in the food and the microbiota composition of the individual. Accordingly, more studies are needed to determine whether these components for longevity are effective when taken through natural foods. In addition, at the point where it is emphasized to be taken as a supplement, there is a need to clarify which strain and which dose should be taken.

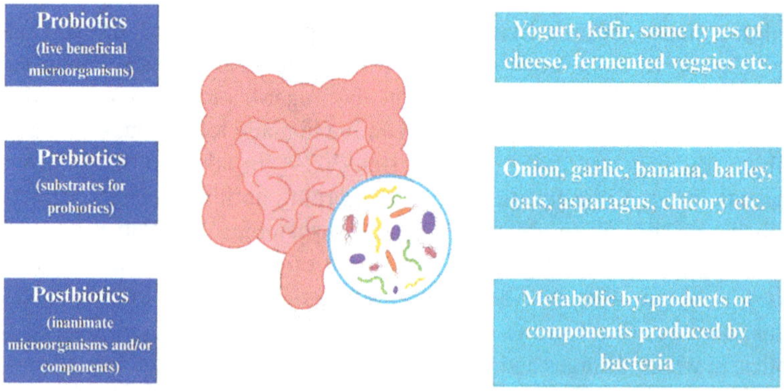

Fig. 3.4 Some natural sources of pro-, pre- and postbiotics

References

Abdelhamid AG, Esaam A, Hazaa MM (2018) Cell free preparations of probiotics exerted antibacterial and antibiofilm activities against multidrug resistant E. coli. Saudi Pharm J 26(5):603–607. https://doi.org/10.1016/j.jsps.2018.03.004

Aggarwal S, Sabharwal V, Kaushik P, Joshi A, Aayushi A, Suri M (2022) Postbiotics: from emerging concept to application. Front Sustain Food Syst 6:887642

Bagher Hashemi SM, Abedi E, Kaveh S, Mousavifard M (2022) Hypocholesterolemic, antidiabetic and bioactive properties of ultrasound-stimulated exopolysaccharide produced by Lactiplantibacillus plantarumstrains. Bioact Carbohydr Diet Fibre 28:100334

Baldi S, Mundula T, Nannini G, Amedei A (2021) Microbiota shaping – the effects of probiotics, prebiotics, and fecal microbiota transplant on cognitive functions: a systematic review. World J Gastroenterol 27(39):6715–6732. https://doi.org/10.3748/wjg.v27.i39.6715

Barbé-Tuana F, Funchal G, Schmitz CRR, Maurmann RM, Bauer ME (2020) The interplay between immunosenescence and age-related diseases. Semin Immunopathol 42(5):545–557. https://doi.org/10.1007/s00281-020-00806-z

Belizário JE, Faintuch J, Garay-Malpartida M (2018) Gut microbiome Dysbiosis and Immunometabolism: new Frontiers for treatment of metabolic diseases. Mediators Inflamm 2018:2037838. https://doi.org/10.1155/2018/2037838

Bergman EN (1990) Energy contributions of volatile fatty acids from the gastrointestinal tract in various species. Physiol Rev 70:567–590

Bevilacqua A, Campaniello D, Speranza B, Racioppo A, Sinigaglia M, Corbo MR (2024) An update on prebiotics and on their health effects. Foods 13(3):446

Biagi E, Franceschi C, Rampelli S, Severgnini M, Ostan R, Turroni S, Consolandi C, Quercia S, Scurti M, Monti D, Capri M, Brigidi P, Candela M (2016) Gut microbiota and extreme longevity. Curr Biol 26(11):1480–1485. https://doi.org/10.1016/j.cub.2016.04.016

Bottacini F, van Sinderen D, Ventura M (2017) Omics of bifidobacteria: research and insights into their health-promoting activities. Biochem J 474(24):4137–4152. https://doi.org/10.1042/bcj20160756

Boyajian JL, Ghebretatios M, Schaly S, Islam P, Prakash S (2021) Microbiome and human aging: probiotic and prebiotic potentials in longevity, skin health and cellular senescence. Nutrients 13(12):4550

Buigues C, Fernández-Garrido J, Pruimboom L, Hoogland AJ, Navarro-Martínez R, Martínez-Martínez M, Verdejo Y, Mascarós MC, Peris C, Cauli O (2016) Effect of a prebiotic formulation on frailty syndrome: a randomized, double-blind clinical trial. Int J Mol Sci 17(6):932. https://doi.org/10.3390/ijms17060932

Busserolles J, Gueux E, Rock E, Demigné C, Mazur A, Rayssiguier Y (2003) Oligofructose protects against the hypertriglyceridemic and pro-oxidative effects of a high fructose diet in rats. J Nutr 133(6):1903–1908. https://doi.org/10.1093/jn/133.6.1903

Che J, Shi J, Fang C, Zeng X, Wu Z, Du Q, Tu M, Pan D (2024) Elimination of pathogen biofilms via Postbiotics from lactic acid bacteria: a promising method in food and biomedicine. Microorganisms 12(4):704. https://doi.org/10.3390/microorganisms12040704

Chen LH, Huang SY, Huang KC, Hsu CC, Yang KC, Li LA, Chan CH, Huang HY (2019) Lactobacillus paracasei PS23 decelerated age-related muscle loss by ensuring mitochondrial function in SAMP8 mice. Aging 11(2):756–770. https://doi.org/10.18632/aging.101782

Cheng LH, Wu CC, Wei YH, Wen PJ, Hsu CC, Tsai YC, Wang S (2024) Anti-aging effects of Lacticaseibacillus paracasei PS117 on cognitive and intestinal health in naturally-aged mice: a focus on senescence-related proteins and microbiota composition. Exp Gerontol 195:112529. https://doi.org/10.1016/j.exger.2024.112529

Choudhary P, Kathuria D, Suri S, Bahndral A, Naveen AK (2023) Probiotics-its functions and influence on the ageing process: a comprehensive review. Food Biosci 52:102389

Claesson MJ, Cusack S, O'Sullivan O, Greene-Diniz R, de Weerd H, Flannery E, Marchesi JR, Falush D, Dinan T, Fitzgerald G, Stanton C, van Sinderen D, O'Connor M, Harnedy N,

O'Connor K, Henry C, O'Mahony D, Fitzgerald AP, Shanahan F, Twomey C et al (2011) Composition, variability, and temporal stability of the intestinal microbiota of the elderly. Proc Natl Acad Sci USA 108(Suppl 1):4586–4591. https://doi.org/10.1073/pnas.1000097107

Clancy R (2003) Immunobiotics and the probiotic evolution. FEMS Immunol Med Microbiol 38(1):9–12

Da M, Sun J, Ma C, Li D, Dong L, Wang LS, Chen F (2024) Postbiotics: enhancing human health with a novel concept. eFood 5(5):e180

Dalile B, Boyle NB, Ruiz FT, Chakrabarti A, Respondek F, Dodd GF et al (2024) Targeting cognitive resilience through prebiotics: a focused perspective. Adv Nutr 16:100343

de Paiva IHR, da Silva RS, Mendonça IP, Duarte-Silva E, Botelho de Souza JR, Peixoto CA (2023) Fructooligosaccharide (FOS) and Galactooligosaccharide (GOS) improve Neuroinflammation and cognition by up-regulating IRS/PI3K/AKT signaling pathway in diet-induced obese mice. J Neuroimmune Pharmacol 18(3):427–447. https://doi.org/10.1007/s11481-023-10069-8

Del Toro-Barbosa M, Hurtado-Romero A, Garcia-Amezquita LE, García-Cayuela T (2020) Psychobiotics: mechanisms of action, evaluation methods and effectiveness in applications with food products. Nutrients 12(12):3896. https://doi.org/10.3390/nu12123896

Fröhlich EE, Farzi A, Mayerhofer R, Reichmann F, Jačan A, Wagner B, Zinser E, Bordag N, Magnes C, Fröhlich E, Kashofer K, Gorkiewicz G, Holzer P (2016) Cognitive impairment by antibiotic-induced gut dysbiosis: analysis of gut microbiota-brain communication. Brain Behav Immun 56:140–155. https://doi.org/10.1016/j.bbi.2016.02.020

Gibson GR, Roberfroid MB (1995) Dietary modulation of the human colonic microbiota: introducing the concept of prebiotics. J Nutr 125:1401–1412. https://doi.org/10.1093/jn/125.6.1401

Guarino MPL, Altomare A, Emerenziani S, Di Rosa C, Ribolsi M, Balestrieri P, Iovino P, Rocchi G, Cicala M (2020) Mechanisms of action of prebiotics and their effects on gastro-intestinal disorders in adults. Nutrients 12(4):1037. https://doi.org/10.3390/nu12041037

Han S, Shin J, Lim S, Ahn HY, Kim B, Cho Y (2019) Dietary effect of lactobacillus plantarum CJLP55 isolated from kimchi on skin pH and its related biomarker levels in adult subjects. J Nutr Health 52(2):149–156

Hernández-Granados MJ, Franco-Robles E (2020) Postbiotics in human health: possible new functional ingredients? Food Res Int 137:109660

Hill C, Guarner F, Reid G, Gibson GR, Merenstein DJ, Pot B et al (2014) The International Scientific Association for Probiotics and Prebiotics consensus statement on the scope and appropriate use of the term probiotic. Nat Rev Gastroenterol Hepatol 11(8):506–514

Hodgkinson K, El Abbar F, Dobranowski P, Manoogian J, Butcher J, Figeys D et al (2023) Butyrate's role in human health and the current progress towards its clinical application to treat gastrointestinal disease. Clin Nutr 42(2):61–75

Jeong JJ, Lee HJ, Jang SE, Han MJ, Kim DH (2018) Lactobacillus plantarum C29 alleviates NF-κB activation and Th17/Treg imbalance in mice with TNBS-induced colitis. Food Agric Immunol 29(1):577–589

Kanner J, Lapidot T (2001) The stomach as a bioreactor: dietary lipid peroxidation in the gastric fluid and the effects of plant-derived antioxidants. Free Radic Biol Med 31(11):1388–1395

Kim DH, Bang E, Ha S, Jung HJ, Choi YJ, Yu BP, Chung HY (2021) Organ-differential roles of Akt/FoxOs Axis as a key metabolic modulator during aging. Aging Dis 12(7):1713–1728. https://doi.org/10.14336/AD.2021.0225

Kumar A, Green KM, Rawat M (2024) A comprehensive overview of postbiotics with a special focus on discovery techniques and clinical applications. Foods 13(18):2937

La Torre D, Verbeke K, Dalile B (2021) Dietary fibre and the gut-brain axis: microbiota-dependent and independent mechanisms of action. Gut Microbiome (Cambridge, England) 2:e3. https://doi.org/10.1017/gmb.2021.3

Landete JM, Gaya P, Rodríguez E, Langa S, Peirotén Á, Medina M, Arqués JL (2017) Probiotic bacteria for healthier aging: immunomodulation and metabolism of phytoestrogens. Biomed Res Int 2017:5939818. https://doi.org/10.1155/2017/5939818

Le Cosquer G, Vergnolle N, Motta JP (2024) Gut microb-aging and its relevance to frailty aging. Microbes Infect 26(3):105309

Lee K, Kim HJ, Kim SA, Park SD, Shim JJ, Lee JL (2021) Exopolysaccharide from lactobacillus plantarum HY7714 protects against skin aging through skin-gut Axis communication. Molecules (Basel, Switzerland) 26(6):1651. https://doi.org/10.3390/molecules26061651

Lee HY, Lee JH, Kim SH, Jo SY, Min KJ (2023) Probiotic Limosilactobacillus Reuteri (Lactobacillus Reuteri) extends the lifespan of drosophila melanogaster through insulin/IGF-1 signaling. Aging Dis 14(4):1407–1424. https://doi.org/10.14336/AD.2023.0122

Lew LC, Hor YY, Jaafar MH, Lau ASY, Ong JS, Chuah LO, Yap KP, Azzam G, Azlan A, Liong MT (2019) Lactobacilli modulated AMPK activity and prevented telomere shortening in ageing rats. Benefic Microbes 10(8):883–892. https://doi.org/10.3920/BM2019.0058

Liguori I, Russo G, Curcio F, Bulli G, Aran L, Della-Morte D, Gargiulo G, Testa G, Cacciatore F, Bonaduce D, Abete P (2018) Oxidative stress, aging, and diseases. Clin Interv Aging 13:757–772. https://doi.org/10.2147/CIA.S158513

Lin X, Xia Y, Wang G, Xiong Z, Zhang H, Lai F, Ai L (2018) Lactobacillus plantarum AR501 alleviates the oxidative stress of D-galactose-induced aging mice liver by upregulation of Nrf2-mediated antioxidant enzyme expression. J Food Sci 83(7):1990–1998. https://doi.org/10.1111/1750-3841.14200

Ling Z, Liu X, Cheng Y, Yan X, Wu S (2022) Gut microbiota and aging. Crit Rev Food Sci Nutr 62(13):3509–3534

Liu Z, Liang Q, Ren Y, Guo C, Ge X, Wang L et al (2023) Immunosenescence: molecular mechanisms and diseases. Signal Transduct Target Ther 8(1):200

López-Otín C, Blasco MA, Partridge L, Serrano M, Kroemer G (2023) Hallmarks of aging: an expanding universe. Cell 186(2):243–278

Mahamud ASU, Tanvir IA, Kabir ME, Samonty I, Chowdhury MAH, Rahman MA (2025) Gerobiotics: exploring the potential and limitations of repurposing probiotics in addressing aging hallmarks and chronic diseases. Probiotics and Antimicrobial Proteins 1–41

Man MQ, Xin SJ, Song SP, Cho SY, Zhang XJ, Tu CX, Feingold KR, Elias PM (2009) Variation of skin surface pH, sebum content and stratum corneum hydration with age and gender in a large Chinese population. Skin Pharmacol Physiol 22(4):190–199. https://doi.org/10.1159/000231524

Menard S, Candalh C, Bambou JC, Terpend K, Cerf-Bensussan N, Heyman M (2004) Lactic acid bacteria secrete metabolites retaining anti-inflammatory properties after intestinal transport. Gut 53:821–828

Mounir M, Ibijbijen A, Farih K, Rabetafika HN, Razafindralambo HL (2022) Synbiotics and their antioxidant properties, mechanisms, and benefits on human and animal health: a narrative review. Biomolecules 12(10):1443. https://doi.org/10.3390/biom12101443

National Council on Aging (2024) Get the facts on health aging. https://www.ncoa.org/article/get-the-facts-on-healthy-aging/. Accessed 18 Mar 2025

Ozma MA, Abbasi A, Akrami S, Lahouty M, Shahbazi N, Ganbarov K, Pagliano P, Sabahi S, Köse Ş, Yousefi M, Dao S, Asgharzadeh M, Hosseini H, Kafil HS (2022) Postbiotics as the key mediators of the gut microbiota-host interactions. Infez Med 30(2):180–193. https://doi.org/10.53854/liim-3002-3

Park SJ, Sharma A, Lee HJ (2023) Postbiotics against obesity: perception and overview based on pre-clinical and clinical studies. Int J Mol Sci 24(7):6414. https://doi.org/10.3390/ijms24076414

Pujari R, Banerjee G (2021) Impact of prebiotics on immune response: from the bench to the clinic. Immunol Cell Biol 99(3):255–273. https://doi.org/10.1111/imcb.12409

Ragonnaud E, Biragyn A (2021) Gut microbiota as the key controllers of "healthy" aging of elderly people. Immun Ageing 18:1–11

Ross FC, Patangia D, Grimaud G, Lavelle A, Dempsey EM, Ross RP, Stanton C (2024) The interplay between diet and the gut microbiome: implications for health and disease. Nat Rev Microbiol 22(11):671–686

Ruiz L, Bottacini F, Boinett CJ, Cain AK, O'Connell-Motherway M, Lawley TD, van Sinderen D (2017) The essential genomic landscape of the commensal Bifidobacterium breve UCC2003. Sci Rep 7(1):5648. https://doi.org/10.1038/s41598-017-05795-y

Salminen S, Collado MC, Endo A, Hill C, Lebeer S, Quigley EM et al (2021) The International Scientific Association of Probiotics and Prebiotics (ISAPP) consensus statement on the definition and scope of postbiotics. Nat Rev Gastroenterol Hepatol 18(9):649–667

Satoh T, Murata M, Iwabuchi N, Odamaki T, Wakabayashi H, Yamauchi K, Abe F, Xiao JZ (2015) Effect of Bifidobacterium breve B-3 on skin photoaging induced by chronic UV irradiation in mice. Benefic Microbes 6(4):497–504. https://doi.org/10.3920/BM2014.0134

Slavin J (2013) Fiber and prebiotics: mechanisms and health benefits. Nutrients 5(4):1417–1435. https://doi.org/10.3390/nu5041417

Sureshkumar S, Jung SK, Kim D, Oh KB, Yang H, Lee HC, Jin JY, Sun LH, Lee S, Byun SJ (2019) Oral administration of lactobacillus reuteriexpressing a 3D8 single-chain variable fragment (scFv) enhances chicken growth and conserves immune homeostasis. 3 Biotech 9(7):282. https://doi.org/10.1007/s13205-019-1811-8

Szydłowska A, Sionek B (2022) Probiotics and postbiotics as the functional food components affecting the immune response. Microorganisms 11(1):104

Theou O, Jayanama K, Fernández-Garrido J, Buigues C, Pruimboom L, Hoogland AJ, Navarro-Martínez R, Rockwood K, Cauli O (2019) Can a prebiotic formulation reduce frailty levels in older people? J Frailty Aging 8(1):48–52. https://doi.org/10.14283/jfa.2018.39

Thevaranjan N, Puchta A, Schulz C, Naidoo A, Szamosi JC, Verschoor CP et al (2017) Age-associated microbial dysbiosis promotes intestinal permeability, systemic inflammation, and macrophage dysfunction. Cell Host Microbe 21(4):455–466

Tsai YC, Cheng LH, Liu YW, Jeng OJ, Lee YK (2021) Gerobiotics: probiotics targeting fundamental aging processes. Biosci Microbiota Food Health 40(1):1–11. https://doi.org/10.12938/bmfh.2020-026

Vaiserman A, Romanenko M, Piven L, Moseiko V, Lushchak O, Kryzhanovska N et al (2020) Differences in the gut Firmicutes to Bacteroidetes ratio across age groups in healthy Ukrainian population. BMC Microbiol 20:1–8

van der Beek CM, Dejong CHC, Troost FJ, Masclee AAM, Lenaerts K (2017) Role of short-chain fatty acids in colonic inflammation, carcinogenesis, and mucosal protection and healing. Nutr Rev 75(4):286–305. https://doi.org/10.1093/nutrit/nuw067

Vulevic J, Juric A, Walton GE, Claus SP, Tzortzis G, Toward RE, Gibson GR (2015) Influence of galacto-oligosaccharide mixture (B-GOS) on gut microbiota, immune parameters and metabonomics in elderly persons. Br J Nutr 114(4):586–595

Wang F, Xu T, Zhang Y, Zheng T, He Y, He F, Jiang Y (2020) Long-term combined administration of Bifidobacterium bifidum TMC3115 and Lactobacillus plantarum 45 alleviates spatial memory impairment and gut dysbiosis in APP/PS1 mice. FEMS Microbiol Lett 367(7):fnaa048

Waworuntu R, Hain H, Chang Q, Thiede L, Hanania T, Berg B (2014) Dietary prebiotics improve memory and social interactions while reducing anxiety when provided early in life to normally developing rodents (637.5). FASEB J 28:637–635

Woo JY, Gu W, Kim KA, Jang SE, Han MJ, Kim DH (2014) Lactobacillus pentosus var. plantarum C29 ameliorates memory impairment and inflammaging in a D-galactose-induced accelerated aging mouse model. Anaerobe 27:22–26. https://doi.org/10.1016/j.anaerobe.2014.03.003

World Health Organization (2024a) World health statistics 2024: monitoring health for the SDGs, sustainable development goals. World Health Organization

World Health Organization (2024b) Ageing and health. https://www.who.int/news-room/fact-sheets/detail/ageing-and-health. Accessed 17 Mar 2025

Xia C, Cao X, Cui L, Liu H, Wang S, Chen T (2020) Anti-aging effect of the combination of Bifidobacterium longum and B. animalis in a d-galactose-treated mice. J Funct Foods 69:103938

Xiao Y, Feng Y, Zhao J, Chen W, Lu W (2025) Achieving healthy aging through gut microbiota-directed dietary intervention: focusing on microbial biomarkers and host mechanisms. J Adv Res 68:179–200

Yang X, Yu D, Xue L, Li H, Du J (2020) Probiotics modulate the microbiota–gut–brain axis and improve memory deficits in aged SAMP8 mice. Acta Pharm Sin B 10(3):475–487

Yang J, Hou L, Wang A, Shang L, Jia X, Xu R, Wang X (2024a) Prebiotics improve frailty status in community-dwelling older individuals in a double-blind, randomized, controlled trial. J Clin Invest 134(18):e176507

Yang J, Luo J, Tian X, Zhao Y, Li Y, Wu X (2024b) Progress in understanding oxidative stress, aging, and aging-related diseases. Antioxidants 13(4):394

Yao Y, Cai X, Fei W, Ye Y, Zhao M, Zheng C (2022) The role of short-chain fatty acids in immunity, inflammation and metabolism. Crit Rev Food Sci Nutr 62(1):1–12. https://doi.org/10.1080/10408398.2020.1854675

Yılmaz T, Şimşek Ö (2020) Potential health benefits of ropy exopolysaccharides produced by Lactobacillus plantarum. Molecules 25(14):3293. https://www.mdpi.com/1420-3049/25/14/3293

Yoo S, Jung SC, Kwak K, Kim JS (2024) The role of prebiotics in modulating gut microbiota: implications for human health. Int J Mol Sci 25(9):4834. https://doi.org/10.3390/ijms25094834

Zhang J, Zhao Y, Sun Z, Sun T (2022) Lacticaseibacillus rhamnosus Probio-M9 extends the lifespan of Caenorhabditis elegans. Commun Biol 5(1):1139. https://doi.org/10.1038/s42003-022-04031-2

Chapter 4
Hormonal Supplements for Health and Longevity

Banteiskhem Kharwanlang and Ramesh Sharma

Abstract Healthy aging usher forth the benefits that old age can provide to society. Healthy aging can be achieved by balancing the optimum neuroendocrine via hormonal supplements, optimum physiology and homeostasis resulting in optimum biological and functional capacity with reduced long-term care in elderly. Hormonal supplements and as well other hormonal interventions such as melatonin, phytomelatonin, dehydroepiandrosterone, phytoestrogen, oral modified insulin-like growth factor-I and dietary arginine and selenium induction of insulin-like growth factor-I plausibly have positive effect on the neuroendocrine but with systematic study on their unwanted effects across different decades in the elderly. Here we propose the role of a theoretical visual representation through a Radar diagram to achieve healthy aging or add life to later years by manoeuvring the macro variables like neuroendocrine, physiology and homeostasis, functional and social capacity, long term care while keeping the average life span constant.

Keywords Aging · Hormones · Melatonin · Insulin-like growth factor-I · DHEA · Estrogen · Testosterone

Abbreviations

AKT	Protein kinase B or PKB
ALP	Alkaline Phosphatase
ATF1	Activating transcription factor-1
BMI	Body Mass Index
CASTOR	Cytosolic arginine sensor for mTORC1 subunit
c-FOS	Proto-oncogene
COL1A1	Collagen Type I Alpha 1 Chai

B. Kharwanlang
Department of Biochemistry, St. Edmund's College, Shillong, Meghalaya, India

R. Sharma (✉)
Department of Biochemistry, North-Eastern Hill University, Shillong, Meghalaya, India

S. I. S. Rattan, C. Gezer (eds.), *Supplements and Drugs for Healthy Longevity*, Healthy Ageing and Longevity 22, https://doi.org/10.1007/978-3-032-11608-6_4

COVID	Coronavirus disease of 2019
CREB	cAMP response element binding protein
c-SRC	cellular sarcoma protein
CTB	Cholera Toxin B subunit
DHEA	Dehydroepiandrosterone
DHEAS	Dehydroepiandrosterone sulfate
DHT	Dihydrotestosterone
DLX5	Distal-less Homeobox 5
DMP1	Dentin matrix protein 1
EEG	Electroencephalography
EGFR	Epidermal growth factor receptor
ERK	Extracellular signal-regulated kinase
FOXO	Forkhead box protein O
GPCR	G- protein coupled receptor
GPER	G protein-coupled estrogen receptor
IGA	Investigator's Global Assessment
IGF	Insulin-like Growth Factor
IGFBP	IGF-binding proteins
IL	Interlukin
JAK	Tyrosine-protein kinase
LDL	Low-density lipoprotein
MAPK	Mitogen activated protein kinase
MMP-1	Matrix metalloproteinase
MT	Melatonin Receptor
mTORC	Mammalian target of rapamycin
NF-κB	Nuclear factor kappa B
NO	Nitric oxide
NrF2	Nuclear factor erythroid 2–related factor 2 rf2
ORNT1	Ornithine Transporter 1
OSX	Osterix
p16INK4a	Cyclin dependent kinase inhibitor 2A
p42/44	Extracellular signal-related kinase or ERK
PAI-1	Plasminogen activator inhibitor type 1
PGC1α	Peroxisome proliferator-activated receptor-γ coactivator 1-α
PI3K	Phosphoinositide 3-kinase
PKC	Protein kinase C
PPARS	Peroxisome proliferator-activated receptor
PXR SXR	Pregnane X receptor
RUNX2	Runt-related transcription factor 2
SIRT1	Silent information regulator sirtuin 1
SOST	Sclerostin
SPARC	secreted protein acidic and rich in cysteine
SPP1	Secreted Phosphoprotein 1
VLDL	Very Low Density Protein
1,25(OH)2D3	1,25-dihydroxyvitamin D3

4.1 Introduction

A world where healthy longevity fosters the growth and development of the global society is worth achieving. Healthy aging can be equated to years of good health that approximate with the lengthened life span. Healthy-lengthened life span brings forth enormous benefits to the global world at various levels. Aged people from different walks of life can usher in their knowledge and experiences to younger generations and vice versa thereby cutting down the time and cost in solving problems arising at any point in future. Healthy aging is a payoff of lifelong consequences which can be achieved by preferably adopting the life-course approach in managing health and disease rather than solely intervening at old age (Fig. 4.1). In addition to various factors, one of the life-course approaches is through hormonal supplementation. Supplement implementation at various levels of a biological being is associated with physical growth and development in early years, whereas at later stages of life prevents functional decline and restores health. Hormonal supplements substantiate dietary nutritional gaps and metabolic deficiency on the production of hormones, play a crucial role in preventing and managing specific health conditions, enhance performance and athletic outcomes, and can help address age-related diseases. However, critical research and knowledge generation on the proper implementation of dietary supplements across various age groups, physiological context and gender is required (Kuh et al. 2014; Archer 2024). Coordinated research on dietary science and its effective application is required according to the ODS Strategic Plan 2025–2029 on Dietary Supplement Research Agenda. Further research must be done on the cellular mechanism, efficacy and metabolism of dietary supplements

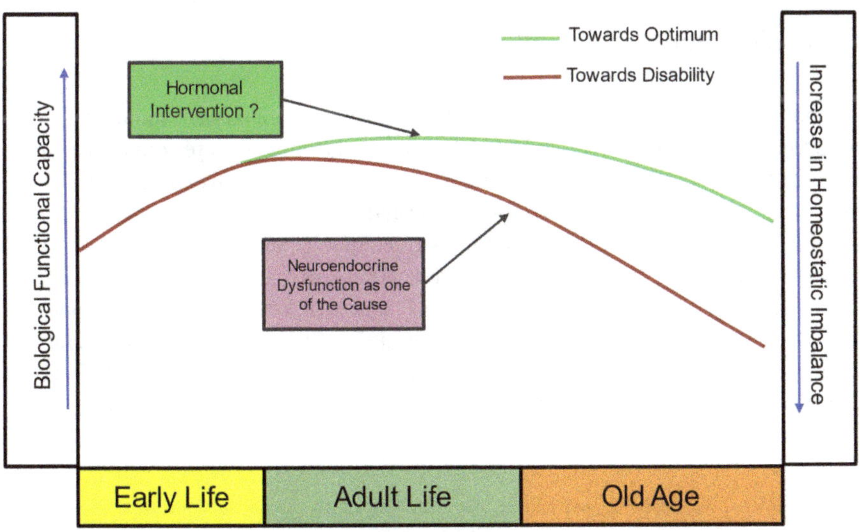

Fig. 4.1 Neuroendocrine dysfunction in elderly. Biological capacity and homeostasis decrease in elderly compared to adults. Hormonal intervention may reverse dysfunction of the neuroendocrine

(Pasiakos and Kuszak 2025). This chapter will focus on the mechanism of action of hormonal supplements or phytohormonal supplements or dietary-induced hormones which mediate their action on health during aging in elderly population and their negating factors that require proper consideration in hormone therapy and supplements (National Academy of Medicine 2022; Coates et al. 2024).

4.1.1 Impaired Neuroendocrine in Aging

Aging is a deteriorative process whereby the cells, organs and systems reach functional impairment, pathological age-related disorder and ultimately death. One of the mechanisms elucidating the decline in the body physiological homeostatic imbalance is the impaired underlying neuroendocrine systemic, cellular communication, coordination and signalling action augmented with diminished response to stress. The various endocrine axis such as hypothalamic-pituitary adrenal/thyroid/gonadal axes, enteroinsular (pancreas and gut) axis and other endocrine glands like pineal, parathyroid and others exhibit a changing pattern in the hormonal secretions associated with various physiological and pathological alterations, detrimental to aged individuals (Fig. 4.1). Glucocorticoids (GCs) signaling impairment with aging like heightened GCs level, disproportionate hydroxysteroid dehydrogenase (HSD) enzymes and reduced glucocorticoid receptor level could have implications in the age-related disorders and may ultimately cause death (Kharwanlang and Sharma 2020).

4.1.2 Hormonal Intervention in Aging

Hormonal replacement therapy (HRT), phytohormones, dietary components having endocrine modulatory effect and off-label use of hormones were found to reverse some of the endocrine function during aging providing improvement in the life of elderly population (Fig. 4.2). Exogenous usage of hormones has some short-term benefits, such as improved body composition, mood, neurocognition, and sexual function, decreased oxidative stress, management of cardiovascular disease and neurodegenerative disease. However, HRT and off label use of hormones are associated with various side effects (Son et al. 2019; Langer et al. 2021; Harper-Harrison et al. 2025). Phytohormones and biological agents seem to offer benefits with fewer side effects (Liu et al. 2020).

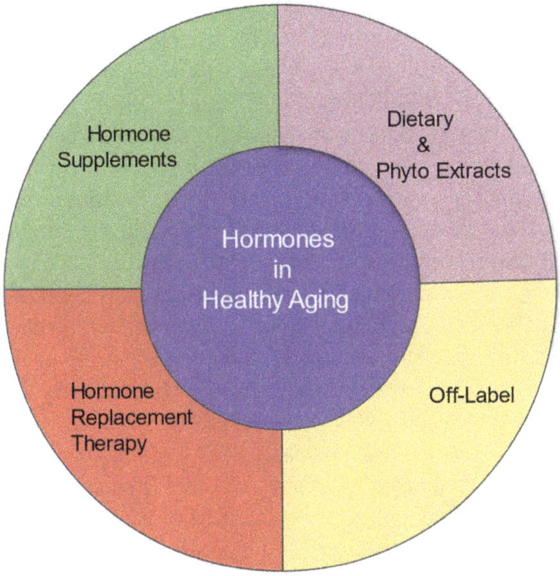

Fig. 4.2 Different roles of hormone for healthy aging. Hormonal intervention may be achieved via hormonal supplements, hormone replacement therapy, natural metabolite having hormonal-like effect such as phytoestrogen or inducing hormonal production by arginine and off label use

4.2 Melatonin

Melatonin, an endogenous indoleamine is the primary hormone secreted by the pineal gland which is crucial for the modulation of the day-night sleep pattern via its action on the suprachiasmatic nucleus (SCN) and as a rhythmic synchronizer via its regulation on the expression of the clock genes in neural and peripheral tissue. Extrapenially, melatonin is synthesized by the mitochondria of peripheral tissues such as retina, gastrointestinal tract, liver, intestine, kidney, the innate immune system and various other tissues. Unlike pineal melatonin, which is endocrine in nature, the extra pineal melatonin is paracrine. Extrapenial melatonin is found to have various physiological effects on many tissues such as modulation of skin pigmentation, immunomodulatory, antioxidant, antiapoptotic, beneficial action on the cardiovascular system, antidepressant and neuro protection (Slominski et al. 2012; Brzezinski et al. 2021). Melatonin is a broad-spectrum antioxidant and free radical scavenger with ability to directly scavenge and neutralize anions, hydroxyl radicals, single oxygen and hydrogen peroxide. Indirectly, melatonin stimulates the cellular antioxidant defense system via elevating glutathione peroxidase, catalase and superoxide dismutase. It has been known to ameliorate age-dependent increase in protein carbonyls caused due to oxidative modification in the cerebral hemispheres and liver of mice (Dkhar and Sharma 2011, 2013). Melatonin decreases lipid peroxidation and malondialdehyde level (Monteiro et al. 2024).

Melatonin exerts its effect by binding to melatonin receptors MT1 and MT2 a family of A G protein-coupled receptors and nuclear melatonin signaling via ROR/RZR (retinoid orphan receptors/retinoid X receptors) groups. The varied multiple biological action of melatonin is attributed to binding with melatonin receptors and its crosstalk with a large diversity of signaling pathways such as PI3K/Akt, PKC and ERK pathways via MT1. MT1 regulation on transcription factors like Nrf2, PGC1α and NF-κB via SIRT1 might underlie the role of melatonin as an antioxidant and in anti-inflammation. The stoichiometry of MT receptors, transducers, effectors, modulatory proteins, crosstalk and interactions with other signaling pathways which vary spatially across tissues, cell types and across dynamic physiological states are important determinants on the various biological actions of melatonin. For example, activation of JAK2, ERK and the Akt/FOXO1 complex are suggested to mediate melatonin-induced cell survival and to modulate pluripotency/differentiation of stem cells, while melatonin-induced inhibition of these pathways is reported in cancer cells (Cecon et al. 2018; Okamoto et al. 2024).

Melatonin decreases with age whereby in elderly the nocturnal peak of this pineal hormone is at its lowest almost 25% of young ones. Decreased production of Melatonin is a consequence of SCN degradation with age. Low melatonin concentration is associated with many diseases in the elderly such as neurological disorder, unable to cope with stress, pain, cardiovascular diseases such as high levels of LDL and coronary heart diseases, physiological conditions such as decreased nighttime blood pressure, metabolic disorder such as type 2 diabetes and others. Elderly people having much lower melatonin production compared to their equivalent healthy age group suffer from sleep disorders (Hardeland 2012; Obayashi et al. 2014; Anghel et al. 2022).

4.2.1 Melatonin in Managing Age-Related Disorder

Reactive oxygen species (ROS) develop in skin cells induced by ultraviolet rays. ROS results in degradation of nucleic acids, proteins and lipids resulting in apoptosis thereby accelerating aging. Melatonin antioxidant property neutralizes the production of ROS as reported that melatonin prevents sun damage of skin cells prior to UV irradiation. Melatonin stabilizes the mitochondrial membrane potential and maintains mitochondrial homeostatic balance in keratinocyte cell culture. Studies on the topical application of melatonin in human do not provide clear demonstration on the efficacy however one study provides significant reduction of the IGA score, reduction in the coarse and fine wrinkle volume after 2 months treatment while another study shows significant improvement on skin tonicity and skin dryness after 3 months of application with melatonin-based cream (Anghel et al. 2022; Milani and Puviani 2018). Suspected differential skin protection was observed in an *ex vivo* study on aged skin cells by downregulation on aging promoting molecules. Intraepidermal mTORC1 pathway and epidermal MMP-1 protein expression is downregulated whereas other key aging biomarkers like SIRT-1, LAMIN-B1, p16INK4 and collagen I remain unchanged (Samra et al. 2023).

Melatonin exerts neuroprotection and positive effects on cognition in elderly. Elderly participants exhibit improved performance in the California Verbal Learning Test compared to placebo. Patients with Alzheimer Disease (AD) showed improvement in Mini Mental State Examination scores. Endogenous high melatonin level is associated with low prevalence mood, cognitive impairment independent of depressive symptoms. Elderly persons with greater levels of melatonin before habitual bedtime positively correlate with hippocampal volume, a brain area critical to memory and vulnerable to damage at early stages of AD (Zhang et al. 2025). Melatonin administration in postmenopausal women with pre-existing sleep disorder improves EEG patterns and subjective sleep quality, bone density and BMI however with increased VLDL and triglycerides (Toffol et al. 2021; Treister-Goltzman and Peleg 2021).

4.2.2 *Phytomelatonin in Managing Age-Related Disorder*

Dietary consumption of tropical fruits like pineapple, orange and banana having trace amounts of phytomelatonin results in a positive correlation of increased melatonin concentration and antioxidant capacity in the serum of healthy volunteers in proportion to serum melatonin levels. Consumption of various food items like grapes, walnut which contain phytomelatonin is associated with increased antioxidant capacity of serum, sleep improvement and cardioprotection (Sae-Teaw et al. 2013; Bonomini et al. 2018). In the human keratinocyte cell line HaCat, phytomelatonin was found to have antioxidants, cell vitality improvement capacity and cytoskeletal remodelling (Ferri et al. 2019). Melatonin extracted from plants inhibits the binding of [^{125}I] iodomelatonin in the brain and their increased level in rabbit plasma (Hattori et al. 1995). Most of the beneficial use of melatonin and phytomelatonin is studied in adults, cell model and animal models. The use of melatonin supplements in older individuals requires intense study since the signaling pathway, cellular and tissue context of an older person is different from adults.

4.2.3 *Plausible Factors to Be Taken into Consideration During Melatonin Hormonal Intervention in Elderly*

Elderly persons who were exposed to significantly less environmental light suffered from diminished nocturnal melatonin secretion especially in insomniac cases (Mishima et al. 2001). Melatonin elimination half-life is 1–2 h and bioavailability is 1–74% depending upon formulation and dosage. Melatonin supplementation on a long- term use comes with concern regarding their side effects such as headache, nausea and dizziness especially in the old people.

Melatonin may increase blood pressure and cause hypothermia. Other side effects include neurological disorders such as convulsion, anxiety and depression, skin disorder such as rashes and digestive problems such as constipation and acute pancreatic. Melatonin stimulates the production of IL-6 a pleiotropic in relation to proinflammatory and antiinflammatory (Savage et al. 2025). Phytomelatonin potential as a 100% natural supplement in place of synthetic melatonin (associated with undesirable contaminants), could be achieved via removing other possible phytochemicals impurity (have some side effects) (Arnao and Hernández-Ruiz 2018).

4.3 GH/IGF-I

The growth hormone (GH)-insulin like growth factors (IGF-I & II) is an important endocrine axis regulating anabolic functions and skeletal growth. Human GH is synthesised and secreted by the somatotrophs of the anterior pituitary. GH secretion is under the control of Growth Hormone Releasing Hormone (GHRH) of the hypothalamus, ghrelin a hunger response hormone and IGF-I. GHRH and ghrelin positively act on the somatotropic cells to secrete GH whereas IGF-I, the primary mediator of GH action, have negative feedback on the somatotrophs and positively regulate the release of somatostatin. Somatostatin inhibits the release of GHRH as well as release of GH. GH activates the GH receptors in hepatocytes, stimulating the gene expression of IGF-I and its secretion from the liver. About 75% of circulating IGF-I is released from the liver acts in an endocrine manner whereas local IGF-I synthesised by tissue acts in an autocrine or paracrine manner (Murray et al. 2004; Khatib et al. 2014).

The *Igf-I* gene comprises six alternatively spliced exons that code for pro-IGF-1A, pro-IGF-1B, and pro-IGF-1C which are further processed to mature IGF-I and E-peptides. Alternative splicing of the *Igf-I* gene and post-translational modifications such as non-glycosylated pro-IGF-I, and glycosylated-pro-IGF-I result in an extracellular milieu comprising of mature IGF-I, non-glycosylated pro-IGF-I, and glycosylated-pro-IGF-I. Layers of complexity on binding to various receptors arise such as action of insulin and IGF-II on the activation of IGF-I receptors (IGF-IR), as well as binding of IGF-I and insulin to the hybrid receptor of IGF-IR and insulin receptor (IR) (Bikle et al. 2015; Poreba and Durzynska 2020).

4.3.1 IGF in Aging

IGF and its related signaling pathway play an important role in normal physiological growth and development across life span and its deficiency is associated with various pathophysiological conditions such as cardiovascular diseases, cancer, diabetes and other endocrinological disorders. IGF-I level decreases with age to about 30–60% which is closely correlated with the low level of GH with aging and appearance of the above listed pathophysiological impairments and diseases (Brinkman et al. 2025).

Reduction in IGF-I is associated with sarcopenia in elderly as a consequence of diminished GH/IGF-I axis ultimately decreasing protein anabolism in skeletal muscle cells resulting in the degeneration of the structure and function of skeletal muscle cells. Age-related sarcopenia, the progressive and age-associated loss of skeletal muscle mass and function, is a major reason for dependency and mortality in aging societies and thus place an enormous economic burden on the healthcare systems (Bian et al. 2020; Jarmusch et al. 2021). When circulating IGF-I levels decrease significantly with age it decreases basal cerebral blood flow and predicts the magnitude of age-related decline in neurovascular coupling responses which contributes to neurovascular aging, impairing cerebral blood flow (CBF) and functional hyperemia in older adults (Toth et al. 2022).

4.3.2 Arginine Supplement in Managing Age-Related Disorder via IGF-I

L-Arginine supplementation in elderly women with or without type 2 diabetes mellitus (T2DM) was able to improve vascular/microvascular health (Costa et al. 2022). Arginine supplementation along with vitamin C increased walking performance, muscle strength, endothelial function and improved fatigue in adults with long COVID (Tosato et al. 2022). Low levels of arginine in elderly women are associated with a higher risk of sarcopenia (Hua et al. 2024). Arginine supplementation led to a significant elevation of serum insulin-like growth factor concentrations (Hurson et al. 1995) whereas some studies suggest no significant change in IGF-I concentration in elderly (Corpas et al. 1993; Nejati et al. 2023). However, in HepG2 cells it was found that the gene expression induction of GH and IGF-I by qPCR assay by L-Arginine whereas IGF binding protein-1 gene expression is reduced (Oh et al. 2017). HepG2 cells lack the urea cycle enzymes ornithine transcarbamoylase, arginase 1 and ORNT1 (Makris et al. 2024), thereby plausibly a threshold arginine concentration is required to induce IGF-I expression. In arginine-depleted medium there is a retention of IGF-I in the endoplasmic reticulum of hepatocytes (Tsugawa et al. 2019).

4.3.3 *Mechanism of Arginine Induction of IGF-I Level via Activating mTORC1/AKT Phosphorylation of FOXO1*

Arginine promotes diverse physiological effects such as immune cell activation, insulin secretion and muscle growth which is maybe largely mediated through activation of mTORC1 via binding to CASTOR1 (Saxton et al. 2016). When mTOR is activated by phosphorylation (p-mTOR), it stimulates FOXO1 phosphorylation (p-FOXO1) via AKT and p-FOXO1 translocate from the nucleus to the cytoplasm (Sun et al. 2019). Stimulation of mTOR activates AKT to inhibit FOXO1 activity (Hsu et al. 2015). The inhibition of mTORC1 led to a significant decrease in FOXO1 protein in cytoplasm accompanied by a subsequent increased nuclear presence (Rao et al. 2012). Interestingly, nuclear FOXO1 activity mediates the inhibition of GHR expression in the liver thereby reducing GH-mediated synthesis of hepatic IGF-I and secretion. FOXO1 induces the synthesis of IGFBI thereby reducing free circulating IGF-I (Melnik and Zouboulis 2013). Therefore, a plausible mechanism in which arginine induces nuclear to cytoplasmic translocation of phosphorylated FOXO1 suppresses the FOXO1 inhibition on GHR and IGF-I gene expression. Arginine may plausibly activate the mTORC1/AKT induced phosphorylation of FOXO1.

4.3.4 *Selenium Supplement in Managing Age-Related Disorder via IGF-I*

Low serum level of selenium is associated with reduced survival in elderly and functional variables (Giovannini et al. 2018) whereas a positive and significant association between serum selenium and IGF-I serum levels in community dwelling older adults could also play a role of selenium in IGF-I hormonal supplementation during aging (Maggio et al. 2010). Yet in another study selenium along with coenzyme Q10 intervention resulted in increased IGF-I level in elderly (Alehagen et al. 2017).

4.3.5 *Soy Dietary Supplement in Managing Age-Related Disorder via IGF-I*

Soy consumption may increase IGF-I plasma levels which is dependent on the intervention dosage, duration and health status of the participants (Zeng et al. 2020). In postmenopausal women, soy intervention showed that concentrations of IGF-I and IGFBP-3 increased and SHBG decreased after consumption of the soy protein supplement (McLaughlin et al. 2011).

4.3.6 IGF-I Oral Supplement

Proliferation of human oral keratinocytes, gingiva-derived mesenchymal stromal cells and mouse osteoblasts by Pro-IGF-I fused with CTB in a dose-dependent manner is observed with promoted osteoblast differentiation through upregulation of ALP, OSX and RUNX2 genes. Oral delivery of bioencapsulated CTB-Pro-IGF-I in femoral fractured diabetic mice results in bone regeneration with increase in bone volume, density and area. This novel IGF-I delivery system will increase affordability and elderly patient compliance during long term care (Park et al. 2020).

4.3.7 Plausible Factors to Be Taken into Consideration During IGF-I Hormonal Intervention in Elderly

Induction of IGF-I level by dietary supplements to intervene age-related disorder should take into account the IGF isoforms, IGF-I interaction with IGF-IR, IR and IGF-IIR, downstream signaling and the crosstalking of IGF-I signaling with other pathways, nuclear localization of pro-IGF-1B, and as well as the broad variability seen at the individual level of IGF-I concentration in elderly aged between 50 and 90 years. IGF-I intervention in cancerous elderly patients should take into consideration that IGFs released following bone resorption facilitated growth of human breast cancer cells in a human adult bone while upregulation of IGF-IR promoted adherence of neuroblastoma cells to human bone marrow endothelial cells (Milani et al. 2004; Rieunier et al. 2019; Poreba and Durzynska 2020). Interestingly, reduction in the GH/IGF-I axis is associated with longevity in exceptionally long-lived individuals. Nonagenarian females with low IGF-I have had significantly longer survival compared with females with levels above 96 mg/ml. In both males and females with a history of cancer, lower IGF-I levels predicted longer survival. A functional mutation in the IGF-I receptor, which confers partial IGF-I resistance, was more prevalent in centenarians, as compared to controls without familial longevity (Milman et al. 2014). Nonagenarians with low IGF-I/IGFBP3 ratio had higher scores for Activities of Daily Living (ADL) and Instrumental ADL scales. IGF-I/IGFBP3 ratio may plausibly be a better indicator of IGF-I bioavailability than total IGF-I (van der Spoel et al. 2015). However, healthy centenarians have plasma IGF-I/IGFBP3 molar ratio greater than aged subjects (Paolisso et al. 1997). A large population-based prospective study showed that individuals' risk of deaths from cancer, CVD, and all causes is associated with lowest and highest levels of circulating IGF-I. For individuals with low IGF-I, the excess risks of death were more pronounced among individuals with liver disorder such as cancer, cirrhosis but were also present among individuals without elevated liver enzymes (Mukama et al. 2023).

A U-shaped relationship between the IGF-I as a biomarker and mortality seems to exist but this relationship is attenuated when participants without liver disease are only included (Mukama et al. 2023). In elderly, moderate levels of IGF-I levels are present with a reduced risk of depression whereas increased risk of depression associated symptoms is conferred with IGF-I values at opposite ends of the continuum, thereby exhibiting a U-shaped relationship (Chigogora et al. 2016). In relation to mortality, a U-shaped relationship is exhibited between IGF-I and mortality risk on either end of the IGF-I continuum whereas IGF-I within the range is associated with the lowest mortality (Rahmani et al. 2022).

4.4 Dehydroepiandrosterone and Dehydroepiandrosterone Sulphate

Dehydroepiandrosterone (DHEA) is mainly synthesised in the adrenal zona reticularis of the adrenal cortex. DHEA sulphate (DHEAS) is an endogenous sulfated product of DHEA in the adrenal cortex whereas in the liver and intestine DHEAS is a sulfated product of exogenous DHEA. DHEA is subsequently converted to testosterone and 5-dihydrotestosterone in the gonads, liver, adrenals and peripheral tissues. Testosterone derived from testicular DHEA accounts for a portion of total testosterone. In post-menopausal women about 90% of estrogens are derived from peripheral DHEAS (Parish et al. 2021; Powrie and Smith 2018).

DHEA and DHEAS binds and activates nuclear receptors such as Androgen receptors, Estrogen receptors, PPARS, PXR/SXR. Activation of plasma membrane GPCR by DHEA activates MAPK, DHEAS activates ERK1/2, c-SRC, ATF1 and CREB. Activation of GPER by DHEA stimulates EGFR, SRC, ER1/2 and PI3K. Activation of N-methyl-D-aspartate receptors by DHEA increased intracellular calcium (Compagnone and Mellon 1998; Prough et al. 2016).

DHEA may be beneficial for bone health through its inhibitory effect on skeletal catabolic IL-6 (Zhou and Glowacki 2018). Interestingly, in humans the increase in serum IL-6 during the process of aging is positively correlated with diminished DHEA leading to age-related diseases such as rheumatoid arthritis, osteoporosis, atherosclerosis, and late-onset B cell neoplasia (Straub et al. 1998). DHEA inhibits the proliferation of human myeloma cells and the production of their Interleukin-6 (Liu et al. 2005).

4.4.1 DHEA in Aging

DHEA and DHEAS decreases by almost 80% in serum between the age of 25 and 75 years with peak concentration earlier in females at puberty. The decline in DHEA and its biological derivatives is associated with various health related disorders such as increased frailty, immune dysfunction, neurodegeneration and the progression of age-related conditions such as osteoporosis, gradual cognitive decline and Alzheimer's disease (Powrie and Smith 2018; Erceg et al. 2025).

4.4.2 DHEA in Managing Age-Related Disorder

DHEA administration in healthy older women enhanced bone mineral density on the lumbar spine which is associated with testosterone, estradiol and insulin-like growth factor (IGF-I) which is absent in male counterpart. IGF-I is associated with improving fracture healing (von Mühlen et al. 2008; Quester et al. 2022). DHEA supplementation in elderly men can induce a small but significant positive effect on body composition via bioactive metabolites such as androgens or estrogens. Appropriate intake dose of DHEA results in elderly serum DHEAS concentrations usually measured in young healthy male adults (Arlt et al. 1999; Corona et al. 2013).

4.4.3 Plausible Factors to Be Taken into Consideration During DHEA Hormonal Intervention in Elderly

DHEA and DHEAS in *in vitro* models of human osteoblastic cells such as MG63 osteosarcoma cell line and normal primary osteoblast-like cells (HOB) did not have independent significant effects on the growth and differentiation of human osteoblastic cells. However, treatment along with 1,25$(OH)_2$D3 resulted in enhancement of specific alkaline phosphatase activity to stimulate human osteoblastic cell differentiation (Scheven and Milne 1998). The *in vitro* stimulation of osteoblastogenesis by DHEA in hMSCs requires the presence of an IGF-I receptor and activation of the PI3K, p38 MAPK, or p42/44 MAPK signaling pathways (Zhou and Glowacki 2018).

4.5 Estrogen

In humans, circulating testosterone (T) and 17β-estradiol (E2) secreted by gonads at sex-specific concentrations are considered male and female sex hormones because they are involved in sexual differentiation and reproduction. During the perimenopausal period, the granulosa cells of the ovaries secrete the major circulating estrogen E2, whereas during menopause the main circulating estrone E1 (produced in adipose tissue from metabolites of adrenal origin) is the main reserve of E2. T is produced by the ovary (25%), the adrenal gland (25%), and in peripheral tissues (50%) following conversion from circulating androstenedione (A4) equally produced by the ovary and the adrenal gland compare to menopause where T of 50% is of ovarian origin, 10% is of adrenal origin and the remaining 40% is produced locally at peripheral tissue from A4 originated from ovarian and adrenals. In males, total circulating T is produced by Leydig cells of the testis. The testes directly produce approximately 20% of circulating E2 whereas the remaining E2 (80%) is formed via aromatization of circulating T in the peripheral tissues (Mauvais-Jarvis and Lindsey 2024).

The biological effects of estrogen are mostly mediated by its binding with ERα and ERβ, which are members of the nuclear receptor superfamily of transcription factors. While ERα and ERβ can bind to most estrogen response elements (ERE) identically, differential tethering of transcription factors on ERE of different target genes has both unique and overlapping estrogenic effects. The above physiological form of estrogen including estriol (E3) can also bind to G Protein-Coupled Estrogen Receptor (GPER). E1 and E2 have been described as agonists for (GPER), while (E3) acts as an antagonist (Hua et al. 2018; Hernández-Silva et al. 2020).

4.5.1 Estrogen and Testosterone in Men

In men, a steady decline in gonadal androgen production is observed with age. Typical testosterone-to-estradiol (T/E) ratios are significantly higher in older versus younger men where the decreased rate of estrogen production is less as compared to testosterone with substantial variation. Both normal aging and hypogonadism are associated with loss of strength and muscle mass, decreased libido and erectile function, decreased cognitive function and loss of bone mineral density. Interestingly, androgen replacement in elderly men increases muscle mass, lean body mass/body fat ratios, well-being, libido and bone density (Orwoll et al. 2006; Leder 2007). Elderly men have increased risk of mortality in parallel with subjects with very low values of both testosterone and estradiol (Tivesten et al. 2009).

4.5.2 Estrogen and Testosterone Supplement in Elderly Men

Another study shows that oral estrogen 17beta-estradiol in elderly men reduces homocysteine, fibrinogen, and PAI-1 and favourably influences VLDL, LDL and HDL subclass levels without changing the markers of thrombotic risk such as thrombin-antithrombin III complex, prothrombin fragment 1.2, D-dimer, antithrombin activity, protein-C and S and von Willebrand factor antigen (Giri et al. 1998). Rather than the absolute value of T and E2, in elderly men, the T/E2 balance plays a key role in the relationship between sex hormones and the risk of cerebrovascular disease. Extremely low T/E2 and inappropriately high T/E2 ratio can be both harmful for the brain blood vessels, while higher free E2 decreased this risk. In managing cerebrovascular disease in elderly men, supplementing with estrogen seems to be worthwhile during testosterone replacement treatment (Gong et al. 2013). In elderly men, beneficial effects of DHEA supplementation at a dose such that it leads to serum DHEAS concentrations usually measured in young healthy adults is observed. Serum testosterone and dihydrotestosterone remained unchanged after the DHEA administration (Arlt et al. 1999). A possible clinical role for estrogenic compounds in cardiovascular risk reduction where a low-dose estrogen supplementation in hypogonadal men is well tolerated, lowers blood pressure and improves vascular function may be through enhancement of basal NO release and attenuation of vasoconstrictor responses to angiotensin II and norepinephrine (Komesaroff et al. 2001).

4.5.3 Plausible Factors to Be Taken while Considering Estrogen and Testosterone Hormonal Intervention in Elderly Men

Variation in E2 serum in elderly men is associated with stroke where subjects in high serum E2 have a higher risk of stroke than those whose estradiol levels were lower. However, testosterone was not related to the risk of stroke (Abbott et al. 2007). Unchanged or maintained testosterone level is correlated with bone health. The U-shaped relationship, of low and high testosterone levels, negatively impacts bone health (Ma et al. 2024) whereas midrange plasma testosterone is associated with proper bone health with lower incidence of any fracture and hip fracture (Yeap et al. 2020).

4.5.4 Estrogen in Females

Estrogen deficiency is characterised by a relatively abrupt decline in the ovarian production of estrogen amounting to the appearance of typical menopausal symptoms such as hot flashes and night sweats, mood changes, sleep disturbance (nocturnal awakening) and cognitive impairment (Birge 2003), with increased cardiovascular disease risk and accelerated vascular aging such as arterial stiffening and endothelial dysfunction. Menopausal transitions can impair the quality of life which extend across all stages of menopause (Hildreth et al. 2018; Todorova et al. 2023). In women above 70 years of age, estrone (E_1), from adipocytes appears to increase in level via aromatization, is an important precursor for peripheral E_2 biosynthesis by 17β-hydroxysteroid dehydrogenase (HSD) enzymes. In addition, the circulating T levels do not differ meaningfully from those of premenopausal women. Intracellular production, action and metabolism of E2 and DHT in the aged women occurs in the same tissue (Davis et al. 2019). Serum E1 level is positively correlated with bone mineral density, and breast cancer risk whereas inversely with colon cancer risk in postmenopausal women (Davis et al. 2020). Higher E2 levels were associated with the prevalence of frailty among women younger than 79 year with increased body mass index and high-sensitivity C-reactive protein (Carcaillon et al. 2012).

The history of hormone replacement therapy (HRT) started in the 1960s reaching the maximum use by the 1990s. HRT was promised to overcome estrogen loss and conserve femininity. However, after findings of the Women's Health Initiative (WHI) in 2002 showing that HRT had more detrimental than beneficial effects created panic among some users. Later a reanalysis of the WHI trial was performed where new studies showed that the use of HRT in younger women or in early postmenopausal women had a beneficial effect on the cardiovascular system, reducing coronary disease and all-cause mortality. A shift from feminine forever to healthy forever in estrogen therapy in women is beneficial with proper dosage, in combination with progesterone, proper guidelines and managing specific diseases in women (Cagnacci and Venier 2019), especially in elderly women (Järvstråt et al. 2015), with lower risk at lower dose, vaginal or transdermal rather than oral preparations and with E2 in place of conjugated estrogen (Baik et al. 2024).

4.5.5 Phytoestrogen in Managing Age-Related Disorder

Notwithstanding HRT newfound beneficial activity with proper guidelines, there has been a rapid increase in consumption of alternative medicines for the treatment of menopausal symptoms such as phytoestrogens, herbs and nutritional supplements (Glazier and Bowman 2001). Phytoestrogens are estrogenic in structure and function, derived from plants mostly and structurally similar to 17β-estradiol with reduced side effects on the breast and endometrium and lower risk of blood clot. During reproductive age, phytoestrogen competes with high endogenous estrogen whereas in postmenopausal mimics or heightened estrogenic activity in lieu of low-level endogenous estrogen (Desmawati and Sulastri 2019). Phytoestrogens reduce the frequency of hot flushes in menopausal women without serious side-effects (Chen et al. 2015). Genistein-like isoflavones have the ability to suppress net bone resorption in postmenopausal women (Weaver et al. 2009). Isoflavones may account for the low incidence of menopausal symptoms, osteoporosis, cardiovascular disease, and breast and endometrial cancer in Asian countries (An et al. 2001).

4.5.6 Isoflavones Mechanism of Action on Estrogen Signaling

One of the proposed schemes to explain the activity of phytoestrogens is their ability to bind and activate human estrogen receptor α (ERα) and human estrogen receptor β

(ERβ). Purified genistein, daidzein, apigenin, and coumestrol exhibit a differential and robust transactivation of ERα- and ERβ-induced transcription showing up to many folds stronger activation via ERβ, whereas equol, naringenin, and kaempferol were weaker agonists when compared to 17β-estradiol (Harris et al. 2005). Genistein is potent ERß agonists triggering transcriptional activation of almost 1000 times compared to the weak agonistic action via Erα, thereby eliciting a distinct action in comparison to estrogen as a result of the differences in their ability to differentially recruit coregulators such as GRIP1 via Genistein-ERs complex. Isoflavones inhibits the expression of TNFα where excessive synthesis of TNFα is associated with osteoporosis in postmenopausal period (An et al. 2001; Manas et al. 2004). Daidzein increases calcium deposition, ALP activity and phosphate mineralization, increased expression of key osteogenic genes like RUNX2, c-FOS, SPARC, DLX5, SPP1, COL1A1, IGF1, SOST, and DMP1 in bone marrow-derived mesenchymal stem cells and adipose-derived stromal/stem cells. The osteogenic differentiation in the presence of 2 g daidzein analog is lost in the presence of ER antagonist (Strong et al. 2014). Daidzein induces the ERβ binding to the estrogen response element (ERE). The effect of daidzein indirectly adapts from a specific activator of ERβ to an activator of both ERα and β in humans since daidzein can be converted to equol (Kostelac et al. 2003).

4.6 Conclusion

Healthy aging is the ability to maintain a dynamic internal state at physiological level that supports life. Maintaining homeostatic balance in elderly requires the feed-forward and feed-back switches coordinated by the positive and negative regulators of the endocrine system. Rescuing the dysregulated endocrine system is plausible to overcome the physiological and pathological age-related disorder as an outcome of an imbalanced homeostasis (Kharwanlang and Sharma 2020). Hormonal supplements along with other phytochemicals and dietary components having hormonal-like effects or modulating effects on the endocrine system could play an important role in achieving healthy aging. The observations such as the U shaped relationship of IGF-I level and testosterone with mortality and bone health respectively, require that the systematic study on the various plausible factors effecting or associated with each hormone need long term studies to evaluate the efficacy and safety of hormonal anti-aging supplement and therapy including off label use across gender, healthy and pathophysiological disorders, ethnic groups (dietary choices are affected by various factors such as culture), financial constraints and others (Fig. 4.3) (Abud et al. 2022; Burns 2023). Effect of hormonal supplements should be studied extensively at molecular action level for example melatonin and plausibly arginine-induce IGF-I secretion activate the AKT/FOXO1 as mentioned before, whereas physiologically

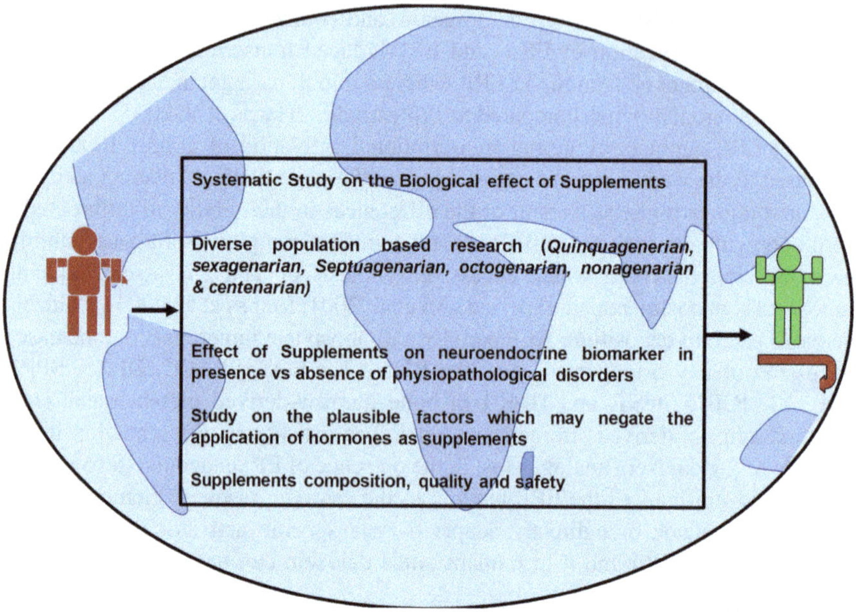

Fig. 4.3 Future systematic study on hormonal supplements for healthy aging. Global study across the later life taking into consideration gender, physiological, pathological, negating factors for safety, composition and efficacy on hormonal intervention for healthy aging

melatonin has a counter effect on some severe arginine induced acute pancreatic (Szabolcs et al. 2006). HRT is well studied however with various side effects so their Phyto counterpart could provide added benefits. Hormonal interventions and supplements could potentially cut off the cost in long term care for elderly. So we propose a theoretical concept as shown in the radar chart (Fig. 4.4), a relationship of optimum neuroendocrine activity via hormonal supplements and therapy with physiological and homeostatic balance leading to reduce long term care for healthy aging with biological and social functional capacity in elderly as opposed to dysregulated neuroendocrine without hormonal intervention with pathophysiological disorders and imbalanced homeostasis leading to high long term care in geriatric health management "Add life to later years" (Elkowitz and Elkowitz 1986; Kharwanlang and Sharma 2020).

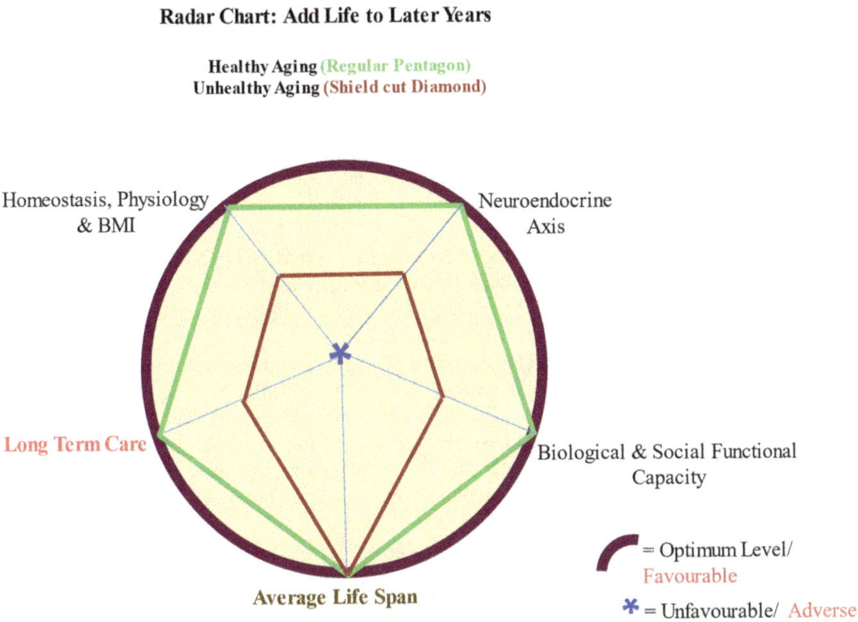

Fig. 4.4 Theoretical radar visual representation on "add life to later years". Each vertex represents the variables which can maintain life like neuroendocrine, homeostasis, functional capacity and long-term care. When the Radar Chart evolves into a Regular Pentagon it represents healthy aging whereas deviation to Shield Cut Diamond represents unhealthy aging. The vertex of age is kept constant as it represents the average life span. Inputting more micro variables such as blood pressure, sleep time and others will change the Regular Pentagon to almost the Circumference of the Radar for healthy aging. A change from Shield Cut Diamond to only a Radius with Age represents extreme frailty in elderly

Acknowledgement BK acknowledges the institutional support provided by St. Edmund's College, Shillong. RS thanks the NEHU, UGC, DST and DBT for the research career support.

References

Abbott RD, Launer LJ, Rodriguez BL, Ross GW, Wilson PW, Masaki KH, Strozyk D, Curb JD, Yano K, Popper JS, Petrovitch H (2007) Serum estradiol and risk of stroke in elderly men. Neurology 68(8):563–568. https://doi.org/10.1212/01.wnl.0000254473.88647.ca

Abud T, Kounidas G, Martin KR, Werth M, Cooper K, Myint PK (2022) Determinants of healthy ageing: a systematic review of contemporary literature. Aging Clin Exp Res 34(6):1215–1223. https://doi.org/10.1007/s40520-021-02049-w. Epub 2022 Feb 8. PMID: 35132578; PMCID: PMC8821855

Alehagen U, Johansson P, Aaseth J, Alexander J, Brismar K (2017) Increase in insulin-like growth factor 1 (IGF-1) and insulin-like growth factor binding protein 1 after supplementation with selenium and coenzyme Q10. A prospective randomized double-blind placebo-controlled trial among elderly Swedish citizens. PLoS One 12(6):e0178614. https://doi.org/10.1371/journal.pone.0178614. PMID: 28609475; PMCID: PMC5469470

An J, Tzagarakis-Foster C, Scharschmidt TC, Lomri N, Leitman DC (2001) Estrogen receptor beta-selective transcriptional activity and recruitment of coregulators by phytoestrogens. J Biol Chem 276(21):17808–17814. https://doi.org/10.1074/jbc.M100953200. Epub 2001 Feb 21

Anghel L, Baroiu L, Popazu CR, Pătraş D, Fotea S, Nechifor A, Ciubara A, Nechita L, Muşat CL, Stefanopol IA, Tatu AL, Ciubara AB (2022) Benefits and adverse events of melatonin use in the elderly (Review). Exp Ther Med 23(3):219. https://doi.org/10.3892/etm.2022.11142. Epub 2022 Jan 14. PMID: 35126722; PMCID: PMC8796282

Archer K (2024) The role of dietary supplements in modern nutrition: benefits, risks, and best practices. Arch Food Nutr 7(4):217

Arlt W, Haas J, Callies F, Reincke M, Hübler D, Oettel M, Ernst M, Schulte HM, Allolio B (1999) Biotransformation of oral dehydroepiandrosterone in elderly men: significant increase in circulating estrogens. J Clin Endocrinol Metab 84(6):2170–2176. https://doi.org/10.1210/jcem.84.6.5789

Arnao MB, Hernández-Ruiz J (2018) The potential of Phytomelatonin as a nutraceutical. Molecules 23(1):238. https://doi.org/10.3390/molecules23010238. PMID: 29361780; PMCID: PMC6017233

Baik SH, Baye F, McDonald CJ (2024) Use of menopausal hormone therapy beyond age 65 years and its effects on women's health outcomes by types, routes, and doses. Menopause 31(5): 363–371. https://doi.org/10.1097/GME.0000000000002335. Epub 2024 Mar 9. PMID: 38595196; PMCID: PMC11465799

Bian A, Ma Y, Zhou X, Guo Y, Wang W, Zhang Y, Wang X (2020) Association between sarcopenia and levels of growth hormone and insulin-like growth factor-1 in the elderly. BMC Musculoskelet Disord 21(1):214. https://doi.org/10.1186/s12891-020-03236-y. PMID: 32264885; PMCID: PMC7140321

Bikle DD, Tahimic C, Chang W, Wang Y, Philippou A, Barton ER (2015) Role of IGF-I signaling in muscle bone interactions. Bone 80:79–88. https://doi.org/10.1016/j.bone.2015.04.036. PMID: 26453498; PMCID: PMC4600536

Birge SJ (2003) The use of estrogen in older women. Clin Geriatr Med 19(3):617–627, viii. https://doi.org/10.1016/s0749-0690(02)00143-x

Bonomini F, Borsani E, Favero G, Rodella LF, Rezzani R (2018) Dietary melatonin supplementation could be a promising preventing/therapeutic approach for a variety of liver diseases. Nutrients 10(9):1135. https://doi.org/10.3390/nu10091135. PMID: 30134592; PMCID: PMC6164189

Brinkman JE, Tariq MA, Leavitt L et al (2025) Physiology, growth hormone. [Updated 2023 May 1]. In: StatPearls [Internet]. Treasure Island (FL), StatPearls Publishing. Available from: https://www.ncbi.nlm.nih.gov/books/NBK482141/

Brzezinski A, Rai S, Purohit A, Pandi-Perumal SR (2021) Melatonin, clock genes, and mammalian reproduction: what is the link? Int J Mol Sci 22(24):13240. https://doi.org/10.3390/ijms222413240. PMID: 34948038; PMCID: PMC8704059

Burns BR (2023) Hormone therapy: aging-related hormone replacement therapy and supplementation. FP Essent 531:22–26

Cagnacci A, Venier M (2019) The controversial history of hormone replacement therapy. Medicina (Kaunas) 55(9):602. https://doi.org/10.3390/medicina55090602. PMID: 31540401; PMCID: PMC6780820

Carcaillon L, García-García FJ, Tresguerres JA, Gutiérrez Avila G, Kireev R, Rodríguez-Mañas L (2012) Higher levels of endogenous estradiol are associated with frailty in postmenopausal women from the Toledo study for healthy aging. J Clin Endocrinol Metab 97(8):2898–2906. https://doi.org/10.1210/jc.2012-1271. Epub 2012 Jun 7. PMID: 22679065

Cecon E, Oishi A, Jockers R (2018) Melatonin receptors: molecular pharmacology and signalling in the context of system bias. Br J Pharmacol. 175(16):3263–3280. https://doi.org/10.1111/bph.13950. Epub 2017 Aug 17. PMID: 28707298; PMCID: PMC6057902

Chen MN, Lin CC, Liu CF (2015) Efficacy of phytoestrogens for menopausal symptoms: a meta-analysis and systematic review. Climacteric 18(2):260–269. https://doi.org/10.3109/13697137.2014.966241. Epub 2014 Dec 1. PMID: 25263312; PMCID: PMC4389700

Chigogora S, Zaninotto P, Kivimaki M, Steptoe A, Batty GD (2016) Insulin-like growth factor 1 and risk of depression in older people: the English longitudinal study of ageing. Transl Psychiatry 6(9):e898. https://doi.org/10.1038/tp.2016.167. PMID: 27648920; PMCID: PMC5048205

Coates PM, Bailey RL, Blumberg JB, El-Sohemy A, Floyd E, Goldenberg JZ, Gould Shunney A, Holscher HD, Nkrumah-Elie Y, Rai D, Ritz BW, Weber WJ (2024) The evolution of science and regulation of dietary supplements: past, present, and future. J Nutr 154(8):2335–2345. https://doi.org/10.1016/j.tjnut.2024.06.017. Epub 2024 Jul 4. PMID: 38971530; PMCID: PMC11375470

Compagnone NA, Mellon SH (1998) Dehydroepiandrosterone: a potential signalling molecule for neocortical organization during development. Proc Natl Acad Sci USA 95(8):4678–4683. https://doi.org/10.1073/pnas.95.8.4678. PMID: 9539798; PMCID: PMC22550

Corona G, Rastrelli G, Giagulli VA, Sila A, Sforza A, Forti G, Mannucci E, Maggi M (2013) Dehydroepiandrosterone supplementation in elderly men: a meta-analysis study of placebo-controlled trials. J Clin Endocrinol Metab 98(9):3615–3626. https://doi.org/10.1210/jc.2013-1358. Epub 2013 Jul 3

Corpas E, Blackman MR, Roberson R, Scholfield D, Harman SM (1993) Oral arginine-lysine does not increase growth hormone or insulin-like growth factor-I in old men. J Gerontol 48(4):M128–M133. https://doi.org/10.1093/geronj/48.4.m128

Costa G, Shushanof M, Bouskela E, Bottino D (2022) Oral L-arginine (5 g/day) for 14 days improves microcirculatory function in healthy young women and healthy and type 2 diabetes mellitus elderly women. J Vasc Res 59(1):24–33. https://doi.org/10.1159/000519428. Epub 2021 Nov 16

Davis SR, Bell RJ, Robinson PJ, Handelsman DJ, Gilbert t, Phung J, Desai R, Lockery JE, Woods RL, Wolfe RS, Reid CM, Nelson MR, Murray AM, McNeil JJ, ASPREE Investigator Group (2019) Testosterone and Estrone increase from the age of 70 years: findings from the sex hormones in older women study. J Clin Endocrinol Metab 104(12):6291–6300. https://doi.org/10.1210/jc.2019-00743

Davis SR, Martinez-Garcia A, Robinson PJ, Handelsman DJ, Desai R, Wolfe R, Bell RJ, ASPREE Investigator Group (2020) Estrone is a Strong predictor of circulating estradiol in women age 70 years and older. J Clin Endocrinol Metab 105(9):e3348–e3354. https://doi.org/10.1210/clinem/dgaa429. PMID: 32614391; PMCID: PMC7394338

Desmawati D, Sulastri D (2019) Phytoestrogens and their health effect. Open Access Maced J Med Sci 7(3):495–499. https://doi.org/10.3889/oamjms.2019.044. PMID: 30834024; PMCID: PMC6390141

Dkhar P, Sharma R (2011) Amelioration of age-dependent increase in protein carbonyls of cerebral hemispheres of mice by melatonin and ascorbic acid. Neurochem Int 59:996–1002. https://doi.org/10.1016/j.neuint.2011.08.019

Dkhar P, Sharma R (2013) Attenuation of age-related increase of protein carbonylation in the liver of mice by melatonin and curcumin. Mol Cell Biochem 380:153–160. https://doi.org/10.1007/s11010-013-1668-9

Elkowitz EB, Elkowitz D (1986) Adding life to later years through exercise. Postgrad Med 80(3): 91–103. https://doi.org/10.1080/00325481.1986.11699513

Erceg N, Micic M, Forouzan E, Knezevic NN (2025) The role of cortisol and Dehydroepiandrosterone in obesity, pain, and aging. Diseases 13(2):42. https://doi.org/10.3390/diseases13020042

Ferri F, Olivieri F, Cannataro R, Caroleo MC, Cione E (2019) Phytomelatonin regulates keratinocytes homeostasis counteracting aging process. Cosmet 6(2):27. https://doi.org/10.3390/cosmetics6020027

Giovannini S, Onder G, Lattanzio F, Bustacchini S, Di Stefano G, Moresi R, Russo A, Bernabei R, Landi F (2018) Selenium concentrations and mortality among community-dwelling older adults: results from IlSIRENTE Study. J Nutr Health Aging 22(5):608–612. https://doi.org/10.1007/s12603-018-1021-9

Giri S, Thompson PD, Taxel P, Contois JH, Otvos J, Allen R, Ens G, Wu AH, Waters DD (1998) Oral estrogen improves serum lipids, homocysteine and fibrinolysis in elderly men. Atherosclerosis 137(2):359–366. https://doi.org/10.1016/s0021-9150(98)00022-7. Erratum in: Atherosclerosis 1998 Jun;138(2):403

Glazier MG, Bowman MA (2001) A review of the evidence for the use of phytoestrogens as a replacement for traditional estrogen replacement therapy. Arch Intern Med 161(9):1161–1172. https://doi.org/10.1001/archinte.161.9.1161

Gong Y, Xiao H, Li C, Bai J, Cheng X, Jin M, Sun B, Lu Y, Shao Y, Tian H (2013) Elevated t/e2 ratio is associated with an increased risk of cerebrovascular disease in elderly men. PLoS One 8(4):e61598. https://doi.org/10.1371/journal.pone.0061598. PMID: 23637864; PMCID: PMC3634802

Hardeland R (2012) Melatonin in aging and disease -multiple consequences of reduced secretion, options and limits of treatment. Aging Dis 3(2):194–225. Epub 2011 Feb 10. PMID: 22724080; PMCID: PMC3377831

Harper-Harrison G, Carlson K, Shanahan MM (2025) Hormone replacement therapy. [Updated 2024 Oct 6]. In: StatPearls [Internet]. StatPearls Publishing, Treasure Island (FL). Available from: https://www.ncbi.nlm.nih.gov/books/NBK493191/

Harris DM, Besselink E, Henning SM, Go VLW, Heber D (2005) Phytoestrogens induce differential estrogen receptor Alpha- or Beta-mediated responses in transfected breast cancer cells. Exp Biol Med (Maywood) 230(8):558–568. https://doi.org/10.1177/153537020523000807

Hattori A, Migitaka H, Iigo M, Itoh M, Yamamoto K, Ohtani-Kaneko R, Hara M, Suzuki T, Reiter RJ (1995) Identification of melatonin in plants and its effects on plasma melatonin levels and binding to melatonin receptors in vertebrates. Biochem Mol Biol Int 35(3):627–634

Hernández-Silva CD, Villegas-Pineda JC, Pereira-Suárez AL (2020) Expression and role of the G protein-coupled estrogen receptor (GPR30/GPER) in the development and immune response in female reproductive cancers. Front Endocrinol (Lausanne) 11:544. https://doi.org/10.3389/fendo.2020.00544. PMID: 32973677; PMCID: PMC7468389

Hildreth KL, Ozemek C, Kohrt WM, Blatchford PJ, Moreau KL (2018) Vascular dysfunction across the stages of the menopausal transition is associated with menopausal symptoms and quality of life. Menopause 25(9):1011–1019. https://doi.org/10.1097/GME.0000000000001112. PMID: 29634636; PMCID: PMC6103796

Hsu CK, Lin CC, Hsiao LD, Yang CM (2015) Mevastatin ameliorates sphingosine 1-phosphate-induced COX-2/PGE2-dependent cell migration via FoxO1 and CREB phosphorylation and translocation. Br J Pharmacol 172(22):5360–5376. https://doi.org/10.1111/bph.13326. Epub 2015 Oct 23. PMID: 26359950; PMCID: PMC5341221

Hua H, Zhang H, Kong Q, Jiang Y (2018) Mechanisms for estrogen receptor expression in human cancer. Exp Hematol Oncol 7:24. https://doi.org/10.1186/s40164-018-0116-7. PMID: 30250760; PMCID: PMC6148803

Hua C, Chen Y, Sun Z, Shi Z, Song Q, Shen L, Lu W, Wang Z, Zang J (2024) Associations of serum arginine acid with sarcopenia in Chinese eldely women. Nutr Metab Lond 21(1):63. https://doi.org/10.1186/s12986-024-00839-3. PMID: 39118134; PMCID: PMC11308234

Hurson M, Regan MC, Kirk SJ, Wasserkrug HL, Barbul A (1995) Metabolic effects of arginine in a healthy elderly population. JPEN J Parenter Enteral Nutr 19(3):227–230. https://doi.org/10.1177/0148607195019003227. Erratum in: JPEN J Parenter Enteral Nutr 1995 Jul-Aug;19(4):329

Jarmusch S, Baber L, Bidlingmaier M, Ferrari U, Hofmeister F, Hintze S, Mehaffey S, Meinke P, Neuerburg C, Schoser B, Tanganelli F, Drey M (2021) Influence of IGF-I serum concentration on muscular regeneration capacity in patients with sarcopenia. BMC Musculoskelet Disord 22(1):807. https://doi.org/10.1186/s12891-021-04699-3. PMID: 34544407; PMCID: PMC8454138

Järvstråt L, Spetz Holm AC, Lindh-Åstrand L, Hoffmann MJ, Fredrikson MG, Hammar ML (2015) Use of hormone therapy in Swedish women aged 80 years or older. Menopause 22(3):275–278. https://doi.org/10.1097/GME.0000000000000294. PMID: 25051289

Kharwanlang B, Sharma R (2020) Aging of the endocrine system. In: Encyclopedia of biomedical gerontology. Academic, pp 159–172. https://doi.org/10.1016/B978-0-12-801238-3.11399-6. ISBN 9780128160763

Khatib N, Gaidhane S, Gaidhane AM, Khatib M, Simkhada P, Gode D, Zahiruddin QS (2014) Ghrelin: ghrelin as a regulatory Peptide in growth hormone secretion. J Clin Diagn Res 8(8): MC13–MC17. https://doi.org/10.7860/JCDR/2014/9863.4767. Epub 2014 Aug 20. PMID: 25302229; PMCID: PMC4190751

Komesaroff PA, Fullerton M, Esler MD, Dart A, Jennings G, Sudhir K (2001) Low-dose estrogen supplementation improves vascular function in hypogonadal men. Hypertension 38(5): 1011–1016. https://doi.org/10.1161/hy1101.095006

Kostelac D, Rechkemmer G, Briviba K (2003) Phytoestrogens modulate binding response of estrogen receptors alpha and beta to the estrogen response element. J Agric Food Chem 51(26):7632–7635. https://doi.org/10.1021/jf034427b

Kuh D, Karunananthan S, Bergman H, Cooper R (2014) A life-course approach to healthy ageing: maintaining physical capability. Proc Nutr Soc 73(2):237–248. https://doi.org/10.1017/S0029665113003923. Epub 2014 Jan 23. PMID: 24456831; PMCID: PMC3981474

Langer RD, Hodis HN, Lobo RA, Allison MA (2021) Hormone replacement therapy – where are we now? Climacteric 24(1):3–10. https://doi.org/10.1080/13697137.2020.1851183. Epub 2021 Jan 6. PMID: 33403881

Leder BZ (2007) Testosterone, estradiol and aromatase inhibitor therapy in elderly men. J Steroid Biochem Mol Biol 106(1–5):162–167. https://doi.org/10.1016/j.jsbmb.2007.05.018. https://www.sciencedirect.com/science/article/pii/S0960076007001197. ISSN 0960-0760

Liu S, Ishikawa H, Li FJ, Ma Z, Otsuyama K, Asaoku H, Abroun S, Zheng X, Tsuyama N, Obata M, Kawano MM (2005) Dehydroepiandrosterone can inhibit the proliferation of myeloma cells and the interleukin-6 production of bone marrow mononuclear cells from patients with myeloma. Cancer Res 65(6):2269–2276. https://doi.org/10.1158/0008-5472.CAN-04-3079

Liu T, Li N, Yan YQ, Liu Y, Xiong K, Liu Y, Xia QM, Zhang H, Liu ZD (2020) Recent advances in the anti-aging effects of phytoestrogens on collagen, water content, and oxidative stress. Phytother Res 34(3):435–447. https://doi.org/10.1002/ptr.6538. Epub 2019 Nov 20. PMID: 31747092; PMCID: PMC7078862

Ma J, Zhao J, Wu N, Han M, Yang Z, Chen H, Zhao Q (2024) Inverted U-shaped association between total testosterone with bone mineral density in men over 60 years old. BMC Endocr Disord 24(1):249. https://doi.org/10.1186/s12902-024-01780-5. PMID: 39558326; PMCID: PMC11572123

Maggio M, Ceda GP, Lauretani F, Bandinelli S, Dall'Aglio E, Guralnik JM, Paolisso G, Semba RD, Nouvenne A, Borghi L, Ceresini G, Ablondi F, Benatti M, Ferrucci L (2010) Association of plasma selenium concentrations with total IGF-1 among older community-dwelling adults: the InCHIANTI study. Clin Nutr 29(5):674–677. https://doi.org/10.1016/j.clnu.2010.03.012. Epub 2010 Apr 22. PMID: 20416996; PMCID: PMC3963695

Makris G, Veit L, Rüfenacht V, Klassa S, Zürcher N, Matsumoto S, Poms M, Häberle J (2024) Expression and function of the urea cycle in widely-used hepatic cellular models. J Inherit Metab Dis 47(6):1228–1238. https://doi.org/10.1002/jimd.12701. Epub 2024 Jan 8

Manas ES, Xu ZB, Unwalla RJ, Somers WS (2004) Understanding the selectivity of genistein for human estrogen receptor-beta using X-ray crystallography and computational methods. Structure 12(12):2197–2207. https://doi.org/10.1016/j.str.2004.09.015. PMID: 15576033

Mauvais-Jarvis F, Lindsey SH (2024) Metabolic benefits afforded by estradiol and testosterone in both sexes: clinical considerations. J Clin Invest 134(17):e180073. https://doi.org/10.1172/JCI180073. PMID: 39225098; PMCID: PMC11364390

McLaughlin JM, Olivo-Marston S, Vitolins MZ, Bittoni M, Reeves KW, Degraffinreid CR, Schwartz SJ, Clinton SK, Paskett ED (2011) Effects of tomato- and soy-rich diets on the IGF-I hormonal network: a crossover study of postmenopausal women at high risk for breast cancer. Cancer Prev Res (Phila) 4(5):702–710. https://doi.org/10.1158/1940-6207.CAPR-

10-0329. Epub 2011 Mar 23. Erratum in: Cancer Prev Res (Phila). 2012 Mar;5(3):498. Dosage error in article text. PMID: 21430071; PMCID: PMC3837570

Melnik BC, Zouboulis CC (2013) Potential role of FoxO1 and mTORC1 in the pathogenesis of Western diet-induced acne. Exp Dermatol 22(5):311–315. https://doi.org/10.1111/exd.12142. PMID: 23614736; PMCID: PMC3746128

Milani M, Puviani M (2018) Anti-aging efficacy of melatonin-based cream: clinical and instrumental skin evaluation. Cosmet 5(4):60. https://doi.org/10.3390/cosmetics5040060

Milani D, Carmichael JD, Welkowitz J, Ferris S, Reitz RE, Danoff A, Kleinberg DL (2004) Variability and reliability of single serum IGF-I measurements: impact on determining predictability of risk ratios in disease development. J Clin Endocrinol Metab 89(5):2271–2274. https://doi.org/10.1210/jc.2003-032150. PMID: 15126552; PMCID: PMC3251209

Milman S, Atzmon G, Huffman DM, Wan J, Crandall JP, Cohen P, Barzilai N (2014) Low insulin-like growth factor-1 level predicts survival in humans with exceptional longevity. Aging Cell 13(4):769–771. https://doi.org/10.1111/acel.12213. Epub 2014 Mar 12. PMID: 24618355; PMCID: PMC4116456

Mishima K, Okawa M, Shimizu T, Hishikawa Y (2001) Diminished melatonin secretion in the elderly caused by insufficient environmental illumination. J Clin Endocrinol Metab 86(1):129–134. https://doi.org/10.1210/jcem.86.1.7097

Monteiro KKAC, Shiroma ME, Damous LL, Simões MJ, Simões RDS, Cipolla-Neto J, Baracat EC, Soares-Jr JM (2024) Antioxidant actions of melatonin: a systematic review of animal studies. Antioxidants (Basel) 13(4):439. https://doi.org/10.3390/antiox13040439. PMID: 38671887; PMCID: PMC11047453

Mukama T, Srour B, Johnson T, Katzke V, Kaaks R (2023) IGF-1 and risk of morbidity and mortality from cancer, cardiovascular diseases, and all causes in EPIC-Heidelberg. J Clin Endocrinol Metab 108(10):e1092–e1105. https://doi.org/10.1210/clinem/dgad212. PMID: 37066827; PMCID: PMC10505533

Murray RD, Kim K, Ren SG, Chelly M, Umehara Y, Melmed S (2004) Central and peripheral actions of somatostatin on the growth hormone-IGF-I axis. J Clin Invest 114(3):349–356. https://doi.org/10.1172/JCI19933. PMID: 15286801; PMCID: PMC484973

National Academy of Medicine; Commission for a Global Roadmap for Healthy Longevity (2022) Global roadmap for healthy longevity. National Academies Press (US), Washington, DC; 2, Vision for Healthy Longevity in 2050. Available from: https://www.ncbi.nlm.nih.gov/books/NBK587286/

Nejati M, Dehghan P, Safari S, Jamilian P, Zarezadeh M (2023) The influence of arginine supplementation on IGF-1: a systematic review and meta-analysis. Clin Nutr ESPEN 55:51–57. https://doi.org/10.1016/j.clnesp.2023.02.024. Epub 2023 Mar 5

Obayashi K, Saeki K, Tone N, Kurumatani N (2014) Relationship between melatonin secretion and nighttime blood pressure in elderly individuals with and without antihypertensive treatment: a cross-sectional study of the HEIJO-KYO cohort. Hypertens Res 37(10):908–913. https://doi.org/10.1038/hr.2014.99. Epub 2014 May 22

Oh HS, Oh SK, Lee JS, Wu C, Lee SJ (2017) Effects of l-arginine on growth hormone and insulin-like growth factor 1. Food Sci Biotechnol 26(6):1749–1754. https://doi.org/10.1007/s10068-017-0236-6. PMID: 30263714; PMCID: PMC6049717

Okamoto HH, Cecon E, Nureki O, Rivara S, Jockers R (2024) Melatonin receptor structure and signaling. J Pineal Res 76(3):e12952. https://doi.org/10.1111/jpi.12952

Orwoll E, Lambert LC, Marshall LM, Phipps K, Blank J, Barrett-Connor E, Cauley J, Ensrud K, Cummings S (2006) Testosterone and estradiol among older men. J Clin Endocrinol Metab 91(4):1336–1344. https://doi.org/10.1210/jc.2005-1830. Epub 2005 Dec 20

Paolisso G, Ammendola S, Del Buono A, Gambardella A, Riondino M, Tagliamonte MR, Rizzo MR, Carella C, Varricchio M (1997) Serum levels of insulin-like growth factor-I (IGF-I) and IGF-binding protein-3 in healthy centenarians: relationship with plasma leptin and lipid concentrations, insulin action, and cognitive function. J Clin Endocrinol Metab 82(7):2204–2209. https://doi.org/10.1210/jcem.82.7.4087

Parish SJ, Simon JA, Davis SR, Giraldi A, Goldstein I, Goldstein SW, Kim NN, Kingsberg SA, Morgentaler A, Nappi RE, Park K, Stuenkel CA, Traish AM, Vignozzi L (2021) International Society for the Study of women's sexual health clinical practice guideline for the use of systemic Testosterone for hypoactive sexual desire disorder in women. J Sex Med 18(5):849–867. https://doi.org/10.1016/j.jsxm.2020.10.009. Epub 2021 Apr 1

Park J, Yan G, Kwon KC, Liu M, Gonnella PA, Yang S, Daniell H (2020) Oral delivery of novel human IGF-1 bioencapsulated in lettuce cells promotes musculoskeletal cell proliferation, differentiation and diabetic fracture healing. Biomaterials 233:119591. https://doi.org/10.1016/j.biomaterials.2019.119591. Epub 2019 Nov 5. PMID: 31870566; PMCID: PMC6990632

Pasiakos SM, Kuszak AJ (2025) Strengthening dietary supplement research: An expanded priority for the NIH Office of Dietary Supplements. Am J Clin Nutr. https://doi.org/10.1016/j.ajcnut.2025.04.001. Epub ahead of print

Poreba E, Durzynska J (2020) Nuclear localization and actions of the insulin-like growth factor 1 (IGF-1) system components: transcriptional regulation and DNA damage response. Mutat Res Rev Mutat Res 784:108307. https://doi.org/10.1016/j.mrrev.2020.108307. Epub 2020 Feb 27

Powrie YSL, Smith C (2018) Central intracrine DHEA synthesis in ageing-related neuroinflammation and neurodegeneration: therapeutic potential? J Neuroinflammation 15(1): 289. https://doi.org/10.1186/s12974-018-1324-0. PMID: 30326923; PMCID: PMC6192186

Prough RA, Clark BJ, Klinge CM (2016) Novel mechanisms for DHEA action. J Mol Endocrinol 56(3):R139–R155. https://doi.org/10.1530/JME-16-0013. Epub 2016 Feb 23

Quester J, Nethander M, Eriksson A, Ohlsson C (2022) Endogenous DHEAS is causally linked with lumbar spine bone mineral density and forearm fractures in women. J Clin Endocrinol Metab 107(5):e2080–e2086. https://doi.org/10.1210/clinem/dgab915. PMID: 34935937; PMCID: PMC9016453

Rahmani J, Montesanto A, Giovannucci E, Zand H, Barati M, Kopchick JJ, Mirisola MG, Lagani V, Bawadi H, Vardavas R, Laviano A, Christensen K, Passarino G, Longo VD (2022) Association between IGF-1 levels ranges and all-cause mortality: A meta-analysis. Aging Cell 21(2): e13540. https://doi.org/10.1111/acel.13540. Epub 2022 Jan 20. PMID: 35048526; PMCID: PMC8844108

Rao RR, Li Q, Gubbels Bupp MR, Shrikant PA (2012) Transcription factor Foxo1 represses T-bet-mediated effector functions and promotes memory CD8(+) T cell differentiation. Immunity 36(3):374–387. https://doi.org/10.1016/j.immuni.2012.01.015. Epub 2012 Mar 15. PMID: 22425248; PMCID: PMC3314246

Rieunier G, Wu X, Macaulay VM, Lee AV, Weyer-Czernilofsky U, Bogenrieder T (2019) Bad to the bone: the role of the insulin-like growth factor Axis in osseous metastasis. Clin Cancer Res 25(12):3479–3485. https://doi.org/10.1158/1078-0432.CCR-18-2697. Epub 2019 Feb 11

Sae-Teaw M, Johns J, Johns NP, Subongkot S (2013) Serum melatonin levels and antioxidant capacities after consumption of pineapple, orange, or banana by healthy male volunteers. J Pineal Res 55(1):58–64. https://doi.org/10.1111/jpi.12025. Epub 2012 Nov 9

Samra T, Gomez-Gomez T, Linowiecka K, Akhundlu A, Lopez de Mendoza G, Gompels M, Lee WW, Gherardini J, Chéret J, Paus R (2023) Melatonin exerts prominent, differential epidermal and dermal anti-aging properties in aged human eyelid skin ex vivo. Int J Mol Sci 24(21):15963. https://doi.org/10.3390/ijms242115963. PMID: 37958946; PMCID: PMC10647640

Savage RA, Zafar N, Yohannan S, Miller JMM (2025) Melatonin. [Updated 2024 Feb 9]. In: StatPearls [Internet]. StatPearls Publishing, Treasure Island (FL). Available from: https://www.ncbi.nlm.nih.gov/books/NBK534823/

Saxton R, Chantranupong L, Knockenhauer K et al (2016) Mechanism of arginine sensing by CASTOR1 upstream of mTORC1. Nature 536:229–233. https://doi.org/10.1038/nature19079

Scheven BA, Milne JS (1998) Dehydroepiandrosterone (DHEA) and DHEA-S interact with 1,25-dihydroxyvitamin D3 (1,25(OH)2D3) to stimulate human osteoblastic cell differentiation. Life Sci 62(1):59–68. https://doi.org/10.1016/s0024-3205(97)01038-2

Slominski RM, Reiter RJ, Schlabritz-Loutsevitch N, Ostrom RS, Slominski AT (2012) Melatonin membrane receptors in peripheral tissues: distribution and functions. Mol Cell Endocrinol

351(2):152–166. https://doi.org/10.1016/j.mce.2012.01.004. Epub 2012 Jan 8. PMID: 22245784; PMCID: PMC3288509

Son DH, Park WJ, Lee YJ (2019) Recent advances in anti-aging medicine. Korean J Fam Med 40(5):289–296. https://doi.org/10.4082/kjfm.19.0087

Straub RH, Konecna L, Hrach S, Rothe G, Kreutz M, Schölmerich J, Falk W, Lang B (1998) Serum dehydroepiandrosterone (DHEA) and DHEA sulfate are negatively correlated with serum interleukin-6 (IL-6), and DHEA inhibits IL-6 secretion from mononuclear cells in man in vitro: possible link between endocrinosenescence and immunosenescence. J Clin Endocrinol Metab 83(6):2012–2017. https://doi.org/10.1210/jcem.83.6.4876

Strong AL, Ohlstein JF, Jiang Q, Zhang Q, Zheng S, Boue SM, Elliott S, Gimble JM, Burow ME, Wang G, Bunnell BA (2014) Novel daidzein analogs enhance osteogenic activity of bone marrow-derived mesenchymal stem cells and adipose-derived stromal/stem cells through estrogen receptor dependent and independent mechanisms. Stem Cell Res Ther 5(4):105. https://doi.org/10.1186/scrt493. PMID: 25168698; PMCID: PMC4355363

Sun H, Shao X, He J, Golos M, Shi B (2019) Role of the mTORFOXO1 pathway in obesity associated renal tubulointerstitial inflammation. Mol Med Rep 19:1284–1293

Szabolcs A, Reiter RJ, Letoha T, Hegyi P, Papai G, Varga I, Jarmay K, Kaszaki J, Sari R, Rakonczay Z Jr, Lonovics J, Takacs T (2006) Effect of melatonin on the severity of L-arginine-induced experimental acute pancreatitis in rats. World J Gastroenterol 12(2): 251–258. https://doi.org/10.3748/wjg.v12.i2.251. PMID: 16482626; PMCID: PMC4066035

Tivesten A, Vandenput L, Labrie F, Karlsson MK, Ljunggren O, Mellström D, Ohlsson C (2009) Low serum testosterone and estradiol predict mortality in elderly men. J Clin Endocrinol Metab 94(7):2482–2488. https://doi.org/10.1210/jc.2008-2650. Epub 2009 Apr 28

Todorova L, Bonassi R, Guerrero Carreño FJ, Hirschberg AL, Yuksel N, Rea C, Scrine L, Kim JS (2023) Prevalence and impact of vasomotor symptoms due to menopause among women in Brazil, Canada, Mexico, and Nordic Europe: a cross-sectional survey. Menopause 30(12): 1179–1189. https://doi.org/10.1097/GME.0000000000002265. Epub 2023 Oct 19. Erratum in: Menopause. 2025;32(5):481. 10.1097/GME.0000000000002550. PMID: 37847872; PMCID: PMC11805475

Toffol E, Kalleinen N, Himanen SL, Partonen T, Haukka J, Polo-Kantola P (2021) Nighttime melatonin secretion and sleep architecture: different associations in perimenopausal and postmenopausal women. Sleep Med 81:52–61. https://doi.org/10.1016/j.sleep.2021.02.011. Epub 2021 Feb 9

Tosato M, Calvani R, Picca A, Ciciarello F, Galluzzo V, Coelho-Júnior HJ, Di Giorgio A, Di Mario C, Gervasoni J, Gremese E et al (2022) Effects of L-arginine plus vitamin C supplementation on physical performance, endothelial function, and persistent fatigue in adults with long COVID: a single-blind randomized controlled trial. Nutrients 14(23):4984. https://doi.org/10.3390/nu14234984

Toth L, Czigler A, Hegedus E, Komaromy H, Amrein K, Czeiter E, Yabluchanskiy A, Koller A, Orsi G, Perlaki G, Schwarcz A, Buki A, Ungvari Z, Toth PJ (2022) Age-related decline in circulating IGF-1 associates with impaired neurovascular coupling responses in older adults. Geroscience 44(6):2771–2783. https://doi.org/10.1007/s11357-022-00623-2. Epub 2022 Jul 23. PMID: 35869380; PMCID: PMC9768079

Treister-Goltzman Y, Peleg R (2021) Melatonin and the health of menopausal women: a systematic review. J Pineal Res 71(2):e12743. https://doi.org/10.1111/jpi.12743. Epub 2021 May 24

Tsugawa Y, Handa H, Imai T (2019) Arginine induces IGF-1 secretion from the endoplasmic reticulum. Biochem Biophys Res Commun 514(4):1128–1132. https://doi.org/10.1016/j.bbrc.2019.05.044. Epub 2019 May 14

van der Spoel E, Rozing MP, Houwing-Duistermaat JJ, Slagboom PE, Beekman M, de Craen AJ, Westendorp RG, van Heemst D (2015) Association analysis of insulin-like growth factor-1 axis parameters with survival and functional status in nonagenarians of the Leiden Longevity Study. Aging (Albany NY) 7(11):956–963. https://doi.org/10.18632/aging.100841. PMID: 26568155; PMCID: PMC4694065 von Mühlen D, Laughlin GA, Kritz-Silverstein D, Bergstrom J, Bettencourt R (2008) Effect of dehydroepiandrosterone supplementation on bone mineral density, bone markers, and body composition in older adults: the DAWN trial. Osteoporos Int 19(5):699–707. https://doi.org/10.1007/s00198-007-0520-z. Epub 2007 Dec 15. PMID: 18084691; PMCID: PMC2435090

Weaver CM, Martin BR, Jackson GS, McCabe GP, Nolan JR, McCabe LD, Barnes S, Reinwald S, Boris ME, Peacock M (2009) Antiresorptive effects of phytoestrogen supplements compared with estradiol or risedronate in postmenopausal women using (41)Ca methodology. J Clin Endocrinol Metab 94(10):3798–3805. https://doi.org/10.1210/jc.2009-0332. Epub 2009 Jul 7. PMID: 19584189; PMCID: PMC2758721

Yeap BB, Alfonso H, Chubb SAP, Center JR, Beilin J, Hankey GJ, Almeida OP, Golledge J, Norman PE, Flicker L (2020) U-shaped Association of Plasma Testosterone, and no Association of Plasma Estradiol, with incidence of fractures in men. J Clin Endocrinol Metab 105(5): dgaa115. https://doi.org/10.1210/clinem/dgaa115

Zeng J, Feng Y, Feng J, Chen X (2020) The effect of soy intervention on insulin-like growth factor 1 levels: a meta-analysis of clinical trials. Phytother Res 34(7):1570–1577. https://doi.org/10.1002/ptr.6630. Epub 2020 Feb 19

Zhang Z, Xue P, Bendlin BB, Zetterberg H, De Felice F, Tan X, Benedict C (2025) Melatonin: a potential nighttime guardian against Alzheimer's. Mol Psychiatry 30(1):237–250. https://doi.org/10.1038/s41380-024-02691-6. Epub 2024 Aug 11. PMID: 39128995; PMCID: PMC11649572

Zhou S, Glowacki J (2018) Dehydroepiandrosterone and Bone. Vitam Horm 108:251–271. https://doi.org/10.1016/bs.vh.2018.01.005. Epub 2018 Feb 24

Chapter 5
NAD$^+$ and Its Precursors for Health and Longevity

Lin Yi

Abstract As the first coenzyme discovered over a century ago, NAD$^+$ (nicotinamide adenine dinucleotide) was found in all living cells from single-celled microorganism, such as bacteria and yeast, to mammals, such as animals and human. It is ubiquitous in all human tissues. Hundreds of very important biological activities in the human body require the participation of NAD$^+$. It is an essential coenzyme to metabolize carbohydrates, fats, and proteins from our diet into energy and other essential nutrients. In addition, it plays an important role in many important biological pathways, such as DNA repair, gene expression, etc. However, NAD$^+$ level decreases significantly during aging. Declining NAD$^+$ level in the human body accelerates aging and contributes to many age-related diseases, such as muscle loss, cardiovascular diseases, neurodegenerative diseases, etc. Replenishing NAD$^+$ has shown to slow the aging process and ameliorate age-related symptoms. NAD$^+$ levels have significant impact on all twelve hallmarks of aging. NR (nicotinamide riboside) and NMN (nicotinamide mononucleotide) are two important NAD$^+$ precursors that effectively raise NAD$^+$ level in both animal studies and human clinical trials. Extensive research has been done during recent years on NR and NMN as NAD$^+$ precursors for their effect and biological mechanism of health and longevity. The dietary supplements of NR and NMN have been commercially available in several countries since 2013. In this chapter, we will give a historical review and current status of NR and NMN as nutritional supplementations on human health and longevity.

Keywords Nicotinamide adenine dinucleotide (NAD$^+$) · Nicotinamide mononucleotide (NMN) · Nicotinamide riboside (NR) · Healthy aging

L. Yi (✉)
Abinopharm, Inc., Shelton, CT, USA
e-mail: lin.yi@abinopharm.com

S. I. S. Rattan, C. Gezer (eds.), *Supplements and Drugs for Healthy Longevity*, Healthy Ageing and Longevity 22, https://doi.org/10.1007/978-3-032-11608-6_5

5.1 The Discovery of NAD^+

The discovery of NAD^+ can be traced back to the scientific studies of fermentation. Although fermentation has been used for producing wine and beer since the beginning of human civilization, it had to wait until mid-1800 when French microbiologist Louis Pasteur found that yeast was responsible for the conversion of sugar into alcohol (Alba-Lois and Segal-Kischinevzky 2010). In 1897, German chemist Eduard Buchner further discovered that the extract from dry and pulverized yeast had the same function of live yeast to ferment sugar into alcohol (Buchner 1897). Buchner named the extract zymase. During 1904–1906, British biochemist Arthur Harden reported that his laboratory found that zymase was, in fact, a mixture of heat-sensitive enzymatic fraction and a heat-stable fraction. Harden found that the two fractions alone were incapable of fermenting sugar, but the mixture of the two had the same power of live yeast to convert sugar into alcohol. He named the heat-stable fraction coferment (Harden 1905; Harden and Young 1906). During the 1910s and 1920s, Swedish biochemist Han von Euler-Chelpin meticulously worked on the chemical nature of the fermentation process, especially the chemical component of the coferment which was then more frequently called cozymase and revealed that the cozymase was a nucleotide consisting of adenine, sugar and phosphate (von Euler-Chelpin 1930). In 1936, German physiologist Otto Warburg figured out that the cozymase also had a pyridine nucleotide (Warburg and Christian 1936). By then, all the chemical components were identified, Warburg gave cozymase a new name diphospho-pyridine nucleotide (DPN) (Slater 2004). Buchner, Harden, von Euler-Chelpin and Warburg were all awarded Nobel Prize for their contributions.

After almost half a century of many scientists' efforts, the structure of cozymase was finally identified, but the several different names (coferment, cozymase, coenzyme, codehydrogenase and DPN) appeared in scientific publications during that time were very confusing (Slater 2004). Eventually coenzyme was adapted by the scientific community and defined as a small, non-protein organic molecule that works with an enzyme to catalyze a biochemical reaction. Since cozymase was the first coenzyme discovered, it was called coenzyme I. From a chemical structure point of view (Fig. 5.1), coenzyme I was a dinucleotide, and its chemical name has been adapted as nicotinamide adenine dinucleotide (NAD^+) since late 1950s. The superscript "+" represented the positive charge on the nitrogen of pyridine ring.

5.2 Historical Therapeutic Usage of NAD^+ and Its Precursors

After the discovery of NAD^+ from yeast fermentation, over a century of scientific studies have confirmed that NAD^+ is one of the most important substances for human health.

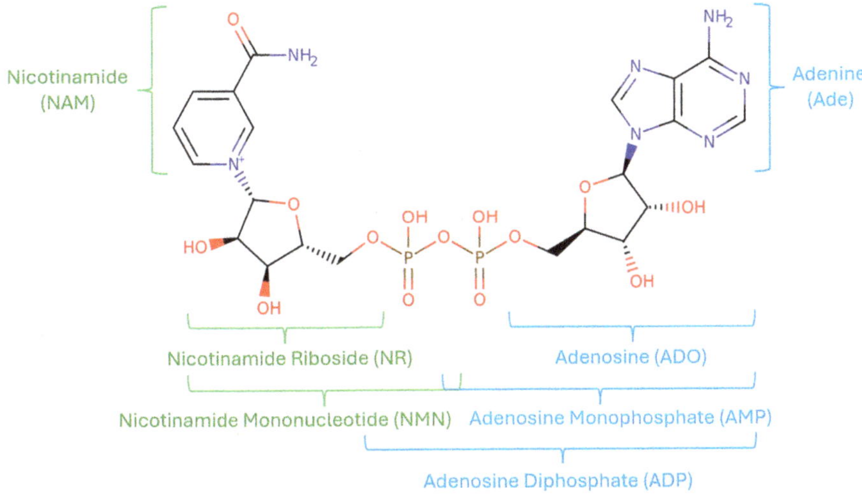

Fig. 5.1 Nicotinamide Adenine Dinucleotide (NAD$^+$) and some of its components

5.2.1 NAD$^+$ and Pellagra

Pellagra was first described as a disease of severe skin rash in 1735 by a Spanish physician, Don Gasper Casal (Jaworek et al. 2021). Pellagra became endemic for nearly two centuries in Southern Europe where corn was the main diet. It was often referred to as "4D diseases": dermatitis, diarrhea, dementia and death. In the US, pellagra became an epidemic for almost 4 decades in southern states with about three million cases and 100,000 deaths after the first case reported in 1902 (Rajakumar 2000). With such an uncurable and widespread deadly disease, the commonly held opinion was that pellagra was an infectious and communicable disease during that time although many observations in Europe and the US indicated it was from certain nutritional deficiency related to maize as the main diet. Dr. Joseph Goldberger was assigned to lead the investigation of pellagra by the US government in 1914. After diligent studies including tests on humans, Goldberger concluded that pellagra was just caused by nutritional deficiency and can be cured by dietary manipulation. By 1926, he eventually found that the heat-stable and water-soluble yeast extract was one of the most effective materials to prevent pellagra (Goldberger et al. 1926). However, he did not realize this yeast extract was the cozymase (now we call NAD$^+$) Harden discovered more than 20 years ago.

In 1937, Conrad Elvehjem and his colleagues identified Goldberger's yeast extract as nicotinic acid (NA) and nicotinic acid amide (now known as nicotinamide, NAM) from liver extract (Koehn and Elvehjem 1937; Elvehjem et al. 1937, 1938). Elvehjem's discovery was quickly confirmed by many other scientists (Spies et al. 1938) and became widely accepted that NA deficiency was the root cause of pellagra. In 1942, flour enrichment with NA began in the US. Pellagra was nearly distinct soon after in the US although there are cases in some undeveloped countries nowadays. Elvehjem and his colleagues further found that tryptophan (Trp) had the same biological functions as NA (Krehl et al. 1945). The connection between NA deficiency and cozymase (NAD^+, coenzyme I) was also reported in 1938–39 (Daft et al. 1938; Axelrod and Elvehjem 1939). Now it is clear that both NA and Trp deficiencies are rooted in NAD^+ deficiency (Feuz et al. 2023). Pellagra caused by NAD^+ deficiency was one of the most significant health disasters of human history.

The United States Government adopted the terms niacin (a shortened form of "nicotinic acid vitamin") as nicotinic acid and niacinamide as nicotinamide in 1942 to avoid the public confusing them with the nearly unrelated (and toxic) nicotine, but later on, niacin has been used as vitamin B3 which includes all NAD^+ precursors: NA, NAM, NR, and NMN. Niacin is also used as the name of a prescription drug with NA as the active pharmaceutical ingredient (details in next section). Therefore, the term, niacin, is used with different meanings in many publications and is confusing.

5.2.2 *Niacin as Medicine*

In 1955, Canadian pathologist Altschul and his colleagues reported that NA in high doses lowered human plasma cholesterol while NAM did not have this effect (Altschul et al. 1955). This landmark discovery led NA to be the first prescription drug for lowering cholesterols since the late 1950s. The name Niacin was specifically used for this prescription drug since then. Soon after Altschul's discovery, Niacin was found to be a miracle broad-spectrum lipid drug: lowering low density lipoprotein cholesterol (LDL-C) and triglyceride (TG) while rising high density lipoprotein cholesterol (HDL) significantly and reducing the progression of atherosclerosis and clinical events/mortality from coronary heart disease (Carlson 2005). Doses of Niacin ranged from 500 mg to 3 g as immediate release (IR) dosage form. However, severe skin flushing was frequently observed during the earlier phase of this drug on the market. In the USA, an extended-release (ER) dosage form (trademark Niaspan®) was approved in 1997 with less frequent flushing. However, two large human clinical trials finished in 2011 and 2014 showed that the combination therapy of Niacin ER with statins, another class of cholesterol lowering drugs, did not provide clinical event benefit over statin alone (Song and FitzGerald 2013; Superko et al. 2017). US FDA withdrew the approval of combination therapy of Niacin ER tablet (Niaspan®) and statins in 2016, but the Niacin immediate release tablet remains on the market for statin intolerant patients.

Although NA is a precursor of NAD^+, its function as a lipid modulating drug has rarely been related to NAD^+ in scientific studies. However, a 2019 publication suggested that the lipid-modulating capability of NA may be rooted to the NAD^+ activation of sirtuin 1 (Romani et al. 2019).

5.3 NAD$^+$ Biosynthesis

Pellagra and many other diseases are ultimately caused by NAD$^+$ deficiency. Efficient NAD$^+$ biosynthesis is obviously one of the most important ways to keep the NAD$^+$ homeostasis. The following discussion mainly focuses on the NAD$^+$ biosynthesis of mammals, especially humans. Other living systems may have slightly different NAD$^+$ biosynthetic pathways.

Right after the discovery of NA treatment of dog's black tong (equivalent to human's pellagra), Elvehjem and his colleagues reported that NA deficiency in dog and pig lowered coenzyme I (NAD$^+$) contents in liver and muscle (Axelrod et al. 1939). They further found that NA supplementation markedly increased NAD$^+$ in human blood and muscle (Axelrod et al. 1941). These results suggested that there were connections between NA and NAD$^+$. In 1940, Philip Handler began to investigate the link of pellagra and the metabolism of NA (Kresge et al. 2004). After more than 15 years of study in this area, Handler and his student Jack Preiss eventually figured out how NA was converted into NAD$^+$ in mammals as disclosed in a series of two research papers in 1958 on the identification of the two intermediates (Preiss and Handler 1958a) and three enzymes (Preiss and Handler 1958b) of the three-step biosynthesis from NA to NAD$^+$ which is now called Preiss-Handler Pathway. Preiss and Handler found that NA was first converted into nicotinic acid mononucleotide (NAMN) by an enzyme Handler later called nicotinic acid mononucleotide pyrophosphorylase (Imsande and Handler 1961), nowadays called nicotinic acid phosphoribosyltransferase (NAPRT) (Galassi et al. 2012); NAMN was then converted to nicotinic acid adenine dinucleotide (NAAD) catalyzed by an enzyme called DPN-pyrophosphorylase, now called nicotinamide mononucleotide adenylyltransferase (NMNAT) (Zhai et al. 2009); finally NAAD was converted to NAD$^+$ by an enzyme called DPN synthetase, now called glutamine-dependent NAD synthetase (NADSYN) (Hara et al. 2003).

In 1945, Elvehjem and his colleagues discovered that Tryptophan (Trp) had similar effect to NA in helping the growth of rats (Krehl et al. 1945). This discovery immediately prompted many studies of Trp metabolism into NA but with little success (Nishizuki and Hayaishi 1963). Instead, these studies eventually led to the deciphering of the NAD$^+$ biosynthetic pathway from Trp by Haiyashi and his colleagues in 1965 (Ikeda et al. 1965), which is now called de novo synthesis. In this NAD$^+$ biosynthetic pathway, Trp was first degraded to N-formylkynurenine by either indoleamine 2,3-dioxygenase (IDO) or tryptophan 2,3-dioxygenase (TDO). Four subsequent enzymatic conversions resulted in the formation of an unstable α-amino-β-carboxymuconate-ε-semialdehyde (ACMS) which can diverge to join TCA cycle or convert to picolinic acid (details in next section) or undergo a non-enzymatic spontaneous cyclization into quinolinic acid (QA). QA is then converted to NAMN by the enzyme quinolinate phosphoribosyltransferase (QPRT) to join the Preiss-Handler pathway and eventually convert into NAD$^+$ (Fukuwatari and Shibata 2013). This de novo NAD$^+$ biosynthesis mainly happens

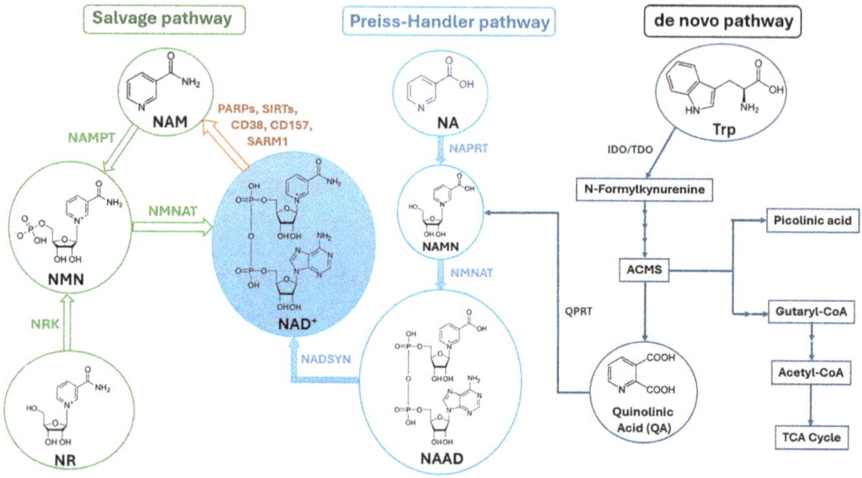

Fig. 5.2 NAD^+ biosynthetic pathways

in the liver and kidney because other tissues do not express some of the enzymes, especially QPRT, involved in this pathway (Xie et al. 2020).

In 1948, Kornberg purified an enzyme from yeast and used it to successfully convert NMN into NAD^+ (DPN at that time) (Kornberg 1948). In 1951, Handler and his colleagues synthesized NMN from NAM in human erythrocytes (Leder and Handler 1951) and later isolated the enzyme responsible for the conversion (Preiss and Handler 1957). These are the earliest reports on another NAD^+ biosynthesis pathway which is now called salvage pathway. However, the exact nature of the two enzymes involved in the salvage pathway was disclosed only recently. Nicotinamide phosphoribosyltransferase (NAMPT) is the enzyme that catalyzes the conversion of NAM to NMN while nicotinamide mononucleotide adenylyltransferase (NMNAT) helps the conversion of NMN into NAD^+(Imai 2009). NAMPT is the rate-limiting enzyme of the salvage pathway (Revollo et al. 2004). NMNAT is the same enzyme that catalyzes the conversion of NAMN to NAAD in the de novo and Preiss-Handler pathways discussed earlier. In 2004, Brenner and colleague discovered that previously known enzyme, nicotinamide riboside kinase (NRK), can convert nicotinamide riboside (NR) into NMN in both yeast and human (Bieganowski and Brenner 2004). Thus, NAM, NR and NMN as precursors of NAD^+ can efficiently be converted into NAD^+ by the salvage pathway.

The three NAD^+ biosynthetic pathways are summarized in Fig. 5.2. Salvage pathway is the most important source of NAD^+ homeostasis in the human body because the two enzymes, NAMPT and NMNAT, involved in this pathway are expressed in most human tissues and the three precursors are readily available from diet, especially for NAM which can also be from intracellular degradation of NAD^+ (details in Sect. 5.4.3.2) and recycled back for NAD^+ biosynthesis (this is why this pathway is called NAD^+ salvage pathway). Liver is also an important source of NAM by metabolism of Trp into NAD^+ via de novo pathway followed by intracellular degradation of NAD$^+$ to NAM which is then released to blood circulation to other tissues (Fukuwatari and Shibata 2013).

5.4 NAD$^+$ and Healthy Aging

NAD$^+$ is found not only to play key roles in redox energy metabolism as a coenzyme but also to be essential for the activities of several aging-related enzymes as a cosubstrate. Due to stress from outside and gradual weakening of internal biological system of human body during aging, NAD$^+$ levels decline with age. Lower NAD$^+$ causes cellular dysfunctions and age-related diseases. NAD$^+$ homeostasis is essential for healthy aging. Shin-ichiro Imai at Washington University in St. Louis studied the roles of NAD$^+$ in aging for almost 30 years and concluded that "the driving force of aging is the systemic decline of NAD$^+$ levels" in his recent review paper titled "NAD World 3.0" (Imai 2025).

5.4.1 NAD$^+$ and Energy Metabolism

Energy metabolism is the intracellular conversion of nutrients (carbohydrates, lipids, and proteins) from our diet into energy in the form of adenosine triphosphate (ATP) for survival and sustainment. Energy metabolism involves many complex electron-transfer of biochemical reactions (redox reactions) with the help of enzymes and coenzymes.

The discovery of NAD$^+$ as a coenzyme for yeast sugar fermentation into alcohol prompted the research of biochemical reactions in the yeast cell. Especially, in 1936, Warburg and colleagues discovered that the pyridine ring of NAD$^+$ played the role of hydrogen-transfer (basically electron transfer since hydrogen carries one electron) during the yeast fermentation of sugar to alcohol (Warburg and Christian 1936). These earlier breakthrough discoveries motivated further studies on the mechanism of NAD$^+$ molecule in living organisms, which eventually brought to light that NAD$^+$ played critical roles on human energy metabolism.

The nutrients in food are mainly carbohydrates, lipids and proteins, but these nutrients are large molecules and cannot be absorbed by the human body directly. Instead, our digestion system (stomach and intestine) breaks down the three classes of nutrients from food into smaller molecules: monosaccharides (mainly glucose) from carbohydrates, long-chain fatty acid from lipids, and small peptides and amino acids from proteins before the nutrients can be absorbed into blood circulation to distribute to various tissues (Da Poian et al. 2010). In the tissue cells, these nutrients go through various metabolic pathways such as glycolysis for glucose, β-oxidation for fatty acid, etc. and eventually converge at TCA cycle in the mitochondria. NAD$^+$, as the most important coenzyme, participates in most of these metabolic pathways and is reduced to NADH by accepting hydride (a negatively charged hydrogen atom)

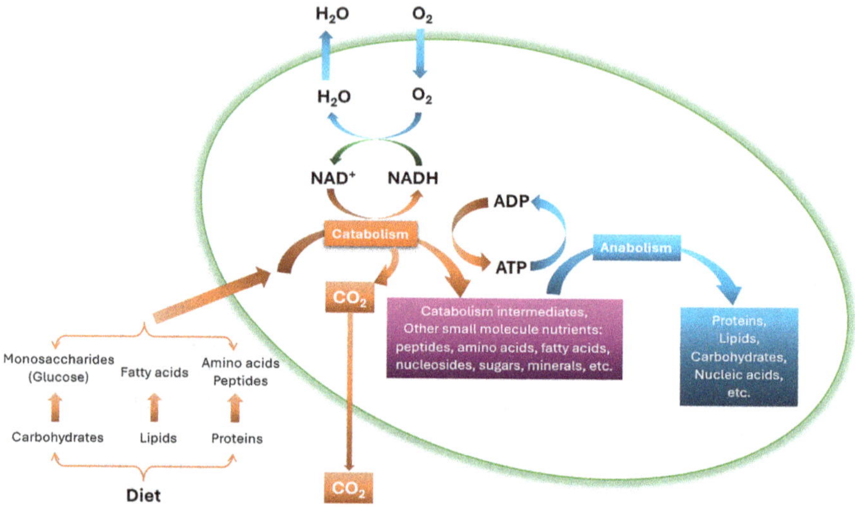

Fig. 5.3 Simplified metabolism diagram

while nutrients are oxidized to release energy in the form of ATP. NADH is oxidized back to NAD$^+$ via oxidative phosphorylation to release more ATP (Arnold and Finley 2023).

The whole metabolism system in the human body consists of catabolism (break down nutrients, energy metabolism is part of the catabolism) in which ATP is produced from ADP as energy reserve, and anabolism (produce biological molecules the body needs, such as DNA, proteins, cholesterol, sugar, etc.) in which ATP is degraded for energy consumption. ATP is constantly degraded to ADP to release energy for sustainment of human life and needs constant replenishment from energy metabolism. On average, an adult needs to produce around 50 kg ATP per day to keep our life functioning. NAD$^+$ and NADH are also constantly recycled back and forth for producing enough ATP the body needs. The half-life of NAD$^+$ varies from 15 min to 15 h depending on the tissues (Liu et al. 2018). During aging, although the NAD$^+$ and NADH pool will not change much, NAD$^+$ levels decline while NADH increases, causing energy metabolism dysregulation and age-associated diseases (McReynolds et al. 2020; Amjad et al. 2021; Liu et al. 2025). Therefore, the NAD$^+$/NADH homeostasis to maintain normal energy metabolism is critical to healthy aging (Fig. 5.3).

5.4.2 NAD$^+$ as Cosubstrate for Signaling

Besides NAD$^+$ as an indispensable coenzyme in energy metabolism, research during the past several decades has revealed that it is an important cosubstrate in several cell signaling pathways that plays critical roles in human aging and age-related diseases.

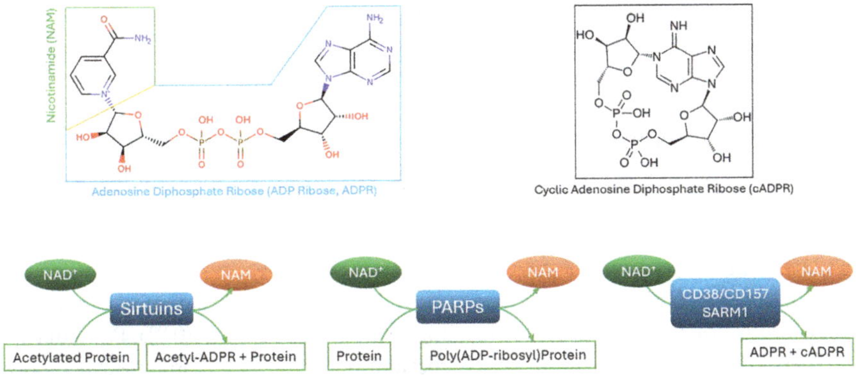

Fig. 5.4 NAD⁺ as the cosubstrate of consumption enzymes

During the first half of twentieth century, the research on NAD⁺ was mostly focused on its discovery, biosynthesis, and roles in metabolism. But, in 1963, Pierre Chambon and colleagues discovered that NAD⁺ was a cosubstrate for an enzyme, now known as poly(ADP-ribose) polymerases (PARPs), which catalyzes the transfer of ADP-ribose from NAD⁺ to proteins, a process called poly-ADP-ribosylation or PARylation (Chambon et al. 1963; Kraus 2015). PARPs belong to a broad family of ADP-ribosyl transferases (ARTs) which involves the enzymatic transfer of the ADP-ribosyl moiety from NAD⁺ to an amino acid residue of target proteins (Suskiewicz et al. 2021). The human PARP protein family has 17 proteins, among which PARP1 is best understood so far and responsible for about 90% of the total PARP activities (Covarrubias et al. 2021). In response to DNA damage due to age or other stressors, PARP1 PARylates itself along with histones and other proteins that act as a scaffold to recruit and activate other DNA repair enzymes and proteins to the damaged DNA site to initiate the repair. The mechanism of action of PARPs is that PARP enzymes firstly cleave the NAD⁺ molecule into NAM and ADP-ribose, in which ADP-ribose is added as single or polymers to PARP itself and other acceptor proteins (Fig. 5.4). Further research on PARPs found its complex role on aging (Mangerich and Bürkle 2012). On one hand, PARPs activities favor healthy longevity by supporting genomic stability via assisting DNA repair, telomere maintenance, and cell cycle control. For example, one study on comparing PARP activities in 132 individuals' mononuclear leukocyte cells from 13 different mammalian species of different lifespans yielded a strong positive correlation between PARP activity and lifespan, with human cells displaying approximately 5 times the activity of rat cells; and, within the same species, PARP activities were found to decline with advancing age (Grube and Bürkle 1992). On the other hand, there are research results that show that PARPs are a driving force of accelerating aging. One such example is that, activated by accumulation of damaged DNA during aging, PARPs compete with Sirtuins (enzymes essential for healthy aging) for NAD⁺ as cosubstrate, which causes the systemic depletion of NAD⁺ and eventually lead to shortening lifespan (Covarrubias et al. 2021). Treatment of mice with Cockayne syndrome (an accelerating aging disorder caused by extensive DNA damage) with PARP inhibitors or NAD^+ supplements promoted lifespan extension and ameliorated the severe phenotypes caused by PARP hyperactivation (Scheibye-Knudsen et al. 2014). Olaparib is an FDA-approved drug for treatment of advanced ovarian cancer and is a selective PARP inhibitor which results in accumulation of DNA double-strand breaks in BRCA1 and BRCA2 mutant cancer cell and leads cancer cell death (Kraus 2020). The activities of PARPs need to be delicately balanced for healthy aging, but it is clear that NAD^+ plays a key role as cosubstrate of PARPs enzymes and supply its ADP-ribose unit for PARylation.

In late 1990s, Guarantee's laboratory at MIT discovered that a family of proteins called sirtuins plays important roles in aging (Naiman and Cohen 2012). Particularly, in 1999, Guarantee and his student Kaeberlein reported that a protein called Sir2 (silent information regulator 2) could extend the lifespan of yeast by as much as 70% when over expressed (Kaeberlein et al. 1999; Zhao et al. 2020). Further research revealed that Sir2 extended the lifespan of worms and flies, and more recently SIRT1, the mammalian ortholog of Sir2, was found to extend lifespan in mice (Satoh et al. 2013; Imai and Guarente 2014). In yeasts there are five homologous proteins called Sir1–5 while, in mammals, there are discovered seven enzymes in the sirtuin family called Sirt 1–7 (Zhao et al. 2020). One of the breakthrough studies on sirtuin enzymes was the discovery by Guarantee and his student Imai in 2000 that the mechanism of Sir2 on yeast lifespan extension was NAD^+-dependent histone deacetylation that removes acetyl group on lysine residues of histone and degraded the NAD^+ cosubstrate, yielding deacetylated and activated histone protein, nicotinamide, and O-acetyl-ADP-ribose (Imai et al. 2000; Landry et al. 2000). Further studies revealed the detailed mechanism of action of the sirtuins is to remove the acetyl or other acyl group from their target proteins to activate their biological activities, in which NAD^+ molecule is firstly cleaved into NAM and ADP-ribose, and secondly, the acetyl or acyl group on the target protein is transferred to ADP-ribose, allowing the formation of the intermediate peptidyl-ADP-ribose. Acetyl-ADP-ribose or other acyl-ADP-ribose is then released from the target enzymes. The freed target enzymes are thus activated to perform their biological activities (Covarrubias et al. 2021) (Fig. 5.4). In 2003, Sinclair and his colleagues reported that a small molecule, resveratrol, was a strong Sir2 allosteric activator and extended the life of yeast by 70% (Howitz et al. 2003). Compounds that activate the activity of sirtuin enzymes are called sirtuin-activating compounds (STAC) which used to be extensively investigated as potential therapeutics for healthy aging or age-related diseases (Bursch et al. 2024). Sirtuin activity plays a crucial role in healthy aging by helping maintain genomic stability, improve mitochondrial function, and reduce cellular senescence (Watroba and Szukiewicz 2021). NAD^+ level decline during aging causes decrease of sirtuin activity and age-related diseases and, vice versa, restoring NAD^+ by supplementing NAD^+ precursors can increase sirtuin activities which, in turn, dramatically ameliorate these age-associated functional defects, counteracting many diseases of aging of tested animals (Imai and Guarente 2014). In addition, the interaction between NAD^+-dependent sirtuins and circadian clock is an exciting emerging area of aging research recently (Zhuang et al. 2024;

Acosta-Rodríguez et al. 2021). In mammals, the circadian system controls daily (~24 h) rhythms in behavior and physiology. Sirtuin activity can influence the expression of the components of circadian clock while, on the other hand, circadian clock controls the release of the key NAD$^+$ biosynthetic enzyme NAMPT in a daily rhythmic way so that the NAD$^+$ levels and sirtuin activity oscillate accordingly. NAD$^+$ decline during age can disrupt the circadian rhythm which leads to the disruption of sleep-wake cycle and other disfunctions while NAD$^+$ precursor supplementation was found to be able to reverse certain age-related declines in circadian rhythm function in mice (Levine et al. 2020). Although sirtuins enzymes has been the subject of extensive research for their roles in healthy aging during the past two decades and the genes that express the sirtuin enzymes are dabbed as longevity gene, there are still a lot of unknown and some of the published conclusions are subjected to disputes (Brenner 2022). But it is now quite clear that NAD$^+$ level is essential for the activities of the seven mammalian sirtuin enzymes which play important roles in human aging and age-related diseases.

The above are just two important examples of NAD$^+$ as cosubstrate in cell signaling of aging biology. It is certainly true that NAD$^+$ plays important roles in other age-related biological pathways which are not covered here. Future research will surely discover new biological pathways of NAD$^+$ on healthy aging. Finally, so far, the studies of NAD$^+$ on aging biological mechanism mostly stay at the animal level. It will be very important to learn how to translate these discoveries into human health and longevity in future research.

5.4.3 NAD$^+$ Level Declines with Age

NAD$^+$ is constantly synthesized, degraded and recycled in the human body to maintain stable NAD$^+$ levels for the various important biological functions it plays, as a coenzyme in energy metabolism and as a cosubstrate in cell signaling for DNA repair, cellular senescence, circadian rhythm maintenance, etc.

The NAD$^+$ levels have been examined from yeast, worms, rats, mice, monkeys, and humans since early 2000s. Among them mice and humans have been studied most extensively on the NAD$^+$ changes during age (Peluso et al. 2021). NAD$^+$ levels in mice were found to decline with age in muscle, liver, adipose, brain, pancreas, spleen, heart, kidney, lung, and cerebellum with a few exceptions (Yoshino et al. 2018). For example, Imai and colleagues reported that, compared to 3–6 months old mice, the NAD$^+$ levels of 25–31 months mice dropped 80–85% in muscle, and 70–75% in adipose, 80–85% in pancreas (Yoshino et al. 2011). Human NAD$^+$ levels were also found decreasing with age in whole blood, plasma, cerebrospinal fluid, muscle and skin with a few exceptions (Iqbal and Nakagawa 2024). One report disclosed that the NAD$^+$ levels in human pelvic skin from 49 participants dropped continuously with age from 0 to 77 years old of both male and female and the decline in male was most significant (Massudi et al. 2012). Another report measured NAD$^+$ concentrations in whole blood from 1518 participants with age between 18 to over

60 years old and found that the NAD$^+$ declined only slightly from about 40–29 mmol/L in males and no statistically significant changes in females (Yang et al. 2022).

From all the test results published so far, the consensus is that NAD$^+$ levels decline with age although there are still discrepancy and disputable test results from different reports (Peluso et al. 2021). The systemic decline in NAD$^+$ level is the driving force of aging (Imai 2025) and NAD$^+$ restoration is currently considered one of the most important healthy aging therapeutics (Guarente et al. 2024). Unfortunately, there is currently no standard on how to measure NAD$^+$ level. All the NAD$^+$ test kits currently on the market are not strictly validated and do not have authority review and approval for clinical use. The analytical methods used in academic institutes require very expensive equipment and are not readily available to the public. Since NAD$^+$ level fluctuates with circadian clock during the day, the time to take test sample can affect the test results. NAD$^+$ can be degraded if the samples are not handled properly. All these factors affect the NAD$^+$ test results and sometimes even give contradictory results of NAD$^+$ levels. Standardization of NAD$^+$ measurement is needed for future therapeutic research and applications of healthy aging.

5.4.3.1 NAD$^+$ Biosynthesis Slowes with Age

The decline of NAD$^+$ level with age is mainly caused by decreasing NAD$^+$ biosynthetic capability and increasing consumption from NAD$^+$ consuming enzymes during aging. So far, there have reported several biological mechanisms that negatively impact NAD$^+$ biosynthesis with age.

The major NAD$^+$ biosynthesis in mammals, the salvage pathway, becomes less efficient during aging because the NAMPT, the enzyme that controls the speed of the whole salvage biosynthesis, is expressed less and less in many tissues during aging (Strømland et al. 2021). Chronic inflammation and the degradation of the circadian rhythm during aging is believed to synergistically reduce NAMPT levels. Li and colleagues reported that the activity of NAMPT in mice gradually decreases with age because of the increase of acetylated NAMPT in old mice, and low-dose nicotine can restore the age-related decline of NAMPT activity through SIRT1 binding and subsequent deacetylation of NAMPT and thus increase NAD$^+$ biosynthesis (Yang et al. 2023).

Another enzyme, nicotinamide N-methyltransferase (NNMT), was found to negatively affect the NAD$^+$ biosynthesis (Covarrubias et al. 2021). NNMT uses S-methyladenosine as a methyl donor to convert NAM to N-methylnicotinamide (MNAM). MNAM can be further oxidized via oxidase to N-Mehtyl-2-pyridone-5-carboxamide (M2PY) and N-methyl-4-pyridone-3-carboxamide (M4PY)), which along with MNAM are excreted in the urine. Study found that NNMT expression increases with age, which, in turn, boosts the metabolization of NAM into M2PY and M4PY to be excreted with urine, thereby depleting the NAM pool for NAD$^+$ biosynthesis and contributing to the decline of NAD$^+$ with age.

5.4.3.2 NAD$^+$ Consumption Increases with Age

As discussed in Sect. 5.4.2, the sirtuins play important roles in healthy longevity by regulating metabolic homeostasis, DNA repair, genomic stability, cell senescence, inflammation, circadian rhythm, etc. NAD$^+$ is the indispensable cosubstrate for sirtuins enzymatic activities by contributing its ADP-ribose for accepting the acetyl or acyl group from target protein while release NAM for NAD$^+$ salvage biosynthesis (Sect. 5.3) (Fig. 5.4). Since the activities of sirtuins are essential for healthy aging, they are the major consumers of NAD$^+$ during aging. SIRT1 and SIRT2 are the most active among the 7 sirtuin enzymes and consume about one third of total NAD$^+$ under basal conditions (Covarrubias et al. 2021).

As discussed in Sect. 5.4.2, NAD$^+$ is an essential cosubstrate for activating PARPs via PARylation to perform their DNA repair function. During PARylation, NAD$^+$ molecule was consumed to transfer its ADP-ribosyl to protein and release NAM. Due to the accumulation of damaged DNA during aging, PARP1 is the major NAD$^+$ consumer of NAD$^+$ in the aged human body and could cause NAD$^+$ depletion under severe conditions. In addition, since PARP1 bounds more tightly than SIRT1 to NAD$^+$, hyperactive PARP1 can consume the majority of NAD$^+$ pool in the old human body, dramatically reducing sirtuin activities and negatively affecting healthy aging.

CD38 and CD157 are both ADP-ribosyl cyclase family of proteins that act as both receptors and ectoenzymes. CD38 is widely expressed in immune cells and other tissues, while CD157 is primarily expressed in myeloid cells and some other tissues. Both involve immune systems and have been implicated in cancer, autoimmune diseases, and other conditions. CD38 and CD157 are also NADase (enzymes that break down NAD$^+$) with glycohydrolase and ADP-ribosyl cyclase activities. Both can exert their glycohydrolytic functions to cleave NAD$^+$ molecule into NAM and ADP-ribose (ADPR), whereas their ADP-ribosyl cyclase activities convert NAD$^+$ into NAM and cyclic ADP-ribose (cADPR) (Covarrubias et al. 2021) (Fig. 5.4). CD38 can also catalyze NAD$^+$ into NAAD. In addition, CD38 uses NMN as an alternative cosubstrate and can degrade extracellular NMN into NAM, while CD157 consumes NR as an alternative cosubstrate and converts extracellular NR into NA. Therefore, CD38 and CD157 can reduce extracellular levels of both NAD$^+$ and its major precursors, NMN and NR. The expression of CD38 was found increasing in response to inflammation during aging and becoming a major consumer of NAD$^+$ in the cardiovascular tissues, macrophages, liver, kidney and skeletal muscle (Kitada et al. 2023). CD38 inhibitor supplementation to aged mice elevate NAD$^+$ levels in multiple tissues and ameliorates age-related diseases (McReynolds et al. 2020). CD157 is less studied for its role in aging although both CD38 and CD157 are mainly expressed in immune cells.

Sterile alpha and Toll/interleukin-1 receptor motif-containing 1 (SARM1) is an enzyme that plays a crucial role in axon degeneration, the process of nerve cell death. It's a key player in the axonal degeneration pathway, particularly after injury or in certain neurodegenerative diseases. SARM1 is only recently found to be a NADase which, like CD38 and CD157, catalyzes the cleavage of NAD^+ into NAM, ADPR and cAPDR, causing axonal NAD^+ depletion, energy metabolism dysfunction and death (Essuman et al. 2017).

Besides the above discussed NADases, other biological processes also found causing the decline of NAD^+ with age. Redox shift (NAD^+ level decrease and NADH increase with age) during energy metabolism and other factors may play minor roles (McReynolds et al. 2020) (Sect. 5.4.1).

5.4.4 NAD⁺ and the Twelve Hallmarks of Aging

Human life expectancy increased from ~50 years in the early 1900s to over 80 years at present (Niccoli and Partridge 2012). Aging is characterized by a progressive loss of physiological integrity, leading to impaired function and increased vulnerability to death. Aging research during the past two decades has shown that the aging rate is controllable. Human health span and lifespan are extendable. In 2013, a group of scientists led by Carlos López-Otín proposed "nine hallmarks of aging" as a guidance of aging research and received tremendous attention in the aging research community (López-Otín et al. 2013). Ten years later, they updated their "hallmarks of aging" to twelve (López-Otín et al. 2023). Here, I would like to briefly check how NAD^+ stands against these twelve hallmarks of aging.

1. Genomic Instability: The accumulation of DNA damage and mutations over time, which can lead to a variety of age-related diseases such as cancer and neurological disorders. NAD^+ assists PARPs to repair damaged DNA via Parylation and activates sirtuins to deacetylate histone, eventually helping genome stability. Sirtuins were even found to interact with PARPS to balance their activities and stabilize DNA (Ruszkiewicz et al. 2022). Supplementations with NAD^+ precursors, NA, NAM and NR, were found increasing or restoring DNA repair and maintaining genome integrity *in vitro*, worms, and mice (Fang et al. 2017; Ruszkiewicz et al. 2022; Xie et al. 2020).
2. Telomere Attrition: Telomeres are the protecting caps at the chromosome ends, which shorten with each cell division, contributing to cellular aging. Shorter telomeres are often associated with a higher risk of age-related diseases and shorter lifespans. Two reports disclosed that intervention of short telomere patients with the NAD^+ precursors, NR, elevated NAD^+ levels, activated PARPs and sirtuin activities, and ameliorated related syndromes (Stock and Liu 2021; Stock et al. 2021). Another publication disclosed that a human clinical trial with NAD^+ precursor, NMN, increased the length of participants' telomeres (Niu et al. 2021).
3. Epigenetic Alterations: Changes in gene expression without altering the DNA sequence itself, impacting cellular function. Changes in epigenetic mechanisms, such as DNA methylation and histone modification, are linked to the aging process and can impact lifespan. NAD^+ acts as a cosubstrate for enzymes like

Sirtuins and PARPs, which mediate epigenetic modifications like histone deacetylation, DNA methylation, circadian rhythm and chromatin structure, ultimately affecting gene expression and cellular functions (Xie et al. 2020; Covarrubias et al. 2021; Imai 2025).

4. Loss of Proteostasis: As organisms age, the efficiency of the proteostasis network declines, leading to the accumulation of misfolded or damaged proteins. This can result in cellular dysfunction and contribute to the development of age-related diseases like neurodegenerative disorders. The human body is estimated to have over 20,000 proteins in various tissues and organs (Ponomarenko et al. 2016). To maintain the proper functions of these proteins is very complex. NAD$^+$ as coenzyme I and cosubstrate of many proteins involves over 500 enzymatic reactions in the body (Rajman et al. 2018). NAD$^+$ as a cosubstrate for sirtuins, which are involved in protein deacetylation, can influence protein stability and aggregation (Ottens et al. 2021). Studies have shown that NAD$^+$ boosting can reduce amyloidosis and improve proteostasis in various tissues, including skeletal muscle and the brain.
5. Disabled Macroautophagy: Macroautophagy, which involves the degradation and recycling of dysfunctional cellular components (damaged organelles and misfolded proteins, cellular waste), is crucial for maintaining cellular health and can contribute to longevity. However, autophagy, which macroautophagy is part of, declines with age, and impaired autophagy is linked to age-related disorders. The relationship between NAD$^+$ and autophagy has been extensively studied. Cells with deficient autophagy has hyperactive PARPs and Sirtuins which deplete NAD$^+$ and cause cell death while NAD$^+$ supplementation with NAM and NR rescues autophagy-deficient cells (Fang et al. 2017; Kataura et al. 2022; Wilson et al. 2023).
6. Deregulated Nutrient-sensing: Impaired ability to sense and respond to nutrient availability negatively impacts metabolism. Various nutrients, like amino acids and glucose, can activate specific nutrient-sensing pathways that influence aging processes. IGF1/PI3K/AKT/mTOR pathway is involved in cell growth and protein synthesis, and AMPK/Sirtuin/PGC1α pathway is involved in energy homeostasis and stress response (Pignatti et al. 2020). Overactivation of these pathways, often seen with diets high in sugars, animal proteins, and unhealthy fats, can accelerate aging. Dietary interventions like caloric restriction have been shown to extend lifespan in many organisms, likely by modulating nutrient-sensing pathways, especially by activation of sirtuins with elevated NAD$^+$ levels during calorie restrict (Wang 2014). Declining NAD$^+$ levels can disrupt nutrient sensing and lead to age-related metabolic diseases, such as obesity, diabetes, dyslipidemia, etc. (Okabe et al. 2019).
7. Mitochondrial Dysfunction: Mitochondria are intracellular powerhouses that produce ATP and carry out diverse functions for cellular energy metabolism. As we age, mitochondria can become less efficient and more prone to damage (Srivastava 2017). To maintain an optimal NAD$^+$/NADH ratio and NAD$^+$ pool are essential for mitochondrial function in energy metabolism as well as mitochondrial biogenesis and mitophagy (Stein and Imai 2012; Fang et al. 2017;

Imai 2025). NAD^+ supplementation was found to restore mitochondria function of aged worms and mice by increasing mitochondria biogenesis and enabling mitochondrial mitophagy.

8. Cellular Senescence: Cellular senescence is a process where cells stop dividing but remain alive. Senescent cells can accumulate with age and contribute to inflammation, tissue damage, and decreased tissue repair, ultimately driving the aging process. NAD^+ deficiency can promote senescence by inducing DNA damage and mitochondrial dysfunction, whereas NAD^+ supplementation can counteract these effects and potentially mitigate senescence (Chini et al. 2024).
9. Stem Cell Exhaustion: A decline in stem cell function and regenerative capacity with age is a significant factor in aging and longevity. NAD^+ is very important for stem cell function, particularly for their ability to renew, repair, and regenerate tissues. NAD^+ is involved in cellular processes like DNA repair and mitochondrial function, and its depletion can lead to stem cell senescence and impaired function (Ji et al. 2023). Zhang and colleagues found that feeding the NAD^+ precursor NR to aging mice prevented muscle degeneration, enhanced muscle function, protected mice from the loss of muscle stem cells, and was also similarly protective of neural and melanocyte stem cells, which may have contributed to the extended lifespan of the NR-treated animals (Zhang et al. 2016).
10. Altered Intercellular Communication: Altered intercellular communication refers to disruptions in the way cells communicate with each other, affecting signal exchange and coordination between cells. Factors contributing to altered communication include a dysfunctional immune system, inflammatory signals from senescent cells, and impaired autophagic responses. The relationship between NAD^+ and intercellular communication has not been well studied. Imai and colleagues proposed extracellular NAMPT (study found that NAMPT exists both intracellularly (iNAMPT) and extracellularly (eNAMPT)) and NMN played a key role in the communication between hypothalamus, small intestine, adipose tissue, and skeletal muscle (Imai 2025).
11. Chronic Inflammation: A prolonged state of immune response, is a significant factor in the aging process and is linked to many age-related diseases. This phenomenon, known as "inflammaging," is characterized by low-grade, persistent inflammation that accelerates tissue damage, cellular senescence, and overall functional decline. Chronic inflammation can lead to a decrease in NAD^+ levels, while increasing NAD^+ may help reduce inflammation. Verdin and colleagues reported that senescent cells promote chronic inflammation and lead to tissue NAD^+ decline during aging via the activation of $CD38^+$ macrophages (Covarrubias et al. 2020). Imai also reported that chronic, low-grade inflammation drives systemic NAD^+ reduction (Imai 2025). A summary of 5 human clinical trials concluded NAD^+ precursor NR can improve some inflammatory measures (Kwon et al. 2022).
12. Dysbiosis: Dysbiosis refers to an imbalance in the community of microorganisms (the "microbiome") living within a specific environment, such as the gut. Dysbiosis is proposed as a primary cause of aging-associated morbidities, and

thus, premature death of elderly people (Ragonnaud and Biragyn 2021). NAD$^+$ deficiency has been associated with gut inflammation and leakage and dysbiosis, whereas, with accumulating evidence, NAD$^+$ elevation may play a crucial role in maintaining the integrity of the gut barrier in animals and humans (Niño-Narvión et al. 2023).

The above brief discussion on the connections of NAD$^+$ with the twelve hallmarks of aging does not mean to be comprehensive and we would like to also caution that, in many cases, the functions of NAD$^+$ may only be a small part of these aging processes. There are many other biological pathways that play crucial roles in each of these aging hallmarks.

5.5 NAD$^+$ Supplementation for Healthy Aging

NAD$^+$ level declines with age in human and other mammals. NAD$^+$ replenishment ameliorates age-related diseases and extends healthspan and lifespan in mammals. Multiple approaches are effective at increasing NAD$^+$ level, such as lifestyle, lowering NAD$^+$-consuming enzyme activities, increasing NAD$^+$-biosynthetic enzymes, or supplementation with NAD$^+$ precursors.

5.5.1 *Lifestyle*

Healthy lifestyles such as healthy diet, fasting, caloric restriction, physical activity, low glucose availability, and heat shocks are reported to have positive influence on NAD$^+$ levels (Poljsak et al. 2020). Among all these healthy lifestyles, calorie restriction was most widely studied to elevate NAD$^+$ levels and extend lifespan in mice. Exercise is also reported to be able to increase NAD$^+$ level by increasing the expression of NAMPT, the rate-limiting enzyme of the salvage NAD$^+$ biosynthesis, in mice skeletal muscle (Brandauer et al. 2013). Many vegetables, milk and meat have small quantities of NAD$^+$ precursors, Trp, NA, NAM, NR, and NMN (Alegre and Pastore 2023). Sleeping quality also affects NAD$^+$ level. Disrupted sleep-wake and circadian rhythms are a feature of aging associated with metabolic disease and reduced levels of NAD$^+$ (Levine et al. 2020).

5.5.2 *NAD$^+$-Consuming Enzyme Inhibitors*

Inhibiting NAD$^+$-consuming enzymes like PARPs, Sirtuins, and CD38 can effectively increase NAD$^+$ levels. These types of substances are mostly for pharmaceuticals, such as a PARP inhibitor, Olaparib, is an anticancer drug (Li et al. 2024).

Natural products apigenin and quercetin are reported to be CD38 inhibitors and can raise NAD^+ levels (Escande et al. 2013).

5.5.3 *NAD^+ Biosynthetic Enzyme Activators*

Activate NAD^+ biosynthetic enzymes can also increase NAD^+ level (Strømland et al. 2021). Natural products, rutin and alpha lipoic acid (ALA), were reported to increase the expression of NAMPT and thus increase NAD^+ level (Henderson et al. 2024). Natural products curcumin, quercetin, luteolin, ginsenosides and salidroside have been shown to activate NAMPT (Guo et al. 2023).

5.5.4 *NAD^+*

Direct supplementation of NAD^+ has been controversial, and very little study has been done although there are both oral NAD^+ supplements and iv infusion currently on the market. Oral NAD^+ cannot be absorbed by intestine but, instead, is degraded into NMN, NR and NAM by intestine cells (She et al. 2021; Canto 2022). One small human clinical study on NAD^+ iv infusion found that plasma NAD^+ level did not increase in the first 2 h which revealed that NAD^+ is rapidly and completely removed from plasma (Grant et al. 2019). However, an *in vitro* study revealed that NAD^+ can pass through plasma cell membrane (Pittelli et al. 2011).

5.5.5 *NAD^+ Precursors*

NAD^+ precursors are those substances which are used to produce NAD^+ in living systems. For humans, NAD^+ precursors include Trp, NA, NAM, NR, and NMN as discussed in Sect. 5.3. These substances are mostly natural products and currently very popular dietary ingredients on the market for NAD^+ restoration. Guarente, Sinclair and Kroemer recently reviewed that NAD^+ precursors are one of the eight promising healthy aging treatments (Guarente et al. 2024).

5.5.5.1 Tryptophan (Trp)

As discussed in Sect. 5.3, Trp is the starting point of mammalian de novo NAD^+ biosynthetic pathway which involves eight enzymes (Ren et al. 2024). Liver and kidney expressed all these enzymes and are the main sites of human de novo NAD^+ biosynthesis. Study revealed that it took about 60 times more Trp to produce the same amount of NAD^+ than niacin (NA or NAM) does, which shows the inefficiency of Trp supplementation for NAD$^+$ restoration (Fukuwatari et al. 2004; Freese and Lysne 2023). Trp is a very poor NAD$^+$ precursor (Cercillieux et al. 2022).

5.5.5.2 Nicotinic Acid (NA)

As discussed in previous sections, NA was widely used for pellagra and dyslipidemia treatment and is a NAD$^+$ precursor in the Preiss-Handler pathway. However, NA supplementation is limited by its common side effect of skin flushing with as lowe as 50 mg dose in as little as 30 min (Poljšak et al. 2022). Study revealed that the flushing is caused by activation of G-protein–coupled receptor, termed GPR109A (Benyó et al. 2005). Preiss-Handler NAD$^+$ biosynthetic pathway also primarily occurs in the liver and kidney since one of the enzymes, NAPRT, is mainly expressed in these two tissues (Cercillieux et al. 2022).

5.5.5.3 Nicotinamide (NAM)

NAM is the NAD$^+$ precursor of salvage pathway, and requires the rate-limiting enzyme, NAMPT, to be converted to NMN which can quickly turn into NAD$^+$ under NMNAT catalysis. The salvage pathway is the most important route of NAD$^+$ biosynthesis in many human tissues where fast NAD$^+$ turnover is required. NAM is a very important NAD$^+$ precursor in that it is readily available endogenously in all cells from the degradation of NAD$^+$ by all the consuming enzymes (PARPs, sirtuins, CD38/CD157 and SARM1) and can be recycled back to NAD$^+$ in a short two-step salvage pathway. NAM as Vitamin B3 is recommended for about 15 mg/day dose for healthy adults. Two factors limit NAM as an exogenous precursor of NAD$^+$ biosynthesis: one is that NAM needs to go through the rate-limiting NAMPT in the 2-step salvage pathway and the express of NAMPT is generally low in aged tissues; the other is that high dose NAM is hepatotoxic and a strong inhibitors of sirtuins and PARPs whose activities are essential for healthy aging (Cercillieux et al. 2022; Imai 2025).

A study of chronic NAM supplementation on mice improved some aspects of healthspan but did not extend lifespan and liver NAD$^+$ level was not increased either (Mitchell et al. 2018). A recent human clinical study found that supplementation with a blend of 210 mg Trp, 4 mg NA and 200 mg NAM daily did not improve mitochondria function and increase skeletal muscle NAD$^+$ level (Connell et al. 2021). Another recent human clinical trial revealed that a single dose of 100 mg NAM supplementation did not raise human blood NAD$^+$ level while a single dose of 500 mg NAM marginally albeit statistical significantly increased NAD$^+$ level after 12 h and 48 h (Ito et al. 2021). Yet a third small clinical trial using a single 200 mg dose of NAM found that blood NAD$^+$ level increased significantly (Ito et al. 2020).

5.5.5.4 Nicotinamide Riboside (NR) and Nicotinamide Mononucleotide (NMN)

As discussed in previous sections, the salvage pathway is the most important route for replenishing NAD$^+$ levels for healthy longevity (Yoshino e al. 2018; Covarrubias et al. 2021; Imai 2025). Trp, NA and NAM all have certain limitations as NAD$^+$ precursors. Besides, both NR and NMN can avoid the NAMPT rate-limiting step in the salvage pathway and thus provide an advantage over NAM for NAD$^+$ restoration. The extensive studies during the last two decades identify NR and NMN as the favorite substances for NAD$^+$ restoration.

5.5.5.4.1 Nicotinamide Riboside (NR)

Introduction: NR molecule is positively charged on the pyridine nitrogen atom and always needs a negative charged counterpart to make a neutral salt, most frequently chloride salt (NR-Cl). Many commercial NR products are actually NR-Cl. NR belongs to a class of chemicals called nucleoside and has two optical isomers: α-NR and β-NR (Fig. 5.5). Only β-NR is biologically active optical isomer for NAD$^+$ biosynthesis. Therefore, when we look for a high potent NR product, we should look for both chemical purity and optical purity. Enzymatic manufacturing NR has better selectivity to produce optically pure NR than chemical synthesis (Huang et al. 2023; Zhang and Sauve 2017). However, currently on the market, the NR ingredient used in dietary supplements are mostly manufactured by chemical synthesis since the technology for industrial scale enzymatic manufacturing of NR is not mature yet (Sharma et al. 2022). NR is chemically unstable because its molecule has a glycosidic bond (the chemical bond that connects nitrogen of nicotinamide with the carbon of D-ribose sugar) which is especially prone to cleavage and certain salts such as chloride and borate help stabilize the substance to a certain extent (Biță et al. 2023).

Historical Background: NR was first identified in 1942 by enzymatic degradation of NAD$^+$ (called cozymase, codehydrogenase I or DPN at that time) (Schlenk 1942).

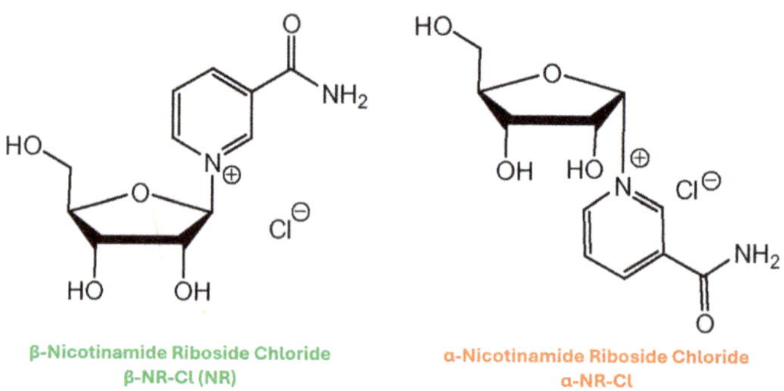

Fig. 5.5 Chemical structure of Nicotinamide Riboside (NR)

There was not much attention to NR after its identification until 2004 when Bieganowski and Brenner found that a class of previously known enzymes called nicotinamide riboside kinases (NRK) from both yeast and human can convert NR to NMN, adding NR as a new NAD$^+$ precursor for the salvage pathway (Bieganowski and Brenner 2004). Brenner also identified that NR is a natural product which can be found in milk. Later studies found that both cow and human milk contain low micromolar NR (Cercillieux et al. 2022). Extensive studies have been done on NR for healthy aging since 2004.

Extension of Lifespan and Healthspan by NR Supplementation in Animal Models: A study found NR extended the lifespan of yeast and nematodes (roundworms) via increasing NAD$^+$ biosynthesis and sirtuin activities, showing similar effect to calorie restriction (Belenky et al. 2007). Another two studies revealed that NR help improve mice healthspan but received mixed results on extending mice lifespan: 24-month old mice supplemented with NR (400 mg/kg/day) lived a ~4% longer lifespan, while a larger study from the NIH interventions testing program (ITP) that fed mice with NR (1 g/kg) failed to find any difference in the lifespan extension (Zhang et al. 2016; Harrison et al. 2021).

Physiological Effects in Animal Models: NR supplementation has shown benefits in preventing or reverse age-related diseases and biological dysfunctions in multiple animal studies (Yoshino et al. 2018; Mehmel et al. 2020; Cercillieux et al. 2022; Sharma et al. 2022): protection against diet-induced body weight gain and glucose intolerance; preventing non-alcoholic fatty liver disease (NAFLD), alcohol-related liver damage/failure and other liver diseases in mice; reversing fatty liver disease in mice; improvement in mouse models of the Cockayne Syndrome (CS) (a progressive neurodegenerative accelerated aging disorder caused by mutations in the CSA or CSB genes which encode proteins involved in DNA repair and transcriptional regulation); attenuation of the complications in multiple mouse models of heart failure; reduction of systemic inflammatory signals; protection against neurodegenerative disorders in Alzheimer's disease (AD) mouse model (cognitive performance improvement and Aβ deposits reduction in the brain cortex tissues); reduction of age-related protein deposits in skeletal muscle and increase of muscle mitochondrial biogenesis; etc. However, some of these results from animal studies of NR treatment are not always consistent. For example, a recent study revealed either very marginal or no effects of NR on glucose tolerance in mice (Cartwright et al. 2021).

Pharmacokinetics (PK)/Pharmacodynamics (PD) in Animal Models: A study confirmed NR increased NAD$^+$ levels in both murine and human cell lines, and in liver and muscle of treated mice (Canto et al. 2012). Interestingly, the same study revealed that NR did not activate GPR109A receptor, the mediator of NA-induced skin flushing. Another study from Brenner's lab revealed that oral NR supplementation of mice elevated liver NAD$^+$ four-fold over baseline (Trammell et al. 2016). This study also concluded that the oral bioavailability of NR is better than that of NA and NAM. The efficiency of NAD$^+$ elevation by NR is tissue specific, and NR is particularly good at increasing liver NAD$^+$ since NRK (the enzyme that catalyzes the conversion of NR to NMN) is mostly expressed in liver (Cercillieux et al. 2022). Previous studies reported that orally administered NR is absorbed into blood circulation via equilibrative nucleoside transporter (ENT) in intestine cell (Kropotov et al. 2021), However, recent studies found that only a small portion of orally administered NR was directly absorbed by the small intestine during the first 30 min of administration, whereas majority of the rest NR underwent gut microbiota-mediated conversion to NAM followed by deamidation to nicotinic acid (NA) which was then transported to liver for NAD^+ biosynthesis via Preiss-Handler pathway (Liu et al. 2018; Cercillieux et al. 2022; Yaku et al. 2025).

Safety in Animal Models: So far, the dosages of NR in most pre-clinical mouse studies generally range between 200 and 500 mg/kg/day with no toxic side effects reported. Toxicology studies in rats revealed that NR is a very well-tolerated molecule, with minor side effects only observed at doses as high as 1 g/kg/day. One report of a very high NR concentration (9 g/kg of diet, approximately 1 g/kg of body weight/day) in mouse led impaired glucose tolerance and promoted white adipose tissue inflammation. These toxicology profiles in rodents are similar to the classic NAD^+ precursor, NAM (Cercillieux et al. 2022).

Human Clinical Trials: Will the extensive preclinical study of NR in animal models be transferred to humans? Human clinical trials will be the litmus. Damgaard and Treeback have published a critical review on all the published human clinical trials on NR from first human clinical trial published in 2016 up to 2023 (Damgaard and Treebak 2023). Here is summary of these human clinical trials on NR: (1) oral daily doses range from 100 to 2000 mg are safe with no mild and severe adverse observations in all studies; (2) Blood NAD^+ level (including whole blood and peripheral blood mononuclear cells (PBMCs)) are convincingly increased and one study claimed that NR helped raise brain NAD^+ level, but NR supplementation did not increase muscle NAD^+ as measured in several studies; (3) there are no effects on muscle function and insulin sensitivity; (4) The common metabolites in various tissues and urine excretas are NAAD, NAM, MNAM, M2PY and M4PY; (5) The reduction of inflammation markers in blood was observed in multiple clinical studies. Since the review was published in 2023, we found three newer human clinical trials, all on brain health, were published with interesting results. The first one was a follow up study of previous clinical trial on Parkinson's Disease (PD) patients with 1000 mg NR daily dose showing encouraging results of brain NAD^+ elevation. This follow-up study was randomized, placebo controlled and double-blinded with 20 PD patients using a much higher daily dose of 3000 mg NR (Berven et al. 2023), which met all prespecified outcomes: no moderate or severe adverse events (especially no skin flushing), very moderate improvement of total MDS-UPDS (the Movement Disorder Society -Unified Parkinson's Disease Rating Scale) scores and five-fold whole blood NAD^+ level increase (brain NAD^+ was not measured). The second human clinical trial was also a randomized placebo-controlled trial with 20 subjects of mild cognitive impairment (MCI) with 1000 mg NR daily dose and achieved a 2.6-fold increase in PBMC blood NAD^+ level with no severe adverse event, but no cognition and physical improvement was observed (Orr et al. 2024). The third human clinical trial on NR was again on MCI patients with 1000 mg NR daily dose, but the trial failed to improve cognition after 8-week supplementation (Wu et al. 2025).

Summary: NR dietary supplements were first introduced onto the US market in 2013 by ChromaDex with their brand "Niagen" and are getting popular for NAD$^+$ boosting and healthy aging. Extensive preclinical and human clinical studies support its safe use. NR has been granted GRAS (generally recognized as safe) status with a daily dose at 180 mg for healthy adults. In Europe, the European Food Safety Authority (EFSA) approved NR as novel food ingredient with daily dosing at 300 mg for healthy adults and 230 mg for pregnant women. NR is also approved as nutritional product ingredient in a few other countries, such as Australia, Japan, Canada, etc. However, there are still many unknowns about its mechanism of actions in different human tissues, PK/PD, etc. (Cercillieux et al. 2022; Canto 2022).

5.5.5.4.2 Nicotinamide Mononucleotide (NMN)

Introduction: NMN molecule is a phosphate ester of NR and belongs to nucleotide class of biochemicals (Fig. 5.6). Unlike NR, NMN molecule is a zwitterion or internal salt. NMN also has two optical isomers: α-NMN and β-NMN. β-NMN is the active optical isomer (literature usually uses NMN for β-NMN). Again, when looking at the quality of NMN products, we should check both chemical purity and optical purity. Unlike NR, very efficient enzymatic manufacturing processes have been widely used for producing NMN on a commercial scale and offer superior optical purity (Rahman et al. 2024).

Historical Background: In 1948, NMN was first isolated in a reasonably pure form by Kornberg and Lindberg who did a series of enzymatic degradation experiments of NAD$^+$ by incubating NAD$^+$ (DPN at that time) with rabbit kidney containing pyrophosphatase (Kornberg and Lindberg 1948) (Fig. 5.2). In 1951, Handler and his colleague enzymatically synthesized NMN from NAM by incubation of human erythrocytes with phosphate, glucose and NAM (Leder and Handler 1951). In 1963, Chambon and colleagues discovered that NMN activated a new enzymatic reaction which, when incubated chicken liver nuclear extracts with ATP and NMN, produced a product called "polyA" (which was later identified as an ADP-ribose Polymer) (Chambon et al. 1963). Their findings led to a series of phenomenal discoveries of poly-ADP-ribose polymerases (PARPs), and we now

Fig. 5.6 Chemical structure of Nicotinamide Mononucleotide (NMN)

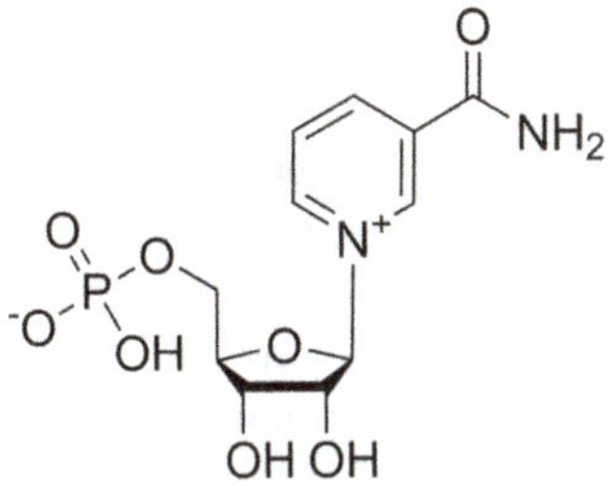

know that it is actually NAD^+ that activates the PARP enzymes to produce PARylated proteins (Fig. 5.4). NMN is just an NAD^+ precursor that was converted into NAD^+ when Chambon and his colleagues added NMN into their chicken liver incubation mixture (refer to Sect. 5.3). After that, there was not much research on NMN until 2007 when Imai and his colleagues found that NMN supplementation can ameliorate the defects of NAD^+ biosynthesis and glucose-stimulated insulin secretion in mice (Revollo et al. 2007). In the next few years, a series of research at Imai's and many other labs further solidified the roles of NMN in NAD^+ elevation and healthy aging. NMN is also a natural product and can be found in many fruits and milk, especially significant high concentrations of NMN (average 9.2 μM) have recently been detected in human breast milk, highest among all other NAD^+ precursors (NAD^+ 2.5 μM and NR 1.1 μM) (Yoshino et al. 2018; Saito et al. 2023; Imai 2025).

Extension of Lifespan and Healthspan by NMN Supplementation: Sinclair and colleagues investigated the effect of long-term NMN administration-from 13 months of age till the natural death- on frailty and lifespan in male and female mice. Their findings showed that chronic NMN treatment delayed frailty, improved gut health, helped male mice metabolic health, and increased median lifespan by 8.5% in female mice although no lifespan extension was observed in male mice (Kane et al. 2024). Another study claimed that NMN supplementation prolonged the lifespan of prematurely aged mice and promoted gut health (Gu et al. 2024).

Physiological Effects in Animal Models: The physiological effects of NMN supplementation in animal models have been comprehensively covered by several reviews (Yoshino et al. 2018; Hong et al. 2020; Nadeeshani et al. 2021). NMN supplementation is particularly effective on metabolic diseases in animal studies: enhancing insulin sensitivity, ameliorating insulin resistance, improving mitochondria function, preventing liver diseases, and reducing age-associated body weight gain. NMN treatment also found therapeutical potential in neuronal functions in brain, hear failure, vascular dysfunction, muscle loss, retinal degeneration, and kidney injury. So many studies have published on the age-related physiological effects of NMN, here we can only do a general summary.

Pharmacokinetics (PK)/Pharmacodynamics (PD) in Animal Models: Sinclair's laboratory found that NAD^+ level of heart tissue dropped over 50% from 6-months to 30-months mice and treatment with NMN 500 mg/kg/day for 7 days increased NAD^+ level about 1.5-fold for 3-months mice and about three-fold for 30-months mice (North et al. 2014). Many other laboratories found that NMN can elevate NAD^+ levels of various animal tissues: pancreas, adipose, heart, skeletal muscle, kidney, eyes, blood vessel, and even hippocampus, and hypothalamus (Yoshino et al. 2018) (Sect. 5.4.3). Recent study further confirmed that oral NMN supplementation of 400 mg/kg raised NAD^+ level over 40% in brain of mice in 45 min (Ramanathan et al. 2022). This result gives another evidence that NMN can penetrate the blood-brain barrier. Imai's lab reported that oral NMN supplementation to mice was rapidly absorbed and plasma NMN quickly increased at 2.5 min, peaked at 5–10 min, went back to original level at 15 min while NAD^+ levels showed a steady increase from 15 to 30 min (Mills et al. 2016). Later, Imai's laboratory identified a specific cell membrane transporter, Slc12a8, for NMN, which explained the rapid oral NMN absorption by intestine cells (Grozio et al. 2019). However, this result was challenged and NMN was claimed to mostly degrade into NR before entering the cell for salvage NAD$^+$ biosynthesis (Schmidt and Brenner 2019; Sauve et al. 2023), but Imai countered the dispute (Imai 2025). Like NR, a few recent studies with substantial evidence revealed that oral NMN administration was mostly degraded to NAM and, in turn, to NA by gut microbiome before being absorbed into blood circulation (Liu et al. 2018; Cercillieux et al. 2022; Yaku et al. 2025). Imai explained that these recent results may come from inaccurate test methods of NMN content and the proper handling of test samples (Imai 2025). Oral NMN is rapidly absorbed into blood circulation within 5–10 min and the samples are recommended to be tested freshly. Imai also disputed the previous reports that NMN is converted to NR by extracellular CD73 and he claimed that there is no significant conversion of NMN to NR considering the rapid absorption of NMN. NMN is very stable in plasma compared to reports that revealed NMN was degraded into NAM.

Safety in Animal Models: Like NR, NMN does not activate the GPR109A receptor, the mediator of NA-induced skin flushing (Canto et al. 2012). In animal studies, NMN has been found to be safe with minimal side effects, even at high doses. Long-term NMN administration in mice, at doses of 100 or 300 mg/kg/day for a year, did not cause significant side effects (Mills et al. 2016). Acute and 90-day sub chronic dosing in Wistar rats and mutagenicity tests concluded that NMN is safe as a new dietary ingredient (Yu et al. 2024). A study revealed that oral administration of NMN at very high doses of 1340 mg/kg/d and 2680 mg/kg/day to mice for 7 days is well tolerated with no deleterious effects, whereas oral administration of 1340 mg/kg/day NMN for 14 days to beagle dogs only resulted in mild increases in creatinine and uric acid (You et al. 2020).

Human Clinical Trials: Human clinical trials are the best way to see how the extensive research on NMN done in the laboratories transfers to human healthy aging. Two recent reviews have summarized the over 20 human clinical trials of NMN (Song et al. 2023; Benjamin and Crews 2024). Here is a summary of the results of the areas revealed by the human clinical trials of NMN conducted so far.

1. NAD$^+$ Elevation: NAD$^+$ level elevation is the single most important biomarker to assess whether NMN supplementation is effective for human healthy longevity. Several human clinical trials were designed primarily for testing whether NMN can increase the human NAD$^+$ levels. The first human clinical trial on NMN revealed that the NAD$^+$ metabolites, M2PY and M4PY, in blood and urine were increased but NAD$^+$ level was not reported (Irie et al. 2020). Another human study evaluating 250 mg daily oral NMN dose (administered as 125 mg twice daily) reported that NAD$^+$ in serum was detectable but could not be quantified due to the low sensitivity of the test method (Katayoshi et al. 2023). In addition, a randomized study administering 300 mg NMN daily, with NAD$^+$ as the primary trial end point, found no meaningfully increase in serum NAD$^+$ level (Huang 2022). In Contrast, a separate randomized human clinical trial assessing daily NMN doses of 300, 600 and 900 mg found that serum NAD$^+$ levels increased

significantly and dose-dependently for all three doses, ranging from 1.5 to 4.7-fold, with 600 mg dose giving the highest 4.7-fold elevation (Yi et al. 2023). Imai and colleagues reported that a trial with 250 mg oral daily NMN dose in prediabetes and obese women increased NAD^+ level in PBMC while NAD^+ level in muscle did not change (Yoshino et al. 2021). A number of human clinical trials using the same oral 250 mg/day NMN reported that NAD^+ levels were significantly raised: two-fold increase of whole blood NAD^+ level (Igarashi et al. 2022), almost doubled in whole blood (Okabe et al. 2022), more than six-fold increase of peak plasma NAD^+ levels over baseline (Yamane et al. 2023), about 1.6-fold increase in PBMC NAD^+ level (Yamaguchi et al. 2024), and two-fold increase of whole blood NAD^+ level (Morifuji et al. 2024). A recent human clinical trial with oral 1000 mg and 2000 mg/day NMN with a unique micro-crystal form (coded as MIB-626) on overweight or obese adults showed about 2–3 fold increases in whole blood NAD^+ level (Pencina et al. 2023a). The same team reported that 2000 mg/day (1000 mg twice daily) NMN (MIB-626) again gave almost two-fold increase of whole blood NAD^+ level but no change in muscle NAD^+ level (Pencina et al. 2023b). From these more than ten human clinical trial studies, there is no doubt oral NMN supplementation can elevate human blood NAD^+ level, but the NAD^+ elevation on other tissues needs to be further investigated. The few exceptions of failed NAD^+ elevation may be attributed to factors such as the quality of NMN materials, blood sample handling, and analytical methods for measuring NAD^+ levels as discussed in Sect. 5.4.3. Importantly, the quality of NMN materials should not only be assessed by chemical purity as many other factors affect the bioavailability of NMN ingredients, such as optical purity, particle size distributions, crystalline form, and dosage form.

2. Safety: From the results of the more than 20 human clinical trials on oral dose of NMN with 100–2000 mg oral daily doses published so far, there are no mild or severe adverse events reported (Song et al. 2023). However, all these human clinical trials are small and short duration. Larger and longer human clinical study is needed for comprehensively assessing the safety of NMN.
3. Physiological Effects on Heathy Aging: A few human clinical studies reported that NMN supplementation improved sleep quality, but two clinical studies did not see any change on sleep quality (Song et al. 2023). The physical improvements are widely recorded among the human clinical studies reported so far (Song et al. 2023). Favorable impact of NMN on metabolic diseases, such as insulin sensitivity and resistance, are also observed in several of the reported human clinical studies (Song et al. 2023). One study revealed NMN supplementation improved muscle insulin sensitivity (Yoshino et al. 2021). Another study saw significant improvement of overall health conditions of mid-age and older adults and lower blood biological age (Yi et al. 2023). In addition, one clinical study saw reductions of body weight and diastolic blood pressure of overweight or obese adults (Pencina et al. 2023b).
4. Blood Biomarkers of Healthy Aging: One human clinical study found that NMN supplementation significantly increased the length of telomer (Niu et al. 2021).

Another study reported that NMN supplementation significantly reduced total LDL, non-HDL cholesterol of overweight or obese adults (Pencina et al. 2023b). Serum insulin secretion improvement was observed in one human clinical trial (Yamane et al. 2023).

Summary NMN dietary supplement was first commercialized in Japan in April 2015 by Shinkowa Pharmaceutical Co., Ltd. under the brand name "Mirailab" which was introduced to the US market in April 2016. Since then, NMN along with NR, has quickly gained popularity as healthy aging ingredient. NMN, a natural product present at low levels in many plants and vegetables and an endogenous substance in mammals, is found safe in all animal and human studies published so far. In 2020, NMN was formally approved as food ingredient in Japan. Later, NMN dietary supplements were approved as natural health products in Canada. In May 2022, the US FDA acknowledged a new dietary ingredient (NDI) application for NMN from SyncoZymes without objection, which means the US FDA approved NMN as a dietary ingredient. However, the US FDA withdrew its NDI approval for NMN as a dietary ingredient in November 2022 because a biotech company, Metro International Biotech, had been developing an NMN product called MIB-626 as a new prescription drug and got FDA authorization for its investigational new drug (IND) application. IND is an authorization from US FDA that allows biopharmaceutical companies to begin to test the potential new drug in human clinical trials. Currently, the Natural Products Association (NPA), a nonprofit organization representing dietary supplement industry, is challenging the FDA's withdrawal of the NDI approval in court and, in November 2024, the court granted NPA's request and issued a joint motion to stay the case, which means NMN dietary supplements can be on the market without punitive and disciplinary penalty until the final outcome of this law suit (Daniells 2024). In Europe, NMN was classified as novel food by the European Food Safety Authority (EFSA) and, so far, NMN has not been authorized by EFSA as novel food.

5.6 Conclusion

Over the past two decades, significant progress of research has been made in understanding the role of NAD$^+$ in health and longevity. It is now a consensus that systemic NAD$^+$ decline is a key contributor to aging and age-related diseases. NAD$^+$ restoration by NAD$^+$ precursors (Trp, NA, NAM, NR and NMN) has been approved to be safe and effective, especially the results from animal and human clinical studies on NR and NMN demonstrate promising safety profile and beneficial effects on healthy aging. However, many unknowns still need to be addressed. What exactly are the PK/PD of these precursors in the human body? What are the optimal doses for each person? How are the human NAD$^+$ levels accurately measured? What is the quality of so many different brands of NAD$^+$ precursor dietary supplements on the market? To answer all these questions and many others, we need more cohesive work from academic, industry and government. Currently, most of the studies on the PK/PD of NR and NMN have been conducted in mice, and the results are often inconsistent or sometime even contradictory. It is still not quite clear how NR and NMN are absorbed, distributed, metabolized and excreted in the humans. NR and NMN dietary supplements have been on the markets for over 10 years, and we have already seen that some people have very good responses on these products while others do not have health improvement at all. How do we know who should take these products and how much should we take? A personalized approach might be the way to go (Kuerec et al. 2024). So far, all the human clinical trials of NR and NMN are small (less than 200 participants) and short term (less than 6 months). Large-scale, multiyear and well-designed human clinical trials are necessary to establish safe and effective dosing regimens. In addition, dosage forms require to be optimized. There are capsule, delayed release capsule, tablet, powder, gummy, injectable and even liposomal dosage forms of NR and NMN dietary supplements on the market, but, except the capsule and tablet, none of them have been carefully studied for the quality, stability, PK/PD (including bioavailability) and safety. Currently, there are many clinical commercial NAD^+ test kits on the market, but none of them have authorization from any government authority. As shown by the drastically different results of NAD^+ levels from the NR and NMN human clinical studies, the measurement of NAD^+ level needs to be standardized. Finally, studies found that some of the NMN dietary supplements on the markets have much lower amount of NMN ingredients than the product labels claim (Sandalova et al. 2024). How do we know the quality of these products? How do we know whether the NMN and NR ingredients are manufactured under cGMP with compliance of government regulations? Health authorities of the countries concerned need to step up the regulation and monitor the manufacturing of these products. After all, there is still a lot of work to do to deliver safe and effective products of NAD^+ precursors from laboratories to consumers for healthy aging.

Conflict of Interest Lin Yi is an employee of Abinopharm, Inc. in USA whose business focuses on healthy longevity and engages in the development, regulatory compliance and commercialization of many of the NAD^+ precursors, Trp, NA, NAM, NR and NMN, as nutritional ingredients or dietary supplements.

References

Acosta-Rodríguez VA, Rijo-Ferreira F, Green CB, Takahashi JS (2021) Importance of circadian timing for aging and longevity. Nat Commun 12(1):2862. https://doi.org/10.1038/s41467-021-22922-6

Alba-Lois L, Segal-Kischinevzky C (2010) yeast fermentation and the making of beer and wine. Nat Educ 3(9):17

Alegre GFS, Pastore GM (2023) NAD+ precursors nicotinamide mononucleotide (NMN) and nicotinamide riboside (NR): potential dietary contribution to health. Curr Nutr Rep 12(3): 445–464. https://doi.org/10.1007/s13668-023-00475-y

Altschul R, Hoffer A, Stephen JD (1955) Influence of nicotinic acid on serum cholesterol in man. Arch Biochem Biophys 54(2):558–559. https://doi.org/10.1016/0003-9861(55)90070-9

Amjad S, Nisar S, Bhat AA, Shah AR, Frenneaux MP, Fakhro K, Haris M, Reddy R, Patay Z, Baur J, Bagga P (2021) Role of NAD$^+$ in regulating cellular and metabolic signaling pathways. Mol Metab 49:101195. https://doi.org/10.1016/j.molmet.2021.101195

Arnold PK, Finley LWS (2023) Regulation and function of the mammalian tricarboxylic acid cycle. J Biol Chem 299(2):102838. https://doi.org/10.1016/j.jbc.2022.102838

Axelrod A, Elvehjem C (1939) Effect of nicotinic acid deficiency on the Cozymase content of tissues. Nature 143:281–282. https://doi.org/10.1038/143281b0

Axelrod A, Madden RJ, Elvehjem C (1939) The effect of a nicotinic acid deficiency upon the coenzyme I content of animal tissues. J Biol Chem 131:85–93. https://doi.org/10.1016/S0021-9258(18)73482-0

Axelrod A, Spies TD, Elvehjem C (1941) The effect of a nicotinic acid deficiency upon the coenzyme I content of the human erythrocyte and muscle. J Biol Chem 138:667–676. https://doi.org/10.1016/S0021-9258(18)51392-2

Belenky P, Racette FG, Bogan KL, McClure JM, Smith JS, Brenner C (2007) Nicotinamide riboside promotes Sir2 silencing and extends lifespan via Nrk and Urh1/Pnp1/Meu1 pathways to NAD+. Cell 129(3):473–484. https://doi.org/10.1016/j.cell.2007.03.024

Benjamin C, Crews R (2024) Nicotinamide mononucleotide supplementation: understanding metabolic variability and clinical implications. Metabolites 14(6):341. https://doi.org/10.3390/metabo14060341

Benyó Z, Gille A, Kero J, Csiky M, Suchánková MC, Nüsing RM, Moers A, Pfeffer K, Offermanns S (2005) GPR109A (PUMA-G/HM74A) mediates nicotinic acid-induced flushing. J Clin Invest 115(12):3634–3640. https://doi.org/10.1172/JCI23626

Berven H, Kverneng S, Sheard E, Søgnen M, Af Geijerstam SA, Haugarvoll K, Skeie GO, Dölle C, Tzoulis C (2023) NR-SAFE: a randomized, double-blind safety trial of high dose nicotinamide riboside in Parkinson's disease. Nat Commun 14(1):7793. https://doi.org/10.1038/s41467-023-43514-6

Bieganowski P, Brenner C (2004) Discoveries of nicotinamide riboside as a nutrient and conserved NRK genes establish a Preiss-Handler independent route to NAD+ in fungi and humans. Cell 117(4):495–502. https://doi.org/10.1016/s0092-8674(04)00416-7

Biță A, Scorei IR, Ciocîlteu MV, Nicolaescu OE, Pîrvu AS, Bejenaru LE, Rău G, Bejenaru C, Radu A, Neamțu J, Mogoşanu GD, Benner SA (2023) Nicotinamide riboside, a promising vitamin B_3 derivative for healthy aging and longevity: current research and perspectives. Molecules 28(16):6078. https://doi.org/10.3390/molecules28166078

Brandauer J, Vienberg SG, Andersen MA, Ringholm S, Risis S, Larsen PS, Kristensen JM, Frøsig C, Leick L, Fentz J, Jørgensen S, Kiens B, Wojtaszewski JF, Richter EA, Zierath JR, Goodyear LJ, Pilegaard H, Treebak JT (2013) AMP-activated protein kinase regulates nicotinamide phosphoribosyl transferase expression in skeletal muscle. J Physiol 591(20):5207–5220. https://doi.org/10.1113/jphysiol.2013.259515

Brenner C (2022) Sirtuins are not conserved longevity genes. Life Metab 1(2):122–133. https://doi.org/10.1093/lifemeta/loac025

Buchner E (1897) Alcoholic fermentation without yeast cells. Ber Dtsch Chem Ges 30:117–124. https://doi.org/10.1002/cber.18970300121

Bursch KL, Goetz CJ, Smith BC (2024) Current trends in Sirtuin activator and inhibitor development. Molecules 29(5):1185. https://doi.org/10.3390/molecules29051185

Canto C, Houtkooper RH, Pirinen E, Youn DY, Oosterveer MH, Cen Y, Fernandez-Marcos PJ, Yamamoto H, Andreux PA, Cettour-Rose P, Gademann K, Rinsch C, Schoonjans K, Sauve AA, Auwerx J (2012) The NAD(+) precursor nicotinamide riboside enhances oxidative metabolism and protects against high-fat diet-induced obesity. Cell Metab 15(6):838–847. https://doi.org/10.1016/j.cmet.2012.04.022

Canto C (2022) NAD$^+$ precursors: a questionable redundancy. Metabolites 12(7):630. https://doi.org/10.3390/metabo12070630

Carlson LA (2005) Nicotinic acid: the broad-spectrum lipid drug. A 50th anniversary review. J Intern Med 258:94–114. https://doi.org/10.1111/j.1365-2796.2005.01528.x

Cartwright DM, Oakey LA, Fletcher RS, Doig CL, Heising S, Larner DP, Nasteska D, Berry CE, Heaselgrave SR, Ludwig C, Hodson DJ, Lavery GG, Garten A (2021) Nicotinamide riboside has minimal impact on energy metabolism in mouse models of mild obesity. J Endocrinol 251(1):111–123. https://doi.org/10.1530/JOE-21-0123

Cercillieux A, Ciarlo E, Canto C (2022) Balancing NAD^+ deficits with nicotinamide riboside: therapeutic possibilities and limitations. Cell Mol Life Sci 79(8):463. https://doi.org/10.1007/s00018-022-04499-5

Chambon P, Weill JD, Mandel P (1963) Nicotinamide mononucleotide activation of new DNA-dependent polyadenylic acid synthesizing nuclear enzyme. Biochem Biophys Res Commun 11:39–43. https://doi.org/10.1016/0006-291x(63)90024-x

Chini CCS, Cordeiro HS, Tran NLK, Chini EN (2024) NAD metabolism: role in senescence regulation and aging. Aging Cell 23(1):e13920. https://doi.org/10.1111/acel.13920

Connell NJ, Grevendonk L, Fealy CE, Moonen-Kornips E, Bruls YMH, Schrauwen-Hinderling VB, de Vogel J, Hageman R, Geurts J, Zapata-Perez R, Houtkooper RH, Havekes B, Hoeks J, Schrauwen P (2021) NAD+-precursor supplementation with L-tryptophan, nicotinic acid, and nicotinamide does not affect mitochondrial function or skeletal muscle function in physically compromised older adults. J Nutr 151(10):2917–2931. https://doi.org/10.1093/jn/nxab193

Covarrubias AJ, Kale A, Perrone R, Lopez-Dominguez JA, Pisco AO, Kasler HG, Schmidt MS, Heckenbach I, Kwok R, Wiley CD, Wong HS, Gibbs E, Iyer SS, Basisty N, Wu Q, Kim IJ, Silva E, Vitangcol K, Shin KO, Lee YM, Riley R, Ben-Sahra I, Ott M, Schilling B, Scheibye-Knudsen M, Ishihara K, Quake SR, Newman J, Brenner C, Campisi J, Verdin E (2020) Senescent cells promote tissue NAD^+ decline during ageing via the activation of $CD38^+$ macrophages. Nat Metab 2(11):1265–1283. https://doi.org/10.1038/s42255-020-00305-3

Covarrubias AJ, Perrone R, Grozio A, Verdin E (2021) NAD^+ metabolism and its roles in cellular processes during ageing. Nat Rev Mol Cell Biol 22(2):119–141. https://doi.org/10.1038/s41580-020-00313-x

Damgaard MV, Treebak JT (2023) What is really known about the effects of nicotinamide riboside supplementation in humans. Sci Adv 9(29):eadi4862. https://doi.org/10.1126/sciadv.adi4862

Daft FS, Fraser HF, Sebrell WH, Pittman M (1938) Treatment of Blacktongue with Cozymase. Science 88(2275):128–129. https://doi.org/10.1126/science.88.2275.128.b

Daniells S (2024) NPA hails "monumental victory" with NMN case. Available at: https://www.nutraingredients-usa.com/Article/2024/11/06/NPA-hails-monumental-victory-with-NMN-case/. Accessed 6 May 2025

Da Poian AT, El-Bacha T, Luz MRMP (2010) Nutrient utilization in humans: metabolism pathways. Nature Education 3(9):11

Elvehjem CA, Madden RJ, Strong FM, Woolley DW (1937) Relation of nicotinic acid and nicotinic acid amide to canine black tongue. J Am Chem Soc 59(9):1767–1768. https://doi.org/10.1021/ja01288a509

Elvehjem CA, Madden RJ, Strong FM, Woolley DW (1938) The isolation and identification of the anti-black tongue factor. J Biol Chem 123:137–149. https://doi.org/10.1016/S0021-9258(18)74164-1

Escande C, Nin V, Price NL, Capellini V, Gomes AP, Barbosa MT, O'Neil L, White TA, Sinclair DA, Chini EN (2013) Flavonoid apigenin is an inhibitor of the NAD+ ase CD38: implications for cellular NAD+ metabolism, protein acetylation, and treatment of metabolic syndrome. Diabetes 62(4):1084–1093. https://doi.org/10.2337/db12-1139

Essuman K, Summers DW, Sasaki Y, Mao X, DiAntonio A, Milbrandt J (2017) The SARM1 toll/Interleukin-1 receptor domain possesses intrinsic NAD^+ cleavage activity that promotes pathological axonal degeneration. Neuron. 93(6):1334–1343.e5. https://doi.org/10.1016/j.neuron.2017.02.022

Fang EF, Lautrup S, Hou Y, Demarest TG, Croteau DL, Mattson MP, Bohr VA (2017) NAD^+ in aging: molecular mechanisms and translational implications. Trends Mol Med 23(10):899–916. https://doi.org/10.1016/j.molmed.2017.08.001

Feuz MB, Meyer-Ficca ML, Meyer RG (2023) Beyond pellagra-research models and strategies addressing the enduring clinical relevance of NAD deficiency in aging and disease. Cells 12(3): 500. https://doi.org/10.3390/cells12030500

Freese R, Lysne V (2023) Niacin—a scoping review for Nordic nutrition recommendations 2023. Food Nutr Res 67. https://doi.org/10.29219/fnr.v67.10299

Fukuwatari T, Ohta M, Kimtjra N, Sasaki R, Shibata K (2004) Conversion ratio of tryptophan to niacin in Japanese women fed a purified diet conforming to the Japanese dietary reference intakes. J Nutr Sci Vitaminol Tokyo 50(6):385–391. https://doi.org/10.3177/jnsv.50.385

Fukuwatari T, Shibata K (2013) Nutritional aspect of tryptophan metabolism. Int J Tryptophan Res 6(Suppl 1). https://doi.org/10.4137/IJTR.S11588

Galassi L, Di Stefano M, Brunetti L, Orsomando G, Amici A, Ruggieri S, Magni G (2012) Characterization of human nicotinate phosphoribosyltransferase: kinetic studies, structure prediction and functional analysis by site-directed mutagenesis. Biochimie 94(2):300–309. https://doi.org/10.1016/j.biochi.2011.06.033

Goldberger J, Wheeler GA, Lillie RD, Rogers LM (1926) A further study of butter, fresh beef, and yeast as pellagra preventives, with consideration of the relation of factor P-P of pellagra (and black tongue of dogs) to vitamin B. Public Health Rep 41(8):297–318. https://doi.org/10.2307/4577789

Grant R, Berg J, Mestayer R, Braidy N, Bennett J, Broom S, Watson J (2019) A pilot study investigating changes in the human plasma and urine NAD+ metabolome during a 6 hour intravenous infusion of NAD. Front Aging Neurosci 11:257. https://doi.org/10.3389/fnagi.2019.00257

Grozio A, Mills KF, Yoshino J, Bruzzone S, Sociali G, Tokizane K, Lei HC, Cunningham R, Sasaki Y, Migaud ME, Imai SI (2019) Slc12a8 is a nicotinamide mononucleotide transporter. Nat Metab 1(1):47–57. https://doi.org/10.1038/s42255-018-0009-4. Epub 2019 Jan 7. Erratum in: Nat Metab. 1(7):743

Grube K, Bürkle A (1992) Poly(ADP-ribose) polymerase activity in mononuclear leukocytes of 13 mammalian species correlates with species-specific life span. Proc Natl Acad Sci USA 89(24):11759–11763. https://doi.org/10.1073/pnas.89.24.11759

Gu Y, Gao L, He J, Luo M, Hu M, Lin Y, Li J, Hou T, Si J, Yu Y (2024) β-Nicotinamide mononucleotide supplementation prolongs the lifespan of prematurely aged mice and protects colon function in ageing mice. Food Funct 15(6):3199–3213. https://doi.org/10.1039/d3fo05221d

Guarente L, Sinclair DA, Kroemer G (2024) Human trials exploring anti-aging medicines. Cell Metab 36(2):354–376. https://doi.org/10.1016/j.cmet.2023.12.007

Guo C, Huang Q, Wang Y, Yao Y, Li J, Chen J, Wu M, Zhang Z, E M, Qi H, Ji P, Liu Q, Zhao D, Su H, Qi W, Li X (2023) Therapeutic application of natural products: NAD$^+$ metabolism as potential target. Phytomedicine 114:154768. https://doi.org/10.1016/j.phymed.2023.154768

Hara N, Yamada K, Terashima M, Osago H, Shimoyama M, Tsuchiya M (2003) Molecular identification of human glutamine- and ammonia-dependent NAD synthetases. Carbon-nitrogen hydrolase domain confers glutamine dependency. J Biol Chem 278(13):10914–10921. https://doi.org/10.1074/jbc.M209203200

Harden A (1905) Zymase and alcoholic fermentation. J Inst Brew 11:2–15. https://doi.org/10.1002/j.2050-0416.1905.tb02115.x

Harden A, Young WJ (1906) The alcoholic ferment of yeast-juice. Part II.—the coferment of yeast-juice. Proc R Soc London: Ser B Contain Pap Biol Charact 78(526):369–375. https://doi.org/10.1098/rspb.1906.0070

Harrison DE, Strong R, Reifsnyder P, Kumar N, Fernandez E, Flurkey K, Javors MA, Lopez-Cruzan M, Macchiarini F, Nelson JF, Markewych A, Bitto A, Sindler AL, Cortopassi G, Kavanagh K, Leng L, Bucala R, Rosenthal N, Salmon A, Stearns TM, Bogue M, Miller RA (2021) 17-a-estradiol late in life extends lifespan in aging UM-HET3 male mice; nicotinamide riboside and three other drugs do not affect lifespan in either sex. Aging Cell 20:e13328. https://doi.org/10.1111/acel.13328

Henderson JD, Quigley SNZ, Chachra SS, Conlon N, Ford D (2024) The use of a systems approach to increase NAD$^+$ in human participants. NPJ Aging 10(1):7. https://doi.org/10.1038/s41514-023-00134-0

Hong W, Mo F, Zhang Z, Huang M, Wei X (2020) Nicotinamide mononucleotide: A promising molecule for therapy of diverse diseases by targeting NAD+ metabolism. Front Cell Dev Biol 8: 246. https://doi.org/10.3389/fcell.2020.00246

Howitz KT, Bitterman KJ, Cohen HY, Lamming DW, Lavu S, Wood JG, Zipkin RE, Chung P, Kisielewski A, Zhang LL, Scherer B, Sinclair DA (2003) Small molecule activators of sirtuins extend Saccharomyces cerevisiae lifespan. Nature 425(6954):191–196. https://doi.org/10.1038/nature01960

Huang H (2022) A multicentre, randomised, double blind, parallel design, placebo controlled study to evaluate the efficacy and safety of Uthever (NMN supplement), an orally administered supplementation in middle aged and older adults. Front Aging 3:851698. https://doi.org/10.3389/fragi.2022.851698

Huang Z, Wang X, Li N, Song F, Zhou J (2023) Systematic engineering of Escherichia coli for efficient production of nicotinamide riboside from nicotinamide and 3-cyanopyridine. Bioresour Technol 377:128953. https://doi.org/10.1016/j.biortech.2023.128953

Igarashi M, Nakagawa-Nagahama Y, Miura M, Kashiwabara K, Yaku K, Sawada M, Sekine R, Fukamizu Y, Sato T, Sakurai T, Sato J, Ino K, Kubota N, Nakagawa T, Kadowaki T, Yamauchi T (2022) Chronic nicotinamide mononucleotide supplementation elevates blood nicotinamide adenine dinucleotide levels and alters muscle function in healthy older men. NPJ Aging. 8(1):5. https://doi.org/10.1038/s41514-022-00084-z

Ikeda M, Tsuji H, Nakamura S, Ichiyama A, Nishizuka Y, Hayaishi O (1965) Studies on the biosynthesis of nicotinamide adenine dinucleotide. II. A role of picolinic carboxylase in the biosynthesis of nicotinamide adenine dinucleotide from tryptophan in mammals. J Biol Chem 240:1395–1401

Imai S, Armstrong CM, Kaeberlein M, Guarente L (2000) Transcriptional silencing and longevity protein Sir2 is an NAD-dependent histone deacetylase. Nature 403(6771):795–800. https://doi.org/10.1038/35001622

Imai S (2009) Nicotinamide phosphoribosyltransferase (Nampt): a link between NAD biology, metabolism, and diseases. Curr Pharm Des 15(1):20–28. https://doi.org/10.2174/138161209787185814

Imai S, Guarente L (2014) NAD+ and sirtuins in aging and disease. Trends Cell Biol 24(8): 464–471. https://doi.org/10.1016/j.tcb.2014.04.002

Imai SI (2025) NAD world 3.0: the importance of the NMN transporter and eNAMPT in mammalian aging and longevity control. Npj Aging 11:1–12. https://doi.org/10.1038/s41514-025-00192-6

Imsande J, Handler P (1961) Biosynthesis of diphosphopyridine nucleotide. III. Nicotinic acid mononucleotide pyrophos-phorylase. J Biol Chem 236:525–530. PMID: 13717627

Iqbal T, Nakagawa T (2024) The therapeutic perspective of NAD$^+$ precursors in age-related diseases. Biochem Biophys Res Commun 702:149590. https://doi.org/10.1016/j.bbrc.2024.149590

Irie J, Inagaki E, Fujita M, Nakaya H, Mitsuishi M, Yamaguchi S, Yamashita K, Shigaki S, Ono T, Yukioka H, Okano H, Nabeshima YI, Imai SI, Yasui M, Tsubota K, Itoh H (2020) Effect of oral administration of nicotinamide mononucleotide on clinical parameters and nicotinamide metabolite levels in healthy Japanese men. Endocr J 67(2):153–160. https://doi.org/10.1507/endocrj.EJ19-0313

Ito TK, Sato T, Hakamato A, Onoda Y, Sato S, Yamazaki F, Horikawa M, Takahashi Y, Kitamoto T, Suzuki M, Uchida S, Odagiri K, Setou M (2020) A nonrandomized study of single oral supplementation within the daily tolerable upper level of nicotinamide affects blood nicotinamide and NAD$^+$ levels in healthy subjects. Transl Med Aging 4:45–54. https://doi.org/10.1016/j.tma.2020.04.002

Ito TK, Sato T, Takanashi Y, Tamannaa Z, Kitamoto T, Odagiri K, Setou M (2021) A single oral supplementation of nicotinamide within the daily tolerable upper-level increases blood NAD+ levels in healthy subjects. Tansl Med Aging 5:43–51. https://doi.org/10.1016/j.tma.2021.09.001

Jaworek AK, Łazarczyk A, Hałubiec P, Leończyk M, Staszewski F, Wojas-Pelc A (2021) The history of pellagra. Dermatology Review 108:554–566. https://doi.org/10.5114/dr.2021.114610

Ji X, Zheng M, Yu T, Kang J, Fan T, Xu B (2023) NAD+-consuming enzymes in stem cell homeostasis. Oxidative Med Cell Longev 2023:4985726. https://doi.org/10.1155/2023/4985726

Kaeberlein M, McVey M, Guarente L (1999) The SIR2/3/4 complex and SIR2 alone promote longevity in Saccharomyces cerevisiae by two different mechanisms. Genes Dev 13:2570–2580. https://doi.org/10.1101/gad.13.19.2570

Kane AE, Chellappa K, S M, Matthew A, Li J, Amorim, JA, Diener C, Zhu DT, Mitchell SJ, Griffin P, Tian X, Petty C, Conway RJ, Walsh K, Shelerud L, Duesing C, Mueller A, Li K, McNamara MS, Shima RT, Mitchell JR, Bonkowski MS, De Cabo R, Gibbons, Sean G, Wu L, Ikeno Y, Baur JA, Rajman LA, Sinclair DA (2024) Long-term NMN treatment increases lifespan and Healthspan in mice in a sex dependent manner. Available at SSRN: https://ssrn.com/abstract=4857461 or https://doi.org/10.2139/ssrn.4857461. Accessed 5 May 2025

Kataura T, Sedlackova L, Otten EG, Kumari R, Shapira D, Scialo F, Stefanatos R, Ishikawa KI, Kelly G, Seranova E, Sun C, Maetzel D, Kenneth N, Trushin S, Zhang T, Trushina E, Bascom CC, Tasseff R, Isfort RJ, Oblong JE, Miwa S, Lazarou M, Jaenisch R, Imoto M, Saiki S, Papamichos-Chronakis M, Manjithaya R, Maddocks ODK, Sanz A, Sarkar S, Korolchuk VI (2022) Autophagy promotes cell survival by maintaining NAD levels. Dev Cell 57(22):2584–2598.e11. https://doi.org/10.1016/j.devcel.2022.10.008

Katayoshi T, Uehata S, Nakashima N, Nakajo T, Kitajima N, Kageyama M, Tsuji-Naito K (2023) Nicotinamide adenine dinucleotide metabolism and arterial stiffness after long-term nicotinamide mononucleotide supplementation: a randomized, double-blind, placebo-controlled trial. Sci Rep 13(1):2786. https://doi.org/10.1038/s41598-023-29787-3

Kitada M, Araki SI, Koya D (2023) The role of CD38 in the pathogenesis of Cardiorenal metabolic disease and aging, an approach from basic research. Cells 12(4):595. https://doi.org/10.3390/cells12040595

Koehn CJ, Elvehjem CA (1937) Further studies on the concentration of the antipellagra factor. J Biol Chem 118(3):693–699. https://doi.org/10.1016/S0021-9258(18)74475-X

Kornberg A, Lindberg O (1948) Diphosphopyridine nucleotide pyrophosphatase. J Biol Chem 176(2):665–677. PMID: 18889922

Kornberg A (1948) The participation of inorganic pyrophosphate in the reversible enzymatic synthesis of diphosphopyridine nucleotide. J Biol Chem 176(3):1475. PMID: 18098602

Kraus WL (2015) PARPs and ADP-Ribosylation: 50 years … and counting. Mol Cell 58(6):902–910. https://doi.org/10.1016/j.molcel.2015.06.006

Kraus WL (2020) PARPs and ADP-ribosylation: 60 years on. Genes Dev 34(5–6):251–253. https://doi.org/10.1101/gad.336420.120

Krehl WA, Teply LJ, Sarma PS, Elvehjem CA (1945) Growth-retarding effect of corn in nicotinic acid-low rations and its counteraction by tryptophan. Science 101(2628):489–490. https://doi.org/10.1126/science.101.2628.489

Kresge N, Simoni RD, Hill RL (2004) Blacktongue and nicotinic acid metabolism: Philip Handler. J Biol Chem 279(48):e8–e10. Available at: https://www.jbc.org/article/S0021-9258(20)67862-0/pdf

Kropotov A, Kulikova V, Nerinovski K, Yakimov A, Svetlova M, Solovjeva L, Sudnitsyna J, Migaud ME, Khodorkovskiy M, Ziegler M, Nikiforov A (2021) Equilibrative nucleoside transporters mediate the import of nicotinamide riboside and nicotinic acid riboside into human cells. Int J Mol Sci 22(3):1391. https://doi.org/10.3390/ijms22031391

Kuerec AH, Wang W, Yi L, Tao R, Lin Z, Vaidya A, Pendse S, Thasma S, Andhalkar N, Avhad G, Kumbhar V, Maier AB (2024) Towards personalized nicotinamide mononucleotide (NMN) supplementation: Nicotinamide adenine dinucleotide (NAD) concentration. Mech Ageing Dev 218:111917. https://doi.org/10.1016/j.mad.2024.111917

Kwon J, Idoine R, Ishtiaq Y, Nkrumah-Elie Y, Rosene M, Shao A (2022) The clinical effects of nicotinamide riboside on inflammatory parameters. Curr Dev Nutr 6(Suppl 1):987. https://doi.org/10.1093/cdn/nzac068.016

Landry J, Slama JT, Sternglanz R (2000) Role of NAD(+) in the deacetylase activity of the SIR2-like proteins. Biochem Biophys Res Commun 278:685–690. https://doi.org/10.1006/bbrc.2000.3854

Leder IG, Handler P (1951) Synthesis of nicotinamide mononucleotide by human erythrocytes in vitro. J Biol Chem 189(2):889–899. https://doi.org/10.1016/S0021-9258(18)44904-6

Levine DC, Hong H, Weidemann BJ, Ramsey KM, Affinati AH, Schmidt MS, Cedernaes J, Omura C, Braun R, Lee C, Brenner C, Peek CB, Bass J (2020) NAD^+ controls circadian reprogramming through PER2 nuclear translocation to counter aging. Mol Cell 78(5):835–849.e7. https://doi.org/10.1016/j.molcel.2020.04.010

Li F, Wu C, Wang G (2024) Targeting NAD metabolism for the therapy of age-related neurodegenerative diseases. Neurosci Bull 40(2):218–240. https://doi.org/10.1007/s12264-023-01072-3

Liu L, Su X, Quinn WJ 3rd, Hui S, Krukenberg K, Frederick DW, Redpath P, Zhan L, Chellappa K, White E, Migaud M, Mitchison TJ, Baur JA, Rabinowitz JD (2018) Quantitative analysis of NAD synthesis-breakdown fluxes. Cell Metab 27(5):1067–1080.e5. https://doi.org/10.1016/j.cmet.2018.03.018

Liu H, Wang S, Wang J, Guo X, Song Y, Fu K, Gao Z, Liu D, He W, Yang LL (2025) Energy metabolism in health and diseases. Signal Transduct Target Ther 10(1):69. https://doi.org/10.1038/s41392-025-02141-x

López-Otín C, Blasco MA, Partridge L, Serrano M, Kroemer G (2013) The hallmarks of aging. Cell 153(6):1194–1217. https://doi.org/10.1016/j.cell.2013.05.039

López-Otín C, Blasco MA, Partridge L, Serrano M, Kroemer G (2023) Hallmarks of aging: an expanding universe. Cell 186(2):243–278. https://doi.org/10.1016/j.cell.2022.11.001

Mangerich A, Bürkle A (2012) Pleiotropic cellular functions of PARP1 in longevity and aging: genome maintenance meets inflammation. Oxidative Med Cell Longev 2012:321653. https://doi.org/10.1155/2012/321653

Massudi H, Grant R, Braidy N, Guest J, Farnsworth B, Guillemin GJ (2012) Age-associated changes in oxidative stress and NAD+ metabolism in human tissue. PLoS One 7(7):e42357. https://doi.org/10.1371/journal.pone.0042357

McReynolds MR, Chellappa K, Baur JA (2020) Age-related NAD^+ decline. Exp Gerontol 134:110888. https://doi.org/10.1016/j.exger.2020.110888

Mehmel M, Jovanović N, Spitz U (2020) Nicotinamide riboside-the current state of research and therapeutic uses. Nutrients 12(6):1616. https://doi.org/10.3390/nu12061616

Mills KF, Yoshida S, Stein LR, Grozio A, Kubota S, Sasaki Y, Redpath P, Migaud ME, Apte RS, Uchida K, Yoshino J, Imai SI (2016) Long-term administration of nicotinamide mononucleotide mitigates age-associated physiological decline in mice. Cell Metab 24(6):795–806. https://doi.org/10.1016/j.cmet.2016.09.013

Mitchell SJ, Bernier M, Aon MA, Cortassa S, Kim EY, Fang EF, Palacios HH, Ali A, Navas-Enamorado I, Di Francesco A, Kaiser TA, Waltz TB, Zhang N, Ellis JL, Elliott PJ, Frederick DW, Bohr VA, Schmidt MS, Brenner C, Sinclair DA, Sauve AA, Baur JA, de Cabo R (2018) Nicotinamide improves aspects of Healthspan, but not lifespan, in mice. Cell Metab 27(3):667–676.e4. https://doi.org/10.1016/j.cmet.2018.02.001

Morifuji M, Higashi S, Ebihara S, Nagata M (2024) Ingestion of β-nicotinamide mononucleotide increased blood NAD levels, maintained walking speed, and improved sleep quality in older adults in a double-blind randomized, placebo-controlled study. GeroScience 46:4671–4688. https://doi.org/10.1007/s11357-024-01204-1

Nadeeshani H, Li J, Ying T, Zhang B, Lu J (2021) Nicotinamide mononucleotide (NMN) as an anti-aging health product – promises and safety concerns. J Adv Res 37:267–278. https://doi.org/10.1016/j.jare.2021.08.003

Naiman S, Cohen HY (2012) The contentious history of sirtuin debates. Rambam Maimonides Med J 3(4):e0022. https://doi.org/10.5041/RMMJ.10093

Niccoli T, Partridge L (2012) Ageing as a risk factor for disease. Curr Biol 22(17):R741–R752. https://doi.org/10.1016/j.cub.2012.07.024

Niño-Narvión J, Rojo-López MI, Martinez-Santos P, Rossell J, Ruiz-Alcaraz AJ, Alonso N, Ramos-Molina B, Mauricio D, Julve J (2023) NAD+ precursors and intestinal inflammation: therapeutic insights involving gut microbiota. Nutrients 15(13):2992. https://doi.org/10.3390/nu15132992

Nishizuki Y, Hayaishi O (1963) Enzymic synthesis of niacin nucleotides from 3-hydroxyanthranilic acid in mammalian tissues. J Biol Chem 238(10):3369–3377. PMID: 14085389

Niu KM, Bao T, Gao L, Ru M, Li Y, Jiang L, Ye C, Wang S, Wu X (2021) The impacts of short-term NMN supplementation on serum metabolism, fecal microbiota, and telomere length in pre-aging phase. Front Nutr 8:756243. https://doi.org/10.3389/fnut.2021.756243

North BJ, Rosenberg MA, Jeganathan KB, Hafner AV, Michan S, Dai J, Baker DJ, Cen Y, Wu LE, Sauve AA, van Deursen JM, Rosenzweig A, Sinclair DA (2014) SIRT2 induces the checkpoint kinase BubR1 to increase lifespan. EMBO J 33(13):1438–1453. https://doi.org/10.15252/embj.201386907

Okabe K, Yaku K, Tobe K, Nakagawa T (2019) Implications of altered NAD metabolism in metabolic disorders. J Biomed Sci 26(1):34. https://doi.org/10.1186/s12929-019-0527-8

Okabe K, Yaku K, Uchida Y, Fukamizu Y, Sato T, Sakurai T, Tobe K, Nakagawa T (2022) Oral administration of nicotinamide mononucleotide is safe and efficiently increases blood nicotinamide adenine dinucleotide levels in healthy subjects. Front Nutr 9:868640. https://doi.org/10.3389/fnut.2022.868640

Orr ME, Kotkowski E, Ramirez P, Bair-Kelps D, Liu Q, Brenner C, Schmidt MS, Fox PT, Larbi A, Tan C, Wong G, Gelfond J, Frost B, Espinoza S, Musi N, Powers B (2024) A randomized placebo-controlled trial of nicotinamide riboside in older adults with mild cognitive impairment. Geroscience 46(1):665–682. https://doi.org/10.1007/s11357-023-00999-9

Ottens F, Franz A, Hoppe T (2021) Build-UPS and break-downs: metabolism impacts on proteostasis and aging. Cell Death Differ 28(2):505–521. https://doi.org/10.1038/s41418-020-00682-y

Peluso A, Damgaard MV, Mori MAS, Treebak JT (2021) Age-dependent decline of NAD+-universal truth or confounded consensus? Nutrients 14(1):101. https://doi.org/10.3390/nu14010101

Pencina KM, Lavu S, Dos Santos M, Beleva YM, Cheng M, Livingston D, Bhasin S (2023a) MIB-626, an Oral formulation of a microcrystalline unique polymorph of β-nicotinamide mononucleotide, increases circulating nicotinamide adenine dinucleotide and its metabolome in middle-aged and older adults. J Gerontol A Biol Sci Med Sci 78(1):90–96. https://doi.org/10.1093/gerona/glac049

Pencina KM, Valderrabano R, Wipper B, Orkaby AR, Reid KF, Storer T, Lin AP, Merugumala S, Wilson L, Latham N, Ghattas-Puylara C, Ozimek NE, Cheng M, Bhargava A, Memish-Beleva-Y, Lawney B, Lavu S, Swain PM, Apte RS, Sinclair DA, Livingston D, Bhasin S (2023b) Nicotinamide adenine dinucleotide augmentation in overweight or obese middle-aged and older adults: a physiologic study. J Clin Endocrinol Metab 108(8):1968–1980. https://doi.org/10.1210/clinem/dgad027

Pignatti C, D'Adamo S, Stefanelli C, Flamigni F, Cetrullo S (2020) Nutrients and pathways that regulate health span and life span. Geriatrics (Basel) 5(4):95. https://doi.org/10.3390/geriatrics5040095

Pittelli M, Felici R, Pitozzi V, Giovannelli L, Bigagli E, Cialdai F, Romano G, Moroni F, Chiarugi A (2011) Pharmacological effects of exogenous NAD on mitochondrial bioenergetics, DNA repair, and apoptosis. Mol Pharmacol 80(6):1136–1146. https://doi.org/10.1124/mol.111.073916

Poljsak B, Kovač V, Milisav I (2020) Healthy lifestyle recommendations: do the beneficial effects originate from NAD+ amount at the cellular level? Oxidative Med Cell Longev 2020:8819627. https://doi.org/10.1155/2020/8819627

Poljšak B, Kovač V, Milisav I (2022) Current uncertainties and future challenges regarding NAD+ boosting strategies. Antioxidants (Basel) 11(9):1637. https://doi.org/10.3390/antiox11091637

Ponomarenko EA, Poverennaya EV, Ilgisonis EV, Pyatnitskiy MA, Kopylov AT, Zgoda VG, Lisitsa AV, Archakov AI (2016) The size of the human proteome: the width and depth. Int J Anal Chem 2016:7436849. https://doi.org/10.1155/2016/7436849

Preiss J, Handler P (1957) Enzymatic synthesis of nicotinamide mononucleotide. J Biol Chem 225(2):759–770. PMID: 13416279

Preiss J, Handler P (1958a) Biosynthesis of diphosphopyridine nucleotide. I. Identification of intermediates. J Biol Chem 233(2):488–492. PMID: 13563526

Preiss J, Handler P (1958b) Biosynthesis of diphosphopyridine nucleotide. II Enzymatic aspects. J Biol Chem 233(2):493–500. PMID: 13563527

Ragonnaud E, Biragyn A (2021) Gut microbiota as the key controllers of "healthy" aging of elderly people. Immun Ageing 18(1):2. https://doi.org/10.1186/s12979-020-00213-w

Rajakumar K (2000) Pellagra in the United States: a historical perspective. South Med J 93(3): 272–277

Rahman SU, Qadeer A, Wu Z (2024) Role and potential mechanisms of nicotinamide mononucleotide in aging. Aging Dis 15(2):565–583. https://doi.org/10.14336/AD.2023.0519-1

Rajman L, Chwalek K, Sinclair DA (2018) Therapeutic potential of NAD-boosting molecules: the in vivo evidence. Cell Metab 27(3):529–547. https://doi.org/10.1016/j.cmet.2018.02.011

Ramanathan C, Lackie T, Williams DH, Simone PS, Zhang Y, Bloomer RJ (2022) Oral administration of nicotinamide mononucleotide increases nicotinamide adenine dinucleotide level in an animal brain. Nutrients 14(2):300. https://doi.org/10.3390/nu14020300

Ren C, Zhang S, Chen Y, Deng K, Kuang M, Gong Z, Zhang K, Wang P, Huang P, Zhou Z, Gong A (2024) Exploring nicotinamide adenine dinucleotide precursors across biosynthesis pathways: unraveling their role in the ovary. FASEB J 38(14):e23804. https://doi.org/10.1096/fj.202400453R

Revollo JR, Körner A, Mills KF, Satoh A, Wang T, Garten A, Dasgupta B, Sasaki Y, Wolberger C, Townsend RR, Milbrandt J, Kiess W, Imai S (2007) Nampt/PBEF/Visfatin regulates insulin secretion in beta cells as a systemic NAD biosynthetic enzyme. Cell Metab 6(5):363–375. https://doi.org/10.1016/j.cmet.2007.09.003

Revollo JR, Grimm AA, Imai S (2004) The NAD biosynthesis pathway mediated by nicotinamide phosphoribosyltransferase regulates Sir2 activity in mammalian cells. J Biol Chem 279(49): 50754–50763. https://doi.org/10.1074/jbc.M408388200

Romani M, Hofer DC, Katsyuba E, Auwerx J (2019) Niacin: an old lipid drug in a new NAD^+ dress. J Lipid Res 60(4):741–746. https://doi.org/10.1194/jlr.s092007

Ruszkiewicz JA, Bürkle A, Mangerich A (2022) Fueling genome maintenance: on the versatile roles of NAD^+ in preserving DNA integrity. J Biol Chem 298(6):102037. https://doi.org/10.1016/j.jbc.2022.102037

Saito Y, Sato K, Jinno S, Nakamura Y, Nobukuni T, Ogishima S, Mizuno S, Koshiba S, Kuriyama S, Ohneda K, Morifuji M (2023) Effect of nicotinamide mononucleotide concentration in human Milk on neurodevelopmental outcome: the Tohoku medical megabank project birth and three-generation cohort study. Nutrients 16(1):145. https://doi.org/10.3390/nu16010145

Sandalova E, Li H, Guan L, Raj SD, Lim TG, Tian E, Kennedy BK, Maier AB (2024) Testing the amount of nicotinamide mononucleotide and urolithin A as compared to the label claim. Geroscience 46(5):5075–5083. https://doi.org/10.1007/s11357-024-01257-2

Satoh A, Brace CS, Rensing N, Cliften P, Wozniak DF, Herzog ED, Yamada KA, Imai S (2013) Sirt1 extends life span and delays aging in mice through the regulation of Nk2 homeobox 1 in the DMH and LH. Cell Metab 18(3):416–430. https://doi.org/10.1016/j.cmet.2013.07.013

Sauve AA, Wang Q, Zhang N, Kang S, Rathmann A, Yang Y (2023) Triple-isotope tracing for pathway discernment of NMN-induced NAD^+ biosynthesis in whole mice. Int J Mol Sci 24(13): 11114. https://doi.org/10.3390/ijms241311114

Scheibye-Knudsen M, Mitchell SJ, Fang EF, Iyama T, Ward T, Wang J, Dunn CA, Singh N, Veith S, Hasan-Olive MM, Mangerich A, Wilson MA, Mattson MP, Bergersen LH, Cogger VC, Warren A, Le Couteur DG, Moaddel R, Wilson DM 3rd, Croteau DL, de Cabo R, Bohr VA

(2014) A high-fat diet and NAD(+) activate Sirt1 to rescue premature aging in cockayne syndrome. Cell Metab 20(5):840–855. https://doi.org/10.1016/j.cmet.2014.10.005

Schlenk F (1942) Identification of the carbohydrate group in the nicotinamide nucleotides. J Biol Chem 146(2):619–626. https://doi.org/10.1016/S0021-9258(18)44986-1

Schmidt MS, Brenner C (2019) Absence of evidence that Slc12a8 encodes a nicotinamide mononucleotide transporter. Nat Metab 1(7):660–661. https://doi.org/10.1038/s42255-019-0085-0

Sharma C, Donu D, Cen Y (2022) Emerging role of nicotinamide riboside in health and diseases. Nutrients 14(19):3889. https://doi.org/10.3390/nu14193889

She J, Sheng R, Qin ZH (2021) Pharmacology and potential implications of nicotinamide adenine dinucleotide precursors. Aging Dis 12(8):1879–1897. https://doi.org/10.14336/AD.2021.0523

Slater EC (2004) Cozymase, coenzyme I (II), D(T)PN and NAD(P). IUBMB Life 56(5):289–290. https://doi.org/10.1080/15216540412331272263

Song WL, FitzGerald GA (2013) Niacin, an old drug with a new twist. J Lipid Res 54(10): 2586–2594. https://doi.org/10.1194/jlr.R040592

Song Q, Zhou X, Xu K, Liu S, Zhu X, Yang J (2023) The safety and antiaging effects of nicotinamide mononucleotide in human clinical trials: an update. Adv Nutr 14(6):1416–1435. https://doi.org/10.1016/j.advnut.2023.08.008

Spies TD, Cooper C, Balnkenhorn MA (1938) The use of nicotinic acid in the treatment of pellagra. JAMA 110(9):622–627. https://doi.org/10.1001/jama.1938.02790090004002

Srivastava S (2017) The mitochondrial basis of aging and age-related disorders. Genes (Basel) 8(12):398. https://doi.org/10.3390/genes8120398

Stein LR, Imai S (2012) The dynamic regulation of NAD metabolism in mitochondria. Trends Endocrinol Metab 23(9):420–428. https://doi.org/10.1016/j.tem.2012.06.005

Stock AJ, Liu Y (2021) NAD-linked metabolism and intervention in short telomere syndromes and murine models of telomere dysfunction. Front Aging 2:785171. https://doi.org/10.3389/fragi.2021.785171

Stock AJ, Ayyar S, Kashyap A, Wang Y, Yanai H, Starost MF, Tanaka-Yano M, Bodogai M, Sun C, Wang Y, Gong Y, Puligilla C, Fang EF, Bohr VA, Liu Y, Beerman I (2021) Boosting NAD ameliorates hematopoietic impairment linked to short telomeres in vivo. Geroscience 45(4):2213–2228. https://doi.org/10.1007/s11357-023-00752-2

Superko HR, Zhao XQ, Hodis HN, Guyton JR (2017) Niacin and heart disease prevention: engraving its tombstone is a mistake. J Clin Lipidol 11(6):1309–1317. https://doi.org/10.1016/j.jacl.2017.08.005

Suskiewicz MJ, Palazzo L, Hughes R, Ahel I (2021) Progress and outlook in studying the substrate specificities of PARPs and related enzymes. FEBS J 288(7):2131–2142. https://doi.org/10.1111/febs.15518

Strømland Ø, Diab J, Ferrario E, Sverkeli LJ, Ziegler M (2021) The balance between NAD$^+$ biosynthesis and consumption in ageing. Mech Ageing Dev 199:111569. https://doi.org/10.1016/j.mad.2021.111569

Trammell S, Schmidt M, Weidemann B, Redpath P, Jaksch F, Dellinger RW, Li Z, Abel ED, Migaud ME, Brenner C (2016) Nicotinamide riboside is uniquely and orally bioavailable in mice and humans. Nat Commun 7:12948. https://doi.org/10.1038/ncomms12948

von Euler-Chelpin H (1930) Fermentation of sugars and fermentative enzymes. Nobel Lecture Viewed May 2, 2025. https://www.nobelprize.org/uploads/2018/06/euler-chelpin-lecture.pdf

Wang Y (2014) Molecular links between caloric restriction and Sir2/SIRT1 activation. Diabetes Metab J 38(5):321–329. https://doi.org/10.4093/dmj.2014.38.5.321

Warburg O, Christian W (1936) Pyridine, the hydrogen transfusing component of fermentative enzymes. Helv Chim Acta 19:79–88. https://doi.org/10.1002/hlca.193601901199

Watroba M, Szukiewicz D (2021) Sirtuins at the service of healthy longevity. Front Physiol 12: 724506. https://doi.org/10.3389/fphys.2021.724506

Wilson N, Kataura T, Korsgen ME, Sun C, Sarkar S, Korolchuk VI (2023) The autophagy-NAD axis in longevity and disease. Trends Cell Biol 33(9):788–802. https://doi.org/10.1016/j.tcb.2023.02.004

Wu CY, Kupferschmid AC, Chen L, McManus AJ, Kivisäkk P, Galler JA, Schwab NA, DesRuisseaux LA, Williams VJ, Gerber J, Riley M, Young C, Guzmán-Vélez E, Dodge HH, Tanzi RE, Singer CM, Arnold SE (2025) Cognitive and Alzheimer's disease biomarker effects of oral nicotinamide riboside (NR) supplementation in older adults with subjective cognitive decline and mild cognitive impairment. Alzheimers Dement (N Y) 11(1):e70023. https://doi.org/10.1002/trc2.70023

Xie N, Zhang L, Gao W, Huang C, Huber PE, Zhou X, Li C, Shen G, Zou B (2020) NAD^+ metabolism: pathophysiologic mechanisms and therapeutic potential. Signal Transduct Target Ther 5(1):227. https://doi.org/10.1038/s41392-020-00311-7

Yaku K, Palikhe S, Iqbal T, Hayat F, Watanabe Y, Fujisaka S, Izumi H, Yoshida T, Karim M, Uchida H, Nawaz A, Tobe K, Mori H, Migaud ME, Nakagawa T (2025) Nicotinamide riboside and nicotinamide mononucleotide facilitate NAD^+ synthesis via enterohepatic circulation. Sci Adv 11(12):eadr1538. https://doi.org/10.1126/sciadv.adr1538

Yamaguchi S, Irie J, Mitsuishi M, Uchino Y, Nakaya H, Takemura R, Inagaki E, Kosugi S, Okano H, Yasui M, Tsubota K, Hayashi K, Yoshino J, Itoh H (2024) Safety and efficacy of long-term nicotinamide mononucleotide supplementation on metabolism, sleep, and nicotinamide adenine dinucleotide biosynthesis in healthy, middle-aged Japanese men. Endocr J 71(2): 153–169. https://doi.org/10.1507/endocrj.EJ23-0431

Yamane T, Imai M, Bamba T, Uchiyama S (2023) Nicotinamide mononucleotide (NMN) intake increases plasma NMN and insulin levels in healthy subjects. Clin Nutr ESPEN 56:83–86. https://doi.org/10.1016/j.clnesp.2023.04.031

Yang F, Deng X, Yu Y, Luo L, Chen X, Zheng J, Qiu Y, Xiao F, Xie X, Zhao Y, Guo J, Hu F, Zhang X, Ju Z, Zhou Y (2022) Association of Human Whole Blood NAD^+ contents with aging. Front Endocrinol (Lausanne) 13:829658. https://doi.org/10.3389/fendo.2022.829658

Yang L, Shen J, Liu C, Kuang Z, Tang Y, Qian Z, Guan M, Yang Y, Zhan Y, Li N, Li X (2023) Nicotine rebalances NAD^+ homeostasis and improves aging-related symptoms in male mice by enhancing NAMPT activity. Nat Commun 14:900. https://doi.org/10.1038/s41467-023-36543-8

Yi L, Maier AB, Tao R, Lin Z, Vaidya A, Pendse S, Thasma S, Andhalkar N, Avhad G, Kumbhar V (2023) The efficacy and safety of β-nicotinamide mononucleotide (NMN) supplementation in healthy middle-aged adults: a randomized, multicenter, double-blind, placebo-controlled, parallel-group, dose-dependent clinical trial. Geroscience 45(1):29–43. https://doi.org/10.1007/s11357-022-00705-1

Yoshino J, Mills KF, Yoon MJ, Imai S (2011) Nicotinamide mononucleotide, a key NAD(+) intermediate, treats the pathophysiology of diet- and age-induced diabetes in mice. Cell Metab 14(4):528–536. https://doi.org/10.1016/j.cmet.2011.08.014

Yoshino J, Baur JA, Imai SI (2018) NAD^+ intermediates: the biology and therapeutic potential of NMN and NR. Cell Metab 27(3):513–528. https://doi.org/10.1016/j.cmet.2017.11.002

Yoshino M, Yoshino J, Kayser BD, Patti GJ, Franczyk MP, Mills KF, Sindelar M, Pietka T, Patterson BW, Imai SI, Klein S (2021) Nicotinamide mononucleotide increases muscle insulin sensitivity in prediabetic women. Science 372(6547):1224–1229. https://doi.org/10.1126/science.abe9985

You Y, Gao Y, Wang H, Li J, Zhang X, Zhu Z, Liu N (2020) Subacute toxicity study of nicotinamide mononucleotide via Oral administration. Front Pharmacol 11:604404. https://doi.org/10.3389/fphar.2020.604404

Yu J, Shen Q, Li J (2024) Toxicology study profile of nicotinamide mononucleotide after acute and 90-day sub chronic dosing in Wistar rats and mutagenicity tests. Curr Res Toxicol 6:100171. https://doi.org/10.1016/j.crtox.2024.100171

Zhai RG, Rizzi M, Garavaglia S (2009) Nicotinamide/nicotinic acid mononucleotide adenylyltransferase, new insights into an ancient enzyme. Cell Mol Life Sci 66(17): 2805–2818. https://doi.org/10.1007/s00018-009-0047-x

Zhang H, Ryu D, Wu Y, Gariani K, Wang X, Luan P, D'Amico D, Ropelle ER, Lutolf MP, Aebersold R, Schoonjans K, Menzies KJ, Auwerx J (2016) NAD^+ repletion improves mitochondrial and stem cell function and enhances life span in mice. Science 352(6292): 1436–1443. https://doi.org/10.1126/science.aaf2693

Zhang N, Sauve AA (2017) Synthesis of β-nicotinamide riboside using an efficient two-step methodology. Curr Protoc Nucleic Acid Chem 71:14.14.1–14.14.9. https://doi.org/10.1002/cpnc.43

Zhao L, Cao J, Hu K, He X, Yun D, Tong T, Han L (2020) Sirtuins and their biological relevance in aging and age-related diseases. Aging Dis 11(4):927–945. https://doi.org/10.14336/AD.2019.0820

Zhuang Y, Zhang Y, Liu C, Zhong Y (2024) Interplay between the circadian clock and Sirtuins. Int J Mol Sci 25(21):11469. https://doi.org/10.3390/ijms252111469

Chapter 6
CoQ_{10} Supplementation for Healthy Longevity

Guillermo López-Lluch

Abstract The form of coenzyme Q in humans, CoQ_{10}, containing a benzene ring linked to a 10 units isoprene tail, plays many essential roles in biochemistry, metabolism and physiology in our organisms. Being the only lipidic member of the mitochondrial electron transport chain it also plays other essential catabolic and anabolic functions by transferring electrons. One of these functions is to act as antioxidant in all the cell membranes and in plasma lipoproteins.

In this chapter the importance of maintaining CoQ_{10} levels during aging is highlighted by the studies that have demonstrated a protective role of cells, tissues and organs against the deterioration associated with aging and age-related diseases. The protective effect can be in part due to prevention of endothelial dysfunction that has been associated with many of the age-related diseases and with the process of aging. The supplementation of CoQ_{10} in elderly people is recommended in the base of many studies indicating its safety and its role as antioxidant and essential compound in mitochondrial function.

Keywords Coenzyme Q_{10} · Healthspan · Mitochondria · Immunity · Chronic diseases · Supplementation · Aging

Abbreviations

AIFM2	Apoptosis-Inducing Factor Mitochondrial 2
CoQ	coenzyme Q
CoQ_{10}	coenzyme Q_{10}
CVD	cardiovascular disease
CRP	C-Reactive Protein
CYB_5R_3	Cytochrome b5 reductase 3

G. López-Lluch (✉)
Department of Physiology, Anatomy and Cell Biology, Andalusian Centre for Developmental Biology (CABD-CSIC-JA), Pablo de Olavide University, Seville, Spain
e-mail: glopllu@upo.es

S. I. S. Rattan, C. Gezer (eds.), *Supplements and Drugs for Healthy Longevity*, Healthy Ageing and Longevity 22, https://doi.org/10.1007/978-3-032-11608-6_6

GPX4	Glutathione Peroxidase 4
GST	Glutathione S-transferase
IL	Interleukine
LDL	Low Density Lipoprotein
mETC	Mitochondrial Electron Transport Chain
mtROS	mitochondrial reactive oxygen species
nSMase	Mg^{2+}-dependent neutral sphingomyelinase
NCPC1L1	Niemann-Pick C1 Like 1
OXPHOS	oxidative phosphorylation
PARL	Presenilin-Associated Rhomboid-Like
PPARα	Peroxisome Proliferator-Activated Receptor
PUFA	Polyunsaturated Fatty Acid
ROS	Reactive Oxygen Species
SAMP1	Senescence-prone
SASP	Senescence associated secretory phenotype
TFN-α	Tumor Necrosis Factor—alfa
VLDL	Very-low density lipoprotein
α-TOC	alpha-tocopherol

6.1 Introduction: Importance of Coenzyme Q in Aging

Coenzyme Q (CoQ, ubiquinone) is a redox-active isoprene lipid involved in fundamental enzymatic and antioxidant roles in mitochondria, all the cell membranes and plasma lipoproteins. CoQ is synthesized by all the aerobic organisms studied to date, from unicellular organisms to mammals indicating its essential role in survival of aerobic organism.

The synthesis of CoQ is located into mitochondria and is catalysed by a combination of different enzymes and regulatory proteins known as the CoQ-synthome (Stefely and Pagliarini 2017). The isoprenoid chain is synthesized at the endoplasmic reticulum from the mevalonate pathway by the activity of COQ1 tetramer and transferred to mitochondria by a yet unknown transporter. In the inner mitochondrial membrane, the isoprenoid chain, with different isoprene units depending on the animal, is linked to the para-hydroxybenzene ring synthesized from tyrosine or phenylalanine (Stefely and Pagliarini 2017; Vargas-Perez et al. 2024) through the activity of COQ2. From there, the benzene ring is modified by the different members of the CoQ-synthome until reaching the functional molecule, CoQ (Wang et al. 2024; Stefely and Pagliarini 2017; Alcázar-Fabra et al. 2016; Wang and Hekimi 2019).

The complexity of the synthesis of CoQ and the essential role of each of the members of the CoQ-synthome imply that any dysfunction of any of the enzymes or regulatory proteins involved in the synthesis produces severe deficiency in the levels of CoQ that is related with dysfunctional mitochondria that is associated with many

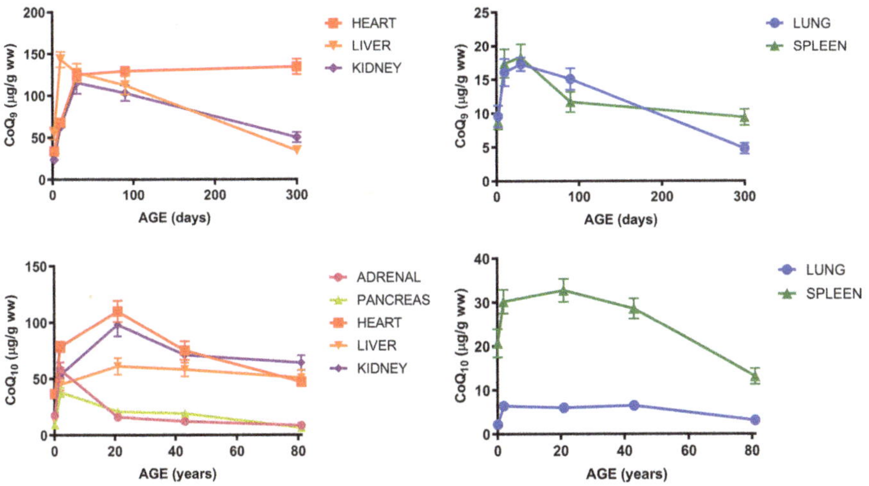

Fig. 6.1 Levels of the main CoQ forms, CoQ_9 (rats) and CoQ_{10} (humans), in different organs along life of the organisms. Data are represented as mean ± SEM from six (rats) or six-eight (human) samples of CoQ in g/g wet weight of the organ. (Figures from data published by, Tables 1 and 2 (Kalen et al. 1989))

diseases and death (Wang et al. 2024; Alcázar-Fabra et al. 2016). Further, decrease in CoQ synthesis can be also associated with the dysfunction of other proteins not directly related with its synthesis but with the activity of mitochondria such as Presenilin-Associated Rhomboid-Like (PARL) protease or mutations in the mtDNA affecting mitochondrial Electron Transport Chain (mtETC) (Radaelli et al. 2023; Spinazzi et al. 2019; Kuhl et al. 2017; Yen et al. 2011). These secondary deficiencies in CoQ have been related with bioenergetics unbalance associated with many diseases and aging (Kuhl et al. 2017; Navas et al. 2021).

This sensible balance between CoQ synthesis and mitochondrial activity suggest that an unbalanced synthesis in CoQ can severely affect mitochondrial activity but, at the same time, mitochondrial dysfunction can importantly affect CoQ synthesis affecting many of the important functions of this molecule as described in the next sections. Probably this vicious cycle explains why a net decrease in CoQ levels is found in rodents and humans in many of the organs after maturity, reaching the lowest levels at elderly ages as was described by Kalen et al. (1989) (Fig. 6.1).

The decrease in CoQ levels during aging many important functions in cells, tissues and organs due to its essential role in bioenergetics and antioxidant protection of cell membranes lipids (Aejmelaeus et al. 1997; Alcázar-Fabra et al. 2016; Arroyo et al. 2000).

6.2 Coenzyme Q Is a Redox Molecule with Many Different Functions

As a redox compound, CoQ accepts electrons from reducing factors and donate them to oxidizing compounds. CoQ is located inside the phospholipidic milieu of all cellular membranes in which CoQ changes between its reduced state (ubiquinol) to the oxidized (ubiquinone). To date, several different enzymatic mechanisms able to regenerate it to its reduced form have been described (Ernster and Dallner 1995). Due to this redox characteristic, CoQ is the only lipidic component of the mitochondrial electron transport chain, mtETC, but also can act as an antioxidant by itself preventing lipid peroxidation chain and in the regeneration of other antioxidants such as ascorbate (Gomez-Diaz et al. 1997; Santos-Ocana et al. 1998) and α-tocopherol (α-TOC) (Bello et al. 2003; Crane and Navas 1997).

Taken into consideration the important role of CoQ in mitochondrial activity not only involved in mtETC but also in the β-oxidation of fatty acids, the synthesis of pyridine nucleotides, and modification of aminoacids, a reduction in the levels of CoQ below a threshold can severely affect mitochondrial activity and contribute to mitochondrial dysfunction. It is known that mitochondrial dysfunction is one of the hallmarks of aging since physiologically active mitochondria maintenance is essential to maintain a plethora of processes in which mitochondria is involved. Thus, its dysfunction must be involved in the general physiological deterioration associated with aging and CoQ decrease (Lopez-Otin et al. 2013).

In fact, consistent accumulated evidence supports that mitochondrial dysfunction plays a key role in shortening lifespan and impairing physiological functions. Deterioration of health is associated with the accumulation of dysfunctional mitochondria contributing to the onset and progression of many age-related diseases including neurodegenerative, metabolic and cardiovascular diseases associated with aging (Srivastava 2017; Sun et al. 2016).

The other main effect of CoQ is the prevention of peroxidation of phospholipids and block of peroxidative chain in cell membranes. In this role, two well-known NAD(P)H enzymes were identified to restore ubiquinol from ubiquinone transferring electrons from cytosolic NAD(P)H that is oxidized. In this function, NQO1 or DT-diaphorase transfers two electrons directly from $NADPH^+$ to ubiquinone (Beyer et al. 1996; Bello et al. 2003) whereas $CytB_5Rase$ transfers electrons one by one from $NADH^+$ (Gomez-Diaz et al. 1997). Other enzymes able to reduce CoQ in cell membranes are lipoamide dehydrogenase, thioredoxin reductase and glutathione reductase (Nordman et al. 2003) indicating a plethora of enzymes able to maintain CoQ in its reduced state in cell membranes. Further, very recently another redox enzyme, previously associated with apoptosis-protection formerly known as apoptosis-inducing factor mitochondrial 2 (AIFM2), demonstrated protection against cell death by oxidative damage of membranes or ferroptosis (Doll et al. 2019; Bersuker et al. 2019). This enzyme protects cells against ferroptosis by a GPX4-independent mechanism by reducing CoQ in cell membranes.

Regarding the role of CoQ in the protection of lipoproteins against oxidation, the maintenance of CoQ_{10} in its reduced form in plasma in which it is the main redox form (Stocker et al. 1991), has been associated with the capacity of oxidoreductases associated with the outer side of plasma membrane of hepatocytes and also red blood cells (Stocker and Suarna 1993; Takahashi et al. 2019).

6.3 Is Coenzyme Q Deficiency a Driver of Mitochondrial Dysfunction in Aging or a Consequence?

The levels of CoQ decrease during aging in an organ-specific manner. Due to its essential functions in bioenergetics and antioxidant protection, this decrease can be considered an accelerating factor in the mitochondrial dysfunction associated with aging (Lopez-Lluch et al. 2010). However, some models of deficiency in the synthesis of CoQ indicate that low levels of CoQ can be associated with longer lifespan (Liu et al. 2017; Rodriguez-Hidalgo et al. 2018). This controversy is probably associated with metabolic adaptations to the conditions imposed by the mutations of some of the genes involved in CoQ synthesis. In fact, studies performed on fibroblasts from seven different patients suffering primary CoQ_{10}-deficiency caused by mutations in any of the genes involved in CoQ_{10} synthesis demonstrate that cells develop a survival phenotype associated with modifications in the epigenome (Fernandez-Ayala et al. 2013). Thus, is it not strange to think that in the mouse models of these mutations a survival phenotype permits longer survival than expected.

It is not clear if CoQ decay during aging is accelerating mitochondrial dysfunction, or age-associated mitochondrial dysfunction causes a decrease in the synthesis of CoQ. In any case, this will be a vicious cycle in which the reduced levels of CoQ in mitochondria severely impairs mitochondrial activities as has been demonstrated in CoQ-deficiency associated rare diseases (Alcázar-Fabra et al. 2016) but, at the same time, mitochondrial dysfunction reduces CoQ synthesis (Radaelli et al. 2023; Spinazzi et al. 2019) impairing yet more mitochondrial activities and also protection against cell death by ferroptosis (Deshwal et al. 2023).

Taken into consideration the decrease in CoQ_{10} levels in many organs and tissues during aging, we can consider this as a secondary deficiency that can be treated with dietary supplements of CoQ_{10}.

6.4 Bioavailability of CoQ, the Main Problem

One of the most important issues in the effective supplementation with CoQ_{10} is the great difference in bioavailability in humans. The range of normal CoQ_{10} levels in plasma is wide, from 400 to 1400 CoQ_{10} pmol/ml blood plasma (Lopez-Lluch et al.

2019). The studies performed with different compounds indicate that even in the case of good incorporation of the compound to blood, the individual response is very different (Lopez-Lluch et al. 2019). This great variability was also found in old population aged more than 65 (Pravst et al. 2020). This fact can affect many of the studies performed about the effect of this compound on diseases, physical activity capacity or simply improvement of physiological functions since many of them do not verify the increase in plasma levels of CoQ_{10} after treatment reaching wrong conclusions about the null effect of the compound. For example, many studies have tried to find if CoQ_{10} can be efficient in the treatment of neurodegenerative diseases such as Alzheimer's disease, Parkinson's disease or Amyotrophic Lateral Sclerosis but it is not clear if CoQ_{10} can access and go through the blood-brain barrier (Mantle and Hargreaves 2025).

We can define bioavailability as the proportion of an ingested substance that reaches the bloodstream following absorption from the digestive tract. The bioavailability of CoQ_{10} is very low, typically of the order of 5% or less, the particular chemical hydrophobic structure of CoQ_{10}. (Mantle and Dybring 2020). To understand the great differences in the uptake of supplemented CoQ_{10} between individuals, the mechanism of CoQ_{10} absorption must be considered.

CoQ_{10} is absorbed in the gut via the same mechanism as any other lipid-soluble substance. Following ingestion and transit through the stomach, CoQ_{10} enters the duodenum where it is micellized (Mantle and Dybring 2020). Micelles are small enough to diffuse between the intestinal villi. Studies based on variable dosage of CoQ_{10} in humans have indicated that there is a finite capacity to absorb CoQ_{10} in a single dose (Singh et al. 2005). This suggests that a carrier is required to facilitate the entry of CoQ_{10} into enterocytes although the carrier has not been definitively identified. Recent data indicate that the cholesterol transporter NCPC1L1 (Niemann-Pick C1 Like 1) can be this carrier (Sahoo et al. 2014; Takekawa et al. 2016). Once inside the enterocytes, CoQ_{10} is incorporated into chylomicrons synthesised in the endoplasmic reticulum and then released from the enterocytes into the lymphatic system reaching further the blood circulation. Chylomicrons in the blood carry CoQ_{10} to the liver. In the liver CoQ_{10} is loaded primarily into low density lipoprotein (LDL) and very low density lipoprotein (VLDL) that transport CoQ_{10} around the body (Mantle and Dybring 2020). Thus, only organs that can get CoQ_{10} from chylomicrons or LDL from blood can incorporate dietary CoQ_{10}. All these processes can explain why humans show a great variability in incorporating CoQ_{10} from diet or supplements.

From a functional point of view, it seems clear that the supplementation with CoQ is not needed when the organism is young and healthy because cell membranes are near saturated at the functional level although membranes can incorporate more exogenous CoQ (Kamzalov et al. 2003). However, CoQ-supplementation become necessary when the organism shows deficiency such as during aging.

Taken into consideration that primary CoQ deficiencies affect mainly neuromuscular and neurological activity, one of the main problems is that the functional incorporation of CoQ into brain or muscle organs is low. In cerebellar ataxia caused by primary CoQ_{10} deficiency affecting also muscle, high doses of CoQ_{10}

administration seems to improve the clinical picture (Musumeci et al. 2001). We can expect that exogenous CoQ_{10} would be taken up by mitochondria in muscles and enhance the capacity of oxidative phosphorylation. In fact, some case studies on myopathic patients showed that muscular deficiency of CoQ_{10} can be at least partially corrected by supplementation (Di Giovanni et al. 2001). Increases in CoQ_{10} levels in muscle after supplementation have been also reported in healthy adults (Karlsson et al. 1991). However, other reports have shown that oral supplementation did not increase muscle and mitochondrial concentration of CoQ_{10} (Svensson et al. 1999). However, we have observed that, both, the brain and muscle incorporated significantly lower amounts of CoQ in comparison with spleen, liver or white blood cells (Bentinger et al. 2003; Turunen et al. 1999), indicating that dietary CoQ reaches these tissues at a very low level and that high doses are needed to significantly increase CoQ levels in these organs.

Probably, when CoQ deficiency is extreme, becase of mutations in the synthesis machinery or muscular deficiency, the amount of CoQ that reaches mitochondria can palliate, at least in part, the deficiency. However, in other cases, such as in aging, this incorporation is not enough or even the deleterious effect of the disease on the cell cannot be reverted by CoQ.

6.5 Functional Nutrients Can Increase CoQ-Uptake or Even Induce Endogenous CoQ-Synthesis

Some studies have indicated that the combination of CoQ with other antioxidants or bioactive compounds can affect the incorporation or the synthesis of CoQ in tissues and organs. In a rodent model, α-TOC deficient diet (available commercially) decreased CoQ levels in liver and blood whereas α-TOC supplemented diet increases uptake of dietary CoQ (Zhang et al. 1996). Perhaps α-TOC not only increases CoQ uptake but also is able to increases endogenous CoQ synthesis; however, this possibility has not been studied in deep. On the contrary, in humans, dietary CoQ_{10} elevates not only CoQ_{10} but also α-TOC in mononuclear cells, whereas CoQ_{10} administration does not lead to increased CoQ_{10} in granulocytes but causes elevation of α-TOC in these cells (Turunen et al. 2002). This relationship indicates a complex relationship between the two main antioxidants in cell membranes probably preserving the equilibrium in the antioxidant process.

Another candidate to increase CoQ uptake is vitamin C, since ascorbate can increase the levels of the reduced form of CoQ, ubiquinol, and it has been considered that this form can be incorporated into cells in an easier way. Vitamin C is one of the main constituents of the supplements containing CoQ_{10} in its reduced form but, to date, the results about the higher availability of the ubiquinol vs. ubiquinone are not clear since no a same lipidic base is used in the comparative studies. (Mantle and Dybring 2020).

It seems that peroxisome proliferator-activated receptor-α (PPARα) is involved in CoQ uptake or in the synthesis of endogenous CoQ. Agonists of PPARα (clofibrate, salicilic acid) increase CoQ in all organs in rats (Turunen et al. 2000). No studies have been carried out to determine the effect of these compounds in COQ levels in human organs, although, in blood, CoQ is not increased (Turunen et al. 2000). It has been demonstrated that both, linolenic and linoleic acids are efficient activators of PPARα, then, they will be excellent nutritional candidates to determine its effect on CoQ levels in both, non-supplemented and CoQ supplemented animals. Moreover, other dietary ligands for this receptor will be also good candidates to increase CoQ synthesis.

6.6 CoQ as an Anti-senescence Factor

The accumulation of senescent cells is nowadays considered a main factor in the progression of aging and age-related diseases (Yamauchi and Takahashi 2025). For this reason, the use of pharmacological compounds that eliminate senescent cells has been proposed as an anti-aging therapy (Zheng et al. 2024).

Mitochondrial dysfunction is an important factor in senescence and in the senescence-associated secretory phenotype (SASP) (Atayik and Cakatay 2023). Senescent cells accumulate during aging and many researchers consider that this accumulation drives the degradation of tissues and organs although many aspects must be yet fixed (Chen et al. 2023). In fact, this relationship has suggested targeting mitochondria as a mechanism to delay or even revert aging by reducing senescence (Atayik and Cakatay 2023; Zheng et al. 2025).

Many years ago, when senescence was not yet a high topic in aging studies, Bliznakow, E.G., already showed that a single administration of CoQ_{10} was able to compensate the suppression of the humoral immune response in old mice indicating a the capacity to reduce senescence of the immune system (Bliznakov 1978). Recent studies also demonstrated that CoQ_{10} supplementation is able to decelerate senescence in SAMP1 (senescence-prone) mice, a model of accelerated senescence (Yan et al. 2006). This effect was mediated by the activation of the PPAR-α, a gene involved in the activation of the metabolism of lipids (Schmelzer et al. 2010). Further studies also demonstrated that the supplementation with CoQ_{10} activated mitochondrial functions delaying senescence in these mice (Tian et al. 2014).

On the other hand, it has been demonstrated in different cultured cells that CoQ_{10} supplementation improves mitochondrial functions and protect cells against senescence-inducers (Chang et al. 2022; Huo et al. 2018; Marcheggiani et al. 2021). Among these inducers, we can find oxidative stress in vascular endothelial cells (Huo et al. 2018), oxidative stress produced by the mutation of presenilin1 in fibroblasts (Ma et al. 2014) in an effect in which the reactivation of autophagy plays an important role (Vegh et al. 2019). Recent studies also demonstrate that supplementation with CoQ_{10} can prevent senescence in many different cells affected by different stressors and aging (Chang et al. 2022; Marcheggiani et al. 2021; Cirilli et al. 2020).

6.7 Supplementation with CoQ$_{10}$ Decelerates Senescence in Animal Models

Some of the studies in animal models have used unbalanced diets or lifestyle to demonstrate an effect of CoQ$_{10}$ in lifespan. For example, lifelong supplementation with CoQ$_{10}$ produced significant increases of mean and maximum life span in rats fed a diet enriched with polyunsaturated fatty acids (PUFA) n-6 (Quiles et al. 2004). We previously showed that lifelong dietary CoQ$_{10}$-supplementation decreased CoQ$_9$/CoQ$_{10}$ ratio in liver membranes, attenuated aging-related decrease of glutathione-dependent antioxidant activities such as cytosolic glutathione-S-transferase (GST) and microsomal Se-independent glutathione peroxidase, increased resistance of membranes against oxidative stress, and abolished aging-dependent increase of Mg^2+-dependent neutral sphingomyelinase (nSMase) in liver plasma membrane (Bello et al. 2005). Proteomics studies have revealed other changes implied in the lifespan extension by dietary CoQ$_{10}$ in rats (Santos-Gonzalez et al. 2007). Serum albumin, which decreases with aging in these animals, was significantly increased by CoQ$_{10}$ supplementation both at 6 and 24 months. Further, lifelong dietary supplementation with CoQ$_{10}$ induced significant decreases of plasma hemopexin, apolipoprotein H and inter-alpha-inhibitor H4P heavy chain (at both 6 and 24 months), preprohaptoglobin, fibrinogen γ-chain precursor, and fetuin-like protein (at 6 months), and alpha-1-antitrypsin precursor and type II peroxiredoxin (at 24 months) (Santos-Gonzalez et al. 2007). On the other hand, CoQ$_{10}$-supplementation resulted in significant increases of serine protease inhibitor 3, vitamin D-binding protein (at 6 months), and Apo A-I (at 24 months). All these results support a putative beneficial role of dietary CoQ$_{10}$ decreasing oxidative stress and cardiovascular risk, and modulating inflammation and osteogenesis during aging.

Other studies have focused the effect of CoQ$_{10}$ supplementation in the progression of aging on a mice model of accelerated senescence such as the SAMP1 mice model (Schmelzer et al. 2010; Yan et al. 2006). In humans, the studies are mainly based on the treatment of chronic diseases although some studies have demonstrated changes in expression profile after supplementation with CoQ$_{10}$. In healthy aged individuals, CoQ10-supplementation induced changes in muscle gene expression affecting 115 genes and in protein levels affecting 174 proteins. CoQ$_{10}$ supplementation also modified the type of muscle fibre in these individuals indicating a regulatory capacity of this molecule (Linnane et al. 2002).

6.8 CoQ in the Treatment of Chronic Diseases

One of the hallmarks of aging is the increase of mitochondrial dysfunction (Lopez-Otin et al. 2013) and it is clear that low levels of CoQ_{10} are associated with mitochondrial dysfunction and with many age-associated chronic diseases (Navas et al. 2021). In fact, treatment with CoQ_{10} has been proposed for many chronic age-related diseases such as cardiovascular disease (CVD) (Pepe et al. 2007; Sue-Ling et al. 2022), kidney chronic disease (Abe et al. 2024; Aziz et al. 2023) neurodegenerative diseases (Ebrahimi et al. 2023; Galpern and Cudkowicz 2007; Yang et al. 2016), sarcopenia (Bolt et al. 2023) and inflammation (Lopez-Lluch 2021; Lopez-Lluch et al. 2010) among others.

The main problem in the studies about the effect of CoQ_{10} in human diseases and aging is that the number of clinical trials is low and that the use of CoQ_{10} is not standardized in dose and time of treatment. This produces a variability of results that mask the putative therapeutic effect of CoQ_{10}. Further, the lack of determinations of CoQ_{10} in plasma before and after the treatment makes impossible to know if null or low effects of CoQ_{10} are due to low bioavailability of the product used in the treatment or to low uptake of CoQ_{10} by many of the participants in the study. Determination of plasma CoQ_{10} levels in these studies must be mandatory.

Fortunately, the number of trials about the effect of CoQ_{10} in human diseases is increasing. For example, recently Vrentzos et al. have demonstrated in a double-blind clinical trial that six-months of supplementation with 240 mg CoQ_{10} improves the symptoms of liver steatosis, endothelial, vascular and myocardial function in sixty patients with metabolic-dysfunction associated with steatotic liver disease (Vrentzos et al. 2024).

Or a recent systematic revision and meta-analysis supplementation with CoQ10 significantly reduced the levels of circulating C-reactive protein (CRP), interleukin (IL)-6 and tumor necrosis factor (TNF)-α. Interestingly, data indicated that the effect on CRP was independent of baseline CRP, treatment duration, dosage, and patients characteristics. On the other hand, in the case of IL-6, a higher baseline was significantly associated with greater effects of CoQ_{10}. However, as in many other cases, these results must be interpreted with caution due to the heterogeneity and the limited number of the assays (Fan et al. 2017).

And in a recent meta-analysis of 20 randomized clinical trials with more than 2500 patients with ovarian aging, the use of antioxidants improved the quality of oocytes and embryos increasing the rate of pregnancy. Among the antioxidants studied, CoQ_{10} was more effective than melatonin, myo-inositol or other vitamins. The use of CoQ_{10} 30 mg/day for 3 months previously to ovarian stimulation cycle produced clear benefits in pregnancy in women older than 35 (Shang et al. 2024).

And in the case of neurodegeneration, recently CoQ_{10} has demonstrated protection against β-amyloid-induced damage of cell membranes (Seychell et al. 2025).

6.9 CoQ in the Prevention of Endothelial Damage: A Key Role in Many Age-Related Diseases

How can CoQ$_{10}$ produce such protective effects in many chronic-diseases if the incorporation of dietary or supplemented CoQ$_{10}$ is low in some organs and tissues? (Bentinger et al. 2003). Probably the answer is in the health of endothelium.

CoQ$_{10}$ supplementation increases the levels of CoQ$_{10}$ in plasma and its can directly affects endothelial function (Yuan et al. 2021) that is associated with many age-associated chronic diseases (Batinac et al. 2024; Borri et al. 2025; Fang et al. 2023; Pacinella et al. 2022). The most important role of CoQ$_{10}$ in these diseases is by preventing endothelial damage that accelerates the development of the disease. For example, in cardiovascular disease, CoQ$_{10}$ not only protects against nitrosative damage but also via inhibition off oxidative damage by oxidized LDL (Belardinelli et al. 2008). In endothelial damage suffered by type-2 diabetes patients, CoQ$_{10}$ can protect cells not only by quenching reactive oxygen species (ROS) levels but also by reducing superoxide production through mitochondria by improving ETC activity (Littarru et al. 2011). These effects can explain why CoQ$_{10}$ supplementation can be associated with significant improvement in arterial endothelial function (Gao et al. 2012).

In the case of neurodegenerative diseases, it is clear that the maintenance of the blood-brain barrier integrity is essential to block or delay the progression of the diseases especially cognitive dysfunction (Andjelkovic et al. 2023; Salmina et al. 2021). Interestingly, CoQ$_{10}$ protects endothelial cells against amyloid angiopathy induced by β-amyloid peptide (Duran-Prado et al. 2014) in an effect that has been associated with the length of the isoprenoid chain being CoQ$_{10}$ the form more effective (Frontinan-Rubio et al. 2021). This protective effect can explain why the reduced form of CoQ$_{10}$, ubiquinol, improves cerebral vasoreactivity and ameliorates chronic inflammation and endothelial function in patients with mild cognitive impairment (Garcia-Carpintero et al. 2021) or the direct relationship of CoQ$_{10}$ levels in plasma with cognitive and executive function in elderly people (Fernandez-Portero et al. 2023).

Another important relationship between CoQ$_{10}$ levels and endothelial health is related with the reduction of chronic inflammation in blood vessels (Pacinella et al. 2022). This effect has been determined in inflammatory markers in overweight/obesity polycystic ovary syndrome patients (Taghizadeh et al. 2021). Even in autoimmune diseases such as anti-phospholipid disease, ubiquinol can modulate the overexpression of inflammatory and thrombotic markers (Perez-Sanchez et al. 2017). This protective effect will reduce the pro-inflammatory cytokines-dependent senescence of vascular endothelial cells critical in the impairment of atherosclerosis (Shang et al. 2023).

The sum of evidence about the role of CoQ$_{10}$ in the protection of endothelial cells against mitochondrial dysfunction and oxidative damage permits to propose the supplementation with this compound in people suffering chronic diseases and in elderly people. This protection impairs the onset and the progression of age-related chronic disease and permit a higher quality of life.

6.10 Conclusions

We can conclude that aging is associated with a decrease in the levels of CoQ_{10} in cells and tissues. Supplementation with dietary CoQ_{10} can help to maintain mitochondrial activity and antioxidant capacity especially in endothelial cells whose activity is essential in many diseases. Maintenance of high CoQ_{10} levels during aging is important to improve cardiovascular function and cognitive capacity, two of the main health dysfunctions during aging. Further, a lifestyle able to maintain endogenous CoQ_{10} production affecting moderate exercise, social activity and balanced diet not only maintain physiological capacity in cells and organs but also can maintain high CoQ_{10} levels improving mitochondrial metabolism and antioxidant protection of cell membranes.

Acknowledgements This work has been financed by studies performed with several grants such the project DEP2012-39985 from the Spanish Government, the Operative Program FEDER Andalucía 2014-2020 grant UPO-1259581, and the grants for the BIO-177 group from the Junta de Andalucía-FEDER funds.

References

Abe Y, Nishiwaki H, Suzuki T, Noma H, Watanabe Y, Ota E, Hasegawa T (2024) Renoprotective effects of coenzyme Q10 supplementation in patients with chronic kidney disease: a protocol for a systematic review. BMJ Open 14:e084088. https://doi.org/10.1136/bmjopen-2024-084088

Aejmelaeus R, Metsa-Ketela T, Laippala P, Solakivi T, Alho H (1997) Ubiquinol-10 and total peroxyl radical trapping capacity of LDL lipoproteins during aging: the effects of Q-10 supplementation. Mol Asp Med 18(Suppl):S113–S120. https://doi.org/10.1016/s0098-2997(97)00002-2

Alcázar-Fabra M, Navas P, Brea-Calvo G (2016) Coenzyme Q biosynthesis and its role in the respiratory chain structure. Biochim Biophys Acta (BBA) – Bioenergetics 1857:1073–1078. https://doi.org/10.1016/j.bbabio.2016.03.010

Andjelkovic AV, Situ M, Citalan-Madrid AF, Stamatovic SM, Xiang J, Keep RF (2023) Blood-brain barrier dysfunction in Normal aging and neurodegeneration: mechanisms, impact, and treatments. Stroke 54:661–672. https://doi.org/10.1161/Strokeaha.122.040578

Arroyo A, Kagan VE, Tyurin VA, Burgess JR, De Cabo R, Navas P, Villalba JM (2000) Nadh and Nadph-dependent reduction of coenzyme Q at the plasma membrane. Antioxid Redox Signal 2: 251–262. https://doi.org/10.1089/ars.2000.2.2-251

Atayik MC, Cakatay U (2023) Mitochondria-associated cellular senescence mechanisms: biochemical and pharmacological perspectives. Adv Protein Chem Struct Biol 136:117–155. https://doi.org/10.1016/bs.apcsb.2023.03.003

Aziz H, Hussein A, Zakari M (2023) Myeloperoxidase and coenzyme Q10 modulated in the chronic kidney disease patients. Georgian Med News:124–128

Batinac T, Baticic L, Krsek A, Knezevic D, Marcucci E, Sotosek V, Curko-Cofek B (2024) Endothelial dysfunction and cardiovascular disease: hyperbaric oxygen therapy as an emerging therapeutic modality? J Cardiovasc Dev Dis 11. https://doi.org/10.3390/jcdd11120408

Belardinelli R, Tiano L, Littarru GP (2008) Oxidative stress, endothelial function and coenzyme Q10. Biofactors 32:129–133. https://doi.org/10.1002/biof.5520320115

Bello RI, Kagan VE, Tyurin V, Navarro F, Alcain FJ, Villalba JM (2003) Regeneration of lipophilic antioxidants by Nad(P)H:quinone oxidoreductase 1. Protoplasma 221:129–135

Bello RI, Gomez-Diaz C, Buron MI, Alcain FJ, Navas P, Villalba JM (2005) Enhanced anti-oxidant protection of liver membranes in long-lived rats fed on a coenzyme Q10-supplemented diet. Exp Gerontol 40:694–706. https://doi.org/10.1016/j.exger.2005.07.003

Bentinger M, Dallner G, Chojnacki T, Swiezewska E (2003) Distribution and breakdown of labeled coenzyme Q10 in rat. Free Radic Biol Med 34:563–575

Bersuker K, Hendricks J, Li Z, Magtanong L, Ford B, Tang PH, Roberts MA, Tong B, Maimone TJ, Zoncu R, Bassik MC, Nomura DK, Dixon SJ, Olzmann JA (2019) The CoQ oxidoreductase Fsp1 acts parallel to Gpx4 to inhibit ferroptosis. Nature. https://doi.org/10.1038/s41586-019-1705-2

Beyer RE, Segura-Aguilar J, Di Bernardo S, Cavazzoni M, Fato R, Fiorentini D, Galli MC, Setti M, Landi L, Lenaz G (1996) The role of Dt-diaphorase in the maintenance of the reduced antioxidant form of coenzyme Q in membrane systems. Proc Natl Acad Sci USA 93:2528–2532. https://doi.org/10.1073/pnas.93.6.2528

Bliznakov EG (1978) Immunological senescence in mice and its reversal by coenzyme Q10. Mech Ageing Dev 7:189–197. https://doi.org/10.1016/0047-6374(78)90065-9

Bolt J, Sandhu S, Mohammadi A (2023) Effect of coenzyme Q10 supplementation on sarcopenia, frailty, and falls: a scoping review. J Nutr Health Aging 27:586–592. https://doi.org/10.1007/s12603-023-1943-8

Borri M, Jacobs ME, Carmeliet P, Rabelink TJ, Dumas SJ (2025) Endothelial dysfunction in the aging kidney. Am J Physiol Renal Physiol 328:F542–F562. https://doi.org/10.1152/ajprenal.00287.2024

Chang TM, Yang TY, Huang HC (2022) Nicotinamide mononucleotide and coenzyme Q10 protects fibroblast senescence induced by particulate matter preconditioned mast cells. Int J Mol Sci 23. https://doi.org/10.3390/ijms23147539

Chen Q, Young L, Barsotti R (2023) Mitochondria in cell senescence: a friend or foe? In: Adv Protein Chem Struct Biol, vol 136, pp 35–91. https://doi.org/10.1016/bs.apcsb.2023.02.019

Cirilli I, Orlando P, Marcheggiani F, Dludla PV, Silvestri S, Damiani E, Tiano L (2020) The protective role of bioactive Quinones in stress-induced senescence phenotype of endothelial cells exposed to cigarette smoke extract. Antioxidants (Basel) 9. https://doi.org/10.3390/antiox9101008

Crane FL, Navas P (1997) The diversity of coenzyme Q function. Mol Aspects Med 18(Suppl):S1–S6

Deshwal S, Onishi M, Tatsuta T, Bartsch T, Cors E, Ried K, Lemke K, Nolte H, Giavalisco P, Langer T (2023) Mitochondria regulate intracellular coenzyme Q transport and ferroptotic resistance via Stard7. Nat Cell Biol 25:246–257. https://doi.org/10.1038/s41556-022-01071-y

Di Giovanni S, Mirabella M, Spinazzola A, Crociani P, Silvestri G, Broccolini A, Tonali P, Di Mauro S, Servidei S (2001) Coenzyme Q10 reverses pathological phenotype and reduces apoptosis in familial CoQ10 deficiency. Neurology 57:515–518

Doll S, Freitas FP, Shah R, Aldrovandi M, Da Silva MC, Ingold I, Grocin AG, Xavier Da Silva TN, Panzilius E, Scheel C, Mourao A, Buday K, Sato M, Wanninger J, Vignane T, Mohana V, Rehberg M, Flatley A, Schepers A, Kurz A, White D, Sauer M, Sattler M, Tate EW, Schmitz W, Schulze A, O'donnel V, Proneth B, Popowicz GM, Pratt D, Angeli JPF, Conrad M (2019) Fsp1 is a glutathione-independent ferroptosis suppressor. Nature 575:293–698. https://doi.org/10.1038/s41586-019-1707-0

Duran-Prado M, Frontinan J, Santiago-Mora R, Peinado JR, Parrado-Fernandez C, Gomez-Almagro MV, Moreno M, Lopez-Dominguez JA, Villalba JM, Alcain FJ (2014) Coenzyme

Q10 protects human endothelial cells from beta-amyloid uptake and oxidative stress-induced injury. PLoS One 9:e109223. https://doi.org/10.1371/journal.pone.0109223
Ebrahimi A, Kamyab A, Hosseini S, Ebrahimi S, Ashkani-Esfahani S (2023) Involvement of coenzyme Q10 in various neurodegenerative and psychiatric diseases. Biochem Res Int 2023: 5510874. https://doi.org/10.1155/2023/5510874
Ernster L, Dallner G (1995) Biochemical, physiological and medical aspects of ubiquinone function. Biochim Biophys Acta 1271:195–204. https://doi.org/10.1016/0925-4439(95)00028-3
Fan L, Feng Y, Chen G-C, Qin L-Q, Fu C-L, Chen L-H (2017) Effects of coenzyme Q10 supplementation on inflammatory markers: a systematic review and meta-analysis of randomized controlled trials. Pharmacol Res 119:128–136. https://doi.org/10.1016/j.phrs.2017.01.032
Fang YC, Hsieh YC, Hu CJ, Tu YK (2023) Endothelial dysfunction in neurodegenerative diseases. Int J Mol Sci 24. https://doi.org/10.3390/ijms24032909
Fernandez-Ayala DJ, Guerra I, Jimenez-Gancedo S, Cascajo MV, Gavilan A, Dimauro S, Hirano M, Briones P, Artuch R, De Cabo R, Salviati L, Navas P (2013) Survival transcriptome in the coenzyme Q10 deficiency syndrome is acquired by epigenetic modifications: a modelling study for human coenzyme Q10 deficiencies. BMJ Open 3. https://doi.org/10.1136/bmjopen-2012-002524
Fernandez-Portero C, Amian JG, Bella R, Lopez-Lluch G, Alarcon D (2023) Coenzyme Q10 levels associated with cognitive functioning and executive function in older adults. J Gerontol A Biol Sci Med Sci 78:1–8. https://doi.org/10.1093/gerona/glac152
Frontinan-Rubio J, Rabanal-Ruiz Y, Duran-Prado M, Alcain FJ (2021) The protective effect of ubiquinone against the amyloid peptide in endothelial cells is isoprenoid chain length-dependent. Antioxidants (Basel) 10. https://doi.org/10.3390/antiox10111806
Galpern WR, Cudkowicz ME (2007) Coenzyme Q treatment of neurodegenerative diseases of aging. Mitochondrion 7(Suppl):S146–S153. https://doi.org/10.1016/j.mito.2007.01.004
Gao L, Mao Q, Cao J, Wang Y, Zhou X, Fan L (2012) Effects of coenzyme Q10 on vascular endothelial function in humans: a meta-analysis of randomized controlled trials. Atherosclerosis 221:311–316. https://doi.org/10.1016/j.atherosclerosis.2011.10.027
Garcia-Carpintero S, Dominguez-Bertalo J, Pedrero-Prieto C, Frontinan-Rubio J, Amo-Salas M, Duran-Prado M, Garcia-Perez E, Vaamonde J, Alcain FJ (2021) Ubiquinol supplementation improves gender-dependent cerebral Vasoreactivity and ameliorates chronic inflammation and endothelial dysfunction in patients with mild cognitive impairment. Antioxidants (Basel) 10. https://doi.org/10.3390/antiox10020143
Gomez-Diaz C, Rodriguez-Aguilera JC, Barroso MP, Villalba JM, Navarro F, Crane FL, Navas P (1997) Antioxidant ascorbate is stabilized by Nadh-coenzyme Q10 reductase in the plasma membrane. J Bioenerg Biomembr 29:251–257. https://doi.org/10.1023/a:1022410127104
Huo J, Xu Z, Hosoe K, Kubo H, Miyahara H, Dai J, Mori M, Sawashita J, Higuchi K (2018) Coenzyme Q10 prevents senescence and dysfunction caused by oxidative stress in vascular endothelial cells. Oxidative Med Cell Longev 2018:3181759. https://doi.org/10.1155/2018/3181759
Kalen A, Appelkvist EL, Dallner G (1989) Age-related changes in the lipid compositions of rat and human tissues. Lipids 24:579–584. https://doi.org/10.1007/bf02535072
Kamzalov S, Sumien N, Forster MJ, Sohal RS (2003) Coenzyme Q intake elevates the mitochondrial and tissue levels of Coenzyme Q and alpha-tocopherol in young mice. J Nutr 133:3175–3180
Karlsson J, Diamant B, Folkers K, Lund B (1991) Muscle fibre types, ubiquinone content and exercise capacity in hypertension and effort angina. Ann Med 23:339–344
Kuhl I, Miranda M, Atanassov I, Kuznetsova I, Hinze Y, Mourier A, Filipovska A, Larsson NG (2017) Transcriptomic and proteomic landscape of mitochondrial dysfunction reveals secondary coenzyme Q deficiency in mammals. elife 6. https://doi.org/10.7554/eLife.30952

Linnane AW, Kopsidas G, Zhang C, Yarovaya N, Kovalenko S, Papakostopoulos P, Eastwood H, Graves S, Richardson M (2002) Cellular redox activity of coenzyme Q10: effect of CoQ10 supplementation on human skeletal muscle. Free Radic Res 36:445–453

Littarru GP, Tiano L, Belardinelli R, Watts GF (2011) Coenzyme Q(10), endothelial function, and cardiovascular disease. Biofactors 37:366–373. https://doi.org/10.1002/biof.154

Liu JL, Yee C, Wang Y, Hekimi S (2017) A single biochemical activity underlies the pleiotropy of the aging-related protein CLK-1. Sci Rep 7:859. https://doi.org/10.1038/s41598-017-00754-z

Lopez-Lluch G (2021) Coenzyme Q homeostasis in aging: response to non-genetic interventions. Free Radic Biol Med 164:285–302. https://doi.org/10.1016/j.freeradbiomed.2021.01.024

Lopez-Lluch G, Rodriguez-Aguilera JC, Santos-Ocana C, Navas P (2010) Is coenzyme Q a key factor in aging? Mech Ageing Dev 131:225–235. https://doi.org/10.1016/j.mad.2010.02.003

Lopez-Lluch G, Del Pozo-Cruz J, Sanchez-Cuesta A, Cortes-Rodriguez AB, Navas P (2019) Bioavailability of coenzyme Q10 supplements depends on carrier lipids and solubilization. Nutrition 57:133–140. https://doi.org/10.1016/j.nut.2018.05.020

Lopez-Otin C, Blasco MA, Partridge L, Serrano M, Kroemer G (2013) The hallmarks of aging. Cell 153:1194–1217. https://doi.org/10.1016/j.cell.2013.05.039

Ma D, Stokes K, Mahngar K, Domazet-Damjanov D, Sikorska M, Pandey S (2014) Inhibition of stress induced premature senescence in presenilin-1 mutated cells with water soluble coenzyme Q10. Mitochondrion 17:106–115. https://doi.org/10.1016/j.mito.2014.07.004

Mantle D, Dybring A (2020) Bioavailability of coenzyme Q(10): an overview of the absorption process and subsequent metabolism. Antioxidants (Basel) 9. https://doi.org/10.3390/antiox9050386

Mantle D, Hargreaves I (2025) Coenzyme Q10 and the Blood–Brain Barrier: An Overview. J Clin Med 14:2748

Marcheggiani F, Kordes S, Cirilli I, Orlando P, Silvestri S, Vogelsang A, Moller N, Blatt T, Weise JM, Damiani E, Tiano L (2021) Anti-ageing effects of ubiquinone and ubiquinol in a senescence model of human dermal fibroblasts. Free Radic Biol Med 165:282–288. https://doi.org/10.1016/j.freeradbiomed.2021.01.032

Musumeci O, Naini A, Slonim AE, Skavin N, Hadjigeorgiou GL, Krawiecki N, Weissman BM, Tsao CY, Mendell JR, Shanske S, De Vivo DC, Hirano M, Dimauro S (2001) Familial cerebellar ataxia with muscle coenzyme Q10 deficiency. Neurology 56:849–855

Navas P, Cascajo MV, Alcazar-Fabra M, Hernandez-Camacho JD, Sanchez-Cuesta A, Rodriguez ABC, Ballesteros-Simarro M, Arroyo-Luque A, Rodriguez-Aguilera JC, Fernandez-Ayala DJM, Brea-Calvo G, Lopez-Lluch G, Santos-Ocana C (2021) Secondary CoQ(10) deficiency, bioenergetics unbalance in disease and aging. Biofactors 47:551–569. https://doi.org/10.1002/biof.1733

Nordman T, Xia L, Bjorkhem-Bergman L, Damdimopoulos A, Nalvarte I, Arner ES, Spyrou G, Eriksson LC, Bjornstedt M, Olsson JM (2003) Regeneration of the antioxidant ubiquinol by lipoamide dehydrogenase, thioredoxin reductase and glutathione reductase. Biofactors 18:45–50. https://doi.org/10.1002/biof.5520180206

Pacinella G, Ciaccio AM, Tuttolomondo A (2022) Endothelial dysfunction and chronic inflammation: the cornerstones of vascular alterations in age-related diseases. Int J Mol Sci 23. https://doi.org/10.3390/ijms232415722

Pepe S, Marasco SF, Haas SJ, Sheeran FL, Krum H, Rosenfeldt FL (2007) Coenzyme Q10 in cardiovascular disease. Mitochondrion 7(Suppl):S154–S167. https://doi.org/10.1016/j.mito.2007.02.005

Perez-Sanchez C, Aguirre MA, Ruiz-Limon P, Abalos-Aguilera MC, Jimenez-Gomez Y, Arias-De La Rosa I, Rodriguez-Ariza A, Fernandez-Del Rio L, Gonzalez-Reyes JA, Segui P, Collantes-Estevez E, Barbarroja N, Velasco F, Sciascia S, Cecchi I, Cuadrado MJ, Villalba JM, Lopez-Pedrera C (2017) Ubiquinol effects on antiphospholipid syndrome Prothrombotic profile: a randomized, placebo-controlled trial. Arterioscler Thromb Vasc Biol 37:1923–1932. https://doi.org/10.1161/atvbaha.117.309225

Pravst I, Rodriguez Aguilera JC, Cortes Rodriguez AB, Jazbar J, Locatelli I, Hristov H, Zmitek K (2020) Comparative bioavailability of different coenzyme Q10 formulations in healthy elderly individuals. Nutrients 12. https://doi.org/10.3390/nu12030784

Quiles JL, Ochoa JJ, Huertas JR, Mataix J (2004) Coenzyme Q supplementation protects from age-related DNA double-strand breaks and increases lifespan in rats fed on a PUFA-rich diet. Exp Gerontol 39:189–194

Radaelli E, Assenmacher CA, Verrelle J, Banerjee E, Manero F, Khiati S, Girona A, Lopez-Lluch G, Navas P, Spinazzi M (2023) Mitochondrial defects caused by PARL deficiency lead to arrested spermatogenesis and ferroptosis. elife 12. https://doi.org/10.7554/eLife.84710

Rodriguez-Hidalgo M, Luna-Sanchez M, Hidalgo-Gutierrez A, Barriocanal-Casado E, Mascaraque C, Acuna-Castroviejo D, Rivera M, Escames G, Lopez LC (2018) Reduction in the levels of CoQ biosynthetic proteins is related to an increase in lifespan without evidence of hepatic mitohormesis. Sci Rep 8:14013. https://doi.org/10.1038/s41598-018-32190-y

Sahoo S, Aurich MK, Jonsson JJ, Thiele I (2014) Membrane transporters in a human genome-scale metabolic knowledgebase and their implications for disease. Front Physiol 5:91. https://doi.org/10.3389/fphys.2014.00091

Salmina AB, Komleva YK, Malinovskaya NA, Morgun AV, Teplyashina EA, Lopatina OL, Gorina YV, Kharitonova EV, Khilazheva ED, Shuvaev AN (2021) Blood-brain barrier breakdown in stress and neurodegeneration: biochemical mechanisms and new models for translational research. Biochemistry (Mosc) 86:746–760. https://doi.org/10.1134/S0006297921060122

Santos-Gonzalez M, Gomez Diaz C, Navas P, Villalba JM (2007) Modifications of plasma proteome in long-lived rats fed on a coenzyme Q10-supplemented diet. Exp Gerontol 42: 798–806

Santos-Ocana C, Villalba JM, Cordoba F, Padilla S, Crane FL, Clarke CF, Navas P (1998) Genetic evidence for coenzyme Q requirement in plasma membrane electron transport. J Bioenerg Biomembr 30:465–475. https://doi.org/10.1023/a:1020542230308

Schmelzer C, Kubo H, Mori M, Sawashita J, Kitano M, Hosoe K, Boomgaarden I, Doring F, Higuchi K (2010) Supplementation with the reduced form of coenzyme Q10 decelerates phenotypic characteristics of senescence and induces a peroxisome proliferator-activated receptor-alpha gene expression signature in SAMP1 mice. Mol Nutr Food Res 54:805–815. https://doi.org/10.1002/mnfr.200900155

Seychell RM, El Saghir A, Farrugia G, Vassallo N (2025) Coenzyme Q10 enhances resilience of mitochondrial-like membranes against Amyloidogenic peptides. Membranes 15:148

Shang D, Liu H, Tu Z (2023) Pro-inflammatory cytokines mediating senescence of vascular endothelial cells in atherosclerosis. Fundam Clin Pharmacol 37:928–936. https://doi.org/10.1111/fcp.12915

Shang Y, Song N, He R, Wu M (2024) Antioxidants and fertility in women with ovarian aging: a systematic review and meta-analysis. Adv Nutr 15:100273. https://doi.org/10.1016/j.advnut.2024.100273

Singh RB, Niaz MA, Kumar A, Sindberg CD, Moesgaard S, Littarru GP (2005) Effect on absorption and oxidative stress of different oral coenzyme Q10 dosages and intake strategy in healthy men. Biofactors 25:219–224. https://doi.org/10.1002/biof.5520250127

Spinazzi M, Radaelli E, Horre K, Arranz AM, Gounko NV, Agostinis P, Maia TM, Impens F, Morais VA, Lopez-Lluch G, Serneels L, Navas P, De Strooper B (2019) PARL deficiency in mouse causes complex III defects, coenzyme Q depletion, and Leigh-like syndrome. Proc Natl Acad Sci USA 116:277–286. https://doi.org/10.1073/pnas.1811938116

Srivastava S (2017) The mitochondrial basis of aging and age-related disorders. Genes (Basel) 8. https://doi.org/10.3390/genes8120398

Stefely JA, Pagliarini DJ (2017) Biochemistry of mitochondrial coenzyme Q biosynthesis. Trends Biochem Sci 42:824–843. https://doi.org/10.1016/j.tibs.2017.06.008

Stocker R, Suarna C (1993) Extrecellular reduction of ubiquinone-1 and -10 by human Hep G2 and blood cells. Biochim Biophys Acta Gen Subj 1158:15–22. https://doi.org/10.1016/0304-4165(93)90090-U

Stocker R, Bowry VW, Frei B (1991) Ubiquinol-10 protects human low density lipoprotein more efficiently against lipid peroxidation than does alpha-tocopherol. Proc Natl Acad Sci USA 88: 1646–1650. https://doi.org/10.1073/pnas.88.5.1646

Sue-Ling CB, Abel WM, Sue-Ling K (2022) Coenzyme Q10 as adjunctive therapy for cardiovascular disease and hypertension: a systematic review. J Nutr 152:1666–1674. https://doi.org/10.1093/jn/nxac079

Sun N, Youle RJ, Finkel T (2016) The mitochondrial basis of aging. Mol Cell 61:654–666. https://doi.org/10.1016/j.molcel.2016.01.028

Svensson M, Malm C, Tonkonogi M, Ekblom B, Sjodin B, Sahlin K (1999) Effect of Q10 supplementation on tissue Q10 levels and adenine nucleotide catabolism during high-intensity exercise. Int J Sport Nutr 9:166–180

Taghizadeh S, Izadi A, Shirazi S, Parizad M, Pourghassem Gargari B (2021) The effect of coenzyme Q10 supplementation on inflammatory and endothelial dysfunction markers in overweight/obese polycystic ovary syndrome patients. Gynecol Endocrinol 37:26–30. https://doi.org/10.1080/09513590.2020.1779689

Takahashi T, Mine Y, Okamoto T (2019) Extracellular coenzyme Q(10) (CoQ(10)) is reduced to ubiquinol-10 by intact Hep G2 cells independent of intracellular CoQ(10) reduction. Arch Biochem Biophys 672:108067. https://doi.org/10.1016/j.abb.2019.108067

Takekawa Y, Sato Y, Yamaki Y, Imai M, Noto K, Sumi M, Takekuma Y, Iseki K, Sugawara M (2016) An Approach to Improve Intestinal Absorption of Poorly Absorbed Water-Insoluble Components via Niemann–Pick C1-Like 1. Biol Pharm Bull 39:301–307. https://doi.org/10.1248/bpb.b15-00359

Tian G, Sawashita J, Kubo H, Nishio SY, Hashimoto S, Suzuki N, Yoshimura H, Tsuruoka M, Wang Y, Liu Y, Luo H, Xu Z, Mori M, Kitano M, Hosoe K, Takeda T, Usami S, Higuchi K (2014) Ubiquinol-10 supplementation activates mitochondria functions to decelerate senescence in senescence-accelerated mice. Antioxid Redox Signal 20:2606–2620. https://doi.org/10.1089/ars.2013.5406

Turunen M, Appelkvist EL, Sindelar P, Dallner G (1999) Blood concentration of coenzyme Q (10) increases in rats when esterified forms are administered. J Nutr 129:2113–2118

Turunen M, Peters JM, Gonzalez FJ, Schedin S, Dallner G (2000) Influence of peroxisome proliferator-activated receptor α on ubiquinone biosynthesis11Edited by J Karn. J Mol Biol 297:607–614. https://doi.org/10.1006/jmbi.2000.3596

Turunen M, Wehlin L, Sjoberg M, Lundahl J, Dallner G, Brismar K, Sindelar PJ (2002) beta2-integrin and lipid modifications indicate a non-antioxidant mechanism for the anti-atherogenic effect of dietary coenzyme Q10. Biochem Biophys Res Commun 296:255–260

Vargas-Perez MLA, Devos DP, Lopez-Lluch G (2024) An AlphaFold structure analysis of COQ2 as key a component of the coenzyme Q synthesis complex. Antioxidants (Basel) 13. https://doi.org/10.3390/antiox13040496

Vegh C, Pupulin S, Wear D, Culmone L, Huggard R, Ma D, Pandey S (2019) Resumption of autophagy by Ubisol-Q(10) in Presenilin-1 mutated fibroblasts and transgenic ad mice: implications for inhibition of senescence and neuroprotection. Oxid Med Cell Longev 2019:7404815. https://doi.org/10.1155/2019/7404815

Vrentzos E, Ikonomidis I, Pavlidis G, Katogiannis K, Korakas E, Kountouri A, Pliouta L, Michalopoulou E, Pelekanou E, Boumpas D, Lambadiari V (2024) Six-month supplementation with high dose coenzyme Q10 improves liver steatosis, endothelial, vascular and myocardial function in patients with metabolic-dysfunction associated steatotic liver disease: a randomized double-blind, placebo-controlled trial. Cardiovasc Diabetol 23:245. https://doi.org/10.1186/s12933-024-02326-8

Wang Y, Hekimi S (2019) The complexity of making ubiquinone. Trends Endocrinol Metab 30: 929–943. https://doi.org/10.1016/j.tem.2019.08.009

Wang Y, Lilienfeldt N, Hekimi S (2024) Understanding coenzyme Q. Physiol Rev 104:1533–1610. https://doi.org/10.1152/physrev.00040.2023

Yamauchi S, Takahashi A (2025) Cellular senescence: mechanisms and relevance to cancer and aging. J Biochem 177:163–169. https://doi.org/10.1093/jb/mvae079

Yan J, Fujii K, Yao J, Kishida H, Hosoe K, Sawashita J, Takeda T, Mori M, Higuchi K (2006) Reduced coenzyme Q10 supplementation decelerates senescence in Samp1 mice. Exp Gerontol 41:130–140. https://doi.org/10.1016/j.exger.2005.11.007

Yang X, Zhang Y, Xu H, Luo X, Yu J, Liu J, Chang RC (2016) Neuroprotection of coenzyme Q10 in neurodegenerative diseases. Curr Top Med Chem 16:858–866. https://doi.org/10.2174/1568026615666150827095252

Yen HC, Chen FY, Chen SW, Huang YH, Chen YR, Chen CW (2011) Effect of mitochondrial dysfunction and oxidative stress on endogenous levels of coenzyme Q(10) in human cells. J Biochem Mol Toxicol 25:280–289. https://doi.org/10.1002/jbt.20387

Yuan S, Hahn SA, Miller MP, Sanker S, Calderon MJ, Sullivan M, Dosunmu-Ogunbi AM, Fazzari M, Li Y, Reynolds M, Wood KC, St Croix CM, Stolz D, Cifuentes-Pagano E, Navas P, Shiva S, Schopfer FJ, Pagano PJ, Straub AC (2021) Cooperation between Cyb5R3 and Nox4 via coenzyme Q mitigates endothelial inflammation. Redox Biol 47:102166. https://doi.org/10.1016/j.redox.2021.102166

Zhang Y, Turunen M, Appelkvist EL (1996) Restricted uptake of dietary coenzyme Q is in contrast to the unrestricted uptake of alpha-tocopherol into rat organs and cells. J Nutr 126:2089–2097

Zheng L, He S, Wang H, Li J, Liu Y, Liu S (2024) Targeting cellular senescence in aging and age-related diseases: challenges, considerations, and the emerging role of Senolytic and Senomorphic therapies. Aging Dis 15:2554–2594. https://doi.org/10.14336/ad.2024.0206

Zheng Y, Yang J, Li X, Qi L, Zheng Z, Kong J, Zhang G, Guo Y (2025) Mitochondria at the crossroads: quality control mechanisms in neuronal senescence and neurodegeneration. Neurobiol Dis 208:106862. https://doi.org/10.1016/j.nbd.2025.106862

Chapter 7
Iodine for Health and Longevity

Johannes Riis, Marco Juul Thomsen, Stine Linding Andersen, and Stig Andersen

Abstract Iodine is a crucial micronutrient that is essential for life. This essential trace element for humans and animals is required for the thyroid gland to produce the thyroid hormones, thyroxine, T4, and triiodothyronine, T3. These hormones regulate development of the body and brain, growth, and metabolism across the lifespan. Deficiency can lead to severe developmental damage in children (cretinism), thyroid swelling (goitre) and thyroid dysfunctions with fatigue, weight gain, and cognitive deficiencies, also known as iodine deficiency disorders, which may in turn influence longevity.

While iodine is present in small amounts in seawater, seaweed has historically been a rich source of iodine. Most inland areas have iodine-poor soils, but iodine may be present in groundwater in some areas depending on the aquifer source rock. This contributes to variations in population iodine intake across geography. Still, iodine intake is often insufficient in many populations globally and many countries have introduced iodine fortification programs to combat iodine deficiency disorders and promote health. The importance of an adequate iodine intake on human health is mediated largely by its essential importance for thyroid hormone production, and iodine intake patterns across the human lifespan may have important implications for promoting longevity.

Keywords Iodine · Thyroid function · Goitre · Longevity · Iodine fortification

J. Riis (✉) · S. Andersen (✉)
Department of Geriatric Medicine, Aalborg University Hospital, Aalborg, Denmark

Department of Clinical Medicine, Aalborg University, Aalborg, Denmark
e-mail: johannes@rn.dk; lasa@rn.dk

M. J. Thomsen · S. L. Andersen
Department of Clinical Medicine, Aalborg University, Aalborg, Denmark

Department of Clinical Biochemistry, Aalborg University Hospital, Aalborg, Denmark

S. I. S. Rattan, C. Gezer (eds.), *Supplements and Drugs for Healthy Longevity*,
Healthy Ageing and Longevity 22, https://doi.org/10.1007/978-3-032-11608-6_7

7.1 History of iodine nutrition Research

7.1.1 The Discovery of Iodine

The element Iodine was discovered during the Napoleonic Wars due to the demand for saltpetre (potassium nitrate) for use in gunpowder production. Bernard Courtois was involved in the production of saltpetre, which required the extraction of potassium compounds from seaweed ash. In 1811, while processing seaweed ash to isolate sodium and potassium compounds, Courtois added excess sulfuric acid to the ash solution while removing sulphur impurities, which resulted in a chemical reaction with the release of a violet-coloured vapor. When the vapour was condensed, shiny, dark crystals were formed. This was the discovery of elemental iodine (Merke 1984; Zimmermann 2008).

The chemical characterisation of iodine was performed by Gay-Lussac and Davy in the early nineteenth century. Gay-Lussac named it from the Greek word "ioeidēs" meaning "violet-coloured", referring to its distinctive purple vapor, "iode" in French and iodine in English. Iodine‘s purple vapor and sublimation from solid to gas made it a subject of particular interest. Following the discoveries of chlorine and bromine, iodine‘s chemical properties suggested it belonged to the halogen group due to its ability to form salts with metals and its oxidizing nature. Both Gay-Lussac and Davy recognized iodine as exhibiting similar reactivity to chlorine and bromine, and its chemical characteristics were consistent with the halogen family (See Table 7.1 for an overview) (Pearce and Zimmermann 2023; Zimmermann 2008).

7.1.2 Early Recognition of the Role in Human Health

Recognition of the importance of iodine for human health emerged in the nineteenth century, primarily through its link to thyroid function and the prevention of enlargement of the thyroid, goitre, and cretinism. Cretinism is a syndrome characterized by stunted growth, intellectual disability, and other symptoms, and is now known to be caused by a severe deficiency of thyroid hormones in infancy (Salisbury 2003). While iodines widespread use in iodized salt as a public health measure came later,

Table 7.1 Timeline of the discovery of Iodine

Year	Event	Facilitator
1811	Courtois discovers iodine (Zimmermann 2008)	Powder production for Napoleons wars
1813	Gay-Lussac and Davy independently confirm and characterise the new element (Zimmermann 2008)	Curiosity about the properties of iodine and the need for systematic classification of elements
1820s	Coindet proposed a link between iodine and the thyroid (Coindet 1820)	Iodine began to be studied and applied in medicine and industry

Table 7.2 Variation in iodine content in the diet by geography in France

Geography	Iodine in diet	Goitre	Cretinism
Paris	100	No	No
Lyon	10	Yes	No
The Alps	1	Yes	Yes

iodine's biological importance was uncovered through a series of studies (Pearce and Zimmermann 2023).

The Swiss physician, Coindet, was among the first to propose a relationship between iodine and the thyroid. In 1820, Coindet observed that the application of iodine tincture could reduce the size of the swelling of the thyroid gland, goitres. He hypothesized that iodine played a role in thyroid health (Coindet 1820).

Areas with iodine-poor soils, such as the Alps and certain inland regions of Europe, were the subject of investigations on iodine deficiency as a direct cause of goitre. In France, Chatin published the hypothesis that iodine deficiency was the cause of goitre based on the observed association between the occurrence of goitre and cretinism depending on variation in iodine content in the diet with geography (See Table 7.2 for overview) (Chatin 1853).

"*The stamping-out of goitre*"programme in Paris was introduced to prevent thyroid disorders. The programme included interventions described in The Lancet (1869) and involved dispensing iodine tablets among 5000 children with goitre, of which 4000 were described as cured.

The biochemical role of iodine for thyroid function was unravelled during the latter half of the nineteenth century. An intervention study by Baumann described the successful treatment of deficient thyroid gland function, hypothyroidism, by use a substrate derived from thyroid tissues, which was found to include 10% iodine. Thus, the authors concluded that ""*Thyroiodine is active after organification*" (Baumann 1896). The organification of iodine has since been demonstrated to be a key step in the synthesis of thyroid hormones within the thyroid gland.

7.2 Iodine, the Thyroid, and Human Health

Today, the link between iodine and thyroid hormones is understood in much more detail. Iodine is mainly absorbed from the gastrointestinal tract as iodide (Josefsson et al. 2002; Nicola et al. 2012). After absorption into plasma, iodide is mainly cleared from plasma by either the thyroid gland or through renal excretion into urine (DeGroot 1966; McConahey et al. 1951). When iodine is absorbed into the thyroid cells, thyrocytes, it is transported through the cell and into colloid in the lumen of the thyroid follicles. Here, thyroid peroxidase (TPO) oxidises iodide which is then bound to tyrosine amino acids (called tyrosyl residues) in a protein called thyroglobulin (TG) forming iodotyrosines. Eventually coupling of iodotyrosines lead to formation of the primary active thyroid hormones, triiodothyronine (T3) and tetraiodothyronine/thyroxine (T4), named after their number of bound iodine atoms, 3 and 4 respectively. T3 and T4 can then be released from the thyroid to peripheral tissues where it increases cell metabolism and basal metabolic rate, and eventually are broken down by deiodination. Resulting iodine from deiodination is then either reused by the thyroid or excreted by the kidneys (Rousset et al. 2015).

As the thyroid hormones thyroxine (T4) and triiodothyronine (T3) produced by the thyroid gland require iodine, lack of iodine impairs the production of these hormones. The hormones T4 and T3 are crucial for regulating growth, development, and metabolism, and it became clear that severe iodine deficiency could cause developmental damage, with physical and mental deficits, a condition termed cretinism in its most severe form.

The findings of the nineteenth-century laid the foundation for iodine supplementation efforts to combat goitre and other iodine deficiency-related disorders. In Switzerland, Bayard and Eggenberger published two reports in the Schweiz Medical Journal in 1923 on the reduction of goitre occurrence and eradication of cretinism by the implementation of an iodine fortification programme (Bayard 1923; Eggenberger 1923). A parallel survey from the mid-western US reported a marked reduction in goitre occurrence from more than 25% in the untreated compared to 0.2% in the group treated by iodine fortification of salt (Marine and Kimbal 1920).

Hetzel and colleagues performed a series of animal studies demonstrating the link between severe iodine deficiency and irreversible neurological developmental deficiencies. In addition to the causal link to severe iodine deficiency, they demonstrated that these deficiencies could be prevented by the addition of iodine (Hertzel and Potter 1983; Hetzel and Mano 1989; Mano et al. 1987; Pharoah et al. 1971).

The link between severe iodine deficiency and irreversible developmental damage in mammals was followed by controlled studies of humans in iodine deficient regions showing that iodine fortification of salt improved cognitive function in the population and eliminated cretinism (Hetzel 1983). The results of such studies have led to recommendations for iodine monitoring and supplementation to deficient populations (World Health Organization: WHO 2014).

7.3 Iodine Intake and the Relation to Human Health

7.3.1 Dietary Sources of iodine and Recommended Intake

The World Health Organisation (WHO) recommends a daily dietary iodine intake of 150 μg for adults and 250 μg for pregnant or lactating women (World Health Organization: WHO 2014). Dietary iodine is naturally occurring in several food sources, however, in modern developed countries it is also highly dependent on processing of food and fortification of iodine. More than 120 countries have implemented salt iodine fortification programmes (UNICEF: United Nations Children's Fund 2008). These include mandatory programmes such as in Denmark (Laurberg et al. 2006), and voluntary iodine fortification of salt such as is the case in the US (Leung et al. 2012). Milk and dairy products may also be a source of iodine to humans as it is widely used to add iodine to cattle in mineral mixes and licks for livestock. This causes dairy products to contain iodine posing a potential source of iodine to humans with an intake of cows' milk (Dineva et al. 2021). Further, iodine content in food may be secondary to other uses in food processing, e.g. in the U.S. iodine has historically been used as a bread conditioner and is still present in dairy due to iodophor cleansers used to clean milk cans and teats (Leung et al. 2012). Iodine is also present in many multivitamins such as prenatal vitamins and in some medications such as amiodarone or iodized contrast.

Aside from iodine intake due to food processing, salt fortification, and supplements, the natural content of iodine is low in terrestrial foods, and dietary sources of iodine are mainly of marine origin (Farebrother et al. 2019; Rasmussen et al. 2002, 2009). Hence, fish and other marine foods have a high iodine content and are reliable sources of iodine as evaluated from populations who rely mainly on marine foods (Andersen et al. 2002, 2005; Johannesen et al. 2023, 2024). In addition, seaweeds are known to contain large amounts of iodine, and the intake may have a marked influence on iodine intake (Andersen et al. 2019). In some areas, particularly in East Asian countries, seaweed products are an important food source, and excessive seaweed consumption associates with thyroid disorders (Müssig 2009). Hence, caution should be taken with e.g. kombu soup (Farebrother et al. 2019). Still, single meals with excessive iodine content from e.g. sushi does not appear to disrupt thyroid (Andersen et al. 2019). Finally, the iodine content of drinking water varies greatly and has been associated with goitre (Andersen and Laurberg 2009). In summary, while the dietary iodine intake varies markedly with the intake of marine foods, marine algae, dairy products, and drinking water, the influence on health is consistent across geographically distinct regions and populations.

7.3.2 *Comparative Studies of Iodine Nutrition and Thyroid Health*

Comparative studies of iodine nutrition based on a difference in iodine intake from natural sources provide an insight into the impact of long-term iodine nutrition in parallel populations (Table 7.3). Comparative studies differ from other studies of iodine nutrition, such as studies monitoring the implementation of iodine supplementation programmes, as natural sources support long-term iodine nutrition at the same level. Therefore, studies comparing natural sources of iodine intake add information on disease occurrence with the potential to influence longevity over long-term—potentially the entire lifespan when the studies address older adults. The table lists such studies of people in Southern, Central and Northern Europe, and China, covering a wide age range, and iodine nutrition ranging from mild iodine deficiency to excessive iodine intakes. These studies illustrate a marked difference in the occurrence of thyroid diseases related to differences in iodine intake levels,

Table 7.3 Geographical variation in iodine intake and associated thyroid dysfunciton and autoimmunity

Geography & Study	Population	Urinary Iodine	Hypo-thyroidism (subclinical)	Hyper-thyroidism (subclinical)	Thyroid antibodies	Goitre	References
Central Europe							
Carpathian basin	$N = 346$ nursing home residents 61–97 year	72 µg/L	0.8 (4.2)%	3.4%	19.3%	39.4%	Szabolcs et al. (1997)
		100 µg/L	1.5 (10.4)%	3.0%	24.4%	16.4%	
		513 µg/L	7.6 (23.9)%	0.0%	22.8%	12.2%	
Northern Europe							
Jutland-Iceland (JUICE)	Jutland, $N = 423$ 68 year	38 µ/L	3.8%	9.7%		12.2 / 3.2% Women / men	Laurberg et al. (1998)
	Iceland, $N = 100$	150 µg/L	18%	1.0%		1.9 /3.2% Women / men	
Denmark (Jutland)	Randers, $N = 212$ 77–78 year	55 µg/d	6.2%	26.2%	42% (TPOab 13%)	26%	Andersen et al. (2009)
	Skagen, $N = 218$ 75–80 year	160 µg/d	12.9%	6.5%	32% (TPOab 19%)	0%	
Denmark (DanThyr)	Aalborg, $N = 2220$ 18–65 year	53 µg/L	8.2/2.1% Women/men	11.1 / 7.2% Women /men	19.4% (all) 17.3% (65 + y)	22.6%	Laurberg et al. (2006)
	Copenhagen, $N = 2429$ 18–65 year	68 µg/L	9.1 / 2.7% Women /men	8.3 / 4.0% Women/men	18.4% (all) 22.1% (65 + y)	15.0%	
Southern Europe							
Greece	Coastal (Chios) $N = 200$	136 µg/L			66%		Giassa et al. (2018)

	Mainland regions, $N = 322$	104 μg/L			27%		
Asia							
China	Panshan, $N = 1103$ 14–88 year	103 μg/L	3.0 (10) %	1.6 (3.7) %	9.2/9.0%	23.2%	Teng et al. (2006)
	Zhangwu, $N = 1584$ 14–95 year	375 μg/L	15 (46) %	2.0 (3.9) %	9.8/9.0%	16.9%	
	Huanghua, $N = 1074$ 14–79 year	615 μg/L	21 (65) %	1.2 (1.1) %	10.5/9.4% (TPOAb/TgAb)	7.6%	

which is seen across all populations. Interestingly, only the study focussing on older people (Andersen et al. 2009) included a follow-up on longevity (Riis et al. 2020).

Table 7.3 displays the comparative studies of iodine intake and prevalences of thyroid disorders. The measure of iodine intake displayed in the table is urinary iodine, which is the gold standard measure for populations and makes sense as this is the method of eventual excretion from the body. However, this measure should be interpreted as an average population measure of iodine nutrition as done in the table rather than for assessment of individual iodine intake. In individuals the measure is extremely variable and rather reflects resent intake of iodine rich foods, something that levels out with a sufficiently large population sample (Andersen et al. 2001; Rasmussen et al. 1999). On a population level, average urinary iodine concentration below 20 μg/L indicate severe iodine deficiency while values of 20–49 μg/L and 50–99 μg/L indicate moderate and mild iodine deficiency, respectively. Values from 100–199 μg/L indicates adequate intake, while values above 200 μg/L indicate excess iodine intake (World Health Organization: WHO 2014).

7.3.3 *Pathophysiology Thyroid Disorders Dependent on iodine Intake*

The comparison of iodine intake shows that hyperthyroidism is more frequent in iodine deficient areas together with goitre, while hypothyroidism is more frequent in iodine replete areas. It can seem counterintuitive that less iodine may cause hyperthyroidism as iodine is required to make thyroid hormones, however, it must be considered that none of the populations had severe iodine deficiency but rather had mild to moderate iodine deficiency. Indeed, in severe iodine deficiency, hypothyroidism is more prevalent than in iodine replete populations (Zimmermann 2009). Mild to moderate iodine deficiency causes more hyperthyroidism, which is mainly due to greater occurrence of multinodular goitre and solitaire adenomas (Carlé et al. 2011). Goitre and nodules develop as the thyroid adapts to less availability of iodine with increased stimulation of the thyroid gland causing goitre and nodules, which in turn can become hyperfunctioning leading to hyperthyroidism (Stubner et al. 1987; Zimmermann 2009). Supplementing iodine to iodine deficient populations may also give a transient rise in hyperthyroidism due to increased function of the thyroid gland especially in people with previously asymptomatic goitre and nodules, which is known as the Jod-Basedow effect (Braverman and Pearce 2024; Roti and Degli Uberti 2001). The association between higher iodine intake and hypothyroidism may in turn be mediated by autoimmunity, especially among people who harbour TPO antibodies (TPOab), where iodine supplementation has been demonstrated to cause hypothyroidism in experimental studies (Chow et al. 1991; Reinhardt et al. 1998).

7.4 Iodine Intake and Longevity

7.4.1 Iodine Deficiency and Longevity

Studies on the direct effect of iodine intake on mortality are scarce. A study based on the National Health and Nutritional Examination Survey III (NHANES) data did not find any association between low spot urine iodine concentration and mortality in a middle-aged adult US population (Inoue et al. 2018). However, the use of spot urine iodine concentrations reflects only short-term iodine exposure from recent dietary intake (Andersen et al. 2008), and it should be kept in mind that the population was iodine replete when interpreting the data among the US population.

A Danish population-based comparative study of older adults covering two regions in Denmark with long-term stable iodine intakes—one being iodine replete and one with moderate iodine insufficiency due to a stable iodine content of drinking water—showed that individuals from the iodine replete area lived longer than those with iodine insufficiency (Riis et al. 2020). The potential link between iodine insufficiency and longevity may primarily be through the distinct importance of iodine for thyroid function and autoimmunity, as described in Table 7.3 and seen in this population. Thus, older people in the iodine replete region had comparably more hypothyroidism and older people in the iodine insufficient region had more hyperthyroidism (Andersen et al. 2009). Hence, these data in older people with long-term follow-up demonstrate an importance of iodine intake for longevity.

The effect of increased population prevalence of hyperthyroidism among adults with iodine insufficiency may have detrimental health effects, especially in older adults (Toft 2001). Hyperthyroidism in this age group is associated with raised morbidity, specifically heart disease and osteoporosis (Collet et al. 2012; Delitala et al. 2020). Thus, hyperthyroidism may be a link between iodine insufficiency and increased mortality.

The finding that insufficient iodine intake may be detrimental to health and thereby shorten life is also supported by a study from Spain of middle-aged adults (Maldonado-Araque et al. 2021). This study used urinary iodine concentrations and showed a dose response relationship between severity of iodine insufficiency and mortality. Interestingly, the association was still present in a sensitivity analysis excluding participants with cardiovascular or thyroid disease. Future thyroid disease may still precipitate the detrimental effect of insufficient iodine intake, but this observation also suggests that other mechanisms may be at play. Here a potential extrathyroidal pathway would be through anti-inflammatory effects of thyroid function, which would support a positive effect raised iodine intake on longevity (Aceves et al. 2013). This is supported by laboratory evidence that iodine might have positive effect on breast and prostate cancers (Aranda et al. 2013; Cann et al. 2000). However, studies on the extrathyroidal effects of iodine are limited and the findings should be corroboration in future research, especially studies including humans to elucidate further health risks of iodine deficiency influencing longevity.

7.4.2 Excessive iodine Intake and Longevity

The large study of NHANES III also reported that a very high iodine intake of three times the recommended daily intake was associated with raised mortality. Excess iodine intake associates with autoimmune thyroiditis and hypothyroidism, as previously described (Laurberg et al. 2006; Teng et al. 2006). Hence, it seems plausible that the effect of excessive intake of iodine on the thyroid may also lead to shorter life expectancy, especially if the exposure occurs in midlife as described in the NHANES III data. It should be noted that only the effects of very high, excessive intakes were associated with increased mortality, and that this was not seen in the group consuming twice the recommended level, but not excessive iodine intake.

Considering this single study association, it should also be noted that the effects of hypothyroidism on health may not be similar at all ages. While the NHANES III sample comprised of a normal adult population, where hypothyroidism may have negative health consequences, it is debated if this is different in the older adult population (Simonsick et al. 2009; Stott et al. 2017; Virgini et al. 2014). In older adults, the effect of even slight hypothyroidism may be beneficial to health and hence increase longevity and support daily functioning without causing symptoms of discomfort or impaired quality of life (Blum et al. 2015; Carlé et al. 2016; Riis et al. 2023; Stott et al. 2017).

7.4.3 Iodine Intake in Early Life Through the Lens of Longevity

Aside from the above evidence on mid- and late life iodine nutrition on longevity, the most researched impact of iodine is on fetal and early life development. As illustrated by the effect of iodine deficiency on intellectual disability and cretinism, iodine is important for brain development. Additionally, a special focus has been on pregnant women as lack or excess of thyroid hormones is important to human foetal brain development (S. L. Andersen et al. 2017). How this eventually affects longevity is yet speculative, as a human cohort study on the subject is unlikely feasible as it would require a lifetime of follow-up, and therefore other sources of evidence are warranted. However, it is a compelling hypothesis that iodine nutrition during pregnancy may be important for longevity mediated by intellectual ability, as it is well established that early life intellectual ability is related to a longer lifespan, and even to less cognitive impairment and better functional ability in old age—in other words healthy longevity (Batty et al. 2007; Poranen-Clark et al. 2016; Richards et al. 2004).

7.4.4 *Considering iodine supplementation to Promote Longevity*

The existing evidence of a direct link between iodine intake and longevity hardly supports making any firm recommendation regarding supplementation with the sole objective of longevity. It is important to consider that the evidence on longevity indicates that complying with the WHO recommendation on iodine intake is the most reasonable approach, and that available evidence suggests that both excessive intake and iodine deficiency does not promote health and longevity. However, considering the wider impact of iodine deficiency on pregnancy, foetal development, and thyroid function there is support for recommending supplementation in iodine deficient regions (Andersen and Laurberg 2016; Bath et al. 2013; Pedersen et al. 1993). On that end, supplementation is not advised for people living in areas with sufficient iodine intake of natural origin, that being from marine foods or drinking water.

The recommendation to supplement iodine in an iodine deficient population should also be viewed as a recommendation on a population level rather than in individuals. Population level supplementation programmes are most effectively done through iodine fortification programmes, which are in places in most iodine deficient populations. These population level programs should consider and balance against the risk of adverse health events with supplementation raising the iodine intake in formerly iodine deficient populations. These programmes cause transient increases in hyperthyroidism with potential influence on mortality (Møllehave et al. 2018; Petersen et al. 2018; Riis et al. 2022; Todd et al. 1995). Thus, care must be taken to balance iodine intake to ensure the effectiveness while observing safety of public supplementation programs through monitoring programmes and evaluations to promote longevity by iodine supplementation.

7.5 Conclusion

There is evidence to support that sufficient iodine intake may be associated with longevity, and that this rise in longevity may be mediated by the importance of iodine for thyroid function, as iodine has several important health benefits throughout the lifespan, especially in pregnancy. Naturally occurring dietary sources of iodine are mainly found in marine foods including fish and seaweed. However, a sudden and lasting increase in iodine intake in individuals with long-term iodine deficiency may have detrimental health effects, as may lasting excessive intake. Thus, individual supplementation of iodine cannot be recommended. Rather iodine fortification programmes to promote health and longevity by ensuring adequate iodine intake is the best strategy for iodine deficient populations.

References

Aceves C, Anguiano B, Delgado G (2013) The Extrathyronine actions of iodine as antioxidant, apoptotic, and differentiation factor in various tissues. Thyroid 23(8):938. https://doi.org/10.1089/THY.2012.0579

Andersen S, Laurberg P (2009) The nature of iodine in drinking water. In: Preedy VR, Burrow GN, Watson RR (eds) Comprehensive handbook of iodine: nutritional, biochemical, pathological and therapeutic aspects, 1st edn. Elsevier

Andersen SL, Laurberg P (2016) Iodine supplementation in pregnancy and the dilemma of ambiguous recommendations. Eur Thyroid J 5(1):35–43. https://doi.org/10.1159/000444254

Andersen S, Pedersen KM, Pedersen IB, Laurberg P (2001) Variations in urinary iodine excretion and thyroid function. A 1-year study in healthy men. Eur J Endocrinol 144(5):461–465. https://doi.org/10.1530/eje.0.1440461

Andersen S, Hvingel B, Laurberg P (2002) Iodine content of traditional Greenlandic food items and tap water in East and West Greenland. Int J Circumpolar Health 61(4):332–340. https://doi.org/10.3402/IJCH.V61I4.17491

Andersen S, Hvingel B, Kleinschmidt K, Jørgensen T, Laurberg P (2005) Changes in iodine excretion in 50-69-y-old denizens of an Arctic society in transition and iodine excretion as a biomarker of the frequency of consumption of traditional inuit foods. Am J Clin Nutr 81(3): 656–663. https://doi.org/10.1093/ajcn/81.3.656

Andersen S, Karmisholt J, Pedersen KM, Laurberg P (2008) Reliability of studies of iodine intake and recommendations for number of samples in groups and in individuals. Br J Nutr 99(4): 813–818. https://doi.org/10.1017/S0007114507842292

Andersen S, Iversen F, Terpling S, Pedersen KM, Gustenhoff P, Laurberg P (2009) More hypothyroidism and less hyperthyroidism with sufficient iodine nutrition compared to mild iodine deficiency–a comparative population-based study of older people. Maturitas 64(2):126–131. https://doi.org/10.1016/j.maturitas.2009.08.007

Andersen SL, Lönn S, Vestergaard P, Törring O (2017) Birth defects after use of antithyroid drugs in early pregnancy: a Swedish nationwide study. Eur J Endocrinol 177(4):369–378. https://doi.org/10.1530/EJE-17-0314

Andersen S, Noahsen P, Rex KF, Florian-Sørensen HC, Mulvad G (2019) Iodine in edible seaweed, its absorption, dietary use, and relation to iodine nutrition in arctic people. J Med Food 22(4): 421–426. https://doi.org/10.1089/JMF.2018.0187

Aranda N, Sosa S, Delgado G, Aceves C, Anguiano B (2013) Uptake and antitumoral effects of iodine and 6-iodolactone in differentiated and undifferentiated human prostate cancer cell lines. Prostate 73(1):31–41. https://doi.org/10.1002/PROS.22536

Bath SC, Steer CD, Golding J, Emmett P, Rayman MP (2013) Effect of inadequate iodine status in UK pregnant women on cognitive outcomes in their children: results from the Avon Longitudinal Study of Parents and Children (ALSPAC). Lancet 382(9889):331–337. https://doi.org/10.1016/S0140-6736(13)60436-5

Batty GD, Deary IJ, Gottfredson LS (2007) Premorbid (early life) IQ and later mortality risk: systematic review. Ann Epidemiol 17(4):278–288. https://doi.org/10.1016/J.ANNEPIDEM.2006.07.010

Baumann F (1896) Über das normale Vorkommen von Jod im Thierkörper. Z Phys Chem 21:319–330

Bayard O (1923) Über das Kropfproblem. Schweiz Med Wochenschr 53:732–737

Blum MR, Bauer DC, Collet T-H, Fink HA, Cappola AR, da Costa BR, Wirth CD, Peeters RP, Åsvold BO, den Elzen WPJ, Luben RN, Imaizumi M, Bremner AP, Gogakos A, Eastell R, Kearney PM, Strotmeyer ES, Wallace ER, Hoff M et al (2015) Subclinical thyroid dysfunction and fracture risk. JAMA 313(20):2055. https://doi.org/10.1001/jama.2015.5161

Braverman KD, Pearce EN (2024) Iodine and hyperthyroidism: a double-edged sword. Endocr Pract 31(3). https://doi.org/10.1016/j.eprac.2024.10.014

Cann SA, Van Netten JP, Van Netten C (2000) Hypothesis: iodine, selenium and the development of breast cancer. Cancer Causes Control 11(2):121–127. https://doi.org/10.1023/A:1008925301459/METRICS

Carlé A, Pedersen IB, Knudsen N, Perrild H, Ovesen L, Rasmussen LB, Laurberg P (2011) Epidemiology of subtypes of hyperthyroidism in Denmark: a population-based study. Eur J Endocrinol 164(5):801–809. https://doi.org/10.1530/EJE-10-1155

Carlé A, Pedersen IB, Knudsen N, Perrild H, Ovesen L, Andersen S, Laurberg P (2016) Hypothyroid symptoms fail to predict thyroid insufficiency in old people: a population-based case-control study. Am J Med 129(10):1082–1092. https://doi.org/10.1016/j.amjmed.2016.06.013

Chatin A (1853) Recherches sur l'iode des eaux douces; de la présence de ce corps dans les plantes at les animaux terrestes. CR Acad Sci

Chow CC, Phillips DIW, Lazarus JH, Parkes AB (1991) Effect of low dose iodide supplementation on thyroid function in potentially susceptible subjects: are dietary iodide levels in Britain acceptable? Clin Endocrinol 34(5):413–416. https://doi.org/10.1111/J.1365-2265.1991.TB00314.X;WGROUP:STRING:PUBLICATION

Coindet JF (1820) Découverte d'un nouveau remède contre le goître. Ann Chim Phys 15:49–59

Collet T-H, Gussekloo J, Bauer DC, den Elzen WPJ, Cappola AR, Balmer P, Iervasi G, Åsvold BO, Sgarbi JA, Völzke H, Gencer B, Maciel RMB, Molinaro S, Bremner A, Luben RN, Maisonneuve P, Cornuz J, Newman AB, Khaw K-T et al (2012) Subclinical hyperthyroidism and the risk of coronary heart disease and mortality. Arch Intern Med 172(10):799–809. https://doi.org/10.1001/archinternmed.2012.402

DeGroot LJ (1966) Kinetic analysis of iodine metabolism. J Clin Endocrinol Metab 26(2):149–173. https://doi.org/10.1210/JCEM-26-2-149

Delitala AP, Scuteri A, Doria C (2020) Thyroid hormone diseases and osteoporosis. J Clin Med 9(4):1034. https://doi.org/10.3390/JCM9041034

Dineva M, Rayman MP, Bath SC (2021) Iodine status of consumers of milk-alternative drinks v. cows' milk: data from the UK National Diet and nutrition survey. Br J Nutr 126(1):28–36. https://doi.org/10.1017/S0007114520003876

Eggenberger H (1923) Die Verhütung des Kropfes und des Kropfrecidives. Schweiz Med Wochenschr 53:245–249

Farebrother J, Zimmermann MB, Andersson M (2019) Excess iodine intake: sources, assessment, and effects on thyroid function. Ann N Y Acad Sci 1446(1):44–65. https://doi.org/10.1111/NYAS.14041

Giassa, Mamali I, Gaki, Kaltsas G, Kouraklis G, Markou, Karatzas T (2018) Iodine intake and chronic autoimmune thyroiditis: a comparative study between coastal and mainland regions in Greece. Hormones 17(4):565–571. https://doi.org/10.1007/S42000-018-0057-X

Hertzel BS, Potter BJ (1983) Iodine deficiency and the role of thyroid hormones in brain development. In: Dreosti IE, Smith RM (eds) Neurobiology of the trace elements. Humana Press, pp 83–133

Hetzel BS (1983) Iodine Deficiency Diosorders (IDD) and their eradication. Lancet 322(8359): 1126–1129. https://doi.org/10.1016/S0140-6736(83)90636-0

Hetzel BS, Mano MT (1989) A review of experimental studies of iodine deficiency during fetal development. J Nutr 119(2):145–151. https://doi.org/10.1093/JN/119.2.145

Inoue K, Leung AM, Sugiyama T, Tsujimoto T, Makita N, Nangaku M, Ritz BR (2018) Urinary iodine concentration and mortality among U.S. Adults. Thyroid 28(7):913–920. https://doi.org/10.1089/thy.2018.0034

Johannesen HL, Skaalum M, Sofía A (2023) The iodine content of Faroese food items to support iodine nutrition in the North Atlantic. Int J Circumpolar Health. https://doi.org/10.1080/22423982.2023.2233145

Johannesen HL, Andersen S, Andersen SL, Hansen S, Petursdóttir J, Weihe P, Strøm M, Petersen MS, Veyhe AS (2024) Iodine nutrition among pregnant women in The Faroe Islands. Br J Nutr 132(4):495–502. https://doi.org/10.1017/S0007114524001697

Josefsson M, Grunditz T, Ohlsson T, Ekblad E (2002) Sodium/iodide-symporter: distribution in different mammals and role in entero-thyroid circulation of iodide. Acta Physiol Scand 175(2): 129–137. https://doi.org/10.1046/J.1365-201X.2002.00968.X

Laurberg P, Pedersen KM, Hreidarsson A, Sigfusson N, Iversen E, Knudsen PR (1998) Iodine intake and the pattern of thyroid disorders: a comparative epidemiological study of thyroid abnormalities in the elderly in Iceland and in Jutland, Denmark. J Clin Endocrinol Metab 83(3): 765–769. https://doi.org/10.1210/JCEM.83.3.4624

Laurberg P, Jørgensen T, Perrild H, Ovesen L, Knudsen N, Pedersen IB, Rasmussen LB, Carlé A, Vejbjerg P (2006) The Danish investigation on iodine intake and thyroid disease, DanThyr: status and perspectives. Eur J Endocrinol 155(2):219–228. https://doi.org/10.1530/EJE.1.02210

Leung AM, Braverman LE, Pearce EN (2012) History of U.S. iodine fortification and supplementation. Nutrients 4(11):1740. https://doi.org/10.3390/NU4111740

Maldonado-Araque C, Valdés S, Badiá-Guillén R, Lago-Sampedro A, Colomo N, Garcia-Fuentes-E, Gutierrez-Repiso C, Goday A, Calle-Pascual A, Castaño L, Castell C, Delgado E, Menendez E, Franch-Nadal J, Gaztambide S, Girbés J, Chaves FJ, Soriguer F, Rojo-Martínez G (2021) Iodine deficiency and mortality in Spanish adults: Di@bet.es study. Thyroid 31(1): 106. https://doi.org/10.1089/THY.2020.0131

Mano MT, Potter BJ, Belling GB, Chavadej J, Hetzel BS (1987) Fetal brain development in response to iodine deficiency in a primate model (Callithrix jacchus jacchus). J Neurol Sci 79(3):287–300. https://doi.org/10.1016/0022-510X(87)90236-X

Marine D, Kimbal O (1920) The prevention of simple goiter in man: fourth paper. Arch Intern Med 25:661–672

McConahey WM, Keating FR, Power MH (1951) An estimation of the renal and extrarenal clearance of radioiodide in man. J Clin Invest 30(7):778–780. https://doi.org/10.1172/JCI102491

Merke F (1984) History and iconography of endemic goitre and cretinism. Kluwer Academic Publishers

Møllehave LT, Linneberg A, Skaaby T, Knudsen N, Jørgensen T, Thuesen BH (2018) Trends in treatments of thyroid disease following iodine fortification in Denmark: a nationwide register-based study. Clin Epidemiol 10:763. https://doi.org/10.2147/CLEP.S164824

Müssig K (2009) Iodine-induced toxic effects due to seaweed consumption. In: Preedy VR, Burrow GN, Watson RR (eds) Comprehensive handbook of iodine : nutritional, biochemical, pathological and therapeutic aspects, 1st edn. Elsevier

Nicola JP, Reyna-Neyra A, Carrasco N, Masini-Repiso AM (2012) Dietary iodide controls its own absorption through post-transcriptional regulation of the intestinal Na+/I− symporter. J Physiol 590(Pt 23):6013. https://doi.org/10.1113/JPHYSIOL.2012.241307

Pearce EN, Zimmermann MB (2023) The prevention of iodine deficiency: a history. Thyroid 33(2): 143–149. https://doi.org/10.1089/THY.2022.0454/ASSET/IMAGES/THY.2022.0454_FIGURE3.JPG

Pedersen KM, Laurberg P, Iversen E, Knudsen PR, Gregersen HE, Rasmussen OS, Larsen KR, Eriksen GM, Johannesen PL (1993) Amelioration of some pregnancy-associated variations in thyroid function by iodine supplementation. J Clin Endocrinol Metab 77(4):1078–1083. https://doi.org/10.1210/jcem.77.4.8408456

Petersen M, Knudsen N, Carlé A, Andersen S, Jørgensen T, Perrild H, Ovesen L, Rasmussen LB, Thuesen BH, Pedersen IB (2018) Increased incidence rate of hypothyroidism after iodine fortification in Denmark. A 20 year prospective population-based study. J Clin Endocrinol Metab. https://doi.org/10.1210/jc.2018-01993

Pharoah POD, Buttfield IH, Hetzel BS (1971) Neurological damage to theFetus resulting from severe iodine deficiency during pregnancy. Lancet 297(7694):308–310. https://doi.org/10.1016/S0140-6736(71)91040-3

Poranen-Clark T, Von Bonsdorff MB, Törmäkangas T, Lahti J, Wasenius N, Räikkönen K, Osmond C, Salonen MK, Rantanen T, Kajantie E, Eriksson JG (2016) Intellectual ability in young adulthood as an antecedent of physical functioning in older age. Age Ageing 45(5):727. https://doi.org/10.1093/AGEING/AFW087

Rasmussen LB, Ovesen L, Christiansen E (1999) Day-to-day and within-day variation in urinary iodine excretion. Eur J Clin Nutr 53(5):401–407. https://doi.org/10.1038/sj.ejcn.1600762

Rasmussen LB, Ovesen L, Bülow I, Jørgensen T, Knudsen N, Laurberg P, Perrild H (2002) Dietary iodine intake and urinary iodine excretion in a Danish population: effect of geography, supplements and food choice. Br J Nutr 87(1):61–69. https://doi.org/10.1079/BJN2001474

Rasmussen LB, Andersen S, Ovesen L, Laurberg P (2009) Iodine intake & food choice. In: Preedy VR, Burrow GN, Watson RR (eds) Comprehensive handbook of iodine: nutritional, biochemical, pathological and therapeutic aspects, 1st edn. Elsevier. https://orbit.dtu.dk/en/publications/iodine-intake-amp-food-choice

Reinhardt W, Luster M, Rudorff KH, Heckmann C, Petrasch S, Lederbogen S, Haase R, Saller B, Reiners C, Reinwein D, Mann K (1998) Effect of small doses of iodine on thyroid function in patients with Hashimoto's thyroiditis residing in an area of mild iodine deficiency. Eur J Endocrinol 139(1):23–28. https://doi.org/10.1530/EJE.0.1390023

Richards M, Shipley B, Fuhrer R, Wadsworth MEJ (2004) Cognitive ability in childhood and cognitive decline in mid-life: longitudinal birth cohort study. BMJ. https://doi.org/10.1136/bmj.37972.513819.EE

Riis J, Pedersen KM, Danielsen MB, Sørensen GVB, Jørgensen MG, Andersen SL, Carlé A, Pedersen IB, Torp-Pedersen C, Andersen S (2020) Long-term Iodine nutrition is associated with longevity in older adults: a 20-year follow-up of the Randers-Skagen study. Br J Nutr:1–15. https://doi.org/10.1017/S0007114520001592

Riis J, Andersen SL, Gade GV, Danielsen MB, Jorgensen MG, Carlé A, Torp-Pedersen C, Andersen S (2022) Raised mortality in old adults with a history of hyperthyroidism following iodine fortification. Clin Endocrinol 96(2):255–262. https://doi.org/10.1111/CEN.14627

Riis J, Kragholm K, Torp-Pedersen C, Andersen S, Riis J, Kragholm K, Torp-Pedersen C (2023) Association between thyroid function, nursing home admission and mortality in community-dwelling adults over 80 years. Arch Gerontol Geriatr 104:104806. https://doi.org/10.1016/J.ARCHGER.2022.104806

Roti E, Degli Uberti E (2001) Iodine excess and hyperthyroidism. Thyroid 11(5):493–500. https://doi.org/10.1089/105072501300176453

Rousset B, Dupuy C, Miot F, Dumont J (2015) Chapter 2: Thyroid hormone synthesis and secretion. In: Endotext. https://www.ncbi.nlm.nih.gov/books/NBK285550/

Salisbury S (2003) Cretinism: the past, present and future of diagnosis and cure. Paediatr Child Health 8(2):105. https://doi.org/10.1093/PCH/8.2.105

Simonsick EM, Newman AB, Ferrucci L, Satterfield S, Harris TB, Rodondi N, Bauer DC (2009) Subclinical hypothyroidism and functional mobility in older adults. Arch Intern Med 169(21): 2011–2017. https://doi.org/10.1001/ARCHINTERNMED.2009.392

Stott DJ, Rodondi N, Kearney PM, Ford I, Westendorp RGJ, Mooijaart SP, Sattar N, Aubert CE, Aujesky D, Bauer DC, Baumgartner C, Blum MR, Browne JP, Byrne S, Collet TH, Dekkers OM, Den Elzen WPJ, Du Puy RS, Ellis G et al (2017) Thyroid hormone therapy for older adults with subclinical hypothyroidism. N Engl J Med 376(26):2534–2544. https://doi.org/10.1056/NEJMoa1603825

Stubner D, Gartner R, Greil W, Gropper K, Brabant G, Permanetter W, Horn K, Pickardt CR (1987) Hypertrophy and hyperplasia during goitre growth and involution in rats – separate bioeffects of TSH and iodine. Acta Endocrinol 116(4):537–548. https://doi.org/10.1530/ACTA.0.1160537

Szabolcs I, Podoba J, Feldkamp J, Dohán O, Farkas I, Sajgó M, Takáts KI, Góth M, Kovács L, Kressinszky K, Hnilica P, Szilágyi G (1997) Comparative screening for thyroid disorders in old age in areas of iodine deficiency, long-term iodine prophylaxis and abundant iodine intake. Clin Endocrinol 47(1):87–92. https://doi.org/10.1046/J.1365-2265.1997.2271040.X

Teng W, Shan Z, Teng X, Guan H, Li Y, Teng D, Jin Y, Yu X, Fan C, Chong W, Yang F, Dai H, Yu Y, Li J, Chen Y, Zhao D, Shi X, Hu F, Mao J et al (2006) Effect of iodine intake on thyroid diseases in China. N Engl J Med 354(26):2783–2793. https://doi.org/10.1056/NEJMOA054022

Todd CH, Allain T, Gomo ZA, Hasler JA, Ndiweni M, Oken E (1995) Increase in thyrotoxicosis associated with iodine supplements in Zimbabwe. Lancet (London, England) 346(8989): 1563–1564. https://doi.org/10.1016/s0140-6736(95)92095-1

Toft AD (2001) Subclinical hyperthyroidism. N Engl J Med 345(7):512–516. https://doi.org/10.1056/NEJMcp010145

UNICEF: United Nations Children's Fund (2008) Sustainable elimination of iodine deficiency. UNICEF

Virgini VS, Wijsman LW, Rodondi N, Bauer DC, Kearney PM, Gussekloo J, den Elzen WPJ, Jukema JW, Westendorp RGJ, Ford I, Stott DJ, Mooijaart SP (2014) Subclinical thyroid dysfunction and functional capacity among elderly. Thyroid 24(2):208. https://doi.org/10.1089/THY.2013.0071

World Health Organization: WHO (2014) Assessment of iodine deficiency disorders and monitoring their elimination, vol 28, Issue 2. World Hearlth Organization. https://apps.who.int/iris/handle/10665/43781

Zimmermann MB (2008) Research on iodine deficiency and goiter in the 19th and early 20th centuries. J Nutr 138(11):2060–2063. https://doi.org/10.1093/jn/138.11.2060

Zimmermann MB (2009) Iodine deficiency. Endocr Rev 30(4):376–408. https://doi.org/10.1210/ER.2009-0011

Part II
Phytochemicals, Extracts, and Dietary Traditions

Chapter 8
Tinospora cordifolia for Health and Longevity

Gurcharan Kaur

Abstract Traditional systems of medicine have long used natural products as dietary supplements for health promotion and for the treatment of myriads of diseases. Several societies like India, China, Japan, and Egypt extensively utilize the herbal medicines in their traditional medicinal systems. Ayurveda, India's age-old traditional system of medicine defines health as body's equilibrium of physical and mental wellbeing. Herbs popularly known as Rasayanas are used in Ayurvedic practices for rejuvenation and treatment of many neurological and physical ailments. *Tinospora cordifolia* is one such popular Rasayana herbs of Ayurveda, commonly known as 'Giloy or Guduchi'. The health promoting effects of *Tinospora cordifolia* are mainly attributed to its traditional claims of therapeutic activity for the treatment of fever, diabetes, anxiety, obesity, immunodeficiency, memory deficit and cognitive impairments. The aim of this chapter is to review and discuss the scientifically validated health benefits of *Tinospora cordifolia* as rejuvenating and therapeutic candidate in view of various pre-clinical and clinical reports from literature. Further, this chapter also aims to present the summarized information about different extracts/decoctions and active phytoconstituents of this medicinal plant which are used in traditional medicine system to treat various ailments as well as for their health promoting potential.

Keywords *Tinospora cordifolia* · Neurological disorders · Neurodegeneration · Neuroinflammation · Neural cancers · Sleep disorders

8.1 Introduction

Traditional herbal medicines have emerged as an important historical component of human health and wellness through trial and error-based development of therapeutics and rejuvenation agents (Li and Weng 2017). Herbal medicines are mostly

G. Kaur (✉)
Emeritus Scientist, Department of Biotechnology, Guru Nanak Dev University, Amritsar, Punjab, India

S. I. S. Rattan, C. Gezer (eds.), *Supplements and Drugs for Healthy Longevity*, Healthy Ageing and Longevity 22, https://doi.org/10.1007/978-3-032-11608-6_8

useful as complementary/adjunct therapies worldwide and account for about 80% of primary healthcare in developing countries like India, China, Nigeria, Ghana etc. (Elufioye et al. 2017). In economically-developed nations such as USA, Canada and European countries, natural products are mainly consumed as nutraceuticals and dietary supplements (Wachtel-Galor and Benzie 2011). "WHO's Global report on Traditional and Complimentary Medicine, 2019" also endorsed the rapid increase in the worldwide acceptance of complementary and traditional medicines. In Ayurveda, the traditional system of medicine in India, clinically approved formulations are usually a concoction of multiple constituents from individual or multiple medicinal plants mixed in specific ratios. The phytoconstituents when tested individually usually are either ineffective or have very low efficacy compared to parent formulations (Li and Weng 2017). The limited efficacy of natural or synthetic single molecules to treat complex diseases like cancers, diabetes, neurodegenerative disorders, autoimmune diseases, at times is benefitted by combinatorial and synergistic actions of natural products in herbal formulations.

Tinospora cordifolia (T. cordifolia), herbaceous climber belonging to Menispermeace family is known by different names in different parts of India like Giloy, Amrita, Guduchi, Gulancha, Ambervel, Gulvel etc. (Upadhyay et al. 2010). This plant is widely distributed pan India, Bangladesh, China, Myanmar and Sri Lanka and has characteristic succulent stem with fleshy aerial branches, heart shaped membranous leaves, tiny green-yellow flowers, cherry like fruits and curved seeds. Stem of *T. cordifolia* undergoes thickness and colour variations according to its age and season and it generally tastes bitter. Traditionally, *T. cordifolia* is prescribed in disease conditions like fever, asthma, dysentery, leprosy, diarrhea, jaundice, skin infections and diabetes by Ayurveda practitioners. Its drug like action has been categorized according to taste, property, digestion and metabolism, potency and mechanism of action (Upadhyay et al. 2010). The therapeutic roles of *T. cordifolia* have been validated by various studies using reverse pharmacological approach which provided the scientific evidence for its different properties like anti-inflammatory, anti-diabetic, anti-cancer, hepatoprotective, immunomodulatory, anti-stress, anti-spasmodic, chemo- and radio-protective, anti-anxiety and neuroprotective properties (Saha and Ghosh 2012; Dhama et al. 2016; Kaur et al. 2017; Ali and Datusalia 2024). *T. cordifolia* earned a special mention not only in Ayurveda but also in Chinese and other traditional systems of medicine. It is commonly considered as 'Heavenly elixir' due to its high medicinal values and rejuvenating potential. Several published reviews elaborate on its phytochemistry, ethnopharmacology and medicinal properties with specific focus on immunomodulatory as well as neurotherapeutic potential (Saha and Ghosh 2012; Dhama et al. 2016; Haque et al. 2017; Saeed et al. 2020; Sharma et al. 2020a). This chapter aims to summarise the preclinical and clinical evidences of health promoting and therapeutic activities of this important medicinal plant and to also provide insights into its conventional role in Ayurvedic practices, folk medicine and other traditional practices.

8.2 Extracts/Decoctions and Active Phytoconstituents of *T. cordifolia*

Recent advances in high throughput technologies such as analytical chromatography-based separation techniques, Mass Spectroscopy, and Nuclear Magnetic Resonance (NMR) techniques have facilitated the screening and isolation of phytoconstituents from herbal products (Harvey et al. 2015). The therapeutic potential of natural products is generally explored using reverse pharmacology which is a scientific approach relating documented clinical and experiential observations to identify lead compounds followed by exploring their effects using trans-disciplinary approaches. This strategy helps to develop these lead molecules into drug candidates after testing their robustness in preclinical and clinical research using advance technological approaches. Moreover, new era use of omics such as metagenomics, metabolomics along with *in silico* approaches and collaboration of biologists with medicinal chemists are a great help to revisit phyto-therapeutics (Li and Weng 2017).

The key pharmacological activities of *T. cordifolia* are attributed to its phytoconstituents, such as alkaloids, glycosides, diterpene lactones, steroids, phenolics, sesquiterpenoids and aliphatic compounds (Saha and Ghosh 2012; Dhama et al. 2016; Haque et al. 2017; Saeed et al. 2020; Sharma et al. 2020a). Extracts containing these compounds either alone or in combinations have been reported using *in vitro* and *in vivo* preclinical model systems to possess various bioactivities as discussed in the following sections. Active compounds from *T. cordifolia* e.g. 11-hydroxymuskatone, N-formylannonain, N-methyle-2-pyrrolidone, cordifolioside A, tinocordioside, magnoflorine and syringin have been identified for immunomodulatory activity and promoting macrophages phagocytic activity, reactive oxygen species generation in human neutrophils and nitric oxide production from splenocytes (Gupta et al. 2024).

8.3 Potential Health Benefits of *T. cordifolia*

8.3.1 T. cordifolia *in Immunomodulation and Brain Health*

Ayurveda acknowledges the importance of healthy brain for holistic health and defines about 80 different types of Central Nervous System (CNS) diseases (Nishteswar et al. 2014). These brain-related ailments have been related with functional and structural changes in prefrontal cortical circuitry, hypothalamic regulation, reticular activation system, limbic system, enteric as well as the autonomous nervous system (Ghodke et al. 2011; Bhalerao et al. 2012; Govindaraj et al. 2015; Travis and Wallace 2015). The brain health promoting Rasayana herbs are known as Medha Rasayanas, which have been scientifically validated for their potential to slow down ageing, neuroregeneration of specific neural tissues as well as their adaptogenic, cognition enhancing and anti-stress potential (Singh et al. 2008). Some of the commonly used Medha Rasayana herbs are *Withania somnifera, T. cordifolia, Ginkgo biloba, Centella asiatica, Glycyrrhiza glabara* and *Convolvulus pluricaulis* (Singh et al. 2008; Kulkarni et al. 2012).

Ageing is a prime risk factor for developing and aggravating neurodegenerative diseases and cognitive dysfunctions due to acceleration of oxidative stress, impairment of mitochondrial function and DNA repair system. Hydro-alcoholic and aqueous extracts of *T. cordifolia* have been shown to ameliorate cyclosporine-induced hippocampal neurodegeneration and cognitive deficits in rats (Agarwal et al. 2002). However, there are limited reports exploring the neurotherapeutic potential of *T. cordifolia*. Neurodegenerative diseases (NDDs) such as AD, PD and other neurodegenerative disorders have a common underlying structural and functional loss of neurons which results in memory impairments and associated cognitive decline (Katsnelson et al. 2016).

Since the hallmarks of inflammation are free radical generation, mitochondrial damage and oxidative stress, so search for the presence of various secondary metabolites in *T. cordifolia* having antioxidative potential to combat inflammation either alone or in combination has been a priority area of research on this plant. Reddi and Tetali (2019) explored the leaf extract of *T. cordifolia* for its antioxidant and anti-inflammatory activities and reported attenuation of Nuclear Factor kappa B (NF-κB) nuclear translocation in activated human monocytic (THP-1) cells. Similarly chloroform and aqueous extracts of *T. cordifolia* were shown to have anti-inflammatory potential against lipopolysaccharide (LPS) mediated stimulation of RAW264.7 macrophage cells (Philip et al. 2018; Ghatpande et al. 2019). These results were further supported by *in vivo* findings on rat model system by comparing the oedema suppressing ability of extract in comparison to indomethacin, positive control anti-inflammatory drug. Interestingly, aqueous extract of *T. cordifolia* has also been shown to recover the iron content of the serum, liver, and spleen in the heat-killed Brucella abortus antigen (HKBA)-related inflammation and anaemic rats (Ghatpande et al. 2019). Similarly aqueous extract of T. cordifolia increased the lymphocyte proliferation and induction of myogenic differentiation of C2C12 myoblast (Sharma et al. 2020).

Recently, Das et al. (2025) reported the anti-inflammatory and wound healing activities *T. cordifolia* extract using a mouse skin wound model. Acceleration of wound healing was confirmed by histopathology along with reduced inflammation, restoration of blood supply, collagen fibers, and swift epidermal closure thus suggesting the potential wound healing activity of *T. cordifolia* as anti-bacterial and anti-inflammatory agent. Metal based nanoparticles are gaining popularity due to their low cost and higher efficacy. Lekkala et al. (2025) carried out green synthesis of silver nanoparticles (AgNPs) using leaf extract of *T. cordifolia* and determined the structural characteristics of these nanoparticles TcAgNPs using advanced techniques. These TcAgNPs showed significant free radical scavenging, anti-microbial and anti-inflammatory activity in a dose-dependent manner (25–100 μg/mL).

Neuro-infections, traumatic injuries, stroke, toxins, neurodegeneration etc. interfere with CNS homeostasis and result in neuroinflammation (Frank-cannon et al.

2009; Kempuraj et al. 2016; Thibaut 2017). Chronic neuroinflammation which persists for long time after an injury or insult becomes self-perpetuating and often, results in reactive microgliosis, which subsequently releases inflammatory molecules and causes oxidative/nitrosative stress (Farooqui et al. 2007; Frank-cannon et al. 2009; Hurley and Tizabi 2013). Chronic neuroinflammation also underlies the initiation and progression of NDDs and age-associated macular degeneration (Frank-cannon et al. 2009; Kempuraj et al. 2016). A study primarily focusing on inflammation associated with PD, reported the beneficial effects of aqueous extract of T. cordifolia against MPTP induced experimental PD by suppressing activation of NF-κB and levels of pro-inflammatory cytokines TNF-α, IL-12, IL-1β along with suppression of gliosis and microgliosis (Jacob et al. 2018; Birla et al. 2019). Similarly neuroprotective activity of ethanolic extract of *T. cordifolia* against 6-Hydroxy Dopamine (6-OHDA) induced Parkinsonism was reported to elevate dopamine levels, increased activity of mitochondrial complex-I, suppressed iron accumulation and improved superior neuromuscular co-ordination in comparison to 6-OHDA alone treated animals (Kosaraju et al. 2014).

A recent study explored the mechanism of action and identified protein targets of *T. cordifolia* in the context of AD using untargeted metabolomics and network pharmacology-based approaches (Amrutha et al. 2025). This *in-silico* study explored the metabolite-protein interactions with the enrichment of 591 human proteins, including receptors for neurotransmitters and some regulatory proteins. Further analysis of database highlighted the enrichment of cell survival pathways along with excitatory synaptic proteins. The docking data suggests that *T. cordifolia* metabolites may be interacting with key AD targets such as Beta-site amyloid precursor protein cleaving enzyme 1 (BACE1) and Monoamine Oxidase-B (MAO-B), suggesting its potential neuroprotective role.

Recently, several studies from our own lab have reported that *T. cordifolia* has rejuvenating and neurotherapeutic potential (Reviewed by Sharma et al. 2020a). Butanol extract of *T. cordifolia* (B-TCE) and an active compound, Tinosporicide, was observed to ameliorate high dose glutamate-induced excitotoxicity in a cell culture model of hippocampal, and cerebellar neurons (Sharma and Kaur 2018; Sharma et al. 2020b). Glutamate as a key excitatory neurotransmitter plays essential role in brain health but excessive accumulation disturbs calcium homeostasis, and causes severe mitochondrial damage which ultimately results in neuronal degeneration. Treatment with B-TCE was observed to prevent glutamate induced degeneration and induce neuronal differentiation targeting expression of structural proteins such as microtubule-associated protein 2 (MAP-2), growth associate protein-43 (GAP-43), neurofilament heavy chain protein (NF-200), downregulating neuroinflammatory molecules like NF-κB, AP-1, pro-inflammatory cytokines, and upregulating cell cycle regulatory proteins PCNA and Cyclin (Sharma and Kaur 2018; Sharma et al. 2019, 2020b). These *in vitro* findings were further validated in rat model system where oral feeding of B-TCE ameliorated glutamate induced anxiety-like behaviour, memory and cognitive deficits and improved neuro-muscular co-ordination in behavioural tasks as well as suppressed astrogliosis (Sharma et al. 2019). *T. cordifolia* in combination with *Phyllanthus emlica* and juice of fresh leaves of *Ocimum sanctum* was observed to improve learning and memory in mice (Malve 2015).

8.3.2 **T. cordifolia** *in Sleep Deprivation and Cognition*

One of the important risk factors of declining health due to unhealthy lifestyle habits is sleep deprivation (SD) in the modern 24/7 society (Shochat et al. 2014). Adequate sleep in day-to-day life plays a pivotal role in maintaining the normal biological functions especially of the nervous system. SD due to hectic lifestyle, sub-optimal work schedules, family commitments and psychological stress conditions has placed a huge burden on our healthcare systems (Hillman and Lack 2013). SD also disrupts neuro-immuno-endocrine system which further aggravates health deterioration. Several pre-clinical studies have used *T. cordifolia* as monotherapy or in combination with some medicinal herbs for treatment of dementia (Agarwal et al. 2002; Malve et al. 2014). SD is associated with anxiety and cognitive deficit and our recent studyhas tested the beneficial effect of TCE and B-TCE extracts for ameliorating anxiety and cognitive dysfunctions induced by acute SD (Mishra et al. 2016; Bajaj et al. 2021). Further exploration of molecular targets revealed that TCE and B-TCE pretreatment modulated the expression of proteins of synaptic plasticity, long-term potentiation (LTP), and neuroinflammation pathways (Mishra et al. 2016; Bajaj and Kaur 2022).

Stress, which is the outcome of physical or emotional burden leading to body's fight and flight response under challenging situations causes physiological changes in the endocrine, immune and nervous systems (Yang et al. 2015). Chronic stress is an important risk factor of anxiety and depression which may exacerbate other neurological health conditions like stroke, epilepsy, post-traumatic stress disorder, NDDs and chronic pain (Streeter et al. 2012). Due to its immune-modulating and nootropic activities, *T. cordifolia* along with some Medha Rasayanas has been reported for depression and stress management (Mutalik and Mutalik 2011; Salve et al. 2015). Changes in neurotransmitters levels such as dopamine, serotonin, norepinephrine, GABA and activation of Hypothalamus-Pituitary-Adrenal (HPA) axis underly the pathophysiology of stress and depression (Naoi et al. 2018). A pre-clinical study on mice model of learned helplessness showed that *T. cordifolia* reduced behavioral anxiety and stress along with reduced levels of MAOA and MAOB monoaminergic metabolism enzymes and the effect was comparable to standard anti-depressant drugs like sertaline and imipramine (Dhingra and Goyal 2008). Similar anti-anxiety activity was observed with fresh leaves of *T. cordifolia* and berberine, one of the active phyto-constituent in another study on stress and depression model system (Peng et al. 2007; Kulkarni and Dhir 2008; Kalabharathi et al. 2014).

The anti-depressant activity of *T. cordifolia* has also been reported in some pilot scale human studies (Table 8.1). Oral intake of encapsulated *T. cordifolia* aqueous extract for 4 weeks was shown to activate the parasympathetic nervous system and

Table 8.1 Relevant clinical trials to validate the mentioned effects, safety, and optimum dose of *T. cordifolia* to facilitate transition from bench to bedside

Activity	Preparation type	Testing subjects	Effects	References
Immunomodulatory activity for the management of Idiopathic Granulomatous Mastitis	Tinosporin tablets (500 mg) an immunomodulator from *T. cordifolia*	91 patients of idiopathic granulomatous mastitis	Effective to suppress inflammation in about 80% of the enrolled patients	Das Sheth et al. (2025)
Management of infections in Chronic Lymphocytic Leukemia (CLL) patient	*T. cordifolia*, Trivrut Avaleha and Glycyrrhiza glabra formulation	A case control study of a 46-year-old male suffering from CLL	Reduction in frequency of infections, improved TLC counts after 10 months of intervention	Prajapati et al. (2025)
Treatment of urinary tract infection (UTI)	Polyherbal formulation of hydro-alcoholic extracts given orally for 42 days	Open-label, single armed clinical trial on UTI patients	Urine cultures negative for E. coli after the completion of therapy	Qureshi et al. (2024)
Management of Hypertriglyceridemia	3 g water extract of *T. cordifolia* for 14 days	24 patients of Hypertriglyceridemia	Significantly decrease in the levels of triglycerides and VLDL and increased HDL levels	Shirolkar et al. (2022)
Analysis of Clinical Trials Registry—India (CTRI) for Ayurvedic drugs	From available datasets between Feb. 2020 and August 2020	COVID-19 related clinical studies involving the healthcare systems of AYUSH, registered on CTRI	T. cordifolia was the most frequently mentioned drug in these CTRI trials	Bhapkar et al. (2020)
Impact of Ayurvedic treatment regime on asymptomatic patients with COVID-19	Formulation of *T. cordifolia*, traditional herbo-minerals, Withania somnifera and Ocimum sanctum orally to the patients twice per day for 7 days	Placebo controlled randomized double-blind pilot clinical trial	Inflammatory markers, hs-CRP, IL-6 and TNF-α in treatment group significantly lower than the placebo group	Devpura et al. (2021)
Innate immunity boosting and restoring respiratory health in patients of COVID-19 infection	Poly-herbal formulation containing *T. cordifolia* approved by Government of India (GoI) for other health conditions	Controlled, exploratory trial for a period of 30 days	Decline in viral load, immune-modulation - TH1, TH2, Th17, IL6, NK cells and CD markers; sIGG and sIGM	Rangnekar et al. (2020)
Impact of chronic and moderate alcohol intake on cholesterol homeostasis and steroid hormone synthesis	Standardized water extract of *T. cordifolia*	Asymptomatic chronic and moderate alcoholics ($n = 12$) without chronic liver disease and healthy volunteers ($n = 14$)	T. cordifolia modulated lipid metabolism by inhibiting cholesterol and glucuronides synthesis	Kumari et al. (2016)

(continued)

Table 8.1 (continued)

Activity	Preparation type	Testing subjects	Effects	References
Testing *T. cordifolia* for adaptogenic activity in healthy volunteers during exercise	150 mg and 300 mg, respectively ($n = 10$ each) of *T. cordifolia* aqueous extract in capsule form once daily in the morning for 28 days	Healthy volunteers during exercise	Significant decrease in mean systolic BP and heart rate on fixed workload exercise compared to placebo	Salve et al. (2015)
Urine metabolic signature of chronic alcoholism *T. cordifolia* aqueous extract	*T. cordifolia* aqueous extract	Patients of chronic alcoholism	Reverted the metabolic signature i.e. tryptophan, fatty acids and pyrimidines metabolism in urine samples of chronic alcoholism patients	Mittal and Dabur (2015)
Anti-tuberculosis treatment induced hepatotoxicity	Combination of *T. cordifolia* and *Curcuma longa*	Patients with active TB diagnosis randomized to a drug control group and a trial group on drugs plus herbal formulation	Herbal formulation significantly prevented hepatotoxicity	Adhvaryu et al. (2008)
Assessment of the safety and efficacy of T. cordifolia in HIV positive patients	*Tinospora cordifolia* extract (TCE)	68 HIV positive participants	60% patients receiving TCE and 20% on placebo reported decrease in symptoms associated with disease	Kalikar et al. (2008)
Generalized immunomodulation in diabetic ulcers	Evaluation of *T. cordifolia* as an adjuvant therapy in diabetic ulcers	A prospective double-blind randomized controlled study lasting for over 18 months in 50 patients	Significant improvement in wound healing, phagocytosis with reduced debridements	Purandare and Supe (2007)
Testing efficacy of *T. cordifolia* extract in patients of allergic rhinitis	A randomized double-blind placebo- controlled trial on patients randomly given either T. cordifolia or placebo for 8 weeks	72 patients of allergic rhinitis	Significant relief from sneezing, nasal discharge, nasal obstruction and nasal pruritus	Badar et al. (2005)

amelioration of physical stress on male subjects (Salve et al. 2015). Further, *T. cordifolia* and Yoga therapy also was observed to exhibit beneficial effects at behavioural and biochemical levels in patients of mental stress (Biswas et al. 2015). The anti-stress and anti-depressant like activity of *T. cordifolia* may be attributed to its anti-inflammatory and anti-oxidant activities (Saha and Ghosh 2012; Sharma and Kaur 2018; Sharma et al. 2019).

8.3.3 **T. cordifolia** *and Metabolic Health*

Several recent reports in literature highlight the medicinal properties of *T. cordifolia*, for the treatment of various metabolic diseases and as a food supplement for the promotion and maintenance of health (Bisht et al. 2025; Mansouri et al. 2025; Singh et al. 2024). Tribal people from different regions of India use either whole plant or decoctions and juice of stem or root for treatment of fever, jaundice, cancer and diarrhea etc. by these tribal people (Upadhyay et al. 2010). In a human pilot scale study, aqueous extract of *T. cordifolia* was tested by Shirolker et al. (2022) to correct dyslipidaemia and for its anti-alcoholic activities. Fourteen days oral intake of extract ameliorated hyperlipidemia and oxidative stress in the urine and blood samples of chronic alcoholics by enhancing levels of dopamine and cAMP activation. This medicinal plant extracts have also been tested for managing neurological outcome of alcoholism or other substances of abuse in other human pilot scale studies (Mittal and Dabur 2015; Kumari et al. 2016).

Another study from our laboratory designed to elucidate the cumulative effects of ageing and/or high fat diet (HFD) induced obesity reported the effects of *T. cordifolia* stem powder (TCP) on metabolic indicators using middle-aged acyclic female rats as the model system of menopause (Singh et al. 2020; Bhandari et al. 2022). Animals were fed on TCP supplemented diet showed improvement in age-associated liver dysfunctions and amelioration of dyslipidemia and oxidative stress, along with normalized blood glucose, insulin, leptin, and secretary pro-inflammatory cytokines. These studies provided evidence that the supplementation of *T. cordifolia* may attenuate obesity-associated physiological dysregulations such as excessive gain in body weight, impairment of liver and kidney functions and increased oxidative stress. TCP supplementation in HFD animals was further observed to correct dyslipidaemia by targeting protein expression of lipid metabolism enzymes and to promote lipolysis and down regulate lipogenesis (Bhandari et al. 2022). These pre-clinical studies provide evidence and suggest that *T. cordifolia* supplementation in diet may prove to be a potential intervention to promote healthy ageing by preventing age-associated hepatic and metabolic functions impairments. Additionally, as an alternative to estrogen replacement therapy, *T. cordifolia* and β-ecdysone, a bioactive compound, have been proposed to act as a selective oestrogen receptor modulator (SERM) that demonstrated pro-stimulatory effects on osteoblast cells and rat primary osteoblasts (Abiramasundari et al. 2018).

The osteoprotective property of β-ecdysone has also been demonstrated by Seidlova-Wuttke et al. (2010) thus suggesting the potential activity of *T. cordifolia* in maintaining the bone health.

8.3.4 Anti-Cancer Activity of T. Cordifolia

It is pertinent to mention that natural medicines including plant extracts have a long history and extensively reported for cancer therapeutics, supported by discovery of some very effective anti-cancer plant-based drugs (Corso and Bindra 2016). Cervical cancer poses a significant public health concern especially for low-and middle-income nations. Vaghasia et al. (2025) explored the dual potential of *T. cordifolia* in combination of some other medicinal plant extracts alongside cisplatin as agents of anti-cancer and immunomodulation against HPV-positive cervical cancer CaSki and HEK 293 cells. These formulations exhibited cytotoxic, genotoxic, anti-migratory, HPV DNA inhibition, and cell cycle modulatory potential thus promoting future studies to combine natural products as adjunct therapy with conventional chemotherapeutics for safer and effective cancer therapies. Several cellular model system studies have reported that *T. cordifolia* extracts suppresses cancers of cervix (Polu et al. 2017), breast (Ansari et al. 2017; Rashmi et al. 2019), and skin (Ali and Dixit 2013).

'Brain cancers' is a term used in clinical practice for a variety of neoplasms developing in the central and peripheral nervous system. They pose heavy burden on health care systems due to their high morbidity and mortality rates in humans. Glioblastoma multiforme (GBM) is the most aggressive and metastasizing intracranial brain cancer. Due to its highly proliferative phenotype, GBM usually does not comply with standard therapeutic measures such as surgical resection, radiation- and chemo-therapy and tends to result in disease relapses and demands advancement towards effective therapies. Anti-brain cancer effects of hydroalcoholic extract, TCE by its apoptosis and senescence promoting effects have been reported from our laboratory using C6-rat glioblastoma and U87MG-Human primary glioblastoma and IMR-32-Human neuroblastoma cells (Mishra and Kaur 2013, 2015). TCE treatment of these brain cancer cells impeded their proliferation rate, induced differentiation and checked metastasis as confirmed by wound-scratch assay and protein expression of extra-cellular matrix cell adhesion molecules as well as cytoskeletal proteins like α-tubulin, MAP-2 and NF-200. Further, chloroform and hexane fractions of TCE were shown to be effective at ~20 folds lower dose than the parent extract (Sharma et al. 2019). These preliminary observations of anti-brain cancer activity of *T. cordifolia* may suggest its role as a candidate for differentiation-based therapy which is more useful approach for treating brain cancers.

8.3.5 T. cordifolia *as Nutraceutical and Functional Foods*

The bitter taste of T. cordifolia is a major limitation factor in its commercial use for preparing functional food-products, which necessitates the refinement of processing techniques and protocols for its active employment in food products (Anjum et al. 2023). A recent study by Singh et al. (2025) reported the nutritional value and biological activities of stem and leaves of this medicinal plant. Ultrasonication, a green extraction technique was observed to promote bioactives extraction from leaves and stem for making tasty drinks and food items with long-shelf life due to their anti-microbial activity. In another recent study, Singh et al. (2024) discussed the potential of incorporating leaves of *T. cordifolia* into a variety of value-added food items like beverages, cookies, as well as herbal products. They further emphasized on the need to combine traditional knowledge with functional foods production using modern scientific techniques to improve the food value and medicinal efficacy of *T. cordifolia* formulations. Tyagi et al. (2020) developed functional food cookies by adding 2–12% of the stem powder of *T. cordifolia*. These cookies after incorporating *T. cordifolia* stem powder had better functional properties with increase in the protein, moisture, total ash, iron, copper, zinc and antioxidant potential and decrease in their fat content.

Recently Kamboj et al. (2023) optimized the protocol to prepare probiotic supplemented herbal wine from T. cordifolia with high yield of ethanol (6.8–10%), total phenols and anti-oxidative capacity. The probiotic supplemented herbal wine safety was then tested on HEK-293 and THP-1 cells, which showed 100% survivability of these cells. The possible use of T. cordifolia as a functional food ingredient has also been tested for preparation of goat milk beverage, and bioactive ingredient for edible films for enhancing shelf life of meat products (Kalem et al. 2018; Sharma et al. 2021; Heena et al. 2021). High nutritional value (292.54 calories/100 gm), richness in dietary fibre, protein content and high levels of mineral elements in T. cordifolia (Kavya et al. 2015) are suggestive of its health promoting and metabolism boosting activities. However, the functional food products developed recently are to be tested and validated for human consumption.

8.4 Future Perspectives

The recent boom in pharmaceutical industry has limited the use of traditional medicines due to issues of their incompatibility with high throughput screening approaches compared to single molecules based synthetic drugs. However, their high bioavailability, metabolite-like properties, and least side-effects are some of the added advantages to promote R&D of natural medicines for health and disease. Natural products generally exhibit diversified stereochemistry and pharmacophore, thus providing better hits for screening of complex drug targets which may involve multiple protein interactions. Proteomic and metabolomic approaches along with other advanced technologies may help to understand the interactive actions of multiple ingredients present in herbal medicines in complicated biological systems to boost future research in the area of phyto-pharmacotherapeutics. Several Food and Drug administration (FDA)-approved plant-based drugs such as morphine from Papaver somniferum, Paclitaxel from Taxus brevifolia, Vincristine (Catharanthus roseus), Camptothecin (Camptotheca acuminata), Artimisinin (Artemisis annua), Homoharringtonine (Cephalotaxus harringtonii), Quinine (Cinchona officinalis), to name a few, are in clinical practice for the treatment of various diseases. Several human pilot scale studies with available formulations/extracts of T. cordifolia are although reported in the literature for its therapeutic and health benefits (Table 8.1), but a much exhaustive and targeted approach is needed in future to develop novel formulations of T. cordifolia and their scientific validation focusing on health and longevity. Moreover, to maintain consistency in the efficacy of T. cordifolia, the important factors such as season and age of the plant at the time of harvesting, and part of plant having appropriate concentrations and ratios of active ingredients needs attention of the herbal industry.

8.5 Conclusion

Based on its multiple pharmacological activities, low hepatotoxicity, and high bio-availability, T. cordifolia has received much attention in Ayurveda and some formulations are available over the counter for human usage. These formulations/extracts of *T. cordifolia* have been tested in several human pilot scale studies which have reported the therapeutic use as well as their health benefits (Table 8.1). Besides being rich in secondary metabolites with known pharmacological activities, high content of protein, dietary fibre and micronutrients like calcium, copper, iron, zinc, phosphorus, manganese etc., makes *T. cordifolia* an ideal supplement for health and longevity. Studies to develop functional foods supplemented with *T. cordifolia* have also been reported recently and their palatability and required dosages in these food products is being explored to ensure their suitability for human consumption. Development of novel formulations from elite varieties of *T. cordifolia* to circumvent the issues of variations in phytoconstituents of their herbal products and future translational research using these standard formulations are some of the existing knowledge gaps.

References

Abiramasundari G, Mohan Gowda CM, Sreepriya M (2018) Selective Estrogen Receptor Modulator and prostimulatory effects of phytoestrogen β-ecdysone in Tinospora cordifolia on osteoblast cells. J Ayurveda Integr Med 9(3):161–168. https://doi.org/10.1016/j.jaim.2017.04.003

Adhvaryu MR, Reddy N, Vakharia BC (2008) Prevention of hepatotoxicity due to anti tuberculosis treatment: a novel integrative approach. World J Gastroenterol 14(30):4753–4762. https://doi.org/10.3748/wjg.14.4753

Agarwal A, Malini S, Bairy KL, Rao MS (2002) Effect of Tinospora cardifolia on learning and memory in normal and memory deficit rats. Indian J Pharmacol 34(5):339–349. http://www.ijp-online.com/temp/IndianJPharmacol345339-2378845_063628.pdf

Ali SA, Datusalia AK (2024) Protective effects of Tinospora cordifolia miers extract against hepatic and neurobehavioral deficits in thioacetamide-induced hepatic encephalopathy in rats via modulating hyperammonemia and glial cell activation. J Ethnopharmacol 323:117700. https://doi.org/10.1016/j.jep.2023.117700

Ali H, Dixit S (2013) Extraction optimization of Tinospora cordifolia and assessment of the anticancer activity of its alkaloid palmatine. Sci World J. https://doi.org/10.1155/2013/376216

Amrutha S, Abhinand CS, Upadhyay SS, Parvaje R, Prasad TSK, Modi PK (2025) Network pharmacology and metabolomics analysis of Tinospora cordifolia reveals BACE1 and MAOB as potential therapeutic targets for neuroprotection in Alzheimer's disease. Sci Rep 15(1):8103. https://doi.org/10.1038/s41598-025-92756-5

Anjum V, Bagale U, Kadi A, Potoroko I, Sonawane SH, Anjum A (2023) Unveiling various facades of Tinospora cordifolia stem in food: medicinal and nutraceutical aspects. Molecules 28(20): 7073. https://doi.org/10.3390/molecules28207073

Ansari JA, Rastogi N, Ahmad MK, Mahdi AA, Khan AR, Thakur R et al (2017) ROS mediated pro-apoptotic effects of Tinospora cordifolia on breast cancer cells. Front Biosci (Elite Ed) 9: 89–100. https://doi.org/10.2741/e788

Badar VA, Thawani VR, Wakode PT, Shrivastava MP, Gharpure KJ, Hingorani LL, Khiyani RM (2005) Efficacy of Tinospora cordifolia in allergic rhinitis. J Ethnopharmacol 6(3):445–449. https://doi.org/10.1016/j.jep.2004.09.034

Bajaj P, Kaur G (2022) Acute sleep deprivation-induced anxiety and disruption of hypothalamic cell survival and plasticity: a mechanistic study of protection by butanol extract of Tinospora cordifolia. Neurochem Res 47(6):1692–1706. https://doi.org/10.1007/s11064-022-03562-8

Bajaj P, Singh H, Kalotra S, Kaur G (2021) Butanol extract of Tinospora cordifolia alleviates acute sleep deprivation-induced impairments in cognitive functions and neuromuscular coordination in middle-aged female rats. NeuroMol Med 2021:8683. https://doi.org/10.1007/s12017-021-08683-x

Bhandari A, Sharma AK, Singh H, Singh A, Kaur G (2022) Aging related changes in metabolic indicators in female rats and their management with Tinospora cordifolia. Biogerontology. https://doi.org/10.1007/s10522-022-09962-1

Bhalerao S, Deshpande T, Thatte U (2012) Prakriti (Ayurvedic concept of constitution) and variations in platelet aggregation. BMC Complement Altern Med 12:248. https://doi.org/10.1186/1472-6882-12-248

Bhapkar V, Sawant T, Bhalerao S (2020) A critical analysis of CTRI registered AYUSH studies for COVID-19. J Ayurveda Integr Med 13(1):100370. https://doi.org/10.1016/j.jaim.2020.10.012

Birla H, Rai SN, Singh SS, Zahra W, Rawat A, Tiwari N et al (2019) Tinospora cordifolia suppresses neuroinflammation in parkinsonian mouse model. NeuroMolecular Med 21:42–53. https://doi.org/10.1007/s12017-018-08521-7

Bisht A, Nayal A, Tewari D, Kumar S, Chandra S (2025) Investigating the aging-modulatory mechanism of Rasayana Churna, an Ayurvedic herbal formulation, using a computational approach. Biogerontology 26(2):53. https://doi.org/10.1007/s10522-025-10188-0

Biswas P, Saha A, Maity LN (2015) Antistress activity of tinospora cordifolia with application of yoga. Int J Ayurvedic Med 6(3):220–224

Corso CD, Bindra RS (2016) Success and failures of combined modalities in glioblastoma multiforme: old problems and new directions. In: Seminars in radiation oncology, vol 26, No. 4. WB Saunders, pp 281–298. https://doi.org/10.1016/j.semradonc.2016.06.003

Das Sheth A, Joshi S, Kumar A, Nair N, Shet T, Sahay A, Thakkar P, Haria P, Katdare A, Parmar V, Desai S, Badwe R (2025) Management of Idiopathic Granulomatous Mastitis: effectiveness of a steroid-free regimen using Tinospora cordifolia-A single-institution experience. Breast J:2997891. https://doi.org/10.1155/tbj/2997891

Das SC, Biswas S, Khan O, Akter R, Azad MAK, Sarkar SK, Masum MA, Bedoura S (2025) Evaluation of anti-inflammatory and wound healing properties of Tinospora cordifolia extract. PLoS One 20(1):e0317928. https://doi.org/10.1371/journal.pone.0317928

Devpura G, Tomar BS, Nathiya D, Sharma A, Bhandari D, Haldar S, Balkrishna A, Varshney A (2021) Randomized placebo-controlled pilot clinical trial on the efficacy of ayurvedic treatment regime on COVID-19 positive patients. Phytomedicine 84:153494. https://doi.org/10.1016/j.phymed.2021.153494

Dhama K, Sachan S, Khandia R, Munjal A, MN Iqbal H, K Latheef S et al (2016) Medicinal and beneficial health applications of Tinospora cordifolia (Guduchi): a miraculous herb countering various diseases/disorders and its immunomodulatory effects. Recent Pat Endocr Metab Immune Drug Discov 10(2):96–111. https://doi.org/10.2174/1872214811666170301105101

Dhingra D, Goyal PK (2008) Evidences for the involvement of monoaminergic and GABAergic systems in antidepressant-like activity of Tinospora cordifolia in mice. Indian J Pharm Sci 70(6): 761–767. https://doi.org/10.4103/0250-474X.49118

Elufioye TO, Berida TI, Habtemariam S (2017) Plants-derived neuroprotective agents: cutting the cycle of cell death through multiple mechanisms. Evid Based Complement Alternat Med:3574012. https://doi.org/10.1155/2017/3574012

Farooqui AA, Horrocks LA, Farooqui T (2007) Modulation of inflammation in brain: a matter of fat. J Neurochem 101(3):577–599. https://doi.org/10.1111/j.1471-4159.2006.04371

Frank-Cannon TC, Alto LT, FE MA, Tansey MG (2009) Does neuroinflammation fan the flame in neurodegenerative diseases? Mol Neurodegener 4(1):47. https://doi.org/10.1186/1750-1326-4-47

Ghatpande NS, Misar AV, Waghole RJ, Jadhav SH, Kulkarni PP (2019) Tinospora cordifolia protects against inflammation associated anemia by modulating inflammatory cytokines and hepcidin expression in male Wistar rats. Sci Rep 9(1):1–11. https://doi.org/10.1038/s41598-019-47458-0

Ghodke Y, Joshi K, Patwardhan B (2011) Traditional medicine to modern pharmacogenomics: Ayurveda Prakriti type and CYP2C19 gene polymorphism associated with the metabolic variability. Evidence-Based Complementary and Alternative Medicine:249528. https://doi.org/10.1093/ecam/nep206

Govindaraj P, Nizamuddin S, Sharath A, Jyothi V, Rotti H, Raval R et al (2015) Genome-wide analysis correlates Ayurveda Prakriti. Sci Rep 5:15786. https://doi.org/10.1038/srep15786

Gupta A, Gupta P, and Bajpai G (2024) Tinospora cordifolia (Giloy): an insight on the multifarious pharmacological paradigms of a most promising medicinal ayurvedic herb. Heliyon 10(4): e26125. https://doi.org/10.1016/j.heliyon.2024.e26125

Haque MA, Jantan I, Bukhari SNA (2017) Tinospora species: an overview of their modulating effects on the immune system. J Ethnopharmacol 207:67–85. https://doi.org/10.1016/j.jep.2017.06.013

Harvey AL, Edrada-Ebel R, Quinn RJ (2015) The re-emergence of natural products for drug discovery in the genomics era. Nat Rev Drug Discov 14(2):111–129. https://doi.org/10.1038/nrd4510

Heena S, Priyanka R, Ashish KS (2021) Fifty years of research on Tinospora cordifolia: from botanical plant to functional ingredient in foods. Trends Food Sci Technol 118:189–206

Hillman DR, Lack LC (2013) Public health implications of sleep loss: the community burden. Med J Aust 199:S7–S10. https://doi.org/10.5694/mja13.10620

Hurley LL, Tizabi Y (2013) Neuroinflammation, neurodegeneration, and depression. Neurotox Res 23(2):131–144. https://doi.org/10.1007/s12640-012-9348-1

Jacob J, Babu BM, Mohan MC, Abhimannue AP, Kumar BP (2018) Inhibition of proinflammatory pathways by bioactive fraction of Tinospora cordifolia. Inflammopharmacology 26(2):531–538. https://doi.org/10.1007/s10787-017-0319-2

Kalabharathi HL, Vaibhavi PS, Naidu S, Pushpa VH, Shruthi SL (2014) Anti-depressant activity of fresh leaves of Tinospora cordifolia miers (Amruthaballi) in albino mice. Int J Pharm Sci Res 5(12):5420–5424

Kalem IK, Bhat ZF, Kumar S, Wang L, Mudiyanselage RJ, Bhat HF (2018) Tinospora cordifolia: a novel bioactive ingredient for edible films for improved lipid oxidative and microbial stability of meat products. J Food Process Preserv 42:e13774

Kalikar MV, Thawani VR, Varadpande UK, Sontakke SD, Singh RP, Khiyani RK (2008) Immunomodulatory effect of Tinospora cordifolia extract in human immuno-deficiency virus positive patients. Indian J Pharmacol 40(3):107–110. https://doi.org/10.4103/0253-7613.42302

Kamboj S, Soni SK, Shukla G (2023) Preparation, characterization, and safety assessment of statistical optimized probiotic supplemented herbal wine from Tinospora cordifolia. 3 Biotech 13(4):118. https://doi.org/10.1007/s13205-023-03501-6

Katsnelson A, De Strooper B, Zoghbi HY (2016) Neurodegeneration: from cellular concepts to clinical applications. Sci Transl Med 8(364):364ps18. https://doi.org/10.1126/scitranslmed.aal2074

Kaur P, Makanjuola VO, Arora R, Singh B, Arora S (2017) Immunopotentiating significance of conventionally used plant adaptogens as modulators in biochemical and molecular signaling pathways in cell mediated processes. Biomed Pharmacother 95:1815–1829. https://doi.org/10.1016/j.biopha.2017.09.081

Kavya B, Kavya N, Ramarao V, Venkateshwarly G (2015) Tinospora cordifolia (Willd.) Miers: nutritional, ethnomedical and therapeutic utility. Int J Res Ayurveda Pharm 6:195–198

Kempuraj D, Thangavel R, Natteru PA, Selvakumar GP, Saeed D, Zahoor H et al (2016) Neuroinflammation induces neurodegeneration. J Neurol Neurosurg Spine 1(1). https://www.ncbi.nlm.nih.gov/pmc/articles/PMC5260818/

Kosaraju J, Chinni S, Roy PD, Kannan E, Antony AS, Kumar MN (2014) Neuroprotective effect of Tinospora cordifolia ethanol extract on 6-hydroxy dopamine induced Parkinsonism. Indian J Pharmacol 46(2):176–180. https://doi.org/10.4103/0253-7613.129312

Kulkarni SK, Dhir A (2008) On the mechanism of antidepressant-like action of berberine chloride. Eur J Pharmacol 589(1–3):163–172. https://doi.org/10.1016/j.ejphar.2008.05.043

Kulkarni R, Girish KJ, Kumar A (2012) Nootropic herbs (MedhyaRasayana) in Ayurveda: an update. Pharmacogn Rev 6(12):147–153. https://doi.org/10.4103/0973-7847.99949

Kumari S, Mittal A, Dabur R (2016) Moderate alcohol consumption in chronic form enhances the synthesis of cholesterol and C-21 steroid hormones, while treatment with Tinospora cordifolia modulate these events in men. Steroids 114:68–77. https://doi.org/10.1016/j.steroids.2016.03.016

Lekkala VDVV, Muktinutalapati AV, Lebaka VR, Lomada D, Korivi M, Li W, Reddy MC (2025) Green synthesis and characterization of silver nanoparticles from Tinospora cordifolia leaf extract: evaluation of their antioxidant, anti-inflammatory, antibacterial, and antibiofilm efficacies. Nanomaterials (Basel) 15(5):381. https://doi.org/10.3390/nano15050381

Li FS, Weng JK (2017) Demystifying traditional herbal medicine with modern approach. Nat Plants 3:17109. https://doi.org/10.1038/nplants.2017.109

Malve HO, Raut SB, Marathe PA, Rege NN (2014) Effect of combination of Phyllanthus emblica, Tinospora cordifolia, and Ocimum sanctum on spatial learning and memory in rats. J Ayurveda Integr Med 5(4):209. https://doi.org/10.4103/0975-9476.146564

Malve HO (2015) Exploring Bhavana samskara using Tinospora cordifolia and Phyllanthus emblica combination for learning and memory in mice. J Ayurveda Integr Med 6(4):233–240. https://doi.org/10.4103/0975-9476.157953

Mansouri M, Imenshahidi M, Rameshrad M, Hosseinzadeh H (2025) Effects of Tinospora cordifolia (giloy) on metabolic syndrome components: a mechanistic review. Naunyn Schmiedeberg's Arch Pharmacol 398(5):4979–5009. https://doi.org/10.1007/s00210-024-03642-2

Mishra R, Kaur G (2013) Aqueous ethanolic extract of Tinospora cordifolia as a potential candidate for differentiation- based therapy of glioblastomas. PLoS One 8:e78764. https://doi.org/10.1371/journal.pone.0078764

Mishra R, Kaur G (2015) Tinospora cordifolia induces differentiation and senescence pathways in Neuroblastoma cells. Mol Neurobiol 52:719–733. https://doi.org/10.1007/s12035-014-8892-5

Mishra R, Manchanda S, Gupta M, Kaur T, Saini V, Sharma A et al (2016) Tinospora cordifolia ameliorates anxiety-like behavior and improves cognitive functions in acute sleep deprived rats. Sci Rep 6:25564. https://doi.org/10.1038/srep25564

Mittal A, Dabur R (2015) Detection of new human metabolic urinary markers in chronic alcoholism and their reversal by aqueous extract of Tinospora cordifolia stem. Alcohol Alcohol 50(3): 271–281. https://doi.org/10.1093/alcalc/agv012

Mutalik M, Mutalik M (2011) Tinospora cordifolia: role in depression, cognition and memory. Aust J Med Herbalism 23(4):168–173

Naoi M, Maruyama W, Shamoto-Nagai M (2018) Type A monoamine oxidase and serotonin are coordinately involved in depressive disorders: from neurotransmitter imbalance to impaired neurogenesis. J Neural Transm 125(1):53–66. https://doi.org/10.1007/s00702-017-1709-8

Nishteswar K, Joshi H, Karra RD (2014) Role of indigenous herbs in the management of Alzheimer's disease. Anc Sci Life 34(1):3–7. https://doi.org/10.4103/0257-7941.150763

Peng WH, Lo KL, Lee YH, Hung TH, Lin YC (2007) Berberine produces antidepressant-like effects in the forced swim test and in the tail suspension test in mice. Life Sci 81(11):933–938. https://doi.org/10.1016/j.lfs.2007.08.003

Philip S, Tom G, Vasumathi AV (2018) Evaluation of the anti-inflammatory activity of Tinospora cordifolia (Willd.) Miers chloroform extract–a preclinical study. J Pharm Pharmacol 70(8): 1113–1125. https://doi.org/10.1111/jphp.12932

Polu PR, Nayanbhirama U, Khan S, Maheswari R (2017) Assessment of free radical scavenging and anti-proliferative activities of Tinospora cordifolia Miers (Willd). BMC Complement Altern Med 17(1):457. https://doi.org/10.1186/s12906-017-1953-3

Prajapati AK, Venate P, Monika T, Huddar VG (2025) Effect of Ayurveda and Siddha interventions in the management of chronic lymphocytic leukemia: a case report. J Ayurveda Integr Med 16(1):101017. https://doi.org/10.1016/j.jaim.2024.101017

Purandare H, Supe A (2007) Immunomodulatory role of Tinospora cordifolia as an adjuvant in surgical treatment of diabetic foot ulcers: a prospective randomized controlled study. Indian J Med Sci 61(6):347–355

Qureshi A, Rahman RU, Shamsi Y (2024) Efficacy and safety of a polyherbal formulation in the management of Escherichia coli urinary tract infection. Drug Metab Pers Ther 39(4):221–230. https://doi.org/10.1515/dmpt-2024-0070

Rangnekar H, Patankar S, Suryawanshi K, Soni P (2020) Safety and efficacy of herbal extracts to restore respiratory health and improve innate immunity in COVID-19 positive patients with mild to moderate severity: a structured summary of a study protocol for a randomised controlled trial. Trials 21(1):943. https://doi.org/10.1186/s13063-020-04906-x

Rashmi KC, Raj MH, Paul M, Girish KS, Salimath BP, Aparna HS (2019) A new pyrrole based small molecule from Tinospora cordifolia induces apoptosis in MDA-MB-231 breast cancer cells via ROS mediated mitochondrial damage and restoration of p53 activity. Chem Biol Interact 299:120–130. https://doi.org/10.1016/j.cbi.2018.12.005

Reddi KK, Tetali SD (2019) Dry leaf extracts of Tinospora cordifolia (Willd.) Miers attenuate oxidative stress and inflammatory condition in human monocytic (THP-1) cells. Phytomedicine 61:152831. https://doi.org/10.1016/j.phymed.2019.152831

Saeed M, Naveed M, Leskovec J, Kakar I, Ullah K, Ahmad F et al (2020) Using Guduchi (Tinospora cordifolia) as an eco-friendly feed supplement in human and poultry nutrition. Poult Sci 99(2):801–811. https://doi.org/10.1016/j.psj.2019.10.051

Saha S, Ghosh S (2012) Tinospora cordifolia: one plant, many roles. Anc Sci Life 31(4):151–159. https://doi.org/10.4103/0257-7941.107344

Salve BA, Tripathi RK, Petare AU, Raut AA, Rege NN (2015) Effect of Tinospora cordifolia on physical and cardiovascular performance induced by physical stress in healthy human volunteers. Ayu 36(3):265–270. https://doi.org/10.4103/0974-8520.182751

Seidlova-Wuttke D, Christel D, Kapur P, Nguyen BT, Jarry H, Wuttke W (2010) Beta-ecdysone has bone protective but no estrogenic effects in ovariectomized rats. Phytomedicine 17(11): 884–889. https://doi.org/10.1016/j.phymed.2010.03.021

Sharma A, Kaur G (2018) Tinospora cordifolia as a potential neuroregenerative candidate against glutamate induced excitotoxicity: an in vitro perspective. BMC Complement Altern Med 18:1–17. https://doi.org/10.1186/s12906-018-2330-6

Sharma P, Dwivedee BP, Bisht D, Dash AK, Kumar D (2019) The chemical constituents and diverse pharmacological importance of Tinospora cordifolia. Heliyon 5:e02437. https://doi.org/10.1016/j.heliyon.2019.e02437

Sharma B, Dutt V, Kaur N, Mittal A, Dabur R (2020) Tinospora cordifolia protects from skeletal muscle atrophy by alleviating oxidative stress and inflammation induced by sciatic denervation. J Ethnopharmacol 254:112720. https://doi.org/10.1016/j.jep.2020.112720

Sharma A, Bajaj P, Bhandari A, Kaur G (2020a) From ayurvedic folk medicine to preclinical neurotherapeutic role of a miraculous herb, Tinospora cordifolia. Neurochem Int 141:104891. https://doi.org/10.1016/j.neuint.2020.104891

Sharma A, Kalotra S, Bajaj P, Singh H, Kaur G (2020b) Butanol extract of Tinospora cordifolia ameliorates cognitive deficits associated with glutamate-induced Excitotoxicity: a mechanistic study using hippocampal neurons. NeuroMolecular Med 22:81–99. https://doi.org/10.1007/s12017-019-08566-2

Sharma H, Singh AK, Deshwal GK, Rao PS, Kumar MD (2021) Functional Tinospora cordifolia (giloy) based pasteurized goat milk beverage: impact of milk protein-polyphenol interaction on bioactive compounds, anti-oxidant activity and microstructure. Food Biosci 42:101101

Shirolkar A, Yadav A, Nale A, Phogat J, Dabur R (2022) Integrated omics analysis revealed the Tinospora cordifolia intervention modulated multiple signalling pathways in hypertriglyceridemia patients-a pilot clinical trial. J Diabetes Metab Disord 21(1):379–397. https://doi.org/10.1007/s40200-022-00985-6

Shochat T, Cohen-Zion M, Tzischinsky O (2014) Functional consequences of inadequate sleep in adolescents: a systematic review. Sleep Med Rev 18(1):75–87. https://doi.org/10.1016/j.smrv.2013.03.005

Singh RH, Narsimhamurthy K, Singh G (2008) Neuronutrient impact of Ayurvedic Rasayana therapy in brain aging. Biogerontology 9(6):369–374. https://doi.org/10.1007/s10522-008-9185-z

Singh H, Sharma AK, Gupta M, Singh AP, Kaur G (2020) Tinospora cordifolia attenuates high fat diet-induced obesity and associated hepatic and renal dysfunctions in rats. PharmaNutrition 13:100189

Singh J, Saxena E, Chaudhary AR, Kaur M, Salotra M, Rasane P, Kaur S, Ercisli S, Durul MS, Bozhuyuk MR, Urusan AH, Ullah R (2024) Immunomodulatory properties of Giloy (Tinospora cordifolia) leaves and its applications in value-added products. Heliyon 11(1):e40948. https://doi.org/10.1016/j.heliyon.2024

Singh SM, Tripathy S, Srivastav PP (2025) Bioactive compound extraction from giloy leaves and steam using ultrasound: bioactivity, antimicrobial, and LC-MS/MS study. Food Sci Biotechnol 34(9):1835–1847. https://doi.org/10.1007/s10068-024-01810-x

Streeter CC, Gerbarg PL, Saper RB, Ciraulo DA, Brown RP (2012) Effects of yoga on the autonomic nervous system, gamma-aminobutyric-acid, and allostasis in epilepsy, depression, and post-traumatic stress disorder. Med Hypotheses 78(5):571–579. https://doi.org/10.1016/j.mehy.2012.01.021

Thibaut F (2017) Neuroinflammation: new vistas for neuropsychiatric research. Dialogues Clin Neurosci 19(1):3. https://www.ncbi.nlm.nih.gov/pmc/articles/PMC5442362/pdf/

Travis FT, Wallace RK (2015) Dosha brain-types: a neural model of individual differences. J Ayurveda Integr Med 6(4):280–285. https://doi.org/10.4103/0975-9476.172385

Tyagi P, Chauhan AK, Singh SN (2020) Sensory acceptability of value-added cookies incorporated with Tinospora cordifolia (TC) stem powder; improvement in nutritional properties and antioxidant potential. J Food Sci Technol 57(8):2934–2940. https://doi.org/10.1007/s13197-020-04325-5

Upadhyay AK, Kumar K, Kumar A, Mishra HS (2010) Tinospora cordifolia (Willd.) Hook. f. and Thoms. (Guduchi)–validation of the Ayurvedic pharmacology through experimental and clinical studies. Int J Ayurveda Res 1(2):112–121. https://doi.org/10.4103/0974-7788.64405

Vaghasia H, Patel R, Prajapati J, Shah K, Saraf M, Rawal RM (2025) Cytotoxic and immunomodulatory properties of Tinospora cordifolia, Boerhaavia diffusa, Berberis aristata, and Ocimum basilicum extracts against HPV-positive cervical cancer cell line. BMC Complement Med Ther 25(1):59. https://doi.org/10.1186/s12906-025-04817-x

Wachtel-Galor S, Benzie IFF (2011) Chapter 1: Herbal medicine: an introduction to its history, usage, regulation, current trends, and research needs. In: Benzie IFF, Wachtel-Galor S (eds) Herbal medicine: biomolecular and clinical aspects, 2nd edn. CRC Press/Taylor & Francis, Boca Raton

Yang L, Zhao Y, Wang Y, Liu L, Zhang X, Li B, Cui R (2015) The effects of psychological stress on depression. Curr Neuropharmacol 13(4):494–504. https://doi.org/10.2174/1570159X1304150831150507

Chapter 9
Neuronutrition: Evaluating the Efficacy of Natural Compounds in Neurodiversity for Brain Health

Anastasiia Badaeva, Sergio Modafferi, Giulia Malaguarnera, Alexey Danilov, Andrey Danilov, Michela Perrone, Beatrice Pranzo, Carlotta Girlando, Francesca Fazzina, Chiara Mieli, Antonio Trapanotto, Damiano Galimberti, Luay Rashan, and Vittorio Calabrese

Abstract Neuronutrition is an emerging field exploring how specific nutrients influence brain health, aging, and neurodegenerative risk. With growing interest in strategies to promote cognitive longevity, targeted compounds such as antioxidants, polyphenols, vitamins, and microbiota-modulating agents are gaining attention for their roles in modulating neuroinflammation, oxidative stress, and the gut–brain axis. This chapter highlights autism spectrum disorder as a case example of how precision neuronutritional interventions can support both neurodevelopment and systemic health. Compounds like glutathione, Boswellia, medicinal mushrooms and vitamins B are examined for their mechanisms of action and translational relevance. The broader goal is to illustrate how neuronutritional interventions can modulate key neurobiological pathways, offering a scientifically grounded approach to enhancing neuronal resilience and extending cognitive healthspan throughout aging.

Keywords Neuronutrition · Hormesis · Vitagene · Autism spectrum disorder

A. Badaeva · A. Danilov · A. Danilov
Department for Nervous Diseases, I.M. Sechenov First Moscow State Medical University of the Ministry of Health of the Russian Federation, Moscow, Russia

S. Modafferi · M. Perrone · B. Pranzo · C. Girlando · F. Fazzina · A. Trapanotto · V. Calabrese (✉)
Department of Biomedical and Biotechnological Sciences, University of Catania, Catania, Italy
e-mail: vittorio.calabrese@unict.it

G. Malaguarnera · C. Mieli
Department of Human Sciences and Quality of Life Promotion, San Raffaele Roma Open University, Rome, Italy

D. Galimberti
International Longevity Science Association (ILSA), Milan, Italy

L. Rashan (✉)
Biodiversity Unit, Dhofar University, Salalah, Oman
e-mail: lrashan@du.edu.om

S. I. S. Rattan, C. Gezer (eds.), *Supplements and Drugs for Healthy Longevity*, Healthy Ageing and Longevity 22, https://doi.org/10.1007/978-3-032-11608-6_9

9.1 Neuronutrition in Aging and Longevity

9.1.1 State of Art

Brain health can be defined as the ability to perform all cognitive, emotional, and behavioral functions effectively and adaptively throughout life, supported by optimal neuronal function, plasticity, and resilience against neurodegeneration and injury. It encompasses maintaining cognitive performance, emotional stability, and neural network integrity, all of which contribute to overall well-being and quality of life. Neuronutrition is an emerging interdisciplinary field that explores how specific nutrients and dietary patterns influence the central nervous system across the lifespan, with growing significance for healthy aging and longevity. It investigates how dietary compounds modulate key molecular and cellular processes such as neurotransmission, mitochondrial function, oxidative metabolism, neuroinflammation, and the gut–brain axis through mechanisms that integrate nutrition science, neurobiology, and biochemistry (Badaeva et al. 2023; Bojang and Manchana 2023). In contrast to traditional dietary models that emphasize general nutritional adequacy, neuronutrition focuses on the targeted use of bioactive compounds including polyphenols, antioxidants, B-vitamins, essential minerals, and omega-3 fatty acids that activate cellular stress response pathways, influence redox and immune signaling, and regulate gene expression to preserve cognitive function and neuronal plasticity (Novikov et al. 2023). These compounds also play a critical role in modulating gut microbiota composition and intestinal barrier integrity, which in turn affect brain health through the gut–brain axis, an essential interface in neuroimmune and neuroendocrine regulation during aging. By supporting cognitive performance, emotional stability, and resilience to neurodegeneration, neuronutrition offers a promising strategy for enhancing brain longevity. Central to this approach is the development of functional foods and personalized interventions aimed at reducing inflammation, oxidative stress, and epigenetic drift as key contributors to age-related cognitive decline and loss of neurological function.

9.1.2 Redoxomics, Vitagenes, and Mitochondria: Molecular Mechanisms

A central mechanism through which neuronutritional interventions exert their effects is the modulation of oxidative and nitrosative stress. Redox imbalance characterized by an overproduction of ROS and/or reduced antioxidant capacity can alter neuronal signaling and synaptic integrity. The brain is particularly susceptible to oxidative damage due to its high lipid content, elevated oxygen consumption, and relatively low antioxidant defenses. Vitagenes are a group of genes that encode cytoprotective proteins, such as heat shock proteins (Hsps), thioredoxins, and sirtuins, which contribute to cellular homeostasis under stress conditions (Calabrese et al.

2010a, b). These proteins act as guardians of mitochondrial function, genomic stability, and intracellular signaling. Specific dietary compounds have demonstrated the ability to activate vitagenes and promote redox balance. For instance, sulforaphane (from broccoli sprouts), resveratrol (from grapes), hydroxytyrosol (from olive oil), acetyl-L-carnitine, and carnosic acid are known to enhance endogenous antioxidant defenses and preserve mitochondrial efficiency (Calabrese et al. 2016a, b; Hayes and Dinkova-Kostova 2006). These compounds also interact with the Nrf2 pathway, a master regulator of the antioxidant response, and contribute to cellular detoxification processes. Mitochondrial dysfunction, tightly linked to oxidative stress, represents another core target in neuronutrition. Nutritional agents capable of improving mitochondrial biogenesis, energy production (via NAD+/NADH modulation), and anti-apoptotic signaling may offer therapeutic advantages in the context of neurodevelopmental and neurodegenerative conditions.

Multiple studies have shown excessive ROS production, impaired antioxidant systems, and elevated lipid peroxidation in various tissues, including the brain. These redox imbalances are associated with mitochondrial dysfunction, impaired neuronal signaling, and disruptions in synaptic connectivity features. Nutritional strategies aimed at restoring redox balance, improving mitochondrial efficiency, and modulating neuroinflammatory pathways (Indika et al. 2023). Unlike pharmacological treatments, which often target symptomatic manifestations, neuronutritional interventions may address the underlying biochemical disturbances, potentially contributing to long-term brain support.

9.1.3 Hormesis in Longevity: General Concepts

Hormesis is a dose-response phenomenon characterized by a biphasic pattern, where low doses of a chemical or environmental stressor induce beneficial or stimulatory effects, while higher doses result in toxicity or inhibition. This response reflects an evolutionarily conserved adaptive mechanism that enables cells, tissues, and organisms to respond to low- level stress by enhancing resilience (Calabrese et al. 2023). Hormetic dose–response curves are often described as biphasic, inverted U-shaped, or inverted J-shaped, and have been observed across all domains of life from bacteria to humans and across diverse cell types. Notably, the quantitative features of the hormetic response appear to be independent of biological model, cell type, endpoint, inducing agent, or underlying mechanism. The hallmark of hormesis is its modest stimulatory amplitude, typically ranging from 30% to 60% above the control. This response is observed across various biological endpoints, including cell proliferation, growth, fecundity, memory, lifespan and others, highlighting its relevance in the context of health, disease prevention, and therapeutic interventions such as neuroprotection (Calabrese 2014).

In the late 1980s, the concept of hormesis was proposed as a mechanism for lifespan extension, with evidence from caloric restriction studies in rodent models suggesting its effectiveness (Boxenbaum et al. 1988). Caloric restriction was shown to induce a stress response, elevating corticosterone levels, which Masoro linked to lifespan extension through a hormetic process (Masoro 2001). These findings contributed to the integration of hormesis into the scientific understanding of aging and biogerontology. More recent studies investigate lifespan extension and stress resistance using invertebrate models, particularly the nematode *Caenorhabditis elegans* (*C. elegans*), due to its short lifespan (20–25 days), ease of rearing, similarity to human-like receptors and biochemical pathways, and genetic amenability. Several hundred compounds have been shown to enhance longevitys in C. *elegans.* These effects are commonly linked to key signaling pathways such as insulin/insulin growth factor (IIS)/AMPK and transcription factors (e.g., Nrf2, Sirt-1, Hsp-1) (Calabrese et al. 2024a, b). Nrf2 belongs to the nuclear factor (erythroid 2)-like (NRF) transcription factors, a family of basic leucine zipper containing transcriptional regulators related to nuclear factor erythroid 2 (NFE2, also called p45), that are characterized by a conserved 43 amino acid Cap'n'collar (CNC) motif. Among CNC family members, the vertebrate NRF subfamily comprising NRF1 (NFE2L1), NRF2 (NFE2L2), and NRF3 (NFE2L3)—plays a central role in transcriptional responses to oxidative and reductive stress (Liu et al. 2019). In basal conditions, Nrf2 is sequestered by cytoplasmic Kelch-like ECH-associated protein 1(Keap1) and is targeted to proteasomal degradation. However, in case of oxidative stress, Keap1 acts as a stress sensor and releases Nrf2, allowing it to translocate to the nucleus, where it heterodimerizes with one of the small musculoaponeurotic fibrosarcoma (sMaf) proteins (Ibrahim et al. 2020). The heterodimers recognize and bind to the antioxidant response elements AREs, regulating the expression of vitagenes, including heat shock proteins (Hsp70), thioredoxin, and sirtuins, crucial for antioxidant defence and redox homeostasis (Calabrese and Mattson 2013).Increasing evidence highlights the neuroprotective role of specific Nrf2 activators as preventive and therapeutic agents in neurodegenerative diseases. Electrophilic compounds, including isocyanate, sulforaphane, and polyphenols, can suppress Keap1 activity, thus activating Nrf2 (Eggler 2013). Several plant-derived compounds such as stilbenoids, quinones, terpenoids, and carotenoids activate the Nrf2 pathway through different mechanisms: (1) Disruption of the Nrf2–Keap1 complex, (2) Covalent modification of Keap1 cysteine residues via oxidation or alkylation, (3) Modulation of kinases such as GSK3β, p38 MAPK, ERK, AMPK, and PI3K/AKT, and (4) Epigenetic regulation involving DNA methylation and histone modification (Thiruvengadam et al. 2021). Moreover, polyphenols exhibit a range of beneficial effects, including antioxidant, anti-inflammatory, anti-amyloidogenic, anti- α-synuclein, and antidepressant properties (Osakabe et al. 2024). In *C. elegans* models of several neurodegenarative diseases, including autism spectrum disorder (ASD), polyphenolic compounds demonstrate neuroprotective effects that are closely linked to hormetic stress response pathways. Hydroxytyrosol, derived from olive extracts, significantly reduces amyloid-β aggregation and β-amyloid-induced paralysis, effects that are abolished upon knockdown of the stress response genes *skn-1/Nrf-2* and *Hsp-16.2* (Romero-Márquez et al. 2022). Similarly, resveratrol from red grapes and phlorizin from apple skins confer protection against proteotoxic stress through mechanisms involving the unfolded protein response, autophagy, and proteasomal activity, all dependent on *skn-1/Nrf-2* signaling (Park and Park 2022). Chlorogenic acid, a major coffee polyphenol, exerts neuroprotective effects in a Parkinson disese's model by inducing autophagy, reducing α-synuclein aggregation, improving locomotor behavior, and lowering reactive oxygen species (ROS) levels (He et al. 2023). These findings highlight the central role of hormetic regulators and autophagic pathways in mediating the beneficial effects of dietary polyphenols against neurodegenerative pathology in genetically tractable *C. elegans* models.

9.1.4 Neurohormesis and Neuroprotection

Neurohormesis is a key concept in understanding how the brain adaptively responds to low-intensity stressors by activating cytoprotective mechanisms that enhance neuronal resilience (Calabrese et al. 2010a, b; Calabrese and Mattson 2017). Universal pharmacological treatments fail in neurodegenerative diseases and this seems to be due to the single mechanism of action of drugs and/or their inability to penetrate neurons (Calabrese et al. 2018; Arendash 2012). Thus, neurohormesis can be considered as a potentially innovative approach in the treatment of neurodegenerative disorder. Following the hormetic paradigm, natural polyphenolic substances at low doses act as antioxidants in a wide range of brain diseases, this occurring through upregulation of Nrf2 signaling pathway and increased expression of vitagenes. This network, by blocking ROS production, plays a fundamental role in the free radical species metabolism, as well as in the biotransformation of xenobiotics, ultimately preventing neuronal cell death associated with neurotoxic damage. The febrile response triggers cellular stress mechanisms and induces the expression of heat shock proteins (Hsps) (Byrne et al. 2022). Various compounds, including sulforaphane and hydroxytyrosol, have been shown to elicit metabolic effects in cellular stress responses similar to fever. Sulforaphane can upregulate Hsps and enhance processes such as synaptic transmission, potentially leading to improved cortical connectivity. Furthermore, the compound has protective impacts against several neurodegenerative disorders via induction of the Nrf2 and heat shock factor 1 (HSF1)-dependent genes.

In this regard, controlled stimulation of hormetic mechanisms through interventions such as moderate physical exercise, intermittent fasting, or nutraceutical supplementation, could represent a promising therapeutic avenue.

Recently, it has been seen that medicinal mushrooms containing polyphenols, counteract oxidative stress, mitochondrial dysfunction, and neuroinflammation closely connected to the initiation and progression of various brain disease conditions. Hericenone, a bioactive compound from *Hericium erinaceus*, has been reported to modulate neuroinflammation and improve social behavior in rodent models of ASD, possibly through the activation of heat shock proteins and attenuation of oxidative damage (Trovato Salinaro et al. 2018). Recent findings have shown that dietary polyphenols are metabolized by gut microbiota resulting in metabolites that are more bioactive as well as possessing more antioxidant capacity

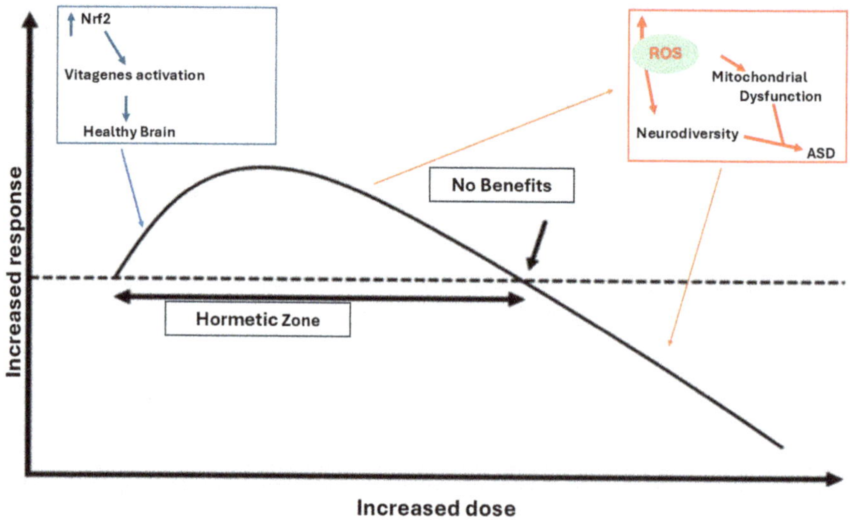

Fig. 9.1 Biphasic response to stress exposure and the concept of hormesis. The hormetic zone represents a dose range in which mild stress induces an adaptive response, leading to beneficial effects such as activation of Nrf2, induction of *vitagenes*, and improved brain resilience. This adaptive phase is characterized by enhanced antioxidant defenses and mitochondrial efficiency. However, when the stressor exceeds the hormetic threshold, the response becomes detrimental, marked by increased reactive oxygen species (ROS), mitochondrial dysfunction, and potential contribution to neurodevelopmental disorders. The diagram illustrates how stress intensity determines the transition from beneficial adaptation to pathological outcomes

than the native form. The gut microbiota break down polyphenols into metabolites that are readily absorbed from the intestine and transported via the blood to the brain where the metabolites exert scavenging properties linked to their chemical structure (Duda-Chodak et al. 2015). Moreover, recent findings indicate that imbalance in gut microbiota or dysbiosis may further compromise the nervous system's hormetic potential. Dysbiosis occurs majorly due to an altered integrity of the intestinal barrier which enhances passage of toxins (produced by pathogenic bacteria) from the gut lumen to the brain. These toxic molecules influence neurotransmitter function in the brain, resulting in abnormalities in behavioural patterns such as impaired socialization, decreased pain response, communication abnormalities and self-abusive or repetitive behaviours, delirium, confusion, which are core symptoms of ASD (Sivamaruthi et al. 2020). This highlights the therapeutic potential of microbiota-targeted interventions, such as dietary modifications or the use of prebiotics and probiotics, to enhance neuroprotective stress responses. Therefore, neurohormesis not only represents a natural defense strategy against neurodegenerative diseases (Calabrese et al. 2016a, b), but also emerges as a potential therapeutic mechanism in neurodevelopmental disorders like autism, where impaired adaptive stress responses are a defining feature (Fig. 9.1).

9.2 Pathophysiological Targets in Neuronutrition

Among the most studied physiological imbalances are oxidative stress, which reflects an imbalance between reactive oxygen species (ROS) and antioxidant defenses, and mitochondrial dysfunction, characterized by impaired bioenergetics, ATP production, and redox buffering capacity (Calabrese et al. 2010a, b). These alterations are not isolated but feed into immune dysregulation, which involves increased circulating cytokines, neuroinflammation, aberrant glial activation, and appear to interact with redox pathways (Rossignol and Frye 2014). Additionally, emerging data on the gut–brain axis suggest that gastrointestinal dysbiosis, altered microbial metabolites, and intestinal barrier dysfunction contribute to both peripheral inflammation and central nervous system signaling disturbances, further compounding metabolic stress and behavioral symptoms. These multidimensional impairments, although variable across individuals, provide a mechanistic rationale for the investigation of nutritional and phytochemical interventions as tools to modulate redox signaling, mitochondrial efficiency, and neuroimmune balance.

9.2.1 Oxidative Stress and Mitochondrial Dysfunction in ASD

Oxidative stress is one of the most consistently documented biochemical hallmarks in individuals with ASD. It arises when the cellular generation of ROS exceeds the capacity of antioxidant defense systems, leading to oxidative damage of lipids, proteins, and nucleic acids. In ASD, this imbalance has been associated with impaired detoxification, mitochondrial injury, and neuroinflammation factors that contribute to both core and associated symptoms. Numerous studies have reported elevated oxidative biomarkers in blood, lymphocytes, and brain tissue of individuals with ASD. These include increased levels of lipid peroxidation products, protein carbonyls, and DNA oxidation, accompanied by significant reductions in endogenous antioxidants. A particularly robust finding is the depletion of reduced glutathione (GSH), which plays a central role in redox regulation and cellular detoxification. The GSH/GSSG ratio is frequently decreased, reflecting compromised antioxidant capacity and increased oxidative burden (Rossignol and Frye 2014). Enzymatic antioxidant defenses are also altered in ASD. Superoxide dismutase (SOD), catalase (CAT), and glutathione peroxidase (GSH-Px) show inconsistent activity patterns, ranging from upregulation as a compensatory mechanism to downregulation due to enzyme depletion or genetic variants. In particular, polymorphisms in genes such as *GSTM1* and *GSTP1*, which encode key glutathione-related detoxifying enzymes, have been linked to increased vulnerability to oxidative insults in ASD. Mitochondria, both sources and targets of ROS, are especially vulnerable in this context. Mitochondrial dysfunction is widely reported in ASD and includes morphological alterations, impaired electron transport chain (ETC) function, reduced ATP production, and increased mitochondrial DNA damage. These abnormalities compromise the brain's energy metabolism and developmental plasticity, particularly during critical periods of synaptogenesis and myelination. To address these imbalances, various nutraceuticals have been studied for their antioxidant and mitochondrial protective properties. Among them, sulforaphane, a bioactive isothiocyanate found in broccoli sprouts, is known to activate the Nrf2 transcription factor and upregulate phase II detoxification enzymes and cytoprotective proteins such as HO-1, NQO1, and Hsp27. Clinical studies have reported improvements in social and behavioral domains following sulforaphane supplementation in ASD populations. Other promising compounds include resveratrol, lipoic acid, and acetyl-L- carnitine, which have demonstrated efficacy in modulating oxidative stress, enhancing mitochondrial respiration, and reducing neuroinflammation through Nrf2-related and vitagene- dependent pathways (Calabrese et al. 2016a, b; Hayes and Dinkova-Kostova 2006).

Taken together, oxidative stress and mitochondrial dysfunction constitute a tightly interlinked pathophysiological axis in ASD. Their early detection and targeted modulation through dietary strategies may offer complementary benefits to increasing quality of life and healthy longevity.

9.2.2 Neuroinflammation and Immune Dysregulation in ASD

Chronic low-grade inflammation and immune abnormalities are well-documented features of ASD. Several studies have reported elevated levels of pro-inflammatory cytokines—including interleukin-1β (IL-1β), interleukin-6 (IL-6), tumor necrosis factor-alpha (TNF-α), and interferon-gamma (IFN-γ)—in blood, cerebrospinal fluid, and brain tissues of individuals with ASD (Rossignol and Frye 2014). These cytokines can cross or compromise the blood–brain barrier, activating microglia and astrocytes, and interfering with neuronal function, potentially contributing to deficits in communication, behavior, and cognition. Microglial activation is considered a hallmark of ASD- associated neuroinflammation. Postmortem studies have revealed increased density and reactivity of microglia in multiple brain regions of individuals with ASD. This hyperactive immune state is accompanied by changes in T-cell populations, elevated immunoglobulins, and the presence of brain-directed autoantibodies (Rossignol and Frye 2014). Pro- inflammatory stimuli, such as IL-1β and TNF-α, have also been shown to disrupt astrocytic glutathione metabolism, decreasing intracellular GSH while increasing extracellular release, further compromising antioxidant defenses (Cordaro et al. 2021). Crucially, neuroinflammation and oxidative stress are tightly interconnected: reactive oxygen species (ROS) can stimulate redox-sensitive transcription factors such as NF-κB, which in turn amplify cytokine production and inflammatory signaling. This bidirectional relationship creates a vicious cycle in which oxidative stress promotes inflammation, and inflammation further depletes antioxidant reserves particularly glutathione (Rossignol and Frye 2014).

From a therapeutic perspective, several natural anti-inflammatory agents have shown promise in modulating immune responses in ASD models. Among them, curcumin, a polyphenol from turmeric, has demonstrated the ability to suppress NF-κB activation and reduce pro- inflammatory cytokines in animal models. Another notable compound is Boswellia serrata, a resinous extract rich in boswellic acids, which inhibits inflammatory mediators such as COX- 2, TNF-α, IL-1β, and MMP-9 (Calabrese et al. 2016a, b). While these agents are still under investigation in ASD-specific contexts, their effects on shared pathways suggest potential as adjunctive therapies. Overall, immune dysregulation in ASD involves a complex network of cytokine imbalance, glial activation, and impaired redox signaling. Understanding these processes is key to developing multimodal interventions that address both neuroinflammation and oxidative stress. Additional metabolic alterations impaired in ASD include one-carbon metabolism, a pathway highly dependent on adequate levels of folate and vitamin B12, contributing to methylation defects and homocysteine accumulation. These imbalances may alter neurotransmitter synthesis, DNA methylation patterns, and redox homeostasis (Schmidt et al. 2012). Homocysteine, in particular, has neurotoxic properties that may further impair the gut barrier and contribute to systemic inflammation. Furthermore, vitamin B12 deficiency has been linked to both gut microbiota imbalance and neurodevelopmental abnormalities. Low levels of B12 can impair the conversion of homocysteine to methionine, disrupt serotonin and dopamine synthesis, and worsen gastrointestinal and behavioral symptoms in ASD (Saad et al. 2016; Zwierz et al. 2025). Taken together, evidence from the existing literature suggests that disruptions in the gut–brain axis, microbial ecology, and associated metabolic pathways are not merely comorbidities but rather integrated components of ASD pathophysiology. These findings support the potential utility of targeted nutritional strategies, including B vitamin supplementation and modulation of the microbiota, in the management of ASD.

9.3 Evidence-Based Nutritional Interventions in ASD

A growing body of evidence has highlighted the potential role of specific nutrients and bioactive compounds in modulating neurobiological mechanisms implicated in ASD. These include oxidative stress, mitochondrial dysfunction, immune activation, altered neurotransmission, and epigenetic regulation. Within a neuronutritional framework, dietary interventions are increasingly viewed as promising complementary strategies to conventional treatments, particularly in addressing the metabolic and physiological imbalances frequently observed in individuals with ASD. Among these bioactive compounds, several have shown encouraging results in both preclinical models and clinical studies. For instance, sulforaphane a potent activator of the Nrf2 pathway (as described in Fig. 9.2) has been associated with improvements in social interaction and verbal communication in a randomized controlled trial involving adolescents and young adults with ASD. Likewise, supplementation with omega-3 fatty acids, including EPA and DHA, has demonstrated benefits in reducing

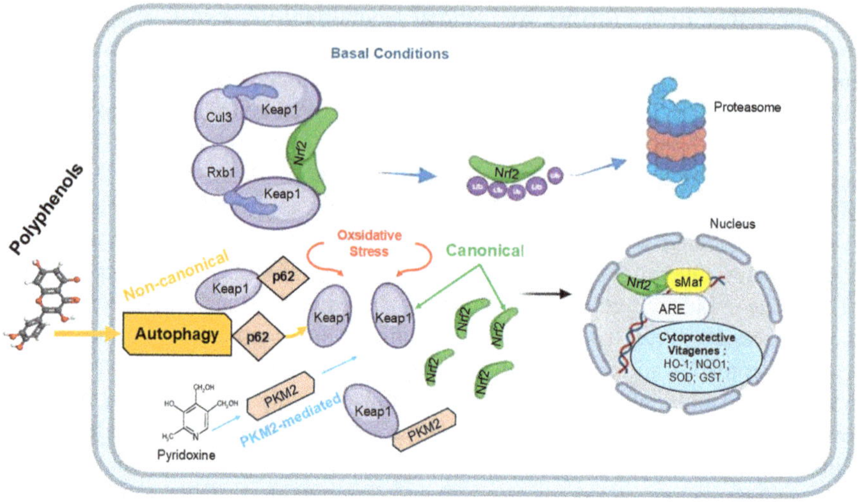

Fig. 9.2 Mechanisms of Nrf2 Activation Under Basal, Oxidative, and Non-Canonical Conditions. Under basal conditions (dark blue), Nrf2 is bound to KEAP1 in the cytoplasm and targeted for degradation. Under oxidative stress (red), Nrf2 translocates to the nucleus and activates ARE-driven genes such as HO-1, GSTs, NQO1, and γ-GCL. In the non-canonical pathway (yellow), polyphenol- induced p62 accumulation competes with KEAP1, activating Nrf2 independently. Pyridoxine (light blue) enhances glutathione synthesis via Nrf2 transactivation involving PKM2

hyperactivity and stereotyped behaviors (Amminger et al. 2006) and enhancing attention and language skills (Richardson et al. 2005; James et al. 2011). In addition, B vitamins, vitamin D, and methyl donors such as folic acid contribute to key pathways in neurodevelopment and redox balance, with accumulating data supporting their utility in ASD symptom management (Schmidt et al. 2012; Saad et al. 2016).

9.3.1 Antioxidants

In the context of ASD, multiple studies have reported significantly reduced GSH levels and GSH:GSSG ratios in brain tissues, blood, and lymphoblastoid cells, supporting the hypothesis of redox imbalance as a core feature of the disorder. These findings are supported by data indicating decreased cysteine availability, the rate- limiting precursor for GSH synthesis, as well as increased exposure to environmental toxicants that may deplete glutathione reserves (Kern et al. 2011). Supplementation with multivitamins, folinic acid, and methylcobalamin has been shown to restore GSH levels and improve clinical parameters including social interaction and communication in children with ASD (Adams et al. 2011; Frye et al. 2013a, b). Biochemically, GSH exerts its antioxidant activity by directly scavenging ROS and reactive nitrogen species (RNS), and by modulating redox-sensitive signaling pathways such as NF-κB and AP-1, which control proinflammatory cytokine expression (Haddad 2005; Biswas 2009). In addition, GSH regulates gene expression, protein function, and cell signaling via thiol-based redox mechanisms (Morris et al. 2014; Aoyama et al. 2013). The depletion of GSH in ASD has also been linked to increased blood- brain barrier permeability and glial cell activation (Li et al. 2012; González-Fraguela et al. 2018). Further mechanistic insights indicate that GSH deficiency promotes neuroinflammation through upregulation of p38 MAPK, JNK, and STAT3 signaling, leading to elevated levels of TNF-α, IL-1β, and IL-6 (Lee et al. 2010; Gavillet et al. 2008). Additionally, GSH catabolism via the γ-glutamyl cycle plays a role in modulating extracellular redox signaling, which in turn can influence immune cell activation and neurotoxicity (Anderson and Meister 1980; Meister 1974, 1991).

Beyond glutathione, other antioxidant molecules have been investigated in ASD. Coenzyme Q10 (CoQ10), a lipid-soluble mitochondrial antioxidant, supports ATP synthesis and reduces oxidative damage in neuronal tissues. Clinical studies suggest that CoQ10 supplementation can improve behavioral symptoms and redox markers in children with ASD (Guevara-Campos et al. 2015). Acetyl-L-carnitine (ALC) is another mitochondrial- targeted antioxidant that enhances fatty acid transport and neurotransmitter production. In individuals with ASD, especially those with mitochondrial dysfunction or Fragile X syndrome, ALC has been associated with improved social behavior and cognitive performance (Malaguarnera and Cauli 2019). In summary, antioxidant therapy—particularly with glutathione-targeted strategies, CoQ10, and ALC—represents a rational adjunct to conventional ASD management, aimed at restoring redox homeostasis and supporting cellular resilience in the face of chronic oxidative stress.

9.3.2 Polyphenols

Polyphenols are a diverse group of plant-derived compounds known for their potent antioxidant, anti-inflammatory, and neuroprotective effects (Behfar et al. 2025). These bioactive molecules modulate key cellular pathways that are commonly altered in ASD, including redox-sensitive signaling cascades (e.g., Nrf2 and NF-κB), mitochondrial function, neuroinflammation, and autophagy. Through these mechanisms, polyphenols contribute to the maintenance of neuronal integrity and immune homeostasis. The growing interest in polyphenols as neuronutritional agents stems from their ability to interact with multiple biological targets involved in the pathophysiology of ASD. Compounds such as sulforaphane, hydroxytyrosol, and boswellic acids, along with polyphenol-rich medicinal mushrooms like *Coriolus versicolor* and *Hericium erinaceus*, have shown promising effects in reducing oxidative stress, modulating cytokine production, and supporting neuronal resilience (Bernini et al. 2015).

9.3.2.1 Boswellia

Boswellia is a genus of resin-producing trees from which frankincense is derived. Among its species, *Boswellia sacra*, *Boswellia serrata*, and *Boswellia carterii* are known for their medicinal properties, particularly due to their content of boswellic acids, a group of pentacyclic triterpenoids with anti-inflammatory, antioxidant, and immunomodulatory activities. These compounds exert their effects by inhibiting key inflammatory enzymes such as 5-lipoxygenase (5-LO), cyclooxygenase-2 (COX-2), and matrix metalloproteinases (e.g., MMP-9), and by reducing proinflammatory cytokines including TNF- α, IL-1β, and IL-6 (Interdonato et al. 2024). Preclinical studies suggest that Boswellia resin extracts can modulate autophagy and apoptosis, two processes implicated in neuroinflammation and neurodegeneration. In a murine model of traumatic brain injury (TBI), oral administration of *Boswellia sacra* gum resin restored antioxidant enzyme levels such as SOD and GSH-Px, reduced lipid peroxidation, and attenuated the activation of apoptotic markers including caspase-3, caspase-8, and cytochrome c (Interdonato et al. 2024). Although direct clinical evidence in ASD is currently lacking, the overlap between the mechanisms of Boswellia and the core pathophysiological features of ASD—namely oxidative stress, mitochondrial dysfunction, and chronic inflammation—provides a strong rationale for its consideration as a neuronutritional agent. Additionally, Boswellia is a component of traditional multi-herbal formulations such as HLXL (Huo-Lou-Xiao-Lin Dan), which has shown efficacy in modulating NF-κB signaling and reducing inflammatory cytokines in animal models (Yang et al. 2011). From a broader perspective, *Boswellia sacra* and related species have been studied for their hepatoprotective, analgesic, and anti-cancer effects, and are valued in both Ayurvedic and Traditional Chinese Medicine for their systemic anti-inflammatory properties. These characteristics make Boswellia a particularly promising candidate for further investigation in neurodevelopmental disorders such as ASD.

9.3.2.2 Medicinal Mushrooms: *Coriolus versicolor* and *Hericium erinaceus*

Edible and medicinal mushrooms are gaining increasing attention in the field of neuronutrition due to their unique combination of antioxidant, immunomodulatory, and neuroprotective properties. Two species, *Coriolus versicolor* (also known as *Trametes versicolor*) and *Hericium erinaceus* (commonly referred to as Lion's Mane), have demonstrated promising biological effects relevant to ASD, including redox regulation, cytokine modulation, and stimulation of neurotrophic factors. *Coriolus versicolor* is rich in β-glucans and other polysaccharides that stimulate the immune system by activating macrophages and natural killer cells. Traditionally used in East Asian medicine for cancer and immunodeficiency conditions, it is now being explored for its capacity to modulate the vitagene network and activate Nrf2–Keap1 signaling pathways (Popiolek-Barczyk et al. 2024; Interdonato et al. 2023).

These mechanisms are of particular interest in ASD, where immune dysregulation and oxidative stress are prevalent. *Hericium erinaceus* has been shown to stimulate the production of nerve growth factor (NGF), a neurotrophin essential for neuronal survival, plasticity, and myelination. Experimental studies indicate that this mushroom supports neurogenesis, enhances mitochondrial function, and exerts anti-inflammatory effects by suppressing proinflammatory cytokines and oxidative damage (Bell et al. 2022; Chang et al. 2015). These effects align with pathological features observed in ASD, particularly impaired synaptic development and neuroinflammation. Both species also contain phenolic compounds and triterpenoids that further contribute to their antioxidant capacity (Li et al. 2021; Amara et al. 2020). Although clinical studies in ASD are still lacking, the strong preclinical rationale, combined with extensive use in traditional medicine, supports their inclusion in future investigations exploring diet-based interventions for neurodevelopmental disorders.

9.3.3 Vitamins

Vitamins are essential micronutrients required for a wide range of metabolic and neurobiological processes, including enzymatic reactions, DNA synthesis and repair, neurotransmitter metabolism, redox regulation, and immune function. Among these, the B- complex vitamins and vitamin D have received significant attention in the context of ASD, given their involvement in pathways related to oxidative stress, mitochondrial function, methylation, and neuroinflammation—all of which are implicated in the disorder's pathophysiology (Morris et al. 2019; Rossignol and Frye 2014). B vitamins, in particular B6, B9 (folate), and B12 (cobalamin), are crucial cofactors in the homocysteine–methionine cycle and are required for one-carbon metabolism and cellular methylation reactions. Aberrant methylation and elevated homocysteine levels have been observed in individuals with ASD, suggesting a functional deficiency in these vitamins despite normal serum concentrations (James et al. 2004; Frye et al. 2013a, b). Clinical studies have demonstrated that supplementation with folinic acid and methyl-B12 may restore redox and methylation capacity, with reported improvements in verbal communication and adaptive behavior (Frye et al. 2013a, b). Vitamin D, on the other hand, plays a dual role as a prohormone and an immunomodulatory agent. It regulates calcium homeostasis and bone development but also influences neurodevelopmental and immune pathways via the vitamin D receptor (VDR) expressed in neurons and glial cells (Christakos et al. 2016). Hypovitaminosis D is frequently reported in individuals with ASD, and meta-analyses confirm a significant reduction of serum vitamin D levels compared to healthy controls. Genetic polymorphisms in the VDR gene may further contribute to altered vitamin D metabolism in ASD populations (Schmidt et al. 2015).

9.4 Discussion and Future Perspectives

Nutritional interventions in ASD are increasingly framed within the paradigm of targeted, mechanism-based supplementation. Rather than adopting a one-size-fits-all model, the efficacy of each compound hinges on its alignment with the individual's pathophysiological profile—particularly in relation to oxidative stress, mitochondrial dysfunction, and neuroinflammation. Among the reviewed agents, SFN emerges as one of the most promising candidates. Acting on both the Nrf2 and NF-κB pathways, SFN enhances antioxidant defenses, modulates inflammation, and supports mitochondrial resilience. Clinical studies have shown encouraging improvements in behavior and communication, justifying its cautious recommendation, especially in patients with biomarkers of redox imbalance. In contrast, GSH, despite being central to cellular redox homeostasis, suffers from limited clinical utility due to low oral bioavailability and inconsistent outcomes (González-Fraguela et al. 2018). However, its precursors, particularly NAC and S- acetylglutathione, show greater promise. These compounds can restore GSH levels and improve oxidative balance, with NAC additionally modulating glutamatergic transmission and inflammation mechanisms particularly relevant in ASD (Bahji et al. 2025). CoQ10, a key player in mitochondrial electron transport and antioxidant defense, has shown beneficial effects in small-scale studies. Its supplementation has been associated with improved behavioral symptoms and biochemical parameters in individuals with mitochondrial dysfunction (Guevara-Campos et al. 2015). ALC also merits attention as a neuroprotective agent, particularly in ASD subgroups with comorbidities such as Fragile X syndrome (Malaguarnera and Cauli 2019). Boswellia extracts and medicinal mushrooms like *Coriolus versicolor* and *Hericium erinaceus* offer compelling preclinical evidence for anti-inflammatory and neuroprotective effects (Trovato Salinaro et al. 2018; Interdonato et al. 2024). However, they lack sufficient standardization, pharmacokinetic data, and human trials to support broader use. These compounds remain promising experimental options, suitable for pilot studies or compassionate use in selected cases. Similarly, hydroxytyrosol demonstrates strong mechanistic rationale for antioxidant and anti-inflammatory activity (Bernini et al. 2015), but awaits clinical validation in ASD. Vitamins, particularly B6, B9 (folate), B12, and D, play indispensable roles in one-carbon metabolism, neurotransmitter synthesis, and immune modulation. Supplementation with methyl donors has yielded clinical benefits in adaptive behavior and verbal communication in ASD (Frye et al. 2013a, b; Saad et al. 2016), especially when guided by metabolic profiling and genetic polymorphisms such as MTHFR (Schmidt et al. 2012; Zwierz et al. 2025). Notably, hyperhomocysteinemia and folate imbalance have also been implicated in vascular and metabolic complications beyond ASD, as shown in diabetes and hepatic cancers (Malaguarnera et al. 2014; Malaguarnera et al. 2015; Malaguarnera et al. 2020), reinforcing the systemic relevance of methylation status.Altogether, this evidence points to a necessary transition toward precision neuronutrition in ASD: interventions must be guided by individualized biological data, not solely by diagnostic category. Combinatorial strategies that integrate antioxidants, mitochondrial, and anti- inflammatory support may be most effective, but must be validated in rigorously controlled, biomarker-informed trials. Only through this integrative, personalized lens can nutritional interventions be meaningfully embedded in the clinical care of ASD. Despite promising signals, current evidence remains limited by significant methodological constraints, including small sample sizes, heterogeneity in clinical endpoints, and lack of standardized biomarkers for patient stratification. Many studies also suffer from a lack of replication and short intervention periods, hindering the generalizability of results. Additionally, variability in supplement formulations, dosages, and outcome measures complicates meta-analysis and guideline development. There is a need for standardization in biomarker selection and interpretation to identify responders, while adverse effects, although rare, are underreported and require more rigorous documentation. Until more robust evidence becomes available, nutritional interventions in ASD should be used as adjunctive tools, integrated into broader multidisciplinary care. Supplementation should be based on documented deficiencies or metabolic impairments, and closely monitored for efficacy and safety. Clinicians should be particularly cautious with high-dose regimens or combinatorial protocols, favoring individualized approaches supported by laboratory assessments and clinical expertise. The advent of induced pluripotent stem cell (iPSC) technology has revolutionized the modeling of neurodevelopmental disorders such as ASD. By enabling the generation of patient-specific neural cells and three-dimensional (3D) cerebral organoids, iPSC-based systems allow researchers to study early brain development in a human-relevant context (Costamagna et al. 2021; Lancaster et al. 2013). Unlike conventional models, these platforms capture the cellular diversity and developmental trajectories of the human brain, providing unprecedented insight into disease mechanisms and therapeutic responses. Over the past decade, iPSC-derived brain models have been effectively utilized to investigate human neurodevelopment and a range of neurodevelopmental disorders (Li et al. 2017; Lancaster et al. 2013; Pașca et al. 2019), as well as to study the effects of prenatal toxicant exposure (Qiao et al. 2018; Meng et al. 2022; Zhong et al. 2020) and gene–environment interactions relevant to early brain development (Modafferi et al. 2021; Umans and Gilad 2024). iPSC-derived brain models hold increasing potential for evaluating natural and dietary polyphenolic compounds, which are gaining attention for their neuroprotective, antioxidant, and anti-inflammatory properties. In particular, neural progenitor cells (NPCs) derived from iPSCs constitute a physiologically relevant in vitro model for the evaluation of bioactive natural compounds, owing to their dependence on oxidative phosphorylation and heightened susceptibility to mitochondrial dysfunction—pathophysiological features frequently observed in ASD (Zink et al. 2021; Frye et al. 2024). In parallel, iPSC-derived cerebral organoids enable researchers to model more complex structural and functional aspects of ASD that are not captured in 2D culture systems (Wang et al. 2023). While multiple studies employing iPSC-based models have demonstrated the ability of pharmacological agents such as insulin-like growth factor 1 (IGF-1), gentamycin, and roscovitine to ameliorate neuronal phenotypes associated with both idiopathic and syndromic forms of ASD (Rossetti et al. 2019; Marchetto et al. 2017; Griesi-Oliveira et al. 2015; Williams et al. 2014), the application of natural compounds within these experimental frameworks remains relatively under-investigated, despite their mechanistic relevance and therapeutic potential. Given their pleiotropic mechanisms and generally favorable safety profiles, polyphenols represent a promising avenue for developing targeted, mechanism-informed therapies (Osakabe and Terao 2018). iPSC-derived neural models provide an ideal platform to explore this potential, offering human-relevant and developmentally complex systems in which to assess both the efficacy of these compounds and their mechanisms of action. As such, they may serve as a critical translational tool for advancing natural compounds toward biologically informed interventions in neurodevelopmental disorders such as ASD.

While several nutraceuticals hold theoretical and early clinical promise, only a subset currently meets the threshold for cautious clinical use. Integrating evidence-based nutraceuticals—like mushrooms, Boswellia and CoQ10—within individualized therapeutic frameworks represents the most grounded path forward. Precision neuronutrition holds the potential to evolve from a supportive measure to a central strategy in ASD management—where food-derived bioactive compounds contribute not only to metabolic and neuroimmune balance but also to the promotion of cognitive resilience, emotional regulation, and healthy longevity across the lifespan.

9.5 Conclusion

Neuronutrition represents a promising frontier in the pursuit of brain health, cognitive resilience, and healthy aging. By targeting key molecular pathways involved in neuroinflammation, oxidative stress, mitochondrial function, and gut–brain communication, specific bioactive nutrients offer the potential to modulate neurological function across the lifespan. The case of ASD illustrates how precision neuronutritional interventions such as sulforaphane, glutathione, B-vitamins, and medicinal mushrooms can support not only neurodevelopmental outcomes but also broader systemic regulation. These findings highlight the translational relevance of neuronutrition beyond early-life disorders, extending into strategies for preserving cognitive function and promoting longevity in aging populations. As evidence grows, the integration of scientifically validated neuronutritional approaches into personalized healthcare may become a cornerstone of preventive neurology and lifespan medicine.

Acknowledgments This research has been conducted with funds from "PIACERI Ricerca di Ateneo 2024/2026, Linea Intervento 1" and from MIUR (PRIN 2022FWB4E).

References

Adams JB, Audhya T, McDonough-Means S, Rubin RA, Quig D, Geis E, Gehn E, Loresto M, Mitchell J, Atwood S (2011) Effect of a vitamin/mineral supplement on children and adults with autism. BMC Pediatr 11:111. https://doi.org/10.1186/1471-2431-11-111

Amara I, Ben Slima S, Ben Saad H, Riahi-Chebbi I, Badri M, Boudawara T (2020) Antioxidant and anti-inflammatory effects of mushroom extracts: therapeutic perspectives. Biomed Pharmacother 129:110419. https://doi.org/10.1016/j.biopha.2020.110419

Amminger GP, Berger GE, Schafer MR, Klier C, Friedrich MH, Feucht M (2006) Omega-3 fatty acids supplementation in children with autism: A double-blind randomized, placebo-controlled pilot study. Biol Psychiatry 61(4):551–553. https://doi.org/10.1016/j.biopsych.2006.01.005

Anderson ME, Meister A (1980) Transport and direct utilization of gamma-glutamylcyst(e)ine for glutathione synthesis. Proc Natl Acad Sci 77(1):217–221. https://doi.org/10.1073/pnas.77.1.217

Aoyama K, Watabe M, Nakaki T (2013) Redox regulation of neuronal survival through glutathione peroxidase and glutathione. Neurochem Int 62(5):469–475. https://doi.org/10.1016/j.neuint.2012.12.012

Arendash GW (2012) Transcranial electromagnetic treatment against Alzheimer's disease: why it has the potential to trump Alzheimer's disease drug development. J Alzheimer's Dis 32(2): 243–266. https://doi.org/10.3233/JAD-2012-120943

Badaeva G, Shumskaya E, Novikova N, Kim I, Klochkova O, Deryabin D (2023) Nutritional psychiatry and neuroinflammation: linking diet to mental health through the gut-brain axis. Int J Mol Sci 24(4):3563. https://doi.org/10.3390/ijms24043563

Bahji A, Forth E, Nasar A, Waqas A, Hawken ER, Ayub M (2025) Navigating agitation in neurodevelopmental disorders: A comparative study of pharmacotherapies via network meta-analysis in children and adults with autism spectrum disorder or intellectual disabilities. J Psychopharmacol 39(3):201–213

Behfar Q, Zeinali S, Mashayekhy E (2025) Role of antioxidants in the management and treatment of autism spectrum: a review of pathways. Neurol Lett 4(1):1–6

Bell V, Valente IM, Gil-Izquierdo A, Rocha CM (2022) Neuroprotective potential of Hericium erinaceus mycelium in aging-related cognitive decline: A systematic review. Front Aging Neurosci 14:878789. https://doi.org/10.3389/fnagi.2022.878789

Bernini R, Merendino N, Romani A, Velotti F (2015) Naturally occurring hydroxytyrosol: synthesis and applications. Food Chem 171:9–19. https://doi.org/10.1016/j.foodchem.2014.08.003

Biswas SK (2009) Does the interdependence between oxidative stress and inflammation explain the antioxidant paradox? Oxidative Med Cell Longev 2(6):319–327. https://doi.org/10.4161/oxim.2.6.9945

Bojang KP, Manchana V (2023) Nutrition and healthy aging: a review. Curr Nutr Rep 3:369–375. https://doi.org/10.1007/s13668-023-00473-0

Boxenbaum H, Neafsey PJ, Fournier DJ (1988) Hormesis, Gompertz functions, and risk assessment. Drug Metab Rev 19(2):195–229. https://doi.org/10.3109/03602538809049623

Byrne K, Zheng S, Bishop S, Boucher J, Ghods S, Kim SH, Lord C (2022) Behavioral responses to fevers and other medical events in children with and without ASD. Autism Res 15(11): 2056–2063. https://doi.org/10.1002/aur.2810. Epub 2022 Sep 26. PMID: 36164255

Calabrese EJ (2014) Hormesis: a fundamental concept in biology. Microb Cell 1(5):145–149. https://doi.org/10.15698/mic2014.05.145. PMID: 28357236; PMCID: PMC5354598

Calabrese EJ, Mattson MP (2013) Hormesis: a fundamental concept with widespread biological and biomedical applications. Environ Toxicol Chem 32(3):499–506. https://doi.org/10.1002/etc.2128

Calabrese EJ, Mattson MP (2017) The role of hormesis in the functional performance and protection of neural systems. Brain Circ 3(1):1–13. https://doi.org/10.4103/2394-8108.203257

Calabrese V, Cornelius C, Stella AM, Calabrese EJ (2010a) Cellular stress responses, mitostress and carnitine insufficiencies as critical determinants in aging and neurodegenerative disorders: role of hormesis and vitagenes. Neurochem Res 35(12):1880–1915. https://doi.org/10.1007/s11064-010-0307-z

Calabrese V, Cornelius C, Dinkova-Kostova AT, Iavicoli I, Di Paola R, Koverech G, Cuzzocrea S, Rizzarelli E, Calabrese EJ (2010b) Cellular stress responses, hormetic phytochemicals and vitagenes in aging and longevity. Biochim Biophys Acta (BBA) – Mol Basis Dis 1822(5): 753–783. https://doi.org/10.1016/j.bbadis.2011.09.002

Calabrese V, Cornelius C, Dinkova-Kostova AT, Calabrese EJ, Rizzarelli E (2016a) Cellular stress responses and hormetic phytochemicals: novel targets for neurodegenerative diseases. Antioxid Redox Signal 25(5):261–308. https://doi.org/10.1089/ars.2016.6757

Calabrese V, Giordano J, Signorile A, Ontario ML, Castorina S, De Pasquale C et al (2016b) Major pathogenic mechanisms in vascular dementia: roles of cellular stress response and hormesis in neuroprotection. J Neurosci Res 94(12):1588–1603. https://doi.org/10.1002/jnr.23925

Calabrese V, Santoro A, Trovato Salinaro A, Modafferi S, Scuto M, Albouchi F et al (2018) Hormetic approaches to the treatment of Parkinson's disease: perspectives and possibilities. J Neurosci Res 96(10):1641–1662. https://doi.org/10.1002/jnr.24244

Calabrese EJ, Osakabe N, Di Paola R, Siracusa R, Fusco R, D'Amico R, Impellizzeri D, Cuzzocrea S, Fritsch T, Abdelhameed AS, Wenzel U, Franceschi C, Calabrese V (2023) Hormesis defines the limits of lifespan. Ageing Res Rev 91:102074. https://doi.org/10.1016/j.arr.2023.102074. Epub 2023 Sep 13

Calabrese EJ, Nascarella M, Pressman P, Hayes AW, Dhawan G, Kapoor R, Calabrese V, Agathokleous E (2024a) Hormesis determines lifespan. Ageing Res Rev 94:102181. https://doi.org/10.1016/j.arr.2023.102181. Epub 2024 Jan 3

Calabrese V, Wenzel U, Piccoli T, Jacob UM, Nicolosi L, Fazzolari G, Failla G, Fritsch T, Osakabe N, Calabrese EJ (2024b) Investigating hormesis, aging, and neurodegeneration: from bench to clinics. Open Med (Wars) 19(1):20240986. https://doi.org/10.1515/med-2024-0986. Erratum in: Open Med (Wars) 2024 Dec 19;19(1):20242000. 10.1515/med-2024-2000. PMID: 38911254; PMCID: PMC11193355

Chang HC, Hsieh CL, Chen YS, Lin CH, Lin JG (2015) Protective effects of Hericium erinaceus mycelium on cognitive impairment and oxidative stress in an Alzheimer's disease model. Evid Based Complement Alternat Med 2015:613063. https://doi.org/10.1155/2015/613063

Christakos S, Dhawan P, Verstuyf A, Verlinden L, Carmeliet G (2016) Vitamin D: metabolism, molecular mechanism of action, and pleiotropic effects. Physiol Rev 96(1):365–408. https://doi.org/10.1152/physrev.00014.2015

Cordaro M, Siracusa R, Impellizzeri D, D'Amico R, Fusco R, Cuzzocrea S, Esposito E (2021) Neuroprotective effects of BDNF in a mouse model of traumatic brain injury. Biomedicine 9(1): 1–17. https://doi.org/10.3390/biomedicines9010032

Costamagna G, Comi GP, Corti S (2021) Advancing drug discovery for neurological disorders using iPSC-derived neural organoids. Int J Mol Sci 22(5):2659. https://doi.org/10.3390/ijms22052659

Duda-Chodak A, Tarko T, Satora P, Sroka P (2015) Interaction of dietary compounds, especially polyphenols, with the intestinal microbiota: a review. Eur J Nutr 54:325–341. https://doi.org/10.1007/s00394-015-0852-y

Eggler AL, Savinov SN (2013) Chemical and biological mechanisms of phytochemical activation of Nrf2 and importance in disease prevention. In: Recent Adv Phytochem, vol 43, pp 121–155. https://doi.org/10.1007/978-3-319-00581-2_7. PMID: 26855455; PMCID: PMC4739799

Frye RE, Delatorre R, Taylor H, Slattery J, Melnyk S, James SJ (2013a) Redox metabolism abnormalities in autistic children: a potential biomarker for treatment. Mol Psychiatry 18(8): 821–828. https://doi.org/10.1038/mp.2012.56

Frye RE, Sequeira JM, Quadros EV, James SJ, Rossignol DA (2013b) Effectiveness of methylcobalamin and folinic acid treatment on adaptive behavior in children with autistic disorder is related to glutathione redox status. Autism Res Treat 2013:609705. https://doi.org/10.1155/2013/609705

Frye RE, Rincon N, McCarty PJ, Brister D, Scheck AC, Rossignol DA (2024) Biomarkers of mitochondrial dysfunction in autism spectrum disorder: A systematic review and meta-analysis. Neurobiol Dis 197:106520. https://doi.org/10.1016/j.nbd.2024.106520

Gavillet M, Allaman I, Magistretti PJ (2008) Modulation of astrocytic glutathione metabolism by proinflammatory cytokines. Glia 56(14):1749–1762. https://doi.org/10.1002/glia.20736

González-Fraguela ME, Ramírez-Lorca R, Álvarez-Mercado AI, Fernández-Arche A, Barranco P, Martínez-Augustin O (2018) Glutathione in autism spectrum disorders: role of redox homeostasis and immune dysregulation. Antioxidants 7(4):85. https://doi.org/10.3390/antiox7040085

Griesi-Oliveira K, Acab A, Gupta AR, Sunaga DY, Chailangkarn T, Nicol X, Nunez Y, Walker MF, Murdoch JD, Sanders SJ et al (2015) Modeling non-syndromic autism and the impact of TRPC6 disruption in human neurons. Mol Psychiatry 20(11):1350–1365. https://doi.org/10.1038/mp.2014.141

Guevara-Campos J, González-Guevara L, Cauli O (2015) Autism spectrum disorders and mitochondrial dysfunction: a systematic review. Neurochem Res 40(10):1986–2000

Haddad JJ (2005) Redox and oxidant-mediated regulation of apoptosis signaling pathways: Immunopharmaco-redox conception of oxidative siege versus cell death commitment. Int Immunopharmacol 5(4):731–752. https://doi.org/10.1016/j.intimp.2004.10.028

Hayes JD, Dinkova-Kostova AT (2006) The Nrf2 regulatory network provides an interface between redox and intermediary metabolism. Trends Biochem Sci 31(1):36–43. https://doi.org/10.1016/j.tibs.2005.11.006

He CL, Tang Y, Wu JM, Long T, Yu L, Teng JF et al (2023) Chlorogenic acid delays the progression of Parkinson's disease via autophagy induction in Caenorhabditis elegans. Nutr Neurosci 26(1):11–24. https://doi.org/10.1080/1028415X.2021.2009993

Ibrahim L, Mesgarzadeh J, Xu I, Powers ET, Wiseman RL, Bollong MJ (2020) Defining the functional targets of Cap'n'collar transcription factors NRF1, NRF2, and NRF3. Antioxidants (Basel) 9(10):1025. https://doi.org/10.3390/antiox9101025. PMID: 33096892; PMCID: PMC7588902

Indika M, Wickramasinghe VP, Dissanayake VHW, Hapangama D (2023) Redox imbalance and mitochondrial dysfunction in autism spectrum disorder: A mechanistic insight into potential nutritional interventions. Nutr Neurosci 26(4):283–295. https://doi.org/10.1080/1028415X.2022.2089155

Interdonato L, Fusco R, Cordaro M, D'Amico R, Siracusa R, Impellizzeri D, Cuzzocrea S (2023) Coriolus versicolor supplementation modulates the Nrf2–Keap1 pathway in a neuroinflammation model. Nutrients 15(7):1430. https://doi.org/10.3390/nu15071430

Interdonato M, Currò M, Allegra A, Molino S, Caccamo D, Ientile R (2024) Clinical effectiveness of N-acetylcysteine and folinic acid in autism spectrum disorders. Neurochem Res 49(1):98–112

James SJ, Cutler P, Melnyk S, Jernigan S, Janak L, Gaylor DW, Neubrander JA (2004) Metabolic biomarkers of increased oxidative stress and impaired methylation capacity in children with autism. Am J Clin Nutr 80(6):1611–1617. https://doi.org/10.1093/ajcn/80.6.1611

James S, Montgomery P, Williams K (2011) Omega-3 fatty acids supplementation for autism spectrum disorders (ASD). Cochrane Database Syst Rev 11:CD007992. https://doi.org/10.1002/14651858.CD007992.pub2

Kern JK, Geier DA, Adams JB, Garver CR, Audhya T, Geier MR (2011) A clinical trial of glutathione supplementation in autism spectrum disorders. Med Sci Monit 17(12):CR677–CR682. https://doi.org/10.12659/MSM.882208

Lancaster MA, Renner M, Martin CA, Wenzel D, Bicknell LS, Hurles ME, Homfray T, Penninger JM, Jackson AP, Knoblich JA (2013) Cerebral organoids model human brain development and microcephaly. Nature 501(7467):373–379. https://doi.org/10.1038/nature12517

Lee HJ, Kang DG, Lee HS (2010) Effect of glutathione depletion on the inflammation-related genes in astrocytes stimulated with LPS. BMC Neurosci 11:121. https://doi.org/10.1186/1471-2202-11-121

Li Q, Chen J, Luo X, Hong Y (2012) Increased blood–brain barrier permeability in autism spectrum disorders. Neurosci Bull 28(6):829–838. https://doi.org/10.1007/s12264-012-1280-z

Li R, Sun L, Fang A, Li P, Wu Q, Wang X (2017) Recapitulating cortical development with organoid culture in vitro and modeling abnormal spindle-like (ASPM related primary) microcephaly disease. Protein Cell 8(11):823–833. https://doi.org/10.1007/s13238-017-0479-2

Li TJ, Li MY, Shen L, Zhang HB, Zuo T (2021) Bioactive components and potential health effects of medicinal mushrooms: a review. Int J Mol Sci 22(19):10362. https://doi.org/10.3390/ijms221910362

Liu P, Kerins MJ, Tian W, Neupane D, Zhang DD, Ooi A (2019) Differential and overlapping targets of the transcriptional regulators NRF1, NRF2, and NRF3 in human cells. J Biol Chem 294(48):18131–18149. https://doi.org/10.1074/jbc.RA119.009591. Epub 2019 Oct 18. PMID: 31628195; PMCID: PMC6885608

Malaguarnera M, Cauli O (2019) Effects of acetyl-L-carnitine in the treatment of fragile X syndrome: a randomized, double-blind, placebo-controlled trial. Nutrients 11(3):525. https://doi.org/10.3390/nu11030525

Malaguarnera G, Gagliano C, Giordano M, Salomone S, Vacante M, Bucolo C, Caraci F, Reibaldi M, Drago F, Avitabile T, Motta M (2014) Homocysteine serum levels in diabetic patients with non proliferative, proliferative and without retinopathy. Biomed Res Int 2014: 191497. https://doi.org/10.1155/2014/191497. Epub 2014 Apr 28. PMID: 24877066; PMCID: PMC4022262

Malaguarnera G, Gagliano C, Salomone S, Giordano M, Bucolo C, Pappalardo A, Drago F, Caraci F, Avitabile T, Motta M (2015) Folate status in type 2 diabetic patients with and without retinopathy. Clin Ophthalmol 9:1437–1442. https://doi.org/10.2147/OPTH.S77538. PMID: 26300625; PMCID: PMC4536839

Malaguarnera G, Catania VE, Latteri S, Borzì AM, Bertino G, Madeddu R, Drago F, Malaguarnera M (2020) Folate levels in hepatocellular carcinoma patients with portal vein thrombosis. BMC Gastroenterol 20(1):375. https://doi.org/10.1186/s12876-020-01525-3. PMID: 33172390; PMCID: PMC7653717

Marchetto MC, Belinson H, Tian Y, Freitas BC, Fu C, Vadodaria K, Beltrao-Braga P, Trujillo CA, Mendes APD, Padmanabhan K et al (2017) Altered proliferation and networks in neural cells derived from idiopathic autistic individuals. Mol Psychiatry 22(6):820–835. https://doi.org/10.1038/mp.2016.95

Masoro EJ (2001) Physiology of aging. Int J Sport Nutr Exerc Metab 11(Suppl):S218–S222. https://doi.org/10.1123/ijsnem.11.s1.s218

Meister A (1974) Glutathione, metabolism and function via the gamma-glutamyl cycle. Biochem Soc Trans 2(3):253–256. https://doi.org/10.1042/bst0020253

Meister A (1991) Glutathione metabolism and its selective modification. J Biol Chem 266(26): 17205–17208. https://doi.org/10.1016/S0021-9258(19)49963-4

Meng Q, Zhang W, Wang X et al (2022) Human forebrain organoids reveal connections between valproic acid exposure and autism risk. Transl Psychiatry 12:130. https://doi.org/10.1038/s41398-022-01898-x

Modafferi S, Zhong X, Kleensang A, Murata Y, Fagiani F, Pamies D, Hogberg HT, Calabrese V, Lachman H, Hartung T, Smirnova L (2021) Gene-environment interactions in developmental neurotoxicity: a case study of synergy between chlorpyrifos and CHD8 knockout in human BrainSpheres. Environ Health Perspect 129(7):077001. https://doi.org/10.1289/EHP8580

Morris G, Puri BK, Walker AJ (2014) The role of neuroglial activation in neuroinflammation and its therapeutic implications. Mol Neurobiol 49(3):1059–1076. https://doi.org/10.1007/s12035-013-8574-0

Morris G, Berk M, Carvalho AF, Maes M, Walker AJ, Puri BK (2019) The role of oxidative stress and antioxidants in autism spectrum disorder: a comprehensive review. Transl Psychiatry 9(1): 15. https://doi.org/10.1038/s41398-018-0340-2

Novikov VN, Badaeva AV, Danilov AB, Vorobyeva YD (2023) Precision Neuronutrition: personalized approaches for optimizing brain health. Biol Life Sci Forum 29:20. https://doi.org/10.3390/IECN2023-15524

Osakabe N, Terao J (2018) Possible mechanisms of postprandial physiological alterations following flavan 3-ol ingestion. Nutr Rev 76:174–186. https://doi.org/10.1093/nutrit/nux070

Osakabe N, Shimizu T, Fujii Y, Fushimi T, Calabrese V (2024) Sensory nutrition and bitterness and astringency of polyphenols. Biomolecules 14(2):234. https://doi.org/10.3390/biom14020234

Park S, Park SK (2022) Anti-oxidant and anti-aging effects of phlorizin are mediated by DAF-16-induced stress response and autophagy in Caenorhabditis elegans. Antioxidants 11(10):1996. https://doi.org/10.3390/antiox11101996

Paşca AM, Park JY, Shin HW, Qi Q, Revah O, Krasnoff R, O'Hara R, Willsey AJ, Palmer TD, Paşca SP (2019) Human 3D cellular model of hypoxic brain injury of prematurity. Nat Med 25(5):784–791. https://doi.org/10.1038/s41591-019-0436-0

Popiolek-Barczyk K, Michalak A, Wawrzczak-Bargiela A, Szydlowska K, Makuch W, Przewlocka B, Mika J (2024) Anti-inflammatory and antioxidant effects of Trametes versicolor in neurological models of chronic stress and pain. J Neuroinflammation 21(1):112. https://doi.org/10.1186/s12974-024-03322-y

Qiao H, Zhang YS, Chen P (2018) Commentary: human brain organoid-on-a-chip to model prenatal nicotine exposure. Front Bioeng Biotechnol 6:138. https://doi.org/10.3389/fbioe.2018.00138

Richardson AJ, Montgomery P, Hall JA (2005) The Oxford-Durham study: a randomized controlled trial of dietary supplementation with fatty acids in children with developmental coordination disorder. Pediatrics 115(5):1360–1366. https://doi.org/10.1542/peds.2004-2167

Romero-Márquez JM, Navarro-Hortal MD, Jiménez-Trigo V, MuñozOllero P, Forbes-Hernández TY, Esteban-Muñoz A et al (2022) An olivederived extract 20% rich in hydroxytyrosol prevents beta-amyloid aggregation and oxidative stress, two features of Alzheimer disease, via SKN-1/NRF2 and HSP-16.2 in Caenorhabditis elegans. Antioxid (Basel) 11(4):629. https://doi.org/10.3390/antiox11040629

Rossetti AC, Koch P, Ladewig J (2019) Drug discovery in psychopharmacology: from 2D models to cerebral organoids. Dialogues Clin Neurosci 21(2):203–224. https://doi.org/10.31887/DCNS.2019.21.2/jladewig

Rossignol DA, Frye RE (2014) Evidence linking oxidative stress, mitochondrial dysfunction, and inflammation in the brain of individuals with autism. Front Physiol 5:150. https://doi.org/10.3389/fphys.2014.00150

Saad K, Abdel-Rahman AA, Elserogy YM, Al-Atram AA, Massoud MA, Al-Ayadhi LY, Bjørklund G (2016) Vitamin B12 supplementation in children with autism spectrum disorder: a randomized controlled trial. Nutrients 8(7):386. https://doi.org/10.3390/nu8070386

Schmidt RJ, Tancredi DJ, Ozonoff S, Hansen RL, Hartiala J, Allayee H, Schmidt LC, Tassone F, Hertz-Picciotto I (2012) Maternal periconceptional folic acid intake and risk of autism spectrum disorders and developmental delay in the CHARGE case-control study. Am J Clin Nutr 96(1):80–89. https://doi.org/10.3945/ajcn.110.004416

Schmidt RJ, Hansen RL, Hartiala J, Allayee H, Schmidt LC, Tancredi DJ, Tassone F, Hertz-Picciotto I (2015) Prenatal vitamins, one-carbon metabolism gene variants, and risk for autism. Epidemiology 22(4):476–485. https://doi.org/10.1097/EDE.0b013e31821d0e30

Sivamaruthi BS, Suganthy N, Kesika P, Chaiyasut C (2020) The role of microbiome, dietary supplements, and probiotics in autism spectrum disorder. Int J Environ Res Public Health 17:2647. https://doi.org/10.3390/ijerph17082647

Thiruvengadam M, Venkidasamy B, Subramanian U, Samynathan R, Ali Shariati M, Rebezov M, Girish S, Thangavel S, Dhanapal AR, Fedoseeva N, Lee J, Chung IM (2021) Bioactive compounds in Oxidative stress-mediated diseases: targeting the NRF2/ARE signaling pathway and epigenetic regulation. Antioxidants (Basel) 10(12):1859. https://doi.org/10.3390/antiox10121859. PMID: 34942962; PMCID: PMC8698417

Trovato Salinaro A, Pennisi M, Di Paola R, Scuto M, Crupi R, Cambria MT et al (2018) Neuroinflammation and neurohormesis in the pathogenesis of Alzheimer's disease and Alzheimer-linked pathologies: modulation by nutritional mushrooms. Immun Ageing 15:8. https://doi.org/10.1186/s12979-017-0108-1

Umans BD, Gilad Y (2024) Oxygen-induced stress reveals context-specific gene regulatory effects in human brain organoids. bioRxiv. https://doi.org/10.1101/2024.09.03.611030

Wang L, Owusu-Hammond C, Sievert D, Gleeson JG (2023) Stem cell-based organoid models of neurodevelopmental disorders. Biol Psychiatry 93(7):622–631. https://doi.org/10.1016/j.biopsych.2023.01.012

Williams EC, Zhong X, Mohamed A, Li R, Liu Y, Dong Q, Ananiev GE, Mok JC, Lin BR, Lu J et al (2014) Mutant astrocytes differentiated from Rett syndrome patient-specific iPSCs have adverse effects on wild-type neurons. Hum Mol Genet 23(11):2968–2980. https://doi.org/10.1093/hmg/ddu008

Yang Y, Zhang J, Liu X, Xu M, Lin W, Zhang S, Sun Q, Liu W (2011) Effect of Boswellia serrata extract on cognitive impairment induced by diabetes in rats. Phytother Res 25(7):1035–1040

Zhong X, Harris G, Smirnova L, Zufferey V, Sá RCDSE, Baldino Russo F, Baleeiro Beltrao Braga PC, Chesnut M, Zurich MG, Hogberg HT, Hartung T, Pamies D (2020) Antidepressant paroxetine exerts developmental neurotoxicity in an iPSC-derived 3D human brain model. Front Cell Neurosci 14:25. https://doi.org/10.3389/fncel.2020.00025

Zink A, Lisowski P, Prigione A (2021) Generation of human iPSC-derived neural progenitor cells (NPCs) as drug discovery model for neurological and mitochondrial disorders. Bio-Protocol 11(5):e3939. https://doi.org/10.21769/BioProtoc.3939

Zwierz M, Suprunowicz M, Mrozek K, Pietruszkiewicz J, Oracz AJ, Konarzewska B, Waszkiewicz N (2025) Vitamin B12 and autism spectrum disorder: A review of current evidence. Nutrients 17(7):1220. https://doi.org/10.3390/nu17071220

Chapter 10
Ayurvedic Supplements for Healthy Brain and Longevity

Himanshi Yadav, Anurag Thapliyal, Shashank Kumar Maurya, and Rajnikant Mishra

Abstract Aging or *Jara*-related changes affect almost all organisms. However, possibilities of long healthy life, age-reversal, or even immortality have either been postulated or experimentally observed based on changes in the molecular cascades and hierarchy from the cell cycle to cell death during development and after birth. Expression of some proteins is autoregulated, and many are multifunctional, like Paired box (Pax) proteins, tumor necrosis factors, interferons, cytokines, chemokines, trophic- or transforming growth factors, and p53. They function as regulator of other genes and proteins that are required for cellular growth, maturation, differentiation, and the creation of cellular identity. This chapter explains the literature and experiment-based importance of available natural and commercial supplements, their bioactive compounds, metabolites, chemical scaffolds, and ayurvedic formulations like *Chyawanprash*, *Triphala Rasayan*, *Brahmi Ghrita*, *Ashwagandharishta*, *Saraswatarishta*, *Amalaki Rasayana*, and *Medhya Rasayana*, as a safe alternative source for a healthy brain and longevity. These compounds and formulations outdo the single-component therapies, offering multi-target and systems-level effects on the management or therapy of stress and inflammation, dysbiosis, synaptic plasticity, repair of genetic components, energy homeostasis, and geriatric care. A comprehensive understanding on a reactive disease management model to a proactive approach would be helpful in the preservation of vitality, healthy systems biology and longevity.

Keywords Aging · Longevity · Ayurvedic formulations · Brain · Pax genes · Microglia · Chyawanprash · Triphala Rasayan · Brahmi Ghrita · Ashwagandharishta · Saraswatarishta · Amalaki Rasayana · Medhya Rasayana

---

H. Yadav · A. Thapliyal · S. K. Maurya
Biochemistry and Molecular Biology Laboratory, Department of Zoology, Faculty of Science, University of Delhi, Delhi, India
e-mail: smaurya1@zoology.du.ac.in

R. Mishra (✉)
Biochemistry and Molecular Biology Laboratory, Department of Zoology, Institute of Science, Banaras Hindu University, Varanasi, Uttar Pradesh, India
e-mail: rmishraa@bhu.ac.in

S. I. S. Rattan, C. Gezer (eds.), *Supplements and Drugs for Healthy Longevity*, Healthy Ageing and Longevity 22, https://doi.org/10.1007/978-3-032-11608-6_10

10.1 Introduction

Leading a healthy life requires respect for sustainable traditions, support, and considerations of biological scientists, modern and alternative medical, psychosocial, and spiritual practitioners. Ayurveda adopts a comprehensive methodology regarding preserving a state of equilibrium, namely, *dhatusamya*, encompassing normal anatomical, physiological, mental, and spiritual wellness. In the Ayurvedic paradigm, *Jara* or aging is conceptualized not as a pathological condition but rather as an intrinsic phenomenon similar to hunger, thirst, or the necessity for sleep. Jara is concomitant with the phenomenon of decay, which is evident through several degenerative alterations. These changes are inherent with time and nature, called Kalaja Vriddhavastha (Natural Aging), and generally perceived as unfavorable (Rao 2018).

In general, aging and age-related ailments are affected mainly by improper utilization of the five senses (*pancha tanmatras*) and discordant impressions of the five sensory organs (*pancha jnanendriyas*), and deleterious transformations of both body and mind (*Parinama*) (Ninivaggi 2010). The mental and physical ailments undermine *dhatusamya* (homeostasis) and result in *Akalaja vriddhavastha* (pathological aging). Consequently, an equilibrated state (*sama*) of tissue (*dhatu*), energy systems (*dosha*), transformative heat (a*gni*), and metabolic byproducts (*mala*) constitutes homeostasis in Ayurveda, which ultimately facilitates healthy aging (*Sukhayu/Kalaja Vriddhavastha*) (Garg et al. 2015). Age-dependent loss of physiological, metabolic, and reproductive capacity increases susceptibility to senescence and mortality. The chronological aging, explained as the precise age of an individual quantified in years, months, and days, is inexorable, immutable, and irreversible. However, physiological or biological aging is considered as an individual's maturation and alterations in specific cellular or molecular criteria. In contrast to chronological aging, biological aging possesses the potential to be reversed or postponed (Booth and Brunet 2016).

Within the framework of natural destruction (*Swabhavoparamavada*), *Charaka* elucidates that there exists a causative factor, and no definitive cause can be attributed to the termination of this existence, as death is perceived as a natural progression subsequent to birth (Dornala and Dornala 2016). The term "*Jara*" encompasses four distinct entities:

1. *Nityaga* signifies the perpetuation of consciousness.
2. *Dhari* refers to the elements inhibiting the onset of Jara/aging.
3. *Jeevitam* embodies the act of sustaining life, and
4. *Anubandha* pertains to the transmigration of the corporeal form (Datta et al. 2009).

Consequently, aging is modulated by various factors influencing *Shareera* (physical), *Bala/Ojas* (immunity), *Agni* (metabolism), *Indriya* (emotional), and *Satwa* (psychic level). Furthermore, *Sharira vriddhikara bhavas* (genotypic and phenotypic characteristics), *Parinama* (cellular transformation), and *Garbhahinivrittikara bhava* (developmental changes induced by pregnancy) play significant roles in shaping an individual's aging trajectory (Singh 2017).

Numerous, diverse, and frequently contradictory theories of aging exist. Many scholars observed or postulate a decline in stem cell proliferation across various tissues, the accumulation of harmful protein aggregates, free radicals, inflammation, compromised mitochondrial function, aggregation of mutations within DNA, defective DNA repair mechanisms, and telomere reduction with manifestations of aging (Aunan et al. 2016). To understand the impacts and mechanisms of sustainable supplements and drugs for health and longevity, the following aspects are being represented with respect to brain functions and mental health.

10.2 Age-Dependent Alterations in Functional and Chemical Anatomy of the Brain and Regulation of Regulators

Specialized central and peripheral nervous system cells lose their supportive roles, express and secrete pro-inflammatory cytokines, and promote excessive phagocytosis and neuronal damage (Fig. 10.1). Alterations in microenvironment, energy homeostasis, composition of genetic sequence elements, functional proteins, and

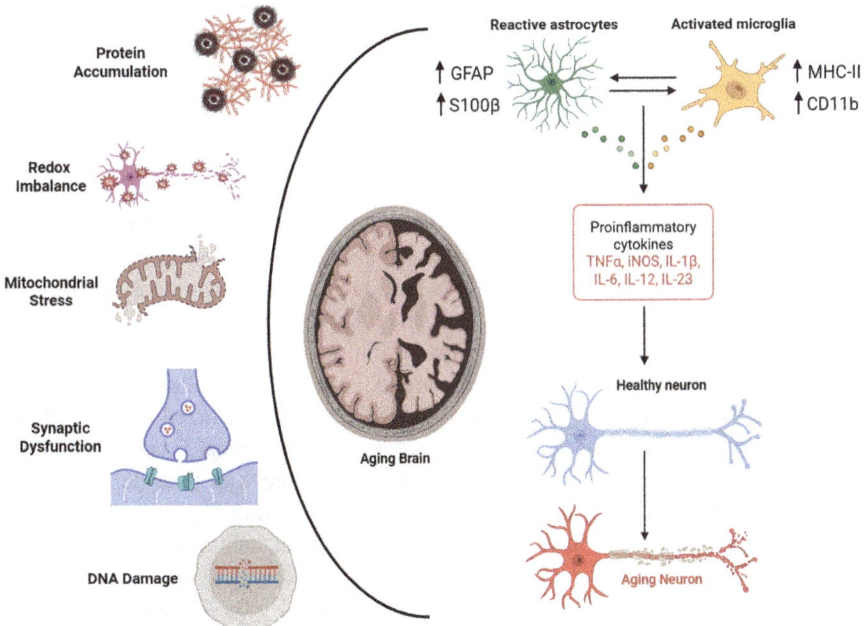

Fig. 10.1 Schematic representation of alterations at cellular and molecular levels during aging in the brain

oxidative stress management are either causes or effects for the neuropathy or neurodegenerative disorders like Major depressive disorder, Alzheimer's (AD), or Parkinson's disease (PD) (Mattson and Magnus 2006; Woodruff-Pak et al. 2010). Changes in the plasticity of cells of the nervous system, synaptic transmission, cytoskeletal frameworks, neurotrophic factors, cholinergic, serotonergic, dopaminergic, glutamatergic neurons, and transcriptional regulators or regulations of regulators have been explained during aging (Boisvert et al. 2018; Angelova and Brown 2019). Most of these pathologies are associated with the neuroinflammatory reactive astrocytes, glia, and senescent microglia (Antignano et al. 2023).

Microglia have a variety of phenotypes depending on their microenvironment. During aging and neurodegenerative disorders, microglia are activated. Disruption of the microenvironment of microglia, injury, or protein accumulation, such as amyloid beta and alpha-synuclein, leads to their activation. During prolonged insults, microglia lead to neuroinflammation, secrete pro-inflammatory cytokines such as tumor necrosis factor-alpha (TNFα), interferon-gamma (INFγ), reactive oxygen and nitrogen species (ROS & RNS), and the generation of oxidative stress due to dysfunctional mitochondria. Expression of Ionized calcium-binding adaptor molecule 1 (IBA1) is increased in activated microglia. Therefore, targeting IBA1 using natural compounds can act as a strategy to combat microglia-induced oxidative stress and neuroinflammation. The colony-stimulating factor 1 receptor (CSF1R) protein and cluster of differentiation 40 ligand (CD40L) are involved in the proliferation and induction of cytokine expression, respectively. Therefore, targeting these proteins using natural compounds might be beneficial for mitigating neuroinflammation. IBA1 increases in the aging brain (Lee et al. 2017). The expression profile of CSF1R, CD40L, and P2RY12 has not been well studied in the aging context, but their role in the microglia pro-inflammatory phenotype has been established, also indicating a theoretical involvement in aging (Kempthorne et al. 2020).

The neuroinflammatory reactive astrocytes contain higher levels of GFAP, calcium-binding protein S100β (Cotrina and Nedergaard 2002), and major histocompatibility complex II (MHC-II) and CD11b in microglia (Frank et al. 2007). The aged microglia also limit neurogenesis by producing pro-inflammatory mediators (McPherson et al. 2011). The IBA1 shows modulation in the distribution pattern of microglia in the aged mouse cortex compared to the young cortex (Harry and Kraft 2012). They also display retracted processes and perinuclear cytoplasm hypertrophy, reminiscent of activated microglia, and influence microglia-mediated immune physiology of the brain (Streit et al. 2004). However, morphological features, distribution, and number of microglial cells progressively change with time during aging, reflecting aging-associated alterations in microglia function (Spittau 2017). The genesis of neurofibrillary tangles (NFT), a pathological hallmark of AD, results from a shift in the expression of neurofilament proteins during aging (Youssef et al. 2016). ROS can activate the nuclear factor Kappa-B (NF-κB) (Reuter et al. 2010). Moreover, the decline of cellular defense mechanisms against oxidative stress has led to neuronal injury and degeneration (Colcombe and Kramer 2003). ROS builds up in mitochondria, producing numerous mitochondrial dysfunctions by damaging cytochrome IV and V complexes, leading to cell oxidative burden (Knott et al. 2008). Excessive ROS accumulation leads to inflammation, cancer, and neurodegeneration (Kawagishi and Finkel 2014). Several dysfunctions caused by ROS include enzyme oxidation, protein aggregation, and membrane rupture, contributing to neuronal death (Kennedy et al. 2012). Similarly, age-dependent regional differences, loss of the stem cells in the hypothalamus, reduction in Paired box (Pax), and co-expressed proteins have been evident (Tripathi and Mishra 2012; Boisvert et al. 2018; Maurya and Mishra 2017, 2018).

10.3 Paired Box (Pax) Proteins as a Regulator of Genes/Proteins During Aging

The mammalian Pax protein family comprises nine members, Pax1–Pax9, having critical roles in organogenesis, tissue regeneration (Li and Eccles 2012), genomic instability, epigenetic drift, mitochondrial dysfunction, stem cell depletion, modulation of oxidative stress, apoptosis, cellular senescence and responses to stress (Mishra 2020). They are required for cellular growth, maturation, differentiation, and the creation of cellular identity (Thompson et al. 2020). They are evolutionarily conserved among orthologs in worms, flies, frogs, fish, and birds (Vorobyov and Horst 2006).

The Pax6 has been proposed to control several downstream regulators as Pax6 knock-out mice fail to develop pituitary, pancreatic alpha-cells, eyes, and central nervous system (Glaser et al. 1994; Tuoc et al. 2009). The Pax2 and Pax8 are associated with renal development and pathophysiology (Zhang et al. 2012), etiology of infertility in both males and females due to the aberrant development and obstruction of reproductive structures (Mittag et al. 2007; Wistuba et al. 2007). Pax3 and Pax7 are crucial for the specification and maintenance of muscle satellite cells, and their age-dependent reduction led to sarcopenia (Buckingham and Relaix 2007; Brzeszczyńska et al. 2016). The symptoms of age-related infertility, cardiomyopathy, and cardiac dysplasia relate to the expression levels of Pax8 (Wu et al. 2016). Pax5 and Pax9 have been observed in immune senescence, and the activity of Pax5 to target genes in B-cells (Landin et al. 2011) is reduced. Pax5 and Pax6 express in neurons and microglia (Tripathi and Mishra 2012; Maurya and Mishra 2017, 2018). Levels of Pax5 is associated with several hematological malignancies, follicular and non-Hodgkin lymphoma (Zhang et al. 2002; Bharti et al. 2016), astrocytoma, medulloblastoma, neuroblastoma (Baumann Kubetzko et al. 2004; Stuart and Gruss 1995; Kozmik et al. 1995), autism spectrum disorder (Iossifov et al. 2012), bipolar disorder (O'Roak et al. 2014) and glioblastoma (Li et al. 2016). The Pax5 and IBA1 co-localize and interact in microglia (Maurya and Mishra 2018). The Pax6 gets expressed in the eye, pancreas, pineal gland, pituitary, and testis. Patients having mutations in Pax6 have symptoms of aging, including loss of synaptic plasticity and memory, mood fluctuation, mental retardation, anxiety, and depression (Tripathi and

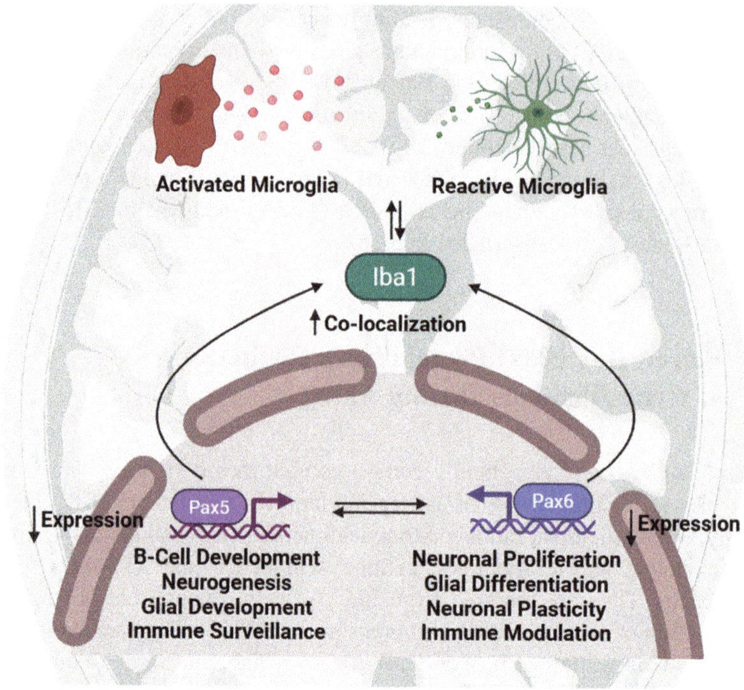

Fig. 10.2 Diagrammatic representation showing binding of Pax5 and Pax6 to promoter sequences of genes related to immunological response, neurogenesis, glial development and differentiation, and synaptic plasticity, and it co-localizes with Iba1 in the microglia of mice, which alter during aging

Mishra 2012). The pro-neurogenic potential of Pax6 has been observed in adult neural progenitors and determines neuronal subtypes in the developing and adult brain (Tuoc et al. 2009). Pax6-dependent neuronal and glial cell proliferation and differentiation are further substantiated by demonstrating molecular cross-talk of Pax6 with SPARC and p53 in the mouse brain. Modulation of Pax6 is suggested to affect the interaction with SPARC and p53, leading to defects in neural plasticity and the p53-mediated cell death pathway during aging (Tripathi and Mishra 2010). Besides aging Pax6 affects neurogenesis (Sisodiya et al. 2001; Tripathi and Mishra 2012). Data mining using the ImmuNet server (http://immunet.princeton.edu/about/) points towards a possible role of Pax6 in brain immunity via Toll-like receptor signaling, T-cell signaling, leukocyte trans-endothelial migration, complement pathway, coagulation cascades, hematopoietic cell lineage, cytosolic DNA sensing pathway, and intestinal immune network for IgA production (Maurya and Mishra 2022a). Also, Pax5 and Pax6 could be crucial in regulating microglia-mediated immunological surveillance during aging (Fig. 10.2).

10.4 Evidence of Interaction of Active Ingredients of Natural Compounds with microglia Proteins, Rasayana Formulations and Clinical Applications

Biodiversity offers us a wide range of bioactive metabolites with varying chemical scaffolds. Curcumin, flavonoids, alkaloids, resveratrol, cannabidiols, and ginsenosides have all been shown to improve cognitive function, reduce oxidative stress, alter inflammatory pathways, and regulate neuroinflammation in neurological illnesses. Considering the "one compound multiple target paradigm" for the treatment of neurological diseases, natural compounds have attracted attention as a safe alternative source (Chen et al. 2022) for healthy aging by providing anti-inflammatory, immunomodulatory, anti-oxidant, and neuroprotective qualities as well as affecting cellular and molecular senescence (Bjørklund et al. 2022).

The mechanisms of action of natural supplements are mostly unknown, but they hold the potential to interact with microglia and regulate the microglia-mediated progression of aging and associated disorders. However, the molecular interaction of IBA1 with curcumin, resveratrol, and cannabidiol indicates their potential impacts through the microglia-mediated redox imbalance during neuroinflammatory conditions (Maurya and Mishra 2022b). Curcumin and resveratrol interactions with CD40L and CSF1R indicate their effects on microglia-related functions and healthy aging. Besides the commonly used natural compounds, marine natural products from algae or seaweeds show the potential to have anti-oxidants, anti-inflammatory, and neuroprotective properties. Upregulation of the IBA1, CSF1R, and CD40L, along with Purinergic Receptor P2Y G-protein coupled 12 (P2RY12), is associated with driving the microglia-mediated neuroinflammation. P2RY12 is an ATP-ADP receptor that senses the ATP gradient formed during neuronal death and induces microglia to move toward the site of degeneration via chemotaxis. The stypoldione also interacts with IBA1 and CSF1R, epitaondiol can inhibit CD40L, and callophycin A can block P2RY12, limiting the proliferation, movement, chemotaxis, and cytokine secretion from microglia during its activation (Thapliyal et al. 2025).

10.5 Key Rasayana Formulations: Mechanisms, Evidence, and Clinical Applications

10.5.1 Chyawanprash

Chyawanprash (CP), originating from India (a polyherbal jam), a traditional Ayurvedic recipe and an ancient formulation, has two fundamental lexemes: "*Chyawan*" and "*Prasha.*" *Chyawan* denotes the name of a revered sage and signifies 'degenerative transformation.' *Prasha* pertains to a substance that is either medicinal or alimentary and is deemed safe for consumption. It was esteemed as one of the most favored consumables for its anti-aging attributes long before the advent of vitamins, minerals, and antioxidant supplements (Parle and Bansal 2006). It has been an integral aspect of the Indian lifestyle as a supplement, holistic 'metabolic' tonic, and medicinal preparation to bolster immunity, enhance overall health, and promote longevity, irrespective of varying societal, political, or scientific contexts. As per ayurvedic formulations, CP is categorized as *Awaleha* due to its textural consistency and method of administration (Kempthorne et al. 2020). Amla, enriched with vitamin C and polyphenols, a citrus fruit recognized as a potent botanical (Bhattacharya et al. 2000), is the principal ingredient along with *Chaturjata* (four aromatic plants), *Dashmula* (ten roots), and *Ashtavarga* (rare medicinal herbs indigenous to the Northwest Himalayas (Balkrishna et al. 2021). An amalgamation of phytocompounds demonstrates superior anti-oxidant efficacy compared to isolated anti-oxidant treatments. CP has adaptogenic attributes because of its anti-aging and anxiolytic properties (Maheshwari et al. 2022), and healing properties by carotenoids, flavonoids, tannins, and phenols (Kumar et al. 2012). The polyphenols in *chyawanprash* are responsible for their neuroprotective, cytoprotective, and anti-oxidant effects (Kasar et al. 2006). Piperine, a bioavailability enhancer, facilitates the healing process against photoaging (Takauji et al. 2016). The withanolides in ashwagandha activate PGC-1α, enhancing ATP production in hippocampal neurons. Amla contains emblicanin, which inhibits NF-κB, reducing TNF-α, and IL-6 levels within microglial cells. Pippali (long pepper) enhances intestinal permeability, facilitating the absorption of neuroprotective polyphenolic compounds (Narayana et al. 2017). In a clinical investigation that spanned 6 months and involved children aged 5–12 years, CP led to a significant enhancement in physical strength, immunity, and vitality as evaluated through KIDSCREEN QOL-27 questionnaires (Gupta et al. 2017). Dendritic cells showed increased TNF-α, and MIP-1 expression and increased IL-1β levels, reduced cortisol levels, enhanced pulmonary function, and improved quality of life (Raghavan et al. 2023). Low spice constituents are widely acknowledged as chemo-preventive, anti-oxidative, anti-inflammatory, anti-mutagenic, and immunomodulatory on gastrointestinal, metabolic, cardiovascular, respiratory, neural, reproductive, and other physiological systems.

10.5.2 *Triphala Rasayan*

It is composed of equal proportions of *Amalaki*, *Bibhitaki*, and Haritaki. It has been demonstrated to promote detoxification (*Ama Pachana*), rejuvenate digestive processes, eye health, and enhance the health of the gut-brain axis. It has also been implicated as scavenger of free radical, anti-oxidant, antipyretic, anti-inflammatory, appetite stimulant, immune-modulating, hepato-protective, reduces gastric hyperacidity, analgesic, anti-cariogenic, anti-mutagenic, wound healing, anti-stress, adaptogenic, hypoglycemic, antibacterial, anticancer, radio-protective, chemo-protective, and has chemo-preventive properties (Dhanalakshmi et al. 2007; Rasool and Sabina 2007; Baliga et al. 2012; Peterson et al. 2017). It contains a wide range of bioactive compounds, such as flavonoids, fatty acids, saponins, anthraquinones, and amino acids. It also serves as a source of vitamin C and flavonoids and increases the activity of anti-oxidant enzymes, such as glutathione-S-transferase, superoxide dismutase (SOD), catalase, and glutathione peroxidase (GPx) (Gupta et al. 2010). *Triphala* displays anti-aging properties (Varma et al. 2016) by reducing lipid peroxidation and renal dysfunction (Baskaran et al. 2015). The active ingredients, like tannins, ellagic acid, gallic acid, and chebulinic acid, are responsible for the immunomodulatory activity of *Triphala* (Belapurkar et al. 2014). *Triphala* treatment was more effective than nonsteroidal anti-inflammatory medicine, indomethacin, in treating inflammatory and arthritic symptoms that reduced TNF-α, interleukin-6 (IL-6), nitric oxide (NO), Monocyte chemoattractant protein-1 (MCP-1), interleukin-1 (IL-1), vascular endothelial growth factor (VEGF), and prostaglandin 2 (PGE2), free radicals (inducible nitric oxide synthase (iNOS) and cyclooxygenase-2 (COX2)), and lysosomal enzymes (Kalaiselvan and Rasool 2016), NF-κB activation, COX-2, IL-17, and receptor activator of nuclear factor-κB ligand (RANKL) (Ahmed et al. 2021). Chebulagic acid has been shown to inhibit COX and 5-lipoxygenase (5-LOX), which are linked to inflammation and cancer (Jantrapirom et al. 2021). The gallic acid in Bibhitaki boosts sirtuin 1 (SIRT1), which promotes mitophagy in aged neurons (Ahmed et al. 2021).

10.5.3 Brahmi Ghrita

An ayurvedic formulation containing *Bacopa monneri*, *Acorus calamus*, *Convolvulos pluricalis, Sassurea lappa*, and Ghrita (ghee) (Sahoo et al. 2014; Yadav et al. 2014) has been very effective for the healthy nervous system. *Brahmi* improves learning, memory, and has anti-epileptic properties (Yadav et al. 2014). Neurons combat oxidative stress through an anti-oxidant system that includes enzymatic and non-enzymatic processes (Turrens 2003). *Brahmi* reduces lipoxygenase activity in the mouse brain and aids in the recovery from oxidative stress (Viji and Helen 2008). Oral administration of *Brahmi* extract in streptozotocin-induced diabetic mice increases SOD and GPx activity while decreasing glutathione (Khan et al. 2015). *Brahmi* extracts also showed considerable free radical scavenging activity (Dwivedi et al. 2013). α-asarone in *Acorus calamus* inhibits acetylcholinesterase (AChE), while scopoletin in Shankhpushpi limits cholinergic action. Bacosides in *Brahmi* and butyrate from ghee trigger the BDNF-TrkB signaling pathway (Fatima et al. 2022).

10.5.4 Ashwagandharishta

It is made up of *Ashwagandha*, *Haritaki*, *Vidanga*, *Dhataki*, and jaggery. It seeks to balance *Vata*, reduce fatigue, and improve cognitive clarity. It is a polyherbal preparation traditionally produced through the fermentation of *Woodfordia fruticosa* flowers. It contains the roots of *Withania somnifera*, a valuable primitive medicine. Yeasts located on the blooms cause the fermentation (Mikulska et al. 2023). *Ashwagandha* has been shown to impede the integration of ribosome 35S into the granulation tissue. *Ashwagandha* was found to affect Mg2+-dependent ATPase activity and reduce succinate dehydrogenase enzyme activity in the mitochondria of granulation tissue (Begum and Sadique 1987). It relieves pain by relaxing the nervous system (Twaij et al. 1989). Ashwagandha's antipyretic, analgesic, and anti-adenoma qualities are also well-established (Singh et al. 1986). *Ashwagandha* (1000 mg/kg/oral) showed an enhanced analgesic effect in rat with thermal analgesia induced by the hot plate method. The maximum analgesic impact of *ashwagandha* was observed at 78.03% after the second hour of administration. The role of pain mediators, 5-hydroxytryptamine and prostaglandins, in analgesic activity was examined by treatment with cyproheptadine (10 mg/kg, i.p.) and paracetamol (100 mg/kg, i.p.). Cyproheptadine significantly improved the analgesic efficacy, whereas paracetamol had little impact, showing that serotonin, not prostaglandins, had a role in the analgesic activity of Ashwagandha (Sakka Mazen et al. 1990). The withanolides in *ashwagandha* reduce cortisol levels while increasing DHEA, offering protection against hippocampal atrophy. The embelin in Vidanga successfully chelates excess iron, reducing oxidative stress in Parkinson's disease. The low glycemic index of jaggery helps to stabilize glucose metabolism in neuronal cells (Singh et al. 2011).

10.5.5 Saraswatarishta

A poly-herbal formulation containing 18 herbs, including *brahmi*, *shankhpushpi*, *ashwagandha*, *kapikachhu* (*Mucuna pruriens*), and *licorice*, has been traditionally used for treating issues such as psychosis, epilepsy, slurred speech, low grasping power, and memory loss (Sharma et al. 2023). It revitalizes the brain, improving verbal fluency, creativity, and cognitive development (Parekar et al. 2014). It also shows an antidepressant-like effect in the animal model of behavioral despair (Parekar et al. 2014). Moreover, daily administration of *Saraswatarishta* has been found to prevent memory impairment caused by diazepam (Rajopadhye and Sahasrabudhe 2020).

10.5.6 Amalaki Rasayana

It is made from *Amalaki* fruits (*Phyllanthus emblica*), used as a traditional medicine for cardiac, brain, and digestive health. This composition is classified as *Vayasthapana rasayanas*, which acts as a tonic for cardiac, brain, and digestive health (Vishwanatha et al. 2016). It is rich in both forms of vitamin C- ascorbic acid and dehydroascorbic acid, and valuable for its anti-oxidant, anti-cancer, and immunomodulatory activities (Kumar et al. 2017). It also contains gallic acid, kaempferol, flavonoids, quercetin, and ellagic acid, which have been shown to be involved in immune activation and anti-oxidant activities (Madhuri et al. 2011). It has been found to promote hematopoiesis (Rajan and Kumari 2016) and enhance telomerase activity in aged individuals (Guruprasad et al. 2017).

10.5.7 Medhya Rasayana

It is also a polyherbal formulation composed of four herbs: *Mandukaparni* (*Centella asiatica Linn.*), *Guduchi* (*Tinospora cordifolia*), *Shankhapushpi* (*Evolvulus alsinoides*), and *Yastimadhu* (*Glycyrrhiza glabra Linn.*), helpful in enhancing memory and intellect (Sarokte and Rao 2013). *Mandukaparni* is categorized as Medhya because its influence enhances cognitive functions, learning, and memory by modulating dopamine, 5-Hydroxytryptamine receptors, and noradrenaline pathways (Kulkarni et al. 2012). Moreover, it alleviates stress, reduces corticosterone levels, and elevates SOD, GPx, and glutathione in mice. *Centella asiatica* extract reduces H_2O_2-induced oxidative stress by inhibiting peroxidation through alterations in anti-oxidant defense mechanisms. It also promotes the formation of neural dendrites, potentially enhancing concentration and short-term memory. Asiaticoside, a bioactive component of *C. asiatica*, has anxiolytic properties that reduce anxiety, which has been found to affect memory adversely. The aqueous extract of *C. asiatica Linn.* alleviates cognitive deficits and oxidative stress (Kataria and Singh 2024). The *Madhura, Shita*, and *Vatapittashamaka* of *Yashtimadhu* provide a calming effect that fosters *Stairya* and *Dhriti*, establishing mental stability and enhancing memory. The roots of *G. glabra Linn.* contain active compounds such as glycyrrhizin, glycoside, isoliquiritin, liquiritin, steroid estrogen, hispaglabridin B, isoliquiritigenin, and paratocarpin B. Reports show that *G. glabra Linn* simultaneously augments blood circulation to the central nervous system and regulates blood glucose levels. The ethanol extract of *G. glabra Linn.* has been shown to confer cerebroprotective effects in hypoxic conditions. Research revealed that giving mice a dose of 150 mg/kg of a water-based licorice extract led to enhancements in both learning and memory, likely due to improved cholinergic transmission in the mouse brain (Muralidharan et al. 2009; Chakravarthi and Avadhani 2013).

10.6 Challenges in Rasayana Research and Clinical Translation

10.6.1 Herb Variability

The concentration of withanolides in ashwagandha exhibits variability contingent upon soil quality and the timing of harvest (Kaur et al. 2018).

10.6.2 Bioavailability Barriers

10.6.2.1 Lipid Solubility

The presence of ghee and Shilajit enhances the bioavailability of fat-soluble active compounds (Kataria and Singh 2024).

10.6.2.2 Nanoformulations

It is pertinent to better understand the solubility and bioavailability issues of herbal formulations. For example, resveratrol shows low solubility in both aqueous and lipophilic solutions, resulting in low bioavailability and inability to reach the target organ/cell and have the intended health benefit, while being a potent polyphenol (Casanova and Santos 2016). To address these issues and limitations, liposomes have been used in drug delivery systems to provide selective targeting, controlled release, and improved bioavailability.

10.6.2.3 Herb-Drug Interactions

The inclusion of turmeric in Rasayanas may potentiate the effects of warfarin, thereby escalating the risk of bleeding (Heck et al. 2000).

10.7 Future Directions: From Tradition to Precision Medicine

10.7.1 AI-Driven Personalized Rasayanas

10.7.1.1 Prakriti Genomics

The correlation between Vata-Pitta-Kapha constitutions and specific single-nucleotide polymorphisms (SNPs) influencing stress response mechanisms (Huang et al. 2022).

10.7.1.2 AI Formulation Engines

Algorithms designed to suggest optimal herb combinations predicated upon biomarkers.

10.7.2 Global Regulatory Harmonization

10.7.2.1 WHO Guidelines

Establishing standardized safety and efficacy protocols for Rasayana products aimed at facilitating international adoption.

10.7.2.2 IPR Protections

Mechanisms for safeguarding traditional knowledge while simultaneously incentivizing research and development efforts (for example, through patenting novel delivery systems).

10.8 Conclusions

Natural compounds and Ayurvedic formulations sold in the markets have a multifaceted role, demonstrating the potential to decelerate the aging process by modulating various critical pathways (Fig. 10.3). These compounds are available as

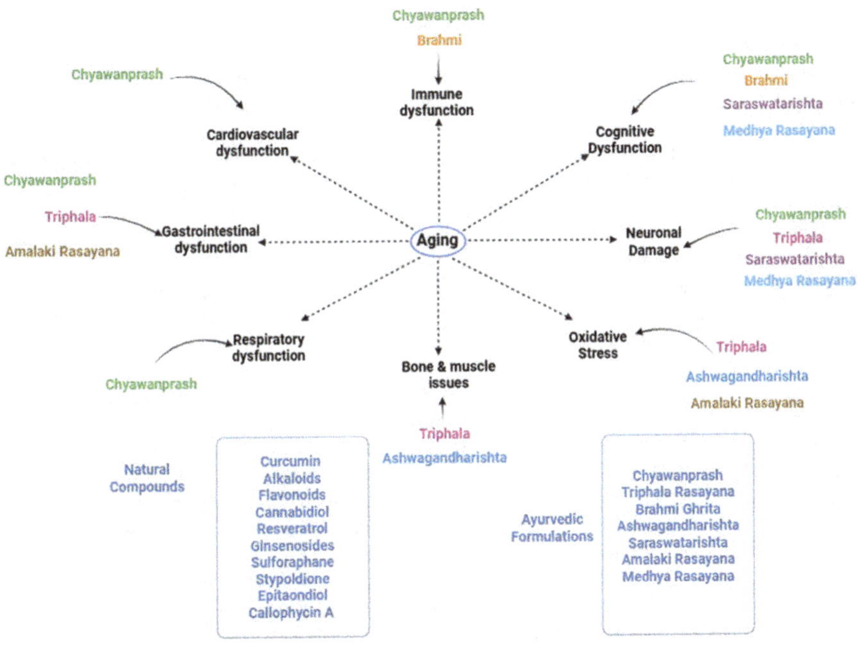

Fig. 10.3 Active molecules and Ayurvedic formulations utilizing a combination of natural compounds hold the potential to regulate age-associated changes in human physiology

commercial products in sufficient quantities for daily consumption in the form of food or supplements. Research data explain that Ayurvedic Rasayanas transcend the limitations of single-component therapies, offer a multi-target, systems-level framework, and their supplementation may facilitate healthy aging. These formulations align with the complex dynamics of aging biology by simultaneously targeting oxidative stress, neuroinflammation, mitochondrial dysfunction, and gut dysbiosis. As the global demographic landscape increasingly shifts towards an aging population, Rasayanas have the potential to transform geriatric care, evolving it from a reactive disease management model to a proactive approach that emphasizes the preservation of vitality, embodying Ayurveda's timeless principle: *Swasthasya Swasthya Rakshanam* (the preservation of health in the healthy). However, additional animal-based basic and clinical studies would prove better to determine the suitable dose of the ayurvedic formulations, constituents, and their influence at cellular and molecular levels in regulating healthy aging.

Acknowledgement Figures were created using Biorender.com.

References

Ahmed S, Ding X, Sharma A (2021) Exploring scientific validation of Triphala Rasayana in Ayurveda as a source of rejuvenation for contemporary healthcare: An update. J Ethnopharmacol 273:113829. https://doi.org/10.1016/j.jep.2021.113829

Angelova DM, Brown DR (2019) Microglia and the aging brain: are senescent microglia the key to neurodegeneration? J Neurochem 151(6):676–688. https://doi.org/10.1111/jnc.14860

Antignano I, Liu Y, Offermann N, Capasso M (2023) Aging microglia. Cell Mol Life Sci 80(5). https://doi.org/10.1007/s00018-023-04775-y

Aunan JR, Watson MM, Hagland HR, Søreide K (2016) Molecular and biological hallmarks of ageing. Br J Surg 103(2):e29–e46. https://doi.org/10.1002/bjs.10053

Baliga MS, Meera S, Mathai B, Rai MP, Pawar V, Palatty PL (2012) Scientific validation of the ethnomedicinal properties of the Ayurvedic drug Triphala: a review. Chin J Integr Med 18(12): 946–954. https://doi.org/10.1007/s11655-012-1299-x

Balkrishna A, Tomer M, Manik M, Srivastava J, Dev R, Haldar S, Varshney A (2021) Chyawanprash, an ancient Indian ayurvedic medicinal food, regulates immune response in zebrafish model of inflammation by moderating inflammatory biomarkers. Front Pharmacol 12. https://doi.org/10.3389/fphar.2021.751576

Baskaran UL, Martin SJ, Mahaboobkhan R, Prince SE (2015) Protective role of Triphala, an Indian traditional herbal formulation, against the nephrotoxic effects of bromobenzene in Wistar albino rats. J Integr Med 13(2):115–121. https://doi.org/10.1016/S2095-4964(15)60165-0

Baumann Kubetzko FB, di Paolo C, Maag C, Meier R, Schäfer BW, Betts DR, Stahel RA, Himmelmann A (2004) The PAX5 oncogene is expressed in N-type neuroblastoma cells and increases tumorigenicity of a S-type cell line. Carcinogenesis 25(10):1839–1846. https://doi.org/10.1093/carcin/bgh190

Begum VH, Sadique J (1987) Effect of Withania somnifera on glycosaminoglycan synthesis in carrageenin-induced air pouch granuloma. Biochem Med Metab Biol 38(3):272–277. https://doi.org/10.1016/0885-4505(87)90091-0

Belapurkar P, Goyal P, Tiwari-Barua P (2014) Immunomodulatory effects of Triphala and its individual constituents: a review. Indian J Pharm Sci 76(6):467–475

Bharti B, Shukla S, Tripathi R, Mishra S, Kumar M, Pandey M, Mishra R (2016) Level of PAX5 in differential diagnosis of non-Hodgkin's lymphoma. Indian J Med Res 143(Suppl 1):S23. https://doi.org/10.4103/0971-5916.191747

Bhattacharya A, Ghosal S, Bhattacharya S (2000) Anti-oxidant activity of tannoid principles of Emblica officinalis (amla) in chronic stress induced changes in rat brain. Indian J Exp Biol 38: 877–880

Bjørklund G, Shanaida M, Lysiuk R, Butnariu M, Peana M, Sarac I, Strus O, Smetanina K, Chirumbolo S (2022) Natural compounds and products from an anti-aging perspective. Molecules 27(20):20. https://doi.org/10.3390/molecules27207084

Boisvert MM, Erikson GA, Shokhirev MN, Allen NJ (2018) The aging astrocyte transcriptome from multiple regions of the mouse brain. Cell Rep 22(1):269–285. https://doi.org/10.1016/j.celrep.2017.12.039

Booth LN, Brunet A (2016) The aging Epigenome. Mol Cell 62(5):728–744. https://doi.org/10.1016/j.molcel.2016.05.013

Brzeszczyńska J, Johns N, Schilb A, Degen S, Degen M, Langen R, Schols A, Glass DJ, Roubenoff R, Greig CA, Jacobi C, Fearon KCH, Ross JA (2016) Loss of oxidative defense and potential blockade of satellite cell maturation in the skeletal muscle of patients with cancer but not in the healthy elderly. Aging (Albany NY) 8(8):1690–1702. https://doi.org/10.18632/aging.101006

Buckingham M, Relaix F (2007) The role of Pax genes in the development of tissues and organs: Pax3 and Pax7 regulate muscle progenitor cell functions. Annu Rev Cell Dev Biol 23:645–673. https://doi.org/10.1146/annurev.cellbio.23.090506.123438

Casanova F, Santos L (2016) Encapsulation of cosmetic active ingredients for topical application—a review. J Microencapsul 33(1):1–17. https://doi.org/10.3109/02652048.2015.1115900

Chakravarthi KK, Avadhani R (2013) Beneficial effect of aqueous root extract of Glycyrrhiza glabra on learning and memory using different behavioral models: an experimental study. J Nat Sci Biol Med 4(2):420–425. https://doi.org/10.4103/0976-9668.117025

Chen X, Pan S, Li F, Xu X, Xing H (2022) Plant-derived bioactive compounds and potential health benefits: involvement of the gut microbiota and its metabolic activity. Biomolecules 12(12):12. https://doi.org/10.3390/biom12121871

Colcombe S, Kramer AF (2003) Fitness effects on the cognitive function of older adults: a meta-analytic study. Psychol Sci 14(2):125–130. https://doi.org/10.1111/1467-9280.t01-1-01430

Cotrina ML, Nedergaard M (2002) Astrocytes in the aging brain. J Neurosci Res 67(1):1–10. https://doi.org/10.1002/jnr.10121

Datta HS, Mitra SK, Paramesh R, Patwardhan B (2009) Theories and management of aging: modern and Ayurveda perspectives. Evid Based Complement Alternat Med 2011(1). https://doi.org/10.1093/ecam/nep005

Dhanalakshmi S, Devi RS, Srikumar R, Manikandan S, Thangaraj R (2007) Protective effect of Triphala on cold stress-induced behavioral and biochemical abnormalities in rats. Yakugaku Zasshi: J Pharm Soc Jpn 127(11):1863–1867. https://doi.org/10.1248/yakushi.127.1863

Dornala SN, Dornala SSN (2016) GERIAYUTRICS–ayurvedic perspective of senescence and rejuvenescence. Ayushdhara 3:574–583

Dwivedi S, Nagarajan R, Hanif K, Siddiqui HH, Nath C, Shukla R (2013) Standardized extract of Bacopa monniera attenuates Okadaic acid induced memory dysfunction in rats: effect on Nrf2 pathway. Evid Based Complement Alternat Med 2013:294501. https://doi.org/10.1155/2013/294501

Fatima U, Roy S, Ahmad S, Ali S, Elkady WM, Khan I, Alsaffar RM, Adnan M, Islam A, Hassan MI (2022) Pharmacological attributes of Bacopa monnieri extract: current updates and clinical manifestation. Front Nutr 9. https://doi.org/10.3389/fnut.2022.972379

Frank MG, Baratta MV, Sprunger DB, Watkins LR, Maier SF (2007) Microglia serve as a neuroimmune substrate for stress-induced potentiation of CNS pro-inflammatory cytokine responses. Brain Behav Immun 21(1):47–59. https://doi.org/10.1016/j.bbi.2006.03.005

Garg KG, Meenakshi S, Ramesh K (2015) Medical and socio-economic problems of old age people and their management. Int Ayurvedic Med J 3:165–175

Glaser T, Jepeal L, Edwards JG, Young SR, Favor J, Maas RL (1994) PAX6 gene dosage effect in a family with congenital cataracts, aniridia, anophthalmia and central nervous system defects. Nat Genet 7(4):463–471. https://doi.org/10.1038/ng0894-463

Gupta SK, Kalaiselvan V, Srivastava S, Agrawal SS, Saxena R (2010) Evaluation of anticataract potential of Triphala in selenite-induced cataract: in vitro and in vivo studies. J Ayurveda Integr Med 1(4):280–286. https://doi.org/10.4103/0975-9476.74425

Gupta A, Kumar S, Dole S, Deshpande S, Deshpande V, Singh S, Sasibhushan V (2017) Evaluation of Cyavanaprāśa on health and immunity related parameters in healthy children: a two arm, randomized, open labeled, prospective, multicenter, Clinical Study. Anc Sci Life 36(3): 141–150. https://doi.org/10.4103/asl.ASL_8_17

Guruprasad KP, Dash S, Shivakumar MB, Shetty PR, Raghu KS, Shamprasad BR, Udupi V, Acharya RV, Vidya PB, Nayak J, Mana AE, Moni R, Sankaran MT, Satyamoorthy K (2017) Influence of Amalaki Rasayana on telomerase activity and telomere length in human blood mononuclear cells. J Ayurveda Integr Med 8(2):105–112. https://doi.org/10.1016/j.jaim.2017.01.007

Harry GJ, Kraft AD (2012) Microglia in the developing brain: a potential target with lifetime effects. Neurotoxicology 33(2):191–206. https://doi.org/10.1016/j.neuro.2012.01.012

Heck AM, DeWitt BA, Lukes AL (2000) Potential interactions between alternative therapies and warfarin. Am J Health-Syst Pharm 57(13):1221–1227. quiz 1228–1230

Huang Z, Chavda VP, Bezbaruah R, Uversky VN, P. S, Patel AB, Chen Z-S (2022) An Ayurgenomics approach: Prakriti-based drug discovery and development for personalized care. Front Pharmacol 13:866827. https://doi.org/10.3389/fphar.2022.866827

Iossifov I, Ronemus M, Levy D, Wang Z, Hakker I, Rosenbaum J, Yamrom B, Lee Y-H, Narzisi G, Leotta A, Kendall J, Grabowska E, Ma B, Marks S, Rodgers L, Stepansky A, Troge J, Andrews P, Bekritsky M et al (2012) De novo gene disruptions in children on the autistic spectrum. Neuron 74(2):285–299. https://doi.org/10.1016/j.neuron.2012.04.009

Jantrapirom S, Hirunsatitpron P, Potikanond S, Nimlamool W, Hanprasertpong N (2021) Pharmacological benefits of Triphala: A perspective for allergic rhinitis. Front Pharmacol 12:628198. https://doi.org/10.3389/fphar.2021.628198

Kalaiselvan S, Rasool MK (2016) Triphala herbal extract suppresses inflammatory responses in LPS-stimulated RAW 264.7 macrophages and adjuvant-induced arthritic rats via inhibition of NF-κB pathway. J Immunotoxicol 13(4):509–525. https://doi.org/10.3109/1547691X.2015.1136010

Kasar RP, Laddha KS, Chaudhary J, Shukla A (2006) Development of quality control methods for poly herbal formulation, Chyawanprash. Nat Prod Radian 5:33–41

Kataria D, Singh G (2024) Health benefits of ghee: review of Ayurveda and modern science perspectives. J Ayurveda Integr Med 15(1):100819. https://doi.org/10.1016/j.jaim.2023.100819

Kaur A, Singh B, Ohri P, Wang J, Wadhwa R, Kaul SC, Pati PK, Kaur A (2018) Organic cultivation of ashwagandha with improved biomass and high content of active Withanolides: use of Vermicompost. PLoS One 13(4):e0194314. https://doi.org/10.1371/journal.pone.0194314

Kawagishi H, Finkel T (2014) Unraveling the truth about anti-oxidants: ROS and disease: finding the right balance. Nat Med 20(7):711–713. https://doi.org/10.1038/nm.3625

Kempthorne L, Yoon H, Madore C, Smith S, Wszolek ZK, Rademakers R, Kim J, Butovsky O, Dickson DW (2020) Loss of homeostatic microglial phenotype in CSF1R-related leukoencephalopathy. Acta Neuropathol Commun 8:72. https://doi.org/10.1186/s40478-020-00947-0

Kennedy KAM, Sandiford SDE, Skerjanc IS, Li SS-C (2012) Reactive oxygen species and the neuronal fate. Cell Mol Life Sci 69(2):215–221. https://doi.org/10.1007/s00018-011-0807-2

Khan MB, Ahmad M, Ahmad S, Ishrat T, Vaibhav K, Khuwaja G, Islam F (2015) Bacopa monniera ameliorates cognitive impairment and neurodegeneration induced by intracerebroventricularstreptozotocin in rat: behavioral, biochemical, immunohistochemical and histopathological evidences. Metab Brain Dis 30(1):115–127. https://doi.org/10.1007/s11011-014-9593-5

Knott AB, Perkins G, Schwarzenbacher R, Bossy-Wetzel E (2008) Mitochondrial fragmentation in neurodegeneration. Nat Rev Neurosci 9(7):505–518. https://doi.org/10.1038/nrn2417

Kozmik Z, Sure U, Rüedi D, Busslinger M, Aguzzi A (1995) Deregulated expression of PAX5 in medulloblastoma. Proc Natl Acad Sci 92(12):5709–5713. https://doi.org/10.1073/pnas.92.12.5709

Kulkarni R, Girish KJ, Kumar A (2012) Nootropic herbs (Medhya Rasayana) in Ayurveda: An update. Pharmacogn Rev 6(12):147–153. https://doi.org/10.4103/0973-7847.99949

Kumar A, Kaur P, Rinwa P (2012) Evaluation of morphological, phytochemical and physicochemical properties of Indian Polyherbal formulation, Chyawanprash for quality evaluation. Asian J Pharm Educ Res 1(2):121–140

Kumar V, Aneesh k A, Kshemada K, Ajith KGS, Binil RSS, Deora N, Sanjay G, Jaleel A, Muraleedharan TS, Anandan EM, Mony RS, Valiathan MS, Santhosh KTR, Kartha CC (2017) Amalaki rasayana, a traditional Indian drug enhances cardiac mitochondrial and contractile functions and improves cardiac function in rats with hypertrophy. Sci Rep 7(1):8588. https://doi.org/10.1038/s41598-017-09225-x

Landin AM, Frasca D, Scallan M, Harrison P, Hagman J, Blomberg B (2011) The role of Pax5 and E47 on AID expression in B lymphocytes of aged BALB/c mice (61.1). J Immunol 186-(1_Supplement):61.1. https://doi.org/10.4049/jimmunol.186.Supp.61.1

Lee AS, Azmitia EC, Whitaker-Azmitia PM (2017) Developmental microglial priming in postmortem autism spectrum disorder temporal cortex. Brain Behav Immun 62:193–202. https://doi.org/10.1016/j.bbi.2017.01.019

Li CG, Eccles MR (2012) PAX genes in cancer; friends or foes? Front Genet 3:6. https://doi.org/10.3389/fgene.2012.00006

Li Y, Min W, Li M, Han G, Dai D, Zhang L, Chen X, Wang X, Zhang Y, Yue Z, Liu J (2016) Identification of hub genes and regulatory factors of glioblastoma multiforme subgroups by RNA-seq data analysis. Int J Mol Med 38(4):1170–1178. https://doi.org/10.3892/ijmm.2016.2717

Madhuri S, Pandey G, Verma KS (2011) Anti-oxidant, immunomodulatory and anticancer activities of Emblica officinalis: an overview. Int Res J Pharm 2(8):38–42

Maheshwari S, Kumar V, Bhadauria G, Mishra A (2022) Immunomodulatory potential of phytochemicals and other bioactive compounds of fruits: a review. Food Front 3(2):221–238. https://doi.org/10.1002/fft2.129

Mattson MP, Magnus T (2006) Ageing and neuronal vulnerability. Nat Rev Neurosci 7(4):278–294. https://doi.org/10.1038/nrn1886

Maurya SK, Mishra R (2017) Pax6 interacts with Iba1 and shows age-associated alterations in brain of aging mice. J Chem Neuroanat 82:60–64. https://doi.org/10.1016/j.jchemneu.2017.05.002

Maurya SK, Mishra R (2018) Co-localization and interaction of Pax5 with Iba1 in brain of mice. Cell Mol Neurobiol 38(4):919–927. https://doi.org/10.1007/s10571-017-0566-1

Maurya SK, Mishra R (2022a) Co-expression and interaction of Pax6 with genes and proteins of immunological surveillance in the brain of mice. Neurotox Res 40(6):2238–2252. https://doi.org/10.1007/s12640-022-00562-y

Maurya SK, Mishra R (2022b) Molecular docking studies of natural immunomodulators indicate their interactions with the CD40L of CD40/CD40L pathway and CSF1R kinase domain of microglia. J Mol Model 28(4):101. https://doi.org/10.1007/s00894-022-05084-0

McPherson CA, Aoyama M, Harry GJ (2011) Interleukin (IL)-1 and IL-6 regulation of neural progenitor cell proliferation with hippocampal injury: differential regulatory pathways in the subgranular zone (SGZ) of the adolescent and mature mouse brain. Brain Behav Immun 25(5):850–862. https://doi.org/10.1016/j.bbi.2010.09.003

Mikulska P, Malinowska M, Ignacyk M, Szustowski P, Nowak J, Pesta K, Szeląg M, Szklanny D, Judasz E, Kaczmarek G, Ejiohuo OP, Paczkowska-Walendowska M, Gościniak A, Cielecka-Piontek J (2023) Ashwagandha (Withania somnifera)-current research on the health-promoting activities: a narrative review. Pharmaceutics 15(4):1057. https://doi.org/10.3390/pharmaceutics15041057

Mishra R (2020) Paired box (Pax) transcription factors and aging. In: Rath PC (ed) Models, molecules and mechanisms in biogerontology: cellular processes, metabolism and diseases. Springer, pp 109–117. https://doi.org/10.1007/978-981-32-9005-1_6

Mittag J, Winterhager E, Bauer K, Grümmer R (2007) Congenital hypothyroid female Pax8-deficient mice are infertile despite thyroid hormone replacement therapy. Endocrinology 148(2):719–725. https://doi.org/10.1210/en.2006-1054

Muralidharan P, Balamurugan G, Babu V (2009) Cerebroprotective effect of Glycyrrhiza glabra Linn. Root extract. Bangladesh J Pharmacol 4(1):Article 1. https://doi.org/10.3329/bjp.v4i1.1052

Narayana DBA, Durg S, Manohar PR, Mahapatra A, Aramya AR (2017) Chyawanprash: a review of therapeutic benefits as in authoritative texts and documented clinical literature. J Ethnopharmacol 197:52–60. https://doi.org/10.1016/j.jep.2016.07.078

Ninivaggi FJ (2010) A comprehensive guide to traditional Indian medicine for the west-health and the disease Process-1, vol 6. Rowman & Littlefield Publishers, p 136

O'Roak BJ, Stessman HA, Boyle EA, Witherspoon KT, Martin B, Lee C, Vives L, Baker C, Hiatt JB, Nickerson DA, Bernier R, Shendure J, Eichler EE (2014) Recurrent de novo mutations implicate novel genes underlying simplex autism risk. Nat Commun 5:5595. https://doi.org/10.1038/ncomms6595

Parekar RR, Jadhav KS, Marathe PA, Rege NN (2014) Effect of Saraswatarishta in animal models of behavior despair. J Ayurveda Integr Med 5(3):141–147. https://doi.org/10.4103/0975-9476.140469

Parle M, Bansal N (2006) Traditional medicinal formulation, Chyawanprash—a review. Ind J Tradit Knowl 5:484–488

Peterson CT, Denniston K, Chopra D (2017) Therapeutic uses of Triphala in Ayurvedic medicine. J Altern Complement Med 23(8):607–614. https://doi.org/10.1089/acm.2017.0083

Raghavan G, Shivanna Y, Gunti P, Bapna A, Chondhekar P, Vyas T (2023) Immunostimulatory activity of a novel ayurvedic propriety formulation based on extracts of herbs used in chyavanprash. Phytomed Plus 3(1):100383. https://doi.org/10.1016/j.phyplu.2022.100383

Rajan S, Kumari S (2016) Amalaki as preventing premature ageing drug-a review. Int J Ayurveda Pharma Res. https://doi.org/10.47070/ijapr.v3i12.63

Rajopadhye DR, Sahasrabudhe AR (2020) Memory enhancing activity of Saraswatarishta in mice. Biomed Pharmacol J 13(4):2033

Rao RV (2018) Ayurveda and the science of aging. J Ayurveda Integr Med 9(3):225–232. https://doi.org/10.1016/j.jaim.2017.10.002

Rasool M, Sabina EP (2007) Anti-inflammatory effect of the Indian Ayurvedic herbal formulation Triphala on adjuvant-induced arthritis in mice. Phytother Res: PTR 21(9):889–894. https://doi.org/10.1002/ptr.2183

Reuter S, Gupta SC, Chaturvedi MM, Aggarwal BB (2010) Oxidative stress, inflammation, and cancer: how are they linked? Free Radic Biol Med 49(11):1603–1616. https://doi.org/10.1016/j.freeradbiomed.2010.09.006

Sahoo S, Rao BCS, Das AK, Das K (2014) Analytical study of Brahmi Ghrita. Res J Pharmacol Pharmacodyn 6(4):193–196

Sakka Mazen E, Pavelescu M, Grigorescu E (1990) Contributions to the pharmacodynamic study of roots of Withania somnifera Dun species of Pakistani origin. Note III: testing of analgesic activity of dichlormethanic and methanolic extract from Withania somnifera roots. Rev Med Chir Soc Med Nat Iasi 94(3–4):603–605

Sarokte AS, Rao MV (2013) Effects of Medhya Rasayana and Yogic practices in improvement of short-term memory among school-going children. Ayu 34(4):383–389. https://doi.org/10.4103/0974-8520.127720

Sharma S, Singh G, Patwardhan K (2023) Effects of two Ayurvedic formulations, Dhanwantaram Kashaya and Saraswatarishta on life history parameters and toxic aggregates in Drosophila models of Huntington's and Alzheimer's disease. J Ethnopharmacol 305:116080. https://doi.org/10.1016/j.jep.2022.116080

Singh K (2017) Yogic life style and geriatric care–a review article. Int Ayurvedic Med J 5:263–268

Singh N, Singh SP, Nath R, Singh DR, Gupta ML, Kohli RP, Bhargava KP (1986) Prevention of urethane—induced lung adenomas by Withania somnifera (L.) Dunal in albino mice. Int J Crude Drug Res 24(2):90–100. https://doi.org/10.3109/13880208609083312

Singh N, Bhalla M, de Jager P, Gilca M (2011) An overview on Ashwagandha: a Rasayana (Rejuvenator) of Ayurveda. Afr J Tradit Complement Altern Med 8(5 Suppl):208–213. https://doi.org/10.4314/ajtcam.v8i5S.9

Sisodiya SM, Free SL, Williamson KA, Mitchell TN, Willis C, Stevens JM, Kendall BE, Shorvon SD, Hanson IM, Moore AT, Van Heyningen V (2001) PAX6 haploinsufficiency causes cerebral malformation and olfactory dysfunction in humans. Nat Genet 28(3):214–216. https://doi.org/10.1038/90042

Spittau B (2017) Aging microglia—phenotypes, functions and implications for age-related neurodegenerative diseases. Front Aging Neurosci 9. https://doi.org/10.3389/fnagi.2017.00194

Streit WJ, Sammons NW, Kuhns AJ, Sparks DL (2004) Dystrophic microglia in the aging human brain. Glia 45(2):208–212. https://doi.org/10.1002/glia.10319

Stuart E, Gruss P (1995) PAX genes: what's new in developmental biology and cancer? Hum Mol Genet 4(suppl_1):1717–1720. https://doi.org/10.1093/hmg/4.suppl_1.1717

Takauji Y, Morino K, Miki K, Hossain M, Ayusawa D, Fujii M (2016) Chyawanprash, a formulation of traditional Ayurvedic medicine, shows a protective effect on skin photoaging in hairless mice. J Integr Med 14(6):473–479. https://doi.org/10.1016/S2095-4964(16)60272-8

Thapliyal A, Singh MB, Kumari K, Dey SK, Maurya SK (2025) Identification of novel inhibitors of microglia-specific proteins responsible for neuroinflammatory response by molecular docking and simulation studies. Int J Model Simul. https://www.tandfonline.com/doi/abs/10.1080/02286203.2025.2484300

Thompson B, Davidson EA, Liu W, Nebert DW, Bruford EA, Zhao H, Dermitzakis ET, Thompson DC, Vasiliou V (2020) Overview of PAX gene family: analysis of human tissue-specific variant expression and involvement in human disease. Hum Genet 140(3):381–400. https://doi.org/10.1007/s00439-020-02212-9

Tripathi R, Mishra R (2010) Interaction of Pax6 with SPARC and p53 in brain of mice indicates Smad3 dependent auto-regulation. J Mol Neurosci 41(3):397–403. https://doi.org/10.1007/s12031-010-9334-0

Tripathi R, Mishra R (2012) Aging-associated modulation in the expression of Pax6 in mouse brain. Cell Mol Neurobiol 32(2):209–218. https://doi.org/10.1007/s10571-011-9749-3

Tuoc TC, Radyushkin K, Tonchev AB, Piñon MC, Ashery-Padan R, Molnár Z, Davidoff MS, Stoykova A (2009) Selective cortical layering abnormalities and behavioral deficits in cortex-specific Pax6 knock-out mice. J Neurosci 29(26):8335–8349. https://doi.org/10.1523/JNEUROSCI.5669-08.2009. PMID: 19571125; PMCID: PMC6665651

Turrens JF (2003) Mitochondrial formation of reactive oxygen species. J Physiol 552(Pt 2):335–344. https://doi.org/10.1113/jphysiol.2003.049478

Twaij HAA, Elisha, Eugenia E, Khalid RM (1989) Analgesic studies on some Iraqi medicinal plants part II. Int J Crude Drug Res 27(2):109–112. https://doi.org/10.3109/13880208909053947

Varma SR, Sivaprakasam TO, Mishra A, Kumar LMS, Prakash NS, Prabhu S, Ramakrishnan S (2016) Protective effects of Triphala on dermal fibroblasts and human keratinocytes. PLoS One 11(1):e0145921. https://doi.org/10.1371/journal.pone.0145921

Viji V, Helen A (2008) Inhibition of lipoxygenases and cyclooxygenase-2 enzymes by extracts isolated from Bacopa monniera (L.) Wettst. J Ethnopharmacol 118(2):305–311. https://doi.org/10.1016/j.jep.2008.04.017

Vishwanatha U, Guruprasad KP, Gopinath PM, Acharya RV, Prasanna BV, Nayak J, Ganesh R, Rao J, Shree R, Anchan S, Raghu KS, Joshi MB, Paladhi P, Varier PM, Muraleedharan K, Muraleedharan TS, Satyamoorthy K (2016) Effect of Amalaki rasayana on DNA damage and repair in randomized aged human individuals. J Ethnopharmacol 191:387–397. https://doi.org/10.1016/j.jep.2016.06.062

Vorobyov E, Horst J (2006) Getting the proto-Pax by the tail. J Mol Evol 63:153–164. https://doi.org/10.1007/s00239-005-0163-7

Wistuba J, Mittag J, Luetjens CM, Cooper TG, Yeung C-H, Nieschlag E, Bauer K (2007) Male congenital hypothyroid Pax8−/− mice are infertile despite adequate treatment with thyroid hormone. J Endocrinol. https://doi.org/10.1677/JOE-06-0054

Woodruff-Pak DS, Foy MR, Akopian GG, Lee KH, Zach J, Nguyen KPT, Comalli DM, Kennard JA, Agelan A, Thompson RF (2010) Differential effects and rates of normal aging in cerebellum and hippocampus. Proc Natl Acad Sci USA 107(4):1624–1629. https://doi.org/10.1073/pnas.0914207107

Wu Y, Zhou X, Huang X, Xia Q, Chen Z, Zhang X, Yang D, Geng Y (2016) Pax8 plays a pivotal role in regulation of cardiomyocyte growth and senescence. J Cell Mol Med 20(4):644–654. https://doi.org/10.1111/jcmm.12779

Yadav KD, Reddy KRC, Kumar V (2014) Beneficial effect of Brahmi Ghrita on learning and memory in normal rat. Ayu 35(3):325–329. https://doi.org/10.4103/0974-8520.153755

Youssef SA, Capucchio MT, Rofina JE, Chambers JK, Uchida K, Nakayama H, Head E (2016) Pathology of the aging brain in domestic and laboratory animals, and animal models of human neurodegenerative diseases. Vet Pathol 53(2):327–348. https://doi.org/10.1177/0300985815623997

Zhang X, Heaney S, Maas RL (2002) Cre-loxp fate-mapping of Pax6 enhancer active retinal and pancreatic progenitors. Genesis 35(1):22–30. https://doi.org/10.1002/gene.10160

Zhang H, Chen H, Luo H, An J, Sun L, Mei L, Feng Y (2012) Functional analysis of Waardenburg syndrome-associated PAX3 and SOX10 mutations: report of a dominant-negative SOX10 mutation in Waardenburg syndrome type II. Hum Genet 131(3):491–503. https://doi.org/10.1007/s00439-011-1098-2

Chapter 11
Natural Healthy-Aging Agents: Wisdom from Traditional Chinese Medicine

Xiaodong Zhang and Yifang Zhang

Abstract Historically, one of the highest ideals of traditional Chinese medicine (TCM) was to use herbs to prevent disease and promote health and longevity. A variety of TCM theories provide a foundation for understanding the physiological basis for the aging process, and in turn therapeutic strategies, practices, and herbs for maximizing our health potential throughout the duration of our lives. There is a plethora of modern research supporting the healthy-aging potential of many botanicals used in TCM. Some contain compounds with antioxidant, anti-inflammatory or immune modulatory activities, all of which are generally positively correlated with healthy aging. Others positively affect telomeres, DNA repair, RNA expression, mitochondrial health, cognitive function, and the microbiome of the gut, all of which are foundational to health aging. This paper addresses the underlying diagnostic constructs for understanding the aging process, introduces readers to some of the primary herbs used in TCM for supporting healthy aging, and provides references to scientific evidence supporting these effects.

Keywords Healthy-aging · TCM theory · Chinese medicinal herbs · Antioxidant · Longevity

11.1 Introduction

With the rapid increase in the global aging population, frailty in the elderly has emerged as a significant public health challenge. While definitions of aging vary, they share common characteristics: aging is associated with phenotypic changes in populations over the average lifespan, including increased susceptibility to age-related diseases and elevated mortality risks (Keshavarz et al. 2023). Strategies to

---

X. Zhang
School of Traditional Chinese Medicine, Nanjing University of Chinese Medicine, Nanjing, China

Y. Zhang (✉)
Shanghai East International Medical Center, Shanghai, China

S. I. S. Rattan, C. Gezer (eds.), *Supplements and Drugs for Healthy Longevity*, Healthy Ageing and Longevity 22, https://doi.org/10.1007/978-3-032-11608-6_11

combat aging and age-related diseases, and to extend healthy lifespan, have garnered increasing attention. The concept of "hallmarks of aging," (López-Otín et al. 2013, 2023) provides a conceptual framework for aging research. To date, 12 hallmarks have been identified: genomic instability, telomere attrition, epigenetic alterations, loss of proteostasis (protein homeostasis), deregulated nutrient sensing, mitochondrial dysfunction, cellular senescence, stem cell exhaustion, altered intercellular communication, impaired autophagy, chronic inflammation, and dysbiosis of the microbiota. This framework has been widely applied in foundational and interventional studies of aging (Tartiere et al. 2024). TCM's understanding of aging is deeply rooted in its theoretical foundations, and the healthy-aging potential of traditional Chinese herbs continues to be discovered and utilized in modern research. This paper integrates traditional and modern evidence of TCM's healthy-aging properties, offering a multifaceted interpretation of its role in combating aging.

11.2 TCM'S Understanding of Aging

TCM's perspective on aging originates from Huangdi Neijing (*The Yellow Emperor's Inner Classic*), the oldest medical text in China, and has evolved into a comprehensive theoretical system. In general, TCM interprets aging through the lens the laws of opposites (yin and yang) of "deficiency" (xu) and "excess" (shi), emphasizing two primary mechanisms: aging caused by deficiency (xu) and aging caused by excess (shi).

11.2.1 Aging Caused by Deficiency

TCM posits that the depletion of qi, blood, yin, and yang is the root cause of aging, closely linked to the "congenital (innate) foundation" (kidney) and "acquired foundation" (spleen). The kidney, considered the "congenital foundation," governs reproduction and development, influencing the entire life cycle from growth to aging and death. *The Yellow Emperor's Inner Classic* describes natural aging processes in men and women, regulated by the waxing and waning of "tian gui" (a vital essence stored in the kidney, promotes the development and maturation of reproductive organs and maintains reproductive function). This suggests that physiological kidney deficiency triggers natural aging. Kidney deficiency often manifests as yin-yang depletion, leading to functional decline, as noted in *The Yellow Emperor's Inner Classic*: "At forty, yin energy diminishes by half, and daily activities weaken." The Tang Dynasty physician Sun Simiao observed, "After fifty, yang energy declines daily." The spleen, as the "acquired foundation," is responsible for transforming food into qi and blood. A weakened spleen leads to qi and blood deficiency, which manifests as signs of aging, such as sleep disturbances mentioned in *The Yellow Emperor's Inner Classic*: "The elderly have depleted qi and blood." In summary, the balance of innate yin-yang and acquired qi-blood governs the aging process. Nourishing deficiency is the fundamental TCM approach to delaying aging, with a focus on regulating qi-blood in the "acquired foundation" and harmonizing yin-yang in the " congenital foundation."

11.2.2 Aging Caused by Excess

With advancing age, the body's spleen and kidney functions weaken, leading to qi, blood, yin, and yang deficiencies. Simultaneously, external factors such as improper diet, irregular lifestyle, lack of exercise, and emotional stress accumulate, contributing to pathogenic factors like qi stagnation, blood stasis, and phlegm-dampness. These "excess factors" can induce or accelerate aging. Epidemiological studies of TCM syndromes in 878 middle-aged and elderly individuals revealed a significant positive correlation between the incidence of qi stagnation, blood stasis, and phlegm-dampness and aging (Yu et al. 1992). These findings support the theory that "excess causes aging" and highlight the importance of regulating qi, activating blood circulation, and resolving dampness and phlegm as auxiliary TCM strategies for healthy aging.

In conclusion, TCM views aging as a process driven by both deficiency and excess, with deficiency as the primary factor and excess as a secondary contributor. However, deficiency and excess interact dynamically: deficiency can lead to excess, and excess can exacerbate deficiency, ultimately forming a complex mechanism of aging characterized by deficiency as the root and excess as the manifestation, with a mixture of both.

11.3 Characteristics of TCM in Healthy-Aging Effects

The first pharmacological treatise in Chinese history, the Shennong Bencao Jing (*Shen Nong's Classic of the Materia Medica*), records numerous herbs with "life-prolonging" or similar effects. A functional classification of 57 herbs still in use today reveals 32 tonic herbs; those that assist in promoting healthy Qi, such as Asian ginseng and Atractylodes macrocephala (qi tonics), Eucommia and Cuscuta seed (yang tonics), Longan fruit (blood tonic), and Goji berries and Dendrobium (yin tonics); and 25 pathogen-expelling herbs, including 13 dampness-phlegm-resolving herbs and 4 blood-activating and stasis-resolving herbs (Cui 1996). Data mining of 530 ancient healthy-aging prescription reveals the usage patterns of herbs, shows that high-frequency herbs (frequency ≥ 53) are predominantly tonifying herbs for deficiency, involving categories such as dampness transforming, blood-quickening, spirit-quieting, interior-warming, securing and astringing, orifice-opening, and exterior-releasing herbs. The top ten herbs by frequency are Poria, Achyranthes, raw and prepared Rehmannia, Goji berries, Asian ginseng, Cuscuta seed, Cistanche,

Dioscorea yam, and Asparagus densiflorus. The most common herb pairing is Asian ginseng and Poria (Zhang et al. 2018). This highlights the central role of tonifying herbs in ancient healthy-aging strategy, complemented by pathogen-expelling herbs for functional regulation, aligning with TCM's understanding of aging pathogenesis. A population cohort study based on the Chinese Longitudinal Healthy Longevity Survey (CLHIS) found that occasional and regular consumption of medicinal plants (e.g., Asian ginseng, Astragalus, Goji berries, Angelica sinensis) correlates with reduced all-cause mortality and cognitive impairment risk (Tao et al. 2023), supporting the potential health benefits of tonifying TCM herbs. Thus, TCM healthy-aging effects emphasizes tonification first, with regulation embedded in tonification, combining both approaches.

11.4 Common Healthy-Aging Chinese Medicinal Herbs

This paper introduces ten common healthy-aging herbs, which are either classified as food-medicine dual-use substances or permitted in supplementing foods in China. Based on the dichotomy of "deficiency" and "excess," they are divided into tonifying herbs (for deficiency) and regulating herbs (for excess). Following the framework of traditional efficacy, historical application, and modern research, this paper explores their healthy-aging properties, aiming to provide readers with a comprehensive understanding of healthy-aging herbs from both traditional and modern perspectives.

11.4.1 Tonifying Herbs

A total of six herbs are included, namely qi-tonifying Asian ginseng and Astragalus; yang-tonifying Cistanche; blood-tonifying Angelica sinensis; and yin-tonifying Goji berries and Dendrobium. Table 11.1 shows their traditional functions, pharmacological activities, healthy-aging components, and mechanisms.

11.4.1.1 Asian Ginseng

Asian ginseng, belonging to the genus Panax in the family Araliaceae, uses roots and rhizomes as medicine. The term *Panax* originates from Greek, where "pan" means "all" and "axos" means "heal," together symbolizing a panacea capable of curing all ailments (Jia and Zhao 2009). Thus, species within the Panax genus are almost universally medicinally valuable. To date, 17 known species of the Panax genus have been identified (Balusamy et al. 2023), among which *Panax ginseng* C.A.Mey, produced in the Far East, is widely used in traditional medicine in China, Korea, and

Table 11.1 Functions, pharmacological activities, healthy-aging ingredients and mechanisms of Chinese tonifying herbs

Herb name	Functions[a]	Pharmacological activities	Healthy-aging ingredients	Healthy-aging mechanisms	Cited references
Asian ginseng (*Ren shen*)	Invigorating primordial qi Restoring pulse and preventing collapse Tonifying spleen and benefiting lung Generating body fluids and nourishing blood Calming the mind and promoting intelligence	Prolonging lifespan Anti-oxidation Protecting cardiovascular Improving nervous system and motor function Protecting skin Regulating immunity Anti-inflammation	Ginsenosides (*protopanaxadiol-type, protopanaxatriol-type*)	Reducing endogenous & exogenous oxidative DNA damage Regulating DNA damage repair Maintaining mitochondrial function Enhancing telomerase activity Promoting proliferation of stem cell Anti-inflammation Regulating intestinal microbes Promoting autophagy	Yang et al. (2017), Su et al. (2023) and Tian et al. (2023)
Astragalus (*Huang qi*)	Tonifying qi and ascending yang Securing the exterior and stopping sweating Promoting diuresis and reducing edema Generating body fluids and nourishing blood Removing stagnation and relieving obstruction Expelling toxins and draining pus	Reducing oxidative stress Regulating immunity Anti-neurodegeneration Protecting vascular Anti-inflammation Anti-tumor	Triterpenoid Saponins (*Astragalosides*) Astragalus flavonoids Astragalus polysaccharides	Anti-oxidation Activating telomerase Inhibiting replicative senescence Reducing glycation end products Inhibiting mitochondrial permeability transition	Liu et al. (2017), Berezutsky et al. (2020) and Li et al. (2012)

(continued)

Table 11.1 (continued)

Herb name	Functions[a]	Pharmacological activities	Healthy-aging ingredients	Healthy-aging mechanisms	Cited references
	Astringing wound and promoting tissue regeneration				
Cistanches (*Rou cong rong*)	Tonifying kidney yang Supplementing the essence and blood Moistening the intestine and relaxing bowels	Extending lifespan Anti-oxidation Enhancing learning and memorization Anti-neurodegeneration Anti-osteoporosis Anti-tumor Modulating immunity	Phenylethanoid (*glycosidesechinacoside, acteoside, isoacteoside*) Polysaccharides	Suppressing oxidation stress Modulating related signaling pathway Regulating mitochondrial function Balancing sex hormone levels Regulating immunity and inhibiting apoptosis Suppressing expression of senescence-related factors Promoting autophagy	Li et al. (2024), Zhang et al. (2023) and Zhu et al. (2025)
Angelica sinensis (*dang gui*)	Nourishing blood and circulating blood Regulating menstruation and alleviating pain Moistening the intestine and relaxing bowels	Improving anemia Anti-oxidation Enhancing immunity Anti-inflammation Anti-fibrosis Anti-tumor Liver protection Anti-radiation	Angelica sinensis polysaccharides (ASP) Ligustilide Ferulic acid	Inhibiting oxidative stress and inflammation Regulating aging-related genes and pathways Modulating telomerase activity Regulating mitophagy and diminishing apoptosis Enhancing immunity	Liu et al. (2019), Guo et al. (2022) and Wang and Peng (2024)

Goji berry (*Gou qi zi*)	Nourishing liver and kidney Replenishing essence and improving vision	Protecting vision Lowering bood sugar and lipid Anti-tumor Modulating immunity Protecting nerves Protecting cardiovascular Anti-oxidation	Lyciumbarbarum polysaccharides (LBP) Carotenoids	Anti-oxidation and anti-apoptotic Reduce DNA fracture damage Blocking aging-related signaling pathway Regulating aging-related gene expression Alleviating telomeres shortening	Shi et al. (2024), Ma et al. (2019) and Ni et al. (2021)
Dendrobium (*Shi hu*)	Benefiting the stomach and generating fluids Nourishing yin and clearing heat	Anti-tumor Anti-fatigue Anti-oxidant Anti-inflammation Liver and gastrointestinal protection Anti-diabetic Modulating immunity Anti-osteoporosis Protecting nerves	Dendrobium Alkaloids Dendrobium polysaccharides	Enhancing antioxidant activity Anti-inflammation Regulating gastrointestinal microecology Activating autophagy Reducing apoptosis Decreasing metabolite accumulation Modulating neuroendocrine immune system	Li et al. (2023), Nie et al. (2020a, b), Bhardwaj et al. (2024) and Shi et al. (2022)

[a]Chinese herbal indications are cited from Chinese pharmacopoeia (2020 edition)

Japan, while *Panax quinquefolius L.*, known as American ginseng, originated from North America and is known as "xiyangshen" in China.

Asian ginseng has been used in China for nearly a thousand years. *The Shen Nong's Classic of the Materia Medica* lists Asian ginseng as a superior herb, stating that it "nourishes life" and "with long-term use, lightens the body and extends longevity," thereby establishing the foundation of ginseng's healthy-aging medicinal value. The Song Dynasty's Bencao Tujing (*Illustrated Classic of Materia Medica*) records a method of testing Ginseng's healthy-fatigue efficacy through competitive walking: two people walk together, one holding Ginseng in the mouth and the other without. After walking three to five time's 550 meter, the person holding Ginseng maintains relaxed breathing, while the other pants heavily. By the Qing Dynasty, the practice of dissolving Ginseng in the mouth was widely used in the imperial court for longevity (Liu et al. 2023). Interestingly, modern clinical trials align with the explorations of ancient Chinese. A randomized, double-blind, placebo-controlled, balanced crossover clinical study found that supplementing with ginsenoside Rg1 1 h before squatting exercise reduces the expression of stem cell senescence markers $p16^{INK4a}$ and inflammatory markers myeloperoxidase (MPO) in skeletal muscle tissue, showing significant differences compared to the placebo group. The study indicates that ginsenoside Rg1 enhances resistance to skeletal muscle stem cell aging and inflammation induced by cellular aging, helping to delay muscle aging (Lee et al. 2021).

Ginseng's healthy-aging effects are inseparable from its powerful qi-tonifying properties. As the "foremost qi-tonifying herb," Ginseng is known for its ability to greatly tonify primal energy, restore pulse and prevent collapse, tonify the spleen and benefit the lung, generate body fluids and nourish blood, and calm the mind and enhance wisdom. The chemical components of Ginseng, such as ginsenosides, polysaccharides, volatile oils, vitamins, sterols, nucleotides, proteins, organic acids, and inorganic elements, form the material basis of its pharmacological effects (Zhang et al. 2025). Among these, ginsenosides, particularly protopanaxadiol-type and protopanaxatriol-type ginsenosides, are key active components (Tian et al. 2023). Modern pharmacological research has revealed the diverse healthy-aging properties of Ginseng, such as extending the lifespan of model organisms (*D. melanogaster, C.elegans*), antioxidant effects, regulating cardiovascular structure and function, improving nervous system and motor function, protecting the skin, modulating immunity, and suppressing inflammatory responses (Yang et al. 2017). The mechanisms of Asian ginseng and its ginsenoside components in healthy-aging primarily involve: ① reducing endogenous and exogenous DNA oxidative damage; ② regulating DNA damage repair; ③ enhancing telomerase activity; ④ promoting mitochondrial autophagy and maintaining mitochondrial function; ⑤ promoting stem cell proliferation; ⑥ anti-inflammatory activity; and ⑦ regulating gut microbiota (Su et al. 2023; Tian et al. 2023). The non-traditional medicinal parts of Ginseng (berries, flowers, leaves) have also been found to have healthy-aging effects, particularly the potential to combat skin aging, but these require further research and development.

Modern research also provides evidence of Ginseng's adaptogenic effects. Adaptogens are defined as substances that are innocuous, cause minimal disorder in the physiological functions of an organism, and have nonspecific actions such as increasing resistance to adverse physical, chemical, and biological stresses. In addition, adaptogens possess normalizing actions (Panossian et al. 2020).

11.4.1.2 Astragalus

Astragalus is the dried root of the leguminous plants *Astragalus membranaceus* (Fisch.) Bge. var. *mongholicus* (Bge.) Hsiao or *Astragalus membranaceus* (Fisch.) Bge., and functions to tonify qi and lift yang, consolidate the exterior and stop sweating, promote diuresis and reduce edema, generate body fluids and nourish blood, resolve stagnation and relieve joint stiffness, support toxin expulsion and promote pus discharge, and heal sores and regenerate tissue. Similar to Asian ginseng, Astragalus is a qi-tonifying herb with a long history. The renowned Ming Dynasty pharmacologist Li Shizhen highly praised the medicinal value of Astragalus, referring to it as the "leader of tonifying herbs" in his work *The Grand Compendium of Materia Medica* (Bencao Gangmu). Another Ming Dynasty herbal text, Bencao Mengquan, *Enlightening Primer of Materia Medica*, praises Astragalus as "the mutton of herbs," implying its potential nutritional value. Mutton contains high levels of selenium (an antioxidant), and Astragalus is a selenium-rich herb. Selenium promotes the accumulation of total polysaccharides in Astragalus (Chen et al. 2023), and the two exhibit good synergistic effects in antioxidation (Guan et al. 2021), contributing to delaying aging. However, research on organic selenium in Astragalus is scarce.

Research on the pharmacologically active components of Astragalus primarily focuses on polysaccharides, flavonoids, and saponins, which form the basis of its healthy-aging effects. Astragalus extracts demonstrate potential in combating aging and age-related diseases through multiple pathways, including reducing oxidative stress, modulating immunity, preventing neurodegeneration, providing vascular protection, and exerting anti-inflammatory, and anti-tumor effects (Liu et al. 2017). Besides its significant antioxidant capacity, the mechanisms of Astragalus in healthy-aging mainly include: ① activating telomerase; ② inhibiting cellular replicative senescence and glycation products; and ③ preventing mitochondrial permeability transition to protect mitochondria (Berezutsky et al. 2020; Li et al. 2012). Particularly in enhancing telomerase activity, Astragalus has distinct advantages. Astragaloside IV is considered the only compound currently known to activate telomerase in the human body (Yu et al. 2018). TA-65, a small-molecule telomerase activator based on Astragalus extract, has shown in cell experiments to increase telomerase activity by twice that of untreated cells (Tsoukalas et al. 2019). Clinical studies have found that TA-65 can safely extend human telomere length. The low-dose TA-65 group showed a significant increase in telomere length over 12 months, while the placebo group showed a significant decrease. The high-dose TA-65 group also showed a trend of improved telomere length, although it did not reach statistical significance (Salvador et al. 2016). Another animal experiment confirmed that Astragalus powder can increase the telomere length of white blood cells in aging-accelerated OXYS rats but cannot delay the accelerated aging of experimental animals, as evidenced by the progressive development of cataracts, age-related macular degeneration (AMD) like retinopathy, and behavioral disorders caused by brain aging (Kozhevnikova et al. 2022). While study results may vary, they all indicate the potential applications of Astragalus in healthy-aging. Currently, as a dietary supplement, Astragalus is sold in the United States in the form of tea, capsules and is available in Asia as tea, beverages, soups, and mixed nut snacks (Shahrajabian et al. 2019).

11.4.1.3 Cistanche

Cistanche is the dried fleshy stem with scale leaves of the Orobanchaceae plant *Cistanche deserticola* Y.C.Ma or *Cistanche tubulosa* (Schenk) Wight. Known as the "desert ginseng" in China due to its growth in arid regions and significant tonic effects, it differs from Asian ginseng which tonifies qi. Cistanche is a traditional yang-tonifying herb, primarily used to tonify kidney yang, enrich essence and blood, and moisten the intestines to relieve constipation. Beyond its medicinal uses, Cistanche has long been a therapeutic food resource in ancient China, particularly in the western regions, where it was commonly cooked with mutton to make soup to "replenish deficiency and fatigue," a practice that continues today. The latest data shows that 71 health foods primarily composed of Cistanche or its extracts, have been registered with China's Special Food Regulatory Bureau (Zhu et al. 2025).

Modern research has discovered that extracts of Cistanche using various solvents (water, ether, methanol) and its main active components, such as phenylethanoid glycosides (echinacoside, acteoside, isoacteoside), and polysaccharides, exhibit a wide range of healthy-aging effects, including extending lifespan, scavenging oxygen free radicals, neuroprotection, enhancing memory and learning, improving cognition, anti-osteoporosis effects, alleviating bone degeneration, anti-tumor, anti-inflammation, anti-fatigue, liver protection, and immunomodulation (Wang et al. 2017).

TCM theory posits that "the kidney governs reproduction" and "the kidney governs bones, produces marrow, and fills the brain." Cistanche with its mild nature and dual use as both medicine and food, is particularly effective in replenishing the essence and blood of the kidney. Therefore, it offers unique preventive and health benefits in maintaining reproductive health, bone health, and brain function. Relevant experimental studies confirm this. ① Reproductive health: extracts of Cistanche can promote testosterone synthesis, improve spermatogenesis; regulate the nuclear factor erythroid 2-related factor 2 (Nrf2) signaling pathway to counteract oxidative stress and reduce testicular oxidative damage; improve ovarian premature failure induced by cisplatin by regulating mitochondrial function; balance sex hormone levels, modulate immunity and cell apoptosis, and promote the repair of ovaries and uterus (Li et al. 2024). ② Bone health: the total glycosides and polysaccharides of

Cistanche can improve reduced bone formation and bone microstructure damage in the senile osteoporosis model senescence-accelerated mouse prone 6 (SAMP6) mice by activating the wingless and integration 1/β-catenin signaling pathway, and regulate the expression of bone morphogenetic protein 2 (BMP-2), osteoprotegerin (OPG), and receptor activator of nuclear factor-kappa B ligand (RANKL). Phenylethanoid glycosides protect osteoblast (OB) by inhibiting p53 expression, indicating their potential for preventing and treating osteoporosis (OP) (Zhang et al. 2023). ③ Brain health: the acteoside in Cistanche can alleviate oxidative stress, promote autophagy, and inhibit NOD-like receptor thermal protein domain associated protein 3 (NLRP3) mediated pyroptosis through multiple pathways by activating the Nrf2/ antioxidant response element (ARE) signaling pathway, thereby exerting neuroprotective effects and showing application value in combating neurodegenerative lesions (Zhu et al. 2025). Extracts of Cistanche can improve learning and memory abilities by promoting neuronal cell differentiation, neurite growth, and the formation of presynaptic structures (Gu et al. 2016), suggesting its potential as a modulator of neurotrophic factors.

Preclinical studies have shown promising healthy-aging effects of Cistanche, but robust clinical evidence for its healthy-aging benefits remains insufficient and requires further research.

11.4.1.4 *Angelica sinensis*

Angelica sinensis is the dried root of the Apiaceae plant *Angelica sinensis* (Oliv.) Diels, and is known for its ability to tonify and invigorate blood, regulate menstruation, relieve pain, and moisten the intestines to relieve constipation. As a classic blood-tonifying herb, Angelica sinensis was praised by the renowned Ming Dynasty physician Zhang Jingyue as the "holy medicine for blood," excelling in "stabilizing the five viscera, strengthening the body, and enhancing mental faculties," implying that Angelica can promote longevity by regulating physical and mental well-being. *The Grand Compendium of Materia Medica* records that Angelica wine can "harmonize blood vessels, strengthen sinews and bones, alleviate various pains, and regulate menstruation," making it a healthful elixir for disease prevention and longevity. As a potential healthy-aging herb, the active components of Angelica primarily include volatile oils (such as ligustilide and Angelica ketone), organic acids (such as ferulic acid and succinic acid), polysaccharides, and flavonoids (Liu et al. 2019). Among these, the healthy-aging effects of Angelica polysaccharides have been extensively studied. Angelica polysaccharides exhibit a variety of pharmacological activities, including improving anemia, antioxidation, immunomodulation, anti-inflammation, hypoglycemic and hypolipidemic effects, anti-tumor properties, liver protection, anti-fibrosis, and anti-radiation (Ren et al. 2025), which can delay aging in multiple systems, including hematopoietic stem cells (HSCs), the liver, kidneys, and brain. The related healthy-aging mechanisms involve: ① inhibiting oxidative stress and inflammatory responses; ② regulating aging-related genes and signaling pathways, such as the two key pathways of cellular senescence, $p16^{INK4a}$/Rb and $p19^{ARF}$/p53/$p21^{Cip1}$; ③ modulating telomerase activity; ④ regulating mitochondrial autophagy, protecting mitochondrial function, and reducing cell apoptosis; and ⑤ enhancing immunomodulation (Guo et al. 2022; Wang and Peng 2024).

Angelica is renowned for its blood-nourishing properties and has been used in China for blood production and deficiency supplementation for millennia. Modern research has confirmed this traditional efficacy. Early studies in animals found that Angelica extracts increase serum iron levels, reduce hepcidin-25 levels, enhance catalase (CAT) expression in the liver, and alleviate oxidative stress, offering preventive and protective effects against exercise-induced anemia (Chang et al. 2016). Human trials have also confirmed that short-term supplementation with Angelica blood-nourishing soup (a blood-nourishing formula composed of Angelica and Astragalus) can shorten running time, suppress exercise-induced hepcidin levels, and promote iron balance recovery, which is beneficial for enhancing exercise performance (Chang et al. 2018). Recent experimental studies have shown that Angelica polysaccharides not only regulate the Sirtuin1/Fork-head box O1 signaling pathway to inhibit oxidative stress and aging in HSCs (Peng et al. 2022) but also reduce reactive oxygen species (ROS) loads in bone marrow stromal cells, improve the hematopoietic microenvironment, and maintain a balance between osteogenesis and adipogenesis. Additionally, Angelica polysaccharides enhance communication between stromal and hematopoietic cells, alter the cytokine environment, and thereby protect HSCs from aging effects (Jing et al. 2022). These studies provide evidence supporting Angelica as a potential hematopoietic protector to maintain blood health and ensure bodily vitality.

Angelica sinensis, with its antioxidant properties, serves as a potential adjunct in healthy-aging effects. However, challenges such as indigestible cell walls and obstacles in macromolecular absorption necessitate the application of fermentation technology to enhance its performance. *In vivo* studies indicate that fermented Angelica more effectively activates the Nrf2 signaling pathway than non-fermented Angelica, further enhancing the activity of antioxidant enzymes such as superoxide dismutase (SOD) and CAT in the liver. It also more effectively regulates gut microbiota and metabolism, thereby alleviating D-galactose-induced hepatic aging in mice (Lu et al. 2023). Clearly, fermentation optimizes the healthy-aging potential of Angelica and merits further investigation.

11.4.1.5 Goji Berries

Goji berries are the dried mature fruits of the Solanaceae plant *Lycium barbarum L.*, are known for their ability to nourish the liver and kidney and improve eyesight. As early as the *Shen Nong's Classic of the Materia Medica*, there are records of goji berries delaying aging, stating that long-term consumption can "strengthen sinews and bones, lighten the body, and prevent aging." *The Grand Compendium of Materia Medica* includes various formulations such as goji wine and goji paste, along with detailed descriptions of methods for consuming goji berries. With the continuous discovery of its nutritional value and health benefits, goji berries have now risen to the status of a "superfood" and are listed as dietary and food supplements in numerous regions, including Caribbean countries, the European Union, North America, Southeast Asia, New Zealand, and Australia (Jagdale et al. 2021).

Goji berries contain a variety of nutrients, with carbohydrates being the primary macronutrient. Among the trace elements, vitamin C is the most prominent, followed by zeaxanthin and β-carotene, precursors to vitamin A. The content of dietary fiber is also considerable, including both soluble and insoluble fibers, which give Goji berries potential prebiotic effects (Shi et al. 2024). Polysaccharides, carotenoids, and phenolic compounds constitute the important bioactive components of Goji berries and are closely related to their health-promoting effects. Experimental studies show that Goji berries possess a wide range of pharmacological properties, including vision protection, hypoglycemic and hypolipidemic effects, anti-tumor properties, immunomodulation, neuroprotection, cardiovascular protection, and antioxidation (Ma et al. 2019), which can prevent and treat aging and age-related diseases through multiple targets and pathways. A systematic review focusing on the healthy-aging effects of goji polysaccharides reveals that they exert anti-aging effects through multiple mechanisms, including: ① antioxidation and anti-apoptosis; ② reducing DNA strand breaks and increasing cell activity; ③ blocking signaling pathways that regulate cell senescence (e.g., the p53 pathway); ④ modulating the expression of aging-related genes; and ⑤ alleviating telomere shortening. Clinical studies have also shown that patients consuming Goji polysaccharides exhibit significantly improved DNA repair rates. Goji polysaccharides can increase the levels of antioxidant enzymes SOD and glutathione peroxidase (GSH-Px) in human serum and reduce the content of oxidative stress metabolite malondialdehyde (MDA), thus demonstrating healthy-aging effects in humans similar to those observed in animal experiments (Ni et al. 2021).

"Improving eyesight" is a characteristic traditional function of Goji berries. Pharmacological research has found that goji polysaccharides can protect the optic nerve through anti-inflammatory, anti-apoptotic, and autophagy-regulating effects, and can modulate the phosphatidylinositol 3-kinase (PI3K) /protein kinase B (AKT) /mammalian target of rapamycin (mTOR) signaling pathway to improve the activity of retinal pigment epithelial cells (RPE), thereby delaying the progression of age-related macular degeneration (Yang et al. 2022). Zeaxanthin, which is rich in goji berries, accumulates in the eyes, preventing oxidative damage and filtering harmful blue light, which is also beneficial for vision health. Clinical randomized trials also confirm that in healthy adults aged 45–65, consuming Goji berries five times a week, 28 grams each time, for 90 days significantly increases macular pigment optical density (MPOD) (Shi et al. 2024). These research findings align well with the traditional use of Goji berries in TCM to improve age-related vision decline in the elderly.

11.4.1.6 Dendrobium

Dendrobium, a traditional Chinese medicinal herb, belongs to the Dendrobium Sw. genus of the Orchidaceae family. In China, 78 species of Dendrobium have been documented, 14 of which are endemic to the country. The 2020 edition of the *Pharmacopoeia of the People's Republic of China* lists five species of Dendrobium for medicinal use, all of which utilize the stem as the medicinal part. The herb known as "Shi Hu" originates from *Dendrobium nobile* Lindl., *Dendrobium huoshanense* C.Z.Tang et S.J.Cheng, *Dendrobium chrysotoxum* Lindl., or *Dendrobium fimbriatum* Hook.; whereas the herb named "Tie Pi Shi Hu" comes from *Dendrobium officinale* Kimura et Migo (Li et al. 2023). All of these species share the therapeutic effects of nourishing the stomach, promoting the production of body fluids, and clearing heat by enriching yin.

The *Shen Nong's Classic of the Materia Medica,* first recorded Dendrobium, describing it as capable of "tonifying the five viscera and relieving deficiency and fatigue" and "extending longevity," classifying it as a superior tonic. *The Grand Compendium of Materia Medica* mentions that Dendrobium can be "boiled in water as a tea, which greatly clears the lung and tonifies the spleen." Ming Dynasty physician Miao Xi Yong suggested that Dendrobium, after being "steamed in wine" and then "brewed as a tea," could "instantly strengthen the legs." Qing Dynasty physician Zhao Xue Min noted that Dendrobium tea "enhances vital energy and strength," and was "widely popular in the south and north of Yangtze river." It is evident that Dendrobium, consumed as a tea, was very popular for health maintenance during the Ming and Qing Dynasties in China. Modern understanding of Dendrobium's health-preserving and healthy-aging functions continues to deepen. At least 450 compounds have been isolated and identified from over 50 species of Dendrobium, with bibenzyls and polysaccharides as the main active components, phenolic compounds and lignans widely distributed, sesquiterpenoids as the most common chemical constituents, and alkaloids serving as quality markers for *Dendrobium nobile*. These components endow Dendrobium with diverse pharmacological activities, such as anti-tumor, anti-fatigue, antioxidation, anti-inflammation, liver protection, gastrointestinal protection, anti-diabetes, neuroprotection, immunomodulation, and anti-osteoporosis (Li et al. 2023), laying the foundation for Dendrobium's healthy-aging and delaying age-related changes in the body effects.

Regarding healthy-aging mechanisms, modern medicine posits that the interplay of insulin resistance and inflammation can accelerate aging. Research indicates that *Dendrobium nobile* can improve insulin resistance through anti-inflammatory, antioxidative, and gut microbiota regulatory effects, thereby delaying aging (Nie et al. 2020a, b). *Dendrobium nobile* Lindl. alkaloids (DNLA) reduces intracellular ROS production, inhibits neuroinflammation, prevents tau protein hyperphosphorylation, enhances autophagy, and reduces neuronal apoptosis, exerting neuroprotective effects that are beneficial for preventing and delaying neurodegenerative diseases (Bhardwaj et al. 2024). Ultrafine powder of *Dendrobium officinale* (DOFP) alleviates physical fatigue and mental depression symptoms in subhealth model mice by reducing metabolic product accumulation, enhancing glycogen storage and antioxidative activity, and regulating abnormalities in the neuroendocrine immune system (Shi et al. 2022), thereby slowing the natural aging process of the body.

Dendrobium is adept at nourishing gastric yin and is a commonly used TCM for stomach and fluid benefits. *In vitro* experiments show that polysaccharides from Dendrobium officinale protect human gastric mucosal epithelial cells by reducing ROS levels and improving nuclear morphology; animal models further indicate that polysaccharides protect the gastric mucosa by downregulating the expression ratio of Bcl-2 associated X protein (Bax)/B-cell lymphoma 2 protein (Bcl-2) in the gastric mucosa, reducing oxidative stress-induced cell apoptosis, and thus protecting the gastric mucosa (Xu et al. 2022). A compound oral solution composed of *Dendrobium nobile*, Goji berries, and *Pueraria lobata* has also been confirmed to reduce ethanol-induced gastric mucosal injury in rats through antioxidative and pro-survival mechanisms (Pei et al. 2021). These studies confirm the potential advantages of Dendrobium in preventing age-related gastric mucosal lesions.

11.4.2 Regulating Herbs

A total of four herbs are included, namely qi-regulating *Rosa rugosa*, dampness-resolving Poria, phlegm-resolving Dried tangerine peel, and blood-activating *Panax notoginseng*. Their traditional functions, pharmacological activities, healthy-aging components, and mechanisms are shown in Table 11.2.

11.4.2.1 Rosa Rugosa

Rosa rugosa, the dried flower buds of the Rosaceae plant *Rosa rugosa* Thunb., functions to regulate qi and resolve depression, harmonize blood, and relieve pain. In China, the earliest record of *Rosa rugosa* as a dual-use medicinal and edible substance dates back to the Ming Dynasty text *Food as Materia Medica* which states that Rosa "benefits the lung and spleen, and supports the liver and gallbladder," and that "consuming it is aromatic, sweet, and refreshing, invigorating the spirit." During the Ming and Qing Dynasties, various methods of consuming Rosa rugosa emerged, including rose tea, rose dew, rose wine, rose sugar, and rose jam. Additionally, ancient people used Rosa for hair cleansing and scenting, leveraging its beautifying effects.

Today, Rosa rugosa continues to hold unique practical value. It contains a rich array of chemical components, with active components primarily consisting of volatile oils, flavonoids, polyphenols, polysaccharides, and pigments. Volatile oils are the main constituents of rose essential oil, with monoterpene compounds such as geraniol, citronellol, nerol, and their acetates being the most prevalent. Rosa rugosa

Table 11.2 Functions, Pharmacological activities, healthy-aging ingredients and Mechanisms of Chinese Regulating herbs

Herb name	Functions[a]	Pharmacological activities	Healthy-aging ingredients	Healthy-aging mechanisms	Cited references
Rose rugosa (*Mei gui hua*)	Regulating qi and relieving stagnation Harmonizing blood Alleviating pain	Anti-oxidation Anti-inflammation Anti-tumor Anti-diabetes Anti-bacterial Reducing hepatotoxicity	Essential oils, flavonoids, polyphenols, polysaccharides	Regulateing antioxidant enzyme activities Inhibit inflammatory mediator production Regulating immunity	Lu and Wang (2018), Guan et al. (2025)
Poria (*Fu ling*)	Promoting urination and draining dampness Strengthening spleen Calming the mind	Cytotoxic effect Anti-invasion Anti-inflammation Anti-hyperglycemia Anti-bacterial and anti-virus Sedative and hypnotic effect Anti-ischemia/reperfusion injury	Triterpene acids (*pachymic acid*) Poria cocos polysaccharide (PCP)	Inducing autophagy Regulating aging-related signaling pathways Regulating neuroinflammation Alleviating oxidative damage Inhibiting mitochondrial permeability transition	Nie et al. (2020a, b), Wei et al. (2022), Lee and Kim (2017) and Hu et al. (2024)
Dried Tangerine Peel (*Chen pi*)	Regulating qi and strengthening spleen Drying dampness and resolving phlegm	Anti-oxidation Anti-inflammation Anti-tumor Anti-obesity Regulating cardiovascular function	Flavones (*Polymethoxyflavones -nobiletin*)	Resisting oxidative stress Regulating inflammation signaling and reducing inflammation mediators Regulating circadian clock Promoting mitochondrial supercomplex (SC) formation	Guan et al. (2024), Chen et al. (2024) and Mileykovskaya et al. (2020)

		Regulating metabolism			
Panax notoginseng (*San qi*)	Dissipating stasis and stopping bleeding Circulating blood and alleviating pain	Cardio-protection Cerebrovascular protection Neuro-protection Anti-tumor Anti-inflammation Hemostasis Anti-coagulation	Panax notoginseng saponins (PNS) Panax notoginseng polysaccharides	Anti-oxidation Regulating immunity Anti-inflammation Regulating neurotransmitter balance Regulating cell signaling pathways Influencing cell-cycle progression	Wei et al. (2024), Zhao et al. (2017)

[a]Chinese herbal indications are cited from Chinese pharmacopoeia (2020 edition)

also exhibits a wide range of pharmacological activities, including antioxidation, anti-inflammation, anti-tumor, anti-diabetes, and antibacterial effects (Lu and Wang 2018).

Related studies indicate that Rosa rugosa can effectively enhance the body's antioxidant stress capacity through multiple pathways, such as improving antioxidant enzyme activity, modulating inflammatory signaling pathways, and strengthening immune defense mechanisms, thereby exerting healthy-aging effects (Guan et al. 2025). Rosa rugosa is commonly used in daily chemical products, and *in vitro* experiments have verified the healthy-aging effects of roses and their phenolic compounds on the skin. Rosarugosides A and D can inhibit the overexpression of ROS in normal human dermal fibroblasts (NHDFs) induced by tumor necrosis factor-alpha (TNF-α); the former effectively suppresses matrix metalloproteinase-1 secretion, while the latter increases procollagen Iα1 expression, both showing potential as protective agents against skin damage (Kim et al. 2024). In the *Caenorhabditis elegans* model, Rosa rugosa aqueous polyphenol (RAP) demonstrated significant healthy-aging effects, extending the lifespan of nematodes and improving their survival rates under oxidative stress and heat shock conditions. Its healthy-aging effects may be related to *in vitro* antioxidation, free radical scavenging, and activation of the endogenous antioxidant defense system (Zhang et al. 2019). Rose essential oil, with its anti- amyloid Beta-protein (Aβ), antioxidative, and antidepressant-like properties, has shown neuroprotective potential in *Caenorhabditis elegans* experiments (Spisni et al. 2023), suggesting that Rosa rugosa and its extracts may also be used to prevent and delay age-related neurodegenerative diseases.

11.4.2.2 Poria

Poria, the dried sclerotium of the pore fungus *Poria cocos* (Schw.) Wolf, is a medicinal fungus primarily known for its diuretic, spleen-strengthening, and tranquillizing effects. The *Shen Nong's Classic of the Materia Medica* first mentioned that long-term consumption of Poria can "calm the soul, nourish the spirit, relieve hunger, and extend longevity." Since then, the health-promoting and longevity-enhancing value of Poria has been highly esteemed. The renowned Northern Song Dynasty literary figure Su Shi had considerable insights into the consumption and health benefits of Poria. He believed that making cakes with Poria, sesame, and honey for daily consumption could keep a person "vigorous and free from disease," which he considered the secret to longevity. Poria cakes were popular in the Qing Dynasty imperial court as a health-preserving food. Later, they spread from the court to the general public, becoming a famous regional specialty.

Poria contains two main types of bioactive components: triterpenoids and polysaccharides. Triterpenoids are typically regarded as the primary chemical constituents and are often used as quality markers for Poria, with poricoic acid being unique to Poria and not found in any other Chinese medicine (Nie et al. 2020a, b). Poricoic acid exhibits various pharmacological activities, including cytotoxicity, anti-invasion, anti-inflammation, hypoglycemic, antiviral, antibacterial, sedative-hypnotic, and anti-ischemia/reperfusion injury effects (Wei et al. 2022). Additionally, poricoic acid possesses healthy-aging properties. *In vitro* experiments have confirmed that poricoic acid can promote autophagy in aging cells and is closely related to the insulin-like growth factor 1 (IGF-1) signaling pathway. Poricoic acid increases autophagosome formation, inhibits mTOR activation, and regulates the expression of proteins related to the IGF-1 signaling pathway, possibly inducing autophagy through phospho-forkhead box protein O1 regulation, thereby delaying cellular aging (Lee and Kim 2017). Poricoic acid can also downregulate key genes in the nuclear factor kappa-B and interleukin-17 signaling pathways, reduce intracellular ROS accumulation, and enhance mitochondrial membrane potential, providing protection against iron ion-induced neuronal cell damage (Hu et al. 2024), suggesting that poricoic acid may also delay cellular aging by promoting resistance to oxidative stress. In addition to poricoic acid, Poria polysaccharides have been found to facilitate the translocation of skn-1 from the cytoplasm to the nucleus, upregulate the expression of related antioxidant genes, and reduce ROS levels in *Caenorhabditis elegans*, thereby slowing the aging process (Song et al. 2024). These studies all confirm the traditional efficacy of Poria in "extending longevity."

In ancient Chinese medicines used for treating memory disorders consistent with the clinical characteristics of Alzheimer's disease (AD), Poria was the most frequently used (May et al. 2016). Modern research has found that Poria ethanol extract can alleviate cognitive dysfunction in amyloid precursor protein/presenilin 1 mice by modulating the balance of Aβ production and clearance, improving gut microbiota, and reversing bile acid metabolic disorders (Sun et al. 2021). Poria essential oil can regulate endogenous antioxidant enzymes, inhibit Aβ accumulation and lower glial fibrillary acidic protein levels, and suppress acetylcholinesterase activity to promote acetylcholine synthesis, thereby improving Aβ-induced learning and memory deficits in rats (Lv et al. 2022). Thus, Poria can be used to prevent brain aging, and the records in the *Shen Nong's Classic of the Materia Medica* regarding its ability to "calm the soul and nourish the spirit" have corresponding scientific basis.

11.4.2.3 Dried Tangerine Peel

Dried tangerine peel, the dried mature peel of the Rutaceae plant *Citrus reticulata* Blanco and its cultivated varieties, is known for its ability to regulate qi and strengthen the spleen, as well as to dry dampness and resolve phlegm. Also known as ju pi, the peel used in traditional Chinese medicine is not fresh but requires aging, a process that promotes the gradual accumulation of flavonoid compounds, thereby significantly enhancing its medicinal efficacy (Fu et al. 2022). The physician Tao Hongjing from the Wei, Jin, and Southern and Northern Dynasties mentioned that dried tangerine peel "when consumed long-term, lightens the body and prolongs life." The term "lightens the body" implies that dried tangerine peel can help maintain a slender figure, which aligns with pharmacological research findings that it has anti-obesity effects; "prolongs life" indicates its potential to extend longevity, a view also supported by modern studies.

The components of dried Tangerine peel are primarily divided into volatile and non-volatile categories, with flavonoids being the most significant among the non-volatile components (Guan et al. 2024). These flavonoids form the material basis of the potential healthy-aging effects of dried Tangerine peel, particularly polymethoxyflavones (PMFs), which have become a recent research focus. The impact of PMFs on the biological processes of aging is reflected in: ① combating ROS through various mechanisms, such as directly scavenging free radicals, chelating metal ions, activating antioxidant enzymes via the Nrf2 pathway, and inhibiting superoxide-generating enzymes; and ② modulating inflammatory signaling pathways to reduce the production of inflammatory mediators like TNF-α and interleukins, thereby preventing age-related pro-inflammatory states (Chen et al. 2024).

Nobiletin, a type of PMFs, possesses various biological activities, including antioxidation, anti-inflammation, anticancer, and metabolic and cardiovascular regulation. Recent studies have shown that nobiletin can enhance the amplitude of the circadian clock, alter its period and phase, and directly bind to and activate the retinoic acid receptor-related orphan receptor α (RORα)/retinoic acid receptor-related orphan receptor γ (RORγ) nuclear receptors. In animal models of metabolic diseases, it inhibits weight gain, improves energy homeostasis, hepatic steatosis, and metabolic gene expression, and enhances insulin sensitivity. In aged mice, it improves behavior and metabolism, extends average lifespan, and regulates metabolic functions of organs such as the liver and pancreas. In other words, nobiletin can prevent aging and metabolism-related diseases caused by circadian rhythm disorders by modulating the circadian clock. Additionally, nobiletin activates the expression of genes encoding components of the respiratory chain complex in skeletal muscle, enhances ATP production, reduces ROS generation, and promotes the formation of mitochondrial supercomplex, which may be one of the key mechanisms for its promotion of healthy aging (Mileykovskaya et al. 2020).

Furthermore, nobiletin has shown promise in improving cognitive impairment in the senescence accelerated mouse prone 8 mouse model of accelerated aging (which can serve as a model for AD) by reducing oxidative stress and tau protein hyperphosphorylation. This suggests its potential for preventing and treating age-related neurodegenerative diseases (Nakajima and Ohizumi 2019). A double-blind, randomized controlled study evaluated the impact of Nobilex® (a test food containing nobiletin) on cognitive function and found that it improved memory function in healthy elderly Japanese individuals (Matsuzaki et al. 2022). The results of this clinical study somewhat support the preclinical evidence that nobiletin may delay the negative effects associated with brain aging.

11.4.2.4 Panax Notoginseng

Panax notoginseng, the dried root and rhizome of the Araliaceae plant *Panax notoginseng* (Burk.) F.H. Chen, is known for its ability to disperse stasis and stop bleeding, as well as to reduce swelling and relieve pain. As a medicinal plant in the Panax genus, the use of Panax notoginseng in China emerged much later than that of Ginseng, with its first recorded use dating back to the Ming Dynasty. Li Shizhen, in his *The Grand Compendium of Materia Medica,* referred to *Panax notoginseng* as "jin bu huan," emphasizing its preciousness. Unlike Ginseng, which is primarily used to tonify qi, Panax notoginseng is distinguished by its ability to stop bleeding and resolve stasis, being described as "able to treat all blood-related diseases." In Yunnan, China, it is customary to braise Panax notoginseng with chicken as a health-preserving recipe. Qing Dynasty physician Zhao Xue Min described this therapeutic method in his writings, suggesting that it "can treat various deficiencies and weaknesses," indicating that Panax notoginseng also possesses tonic and strengthening effects. With the advancement of modern research, the application of Panax notoginseng in health preservation for the elderly in China has become increasingly widespread.

The bioactive components of Panax notoginseng include saponins, amino acids, polysaccharides, and flavonoid compounds, the types and content of which are influenced by region, tissue, and developmental age. Saponins are the most important components, categorized into dammarane, oleanane, and ocotillol types, primarily including ginsenosides Rb1, Rc, Rb2, Rg1, Re, Ro, Rd., notoginsenoside R1, and ginsenoside R2. Relevant studies indicate that the total saponins of Panax notoginseng possess a range of pharmacological activities, such as cardiac protection, cerebrovascular protection, neuroprotection, anti-tumor, anti-inflammatory, hemostatic, and anti-coagulation effects (Wei et al. 2024). Research on model organisms has revealed potential healthy-aging functions of Panax notoginseng. While Panax notoginseng extract does not extend the lifespan of *D. melanogaster*, it effectively enhances their resistance to various oxidative stressors and endoplasmic reticulum (ER) stress, and reduces cardiac cycle in aging *D. melanogaster,* improving cardiac function (Teseo et al. 2021). Moderate amounts of Panax notoginseng saponins can increase antioxidant enzyme activity, significantly extend the lifespan of *D. melanogaster*, and protect their reproductive capacity (Zou et al. 2023). Two polysaccharides (MRP5 and MRP5A) from Panax notoginseng, though weak *in vitro* antioxidant capacity, exhibit significant antioxidant stress and lifespan-extending effects in *Caenorhabditis elegans* (Feng et al. 2019). Further studies show that extracts and bioactive components of Panax notoginseng hold multifaceted potential in preventing aging and cell senescence-related diseases, including extending lifespan, anti-vascular aging, anti-neurodegeneration, and anti-tumor effects. Different components function through various mechanisms, such as antioxidation, immune function modulation, anti-inflammation, neurotransmitter balance regulation, cell signaling pathway modulation, and cell cycle influence (Zhao et al. 2017).

In ancient times, Panax notoginseng was first applied to fractures and other orthopedic conditions. Experimental studies have revealed that moderate amounts of Panax notoginseng saponins promote angiogenesis and prevent bone loss in mouse osteoporosis models, making it a candidate drug for preventing postmenopausal osteoporosis in women (Hu et al. 2020). Panax notoginseng saponins may also delay joint cartilage degradation by inhibiting the PI3K-AKT pathway, thereby protecting chondrocytes in osteoarthritis from aging and apoptosis (Zhang et al. 2020). Thus, Panax notoginseng holds promise in preventing bone loss due to aging and promoting bone health.

11.5 Conclusion

TCM posits that human aging stems from deficiencies in qi, blood, yin, and yang, further exacerbated by "excess" pathological influences such as stagnant qi, phlegm-dampness, and blood stasis. By appropriately employing tonifying and regulating herbs, one can "nourish deficiencies" and "eliminate excesses" to delay aging and promote healthy longevity. This paper reviews the characteristics of commonly used healthy-aging herbs based on traditional applications and modern research. Under the modern paradigm of healthy-aging research, the material basis, effects, and mechanisms of TCM's healthy-aging properties are continually revealed and verified, with promising results.

Chinese Medicinal Herbs are derived from natural foods, plants and botanical sources, with their efficacy systematically documented through thousands of years of human applications. These sources are utilized in various forms: sole herbs, compound formulas, isolated active compounds, or standardized extracts. From the TCM perspectives, herbs are primarily administered as compound formulas and occasionally single formula, e.g. Ginseng and other botanicals both being applied as holistic multi-component systems and sole use. Asian ginseng, Astragalus, Goji berries, Angelica sinensis and Rosa rugosa have been gaining international market presence. When using isolated active compounds or standardized extracts, optimal therapeutic outcomes are requiring integration with TCM theory, particularly through pattern differentiation of Yin-Yang and deficiency-excess states in clinical practice.

However, most current research remains in the preclinical stage, lacking large-scale, rigorously designed clinical trials targeting healthy-aging TCM herbs. Additionally, the connection between TCM's healthy-aging effects and its traditional functions, as well as its integration with TCM theories on aging, remains underexplored. Incorporating TCM theories into the existing framework of aging research presents a challenge and an opportunity. It may foster the generation of new scientific evidence for TCM's healthy-aging properties and further advance the field of healthy-aging research.

References

Balusamy SR et al (2023) A comprehensive and systemic review of ginseng-based nanomaterials: synthesis, targeted delivery, and biomedical applications. Med Res Rev 43(5):1374–1410. https://doi.org/10.1002/med.21953

Berezutsky MA, Durnova NA, Vlasova IA (2020) Experimental and clinical studies of mechanisms of the antiaging effects of chemical compounds in astragalus membranaceus (review). Adv Gerontol 10(2):142–149. https://doi.org/10.1134/s2079057020020046

Bhardwaj K et al (2024) Dendrobine: a neuroprotective sesquiterpenic alkaloid for the prevention and treatment of diseases: a review. Mini-Rev Med Chem 24(15):1395–1408. https://doi.org/10.2174/0113895575274314240125105120

Chang C-W et al (2016) Protective effects of the roots of Angelica sinensis on strenuous exercise-induced sports anemia in rats. J Ethnopharmacol 193:169–178. https://doi.org/10.1016/j.jep.2016.08.010

Chang C-W et al (2018) Repressed exercise-induced hepcidin levels after Danggui Buxue tang supplementation in male recreational runners. Nutrients 10(9):1318. https://doi.org/10.3390/nu10091318

Chen L et al (2023) Research and Development of selenium-rich Chinese medicinal materials. Cent South Agric Sci Technol 44(10):214–217, 235

Chen X et al (2024) Citrus polymethoxyflavones regulate against aging-associated diseases: advances in biological mechanisms responsible for their regulation. J Agric Food Chem 72(51):28209–28224. https://doi.org/10.1021/acs.jafc.4c08004

Cui Y (1996) A preliminary exploration on the medicines for anti-aging and prolonging life in Shennong's herbal classic and its anti-aging thoughts. Henan Traditional Chin Med J 11(1): 12–14

Feng S et al (2019) Antioxidant and anti-aging activities and structural elucidation of polysaccharides from Panax Notoginseng root. Process Biochem 78:189–199. https://doi.org/10.1016/j.procbio.2019.01.007

Fu M et al (2022) Research Progress on the surface microorganisms of Citri Reticulatae Pericarpium and derived microbial transformation of flavonoids. Modern Food Sci Technol 38(4):282–291

Gu C, Yang X, Huang L (2016) Cistanches Herba: a neuropharmacology review. Front Pharmacol 7. https://doi.org/10.3389/fphar.2016.00289

Guan L et al (2021) Effects of selenium-enriched Astragalus polysaccharides on antioxidant function, lipid metabolism, tissue selenium content, and immunity of Bashang long-tailed chickens. Southwest China J Agric Sci 34(11):2531–2538

Guan X et al (2024) Research Progress on chemical constituents and pharmacological effects of Chenpi(dried tangerine Peel). Chin Arch Tradit Chin Med 42(6):41–49, 266

Guan Y et al (2025) Research Progress on the anti-aging mechanism of Rosa rugosa. Chin Arch Tradit Chin Med:1–14. Available at: http://kns.cnki.net/kcms/detail/21.1546.R.20240904.1245.002.html. Accessed 26 Mar 2025

Guo X et al (2022) The role and mechanism of polysaccharides in anti-aging. Nutrients 14(24): 5330. https://doi.org/10.3390/nu14245330

Hu H et al (2020) Panax notoginseng saponins prevent bone loss by promoting angiogenesis in an osteoporotic mouse model. Biomed Res Int 2020(1). https://doi.org/10.1155/2020/8412468

Hu S et al (2024) RNA-seq analysis reveals potential neuroprotective mechanisms of pachymic acid toward iron-induced oxidative stress and cell death. Cell Transplant 33. https://doi.org/10.1177/09636897231218382

Jagdale YD et al (2021) Nutritional profile and potential health benefits of super foods: a review. Sustainability 13(16):9240. https://doi.org/10.3390/su13169240

Jia L, Zhao Y (2009) Current evaluation of the millennium phytomedicine- ginseng (i): etymology, pharmacognosy, phytochemistry, market and regulations. Curr Med Chem 16(19):2475–2484. https://doi.org/10.2174/092986709788682146

Jing P et al (2022) Angelica sinensis polysaccharides prevents hematopoietic regression in D-galactose-induced aging model via attenuation of oxidative stress in hematopoietic microenvironment. Mol Biol Rep 50(1):121–132. https://doi.org/10.1007/s11033-022-07898-w

Keshavarz M et al (2023) Aging—what it is and how to measure it. Mech Ageing Dev 213:111837. https://doi.org/10.1016/j.mad.2023.111837

Kim KS et al (2024) Rosarugosides A and D from Rosa rugosa flower buds: their potential anti-skin-aging effects in TNF-α-induced human dermal fibroblasts. Plants 13(9):1266. https://doi.org/10.3390/plants13091266

Kozhevnikova OS et al (2022) Astragalus membranaceus increases leukocyte telomere length, but does not suppress development of accelerated senescence signs in oxys rats. Adv Gerontol 12(2):128–134. https://doi.org/10.1134/s2079057022020114

Lee S-G, Kim M-M (2017) Pachymic acid promotes induction of autophagy related to IGF-1 signaling pathway in wi-38 cells. Phytomedicine 36:82–87. https://doi.org/10.1016/j.phymed.2017.09.020

Lee TX et al (2021) Reduced stem cell aging in exercised human skeletal muscle is enhanced by ginsenoside RG1. Aging 13(12):16567–16576. https://doi.org/10.18632/aging.203176

Li X-T et al (2012) Mitochondrial protection and anti-aging activity of astragalus polysaccharides and their potential mechanism. Int J Mol Sci 13(2):1747–1761. https://doi.org/10.3390/ijms13021747

Li P-Y, Li L, Wang Y-Z (2023) Traditional uses, chemical compositions and pharmacological activities of dendrobium: a review. J Ethnopharmacol 310:116382. https://doi.org/10.1016/j.jep.2023.116382

Li Z et al (2024) The role of cistanches herba and its ingredients in improving reproductive outcomes: a comprehensive review. Phytomedicine 129:155681. https://doi.org/10.1016/j.phymed.2024.155681

Liu M et al (2023) Application of spleen-stomach regulation methods in health-preserving prescriptions of the Qing dynasty Imperial court. J Shaanxi Univ Chin Med 46(4):73–76

Liu P, Zhao H, Luo Y (2017) Anti-aging implications of astragalus membranaceus (Huangqi): a well-known Chinese tonic. Aging Dis 8(6):868. https://doi.org/10.14336/ad.2017.0816

Liu Y et al (2019) New insights for cellular and molecular mechanisms of aging and aging-related diseases: herbal medicine as potential therapeutic approach. Oxidative Med Cell Longev 2019:1–25. https://doi.org/10.1155/2019/4598167

López-Otín C et al (2013) The hallmarks of aging. Cell 153(6):1194–1217. https://doi.org/10.1016/j.cell.2013.05.039

López-Otín C et al (2023) Hallmarks of aging: an expanding universe. Cell 186(2):243–278. https://doi.org/10.1016/j.cell.2022.11.001

Lu J, Wang C (2018) Medicinal components and pharmacological effects of Rosa rugosa. Rec Nat Prod 12(6):535–543. https://doi.org/10.25135/rnp.60.17.12.191

Lu X et al (2023) Fermentedangelica sinensisactivates Nrf2 signaling and modulates the gut microbiota composition and metabolism to attenuated-gal induced liver aging. Food Funct 14(1):215–230. https://doi.org/10.1039/d2fo01637k

Lv Q et al (2022) Neuroprotective effects of Poria Cocos (agaricomycetes) essential oil on Aβ1-40-Induced learning and memory deficit in rats. Int J Med Mushrooms 24(10):73–82. https://doi.org/10.1615/intjmedmushrooms.2022045031

Ma ZF et al (2019) Goji berries as a potential natural antioxidant medicine: an insight into their molecular mechanisms of action. Oxidative Med Cell Longev 2019:1–9. https://doi.org/10.1155/2019/2437397

Matsuzaki K et al (2022) A narrative review of the effects of citrus peels and extracts on human brain health and metabolism. Nutrients 14(9):1847. https://doi.org/10.3390/nu14091847

May BH et al (2016) Memory impairment, dementia, and alzheimer's disease in classical and contemporary traditional Chinese medicine. J Altern Complement Med 22(9):695–705. https://doi.org/10.1089/acm.2016.0070

Mileykovskaya E et al (2020) Nobiletin: targeting the circadian network to promote bioenergetics and healthy aging. Biochem Mosc 85(12–13):1554–1559. https://doi.org/10.1134/s000629792012007x

Nakajima A, Ohizumi Y (2019) Potential benefits of nobiletin, a citrus flavonoid, against alzheimer's disease and parkinson's disease. Int J Mol Sci 20(14):3380. https://doi.org/10.3390/ijms20143380

Ni J et al (2021) Lycium barbarum polysaccharides in ageing and its potential use for prevention and treatment of osteoarthritis: a systematic review. BMC Complement Med Ther 21(1). https://doi.org/10.1186/s12906-021-03385-0

Nie A et al (2020a) Phytochemistry and pharmacological activities of Wolfiporia Cocos (F.A. Wolf) ryvarden & Gilb. Front Pharmacol 11. https://doi.org/10.3389/fphar.2020.505249

Nie X et al (2020b) Anti-aging properties of dendrobium Nobile Lindl.: from molecular mechanisms to potential treatments. J Ethnopharmacol 257:112839. https://doi.org/10.1016/j.jep.2020.112839

Panossian AG et al (2020) Evolution of the adaptogenic concept from traditional use to medical systems: pharmacology of stress-and aging-related diseases. Med Res Rev 41(1):630–730. https://doi.org/10.1002/med.21743

Pei Z et al (2021) Development of a compound oral liquid containing herbal extracts and its effect on immunity and gastric mucosa. J Food Sci 86(6):2684–2699. https://doi.org/10.1111/1750-3841.15761

Peng L et al (2022) Angelica sinensis polysaccharide suppresses the aging of hematopoietic stem cells through SIRT1/FOXO1 signaling. Clin Lab 68(05/2022). https://doi.org/10.7754/clin.lab.2021.210731

Ren C et al (2025) Pharmacological action of Angelica sinensis polysaccharides: a review. Front Pharmacol 15:1–18. https://doi.org/10.3389/fphar.2024.1510976

Salvador L et al (2016) A natural product telomerase activator lengthens telomeres in humans: a randomized, double blind, and placebo controlled study. Rejuvenation Res 19(6):478–484. https://doi.org/10.1089/rej.2015.1793

Shahrajabian MH et al (2019) A review of astragalus species as foodstuffs, dietary supplements, a traditional Chinese medicine and a part of Modern Pharmaceutical Science. Appl Ecol Environ Res 17(6):10.15666/aeer/1706_1337113382

Shi M-L et al (2022) Effects of dendrobium officinale ultrafine powder on sub-health mice induced by unhealthy lifestyle based on neuroendocrine immune system. Food Funct 13(23): 12436–12450. https://doi.org/10.1039/d2fo02158g

Shi X et al (2024) Advances in the study of bioactive compounds and nutraceutical properties of Goji Berry (Lycium Barbarum L.). Appl Sci 15(1):262. https://doi.org/10.3390/app15010262

Song Z et al (2024) Antioxidant and anti-aging effects and mechanism of Poria cocos polysaccharide in Caenorhabditis elegans. Chin Tradit Herb Drug 55(4):1133–1144

Spisni E, Valerii MC, Massimino ML (2023) Essential oil molecules can break the loop of oxidative stress in neurodegenerative diseases. Biology 12(12):1504. https://doi.org/10.3390/biology12121504

Su J et al (2023) Research progress on the anti-aging potential of the active components of ginseng. Nutrients 15(15):3286. https://doi.org/10.3390/nu15153286

Sun Y et al (2021) Poria cocos could ameliorate cognitive dysfunction in app/ps1 mice by restoring imbalance of AB production and clearance and gut microbiota dysbiosis. Phytother Res 35(5): 2678–2690. https://doi.org/10.1002/ptr.7014

Tao L et al (2023) Association of medicinal plant consumption with all-cause mortality and cognitive impairment in older adult: a large prospective cohort study. Phytomedicine 119: 154995. https://doi.org/10.1016/j.phymed.2023.154995

Tartiere AG, Freije JM, López-Otín C (2024) The hallmarks of aging as a conceptual framework for health and longevity research. Front Aging 5. https://doi.org/10.3389/fragi.2024.1334261

Teseo S et al (2021) G. Sinense and P. Notoginseng extracts improve Healthspan of aging flies and provide protection in a Huntington disease model. Aging Dis 12(2):425. https://doi.org/10.14336/ad.2020.0714-1

Tian T et al (2023) The anti-aging mechanism of ginsenosides with medicine and food homology. Food Funct 14(20):9123–9136. https://doi.org/10.1039/d3fo02580b

Tsoukalas D et al (2019) Discovery of potent telomerase activators: unfolding new therapeutic and anti-aging perspectives. Mol Med Rep. https://doi.org/10.3892/mmr.2019.10614

Wang L, Peng W (2024) Research progress on anti-aging effects of Angelica polysaccharide. Pract Geriatr 38(3):232–235

Wang N et al (2017) Herba cistanches: anti-aging. Aging Dis 8(6):740. https://doi.org/10.14336/ad.2017.0720

Wei C et al (2022) Pharmacological profiles and therapeutic applications of pachymic acid (review). Exp Ther Med 24(3). https://doi.org/10.3892/etm.2022.11484

Wei G et al (2024) Panax notoginseng: Panoramagram of phytochemical and pharmacological properties, biosynthesis, and regulation and production of ginsenosides. Horticult Res 11(8). https://doi.org/10.1093/hr/uhae170

Xu X et al (2022) Bioactivities and mechanism of actions of dendrobium officinale: a comprehensive review. Oxidative Med Cell Longev 2022:1–21. https://doi.org/10.1155/2022/6293355

Yang Y et al (2017) Ginseng: an nonnegligible natural remedy for healthy aging. Aging Dis 8(6): 708. https://doi.org/10.14336/ad.2017.0707

Yang C et al (2022) Effects of lycium Barbarum L. polysaccharides on vascular retinopathy: an insight review. Molecules 27(17):5628. https://doi.org/10.3390/molecules27175628

Yu Z et al (1992) Effect of aging on the syndrome of essential debility and evil reality—an exploration on mechanism of senility by traditional Chinese medicine. Chin J Integr Traditional West Med 12(2):80–82

Yu Y et al (2018) Cycloastragenol: an exciting novel candidate for age-associated diseases (review). Exp Ther Med. https://doi.org/10.3892/etm.2018.6501

Zhang M, Zhao J, Zhang X (2018) Analysis on composition principles of ancient anti-senile prescriptions using traditional Chinese medicine inheritance support system. J Basic Chin Med 24(3):393–396

Zhang J et al (2019) An aqueous polyphenol extract from rosa rugosa tea has antiaging effects on caenorhabditis elegans. J Food Biochem 43(4). https://doi.org/10.1111/jfbc.12796

Zhang Y et al (2020) Panax notoginseng saponins prevent senescence and inhibit apoptosis by regulating the pi3k-akt-mtor pathway in osteoarthritic chondrocytes. Int J Mol Med. https://doi.org/10.3892/ijmm.2020.4491

Zhang Y et al (2023) Targeting cellular senescence in senile osteoporosis: therapeutic potential of traditional Chinese medicine. Front Med 10. https://doi.org/10.3389/fmed.2023.1288993

Zhang Z-B et al (2025) The history, beneficial ingredients, mechanism, processing, and products of panax ginseng for medicinal and edible value. Food Med Homol. https://doi.org/10.26599/fmh.2025.9420059

Zhao H et al (2017) Therapeutic potential and cellular mechanisms of Panax Notoginseng on prevention of aging and cell senescence-associated diseases. Aging Dis 8(6):721. https://doi.org/10.14336/ad.2017.0724

Zhu G et al (2025) Medicinal and food plant: Cistanches Herba, a potential therapeutic hope for Parkinson's disease and related complications. J Agric Food Res 19:101671. https://doi.org/10.1016/j.jafr.2025.101671

Zou Y et al (2023) Anti-aging effect of Panax Notoginseng saponins via protecting endogenous antioxidant in fruit flies. Indian J Pharm Educ Res 57(3):797–801. https://doi.org/10.5530/ijper.57.3.97

Chapter 12
African Medicinal Formulations for Health and Longevity

E. I. O. Ajayi, J. O. Oladele, I. O. Babatunde, O. U. H. Imere, and A. T. Adewole

Abstract African medicinal formulations, rooted in centuries of ethnomedical knowledge and biodiversity, represent a treasure trove of therapeutic agents with

E. I. O. Ajayi (✉)
Diabetes Complications and Infectious Diseases-Related Movement Disorders Research Group, Membrane Biochemistry and Nanobiotechnology Unit, Department of Biochemistry, Faculty of Basic and Applied Sciences, College of Science, Engineering and Technology, Osun State University, Oke Baale, Osogbo, Nigeria

The STREEAAMMIE Centre, Ebelola Bioenergetic Solutions, Kajola Ijesa, Osun State, Nigeria
e-mail: ebenezer.ajayi@uniosun.edu.ng

J. O. Oladele
Phytochemistry and Toxicology Unit, Royal Scientific Research Institute, Osogbo, Osun State, Nigeria

Department of Veterinary Physiology and Pharmacology, College of Veterinary Medicine and Biomedical Sciences, Texas A&M University, College Station, TX, USA

I. O. Babatunde
Toxicology and Pharmacology Unit, Department of Biochemistry, Faculty of Basic and Applied Sciences, College of Science, Engineering and Technology, Osun State University, Oke Baale, Osogbo, Nigeria

O. U. H. Imere
Diabetes Complications and Infectious Diseases-Related Movement Disorders Research Group, Membrane Biochemistry and Nanobiotechnology Unit, Department of Biochemistry, Faculty of Basic and Applied Sciences, College of Science, Engineering and Technology, Osun State University, Oke Baale, Osogbo, Nigeria

Toxicology and Pharmacology Unit, Department of Biochemistry, Faculty of Basic and Applied Sciences, College of Science, Engineering and Technology, Osun State University, Oke Baale, Osogbo, Nigeria

A. T. Adewole
Diabetes Complications and Infectious Diseases-Related Movement Disorders Research Group, Membrane Biochemistry and Nanobiotechnology Unit, Department of Biochemistry, Faculty of Basic and Applied Sciences, College of Science, Engineering and Technology, Osun State University, Oke Baale, Osogbo, Nigeria

Nutrition Unit, Department of Biochemistry, Faculty of Basic and Applied Sciences, College of Science, Engineering and Technology, Osun State University, Oke Baale, Osogbo, Nigeria

S. I. S. Rattan, C. Gezer (eds.), *Supplements and Drugs for Healthy Longevity*, Healthy Ageing and Longevity 22, https://doi.org/10.1007/978-3-032-11608-6_12

promising applications for promoting health and longevity. Whereas some are underutilized, natural products rich in bioactive phytoconstituents and promising medicinal bioresources contribute to antioxidant, anti-inflammatory, adaptogenic, neuroprotective, and anti-aging properties. These play roles in wellness and longevity through mechanisms involving redox modulation, cellular stress responses, and mitochondrial function, which align with emerging paradigms in geroscience. Using an integrative approach that combines ethnobotanical documentation, phytochemical profiling, and preclinical evaluation in models of metabolic dysfunction, immunosenescence, and neurodegeneration, evidence support the role of many African medicinal formulations in extending healthspan. Furthermore, we highlight the possibility of synergistic interactions among plant constituents (polyherbal formulations) and animal bioresources and the relevance of traditional preparation methods, such as prolonged decoction, fermentation, and pounded admixtures, and addition of edible mineral-ladden clays in enhancing bioavailability and efficacy. While noting the challenges associated with standardization, quality control, and safety assessment, there are opportunities for translational research through omics technologies, artificial intelligence-driven screening, and biotechnological enhancement that tripartite collaboration among indigenous practitioners, academic researchers and the biotechnology industry can drive. Importantly, sociocultural frameworks sustain traditional medicine practices, however there is a need for a model that will harmonize indigenous knowledge with modern drug development pipelines. African medicinal formulations are a viable and underexplored component of global strategies for preventive healthcare and healthy aging. They warrant recognition, preservation, and scientific validation as Africa's medicinal heritage.

Keywords African traditional medicine · Ethnopharmacology · Animal biorsources · Polyherbal formulations · Healthspan extension

12.1 Overview of African Traditional Medicine (ATM)

The ATM is a core aspect of folk medicine that uses a holistic approach to bring about wellness to health (Chaitanya et al. 2022). The practice of ATM has evolved over many centuries and has been practiced by many diverse cultures and regions. It is a holistic healing system richly encompassed with cultural, spiritual, belief, physical, and social elements. The ATM has surfaced over a long time, as old as humans, and remains the most easily reached therapeutic system to mankind (Ebu et al. 2021). This healthcare practice system remains sustainable to Africans because Western or conventional treatment remains inaccessible to most African people due to high costs or limited healthcare (Baily 2022). Most importantly, Western medicine is still not efficient enough to effectively eradicate some deadly global diseases like HIV/AIDs and cancer. Thirdly, diseases like malaria, which are common and widely affecting Africa, have yet to be completely eradicated.

Many traditional healers or practitioners use empirical knowledge via observation, experience, experimental knowledge, and cultural traditions to diagnose and treat ailments (Zuma et al. 2016). Traditional healers are usually with or without formal education, and they transfer or gain knowledge via apprenticeship, community-based practices, and oral tradition. Methods of divination and incantations might be used alongside herbal treatments for treatment and diagnosis.

ATM integrates herbalism, spirituality, cultural significance, and diverse practices (Chaitanya et al. 2022). Herbalism, which is mostly common practice, involves the use of plant parts such as roots, leaves, stems, and bark to treat ailments. The belief of the linkage between the body (physical), spiritual, mind and emotion make up the spirituality aspect. Rituals, rites, ceremonies and other community-based practices make up the cultural significance, while the differences in region and communities practices make up the diverse practices.

The most common practice among the different aspects of ATM is herbal treatments or herbalism. Over 70% of Africa relies on this as primary health care. This is because Africa is richly blessed with medicinal plants that are easily accessible to people (WHO 2025). Over 5000 species of plants are used for medicinal purposes. The plant has many phytochemicals as secondary metabolites, which can either act alone or synergistically to improve treatment or management. These, coupled with their natural origin, make them less toxic to humans.

Despite the promising potential of traditional African medicine, its usage is still faced with challenges, among which include difficulty in integration with modern medicines, Skepticism and reluctance in acceptability, usage, and dosage regulation and standardization (Sifuna 2022).

12.2 History and Significance

The existence of ATM historically has been for over thousands of years. Its emergence can be traced to the use of ancient roots, traditional knowledge, spiritual and cultural usage. The historical emergence of ATM can be traced back to pre-colonial, colonial and modern or post-colonial eras (Abdullahi 2011; Chaitanya et al. 2022) (Fig. 12.1).

At ancient roots period, the holistic approach is the integration of herbal medicine with spiritual beliefs, ancestral power and divination. Where traditional practitioners or diviners spoke with the ancestral spirit for provision of diagnosis and treatment. At the pre-colonial era, the used of herbalism and natural remedies as community and traditional –based approach to treat various illness and maintain the community health surfaced. At this time, herbal medicine had surfaced but had not adequately research. Orally transfer of information leads to misuse of plants or misidentification of medicinal plants which leads to health complication and serious adverse (Abdullahi 2011).

Owing to these, coupled with unreal or magical practices and superstitious belief collectively termed as witchcraft as seen by the Europeans or colonial authorities, the traditional healers and diviners were prohibited during the colonial era. These lead to suppression and marginalization of ATM. Although attempt was made to re-educate

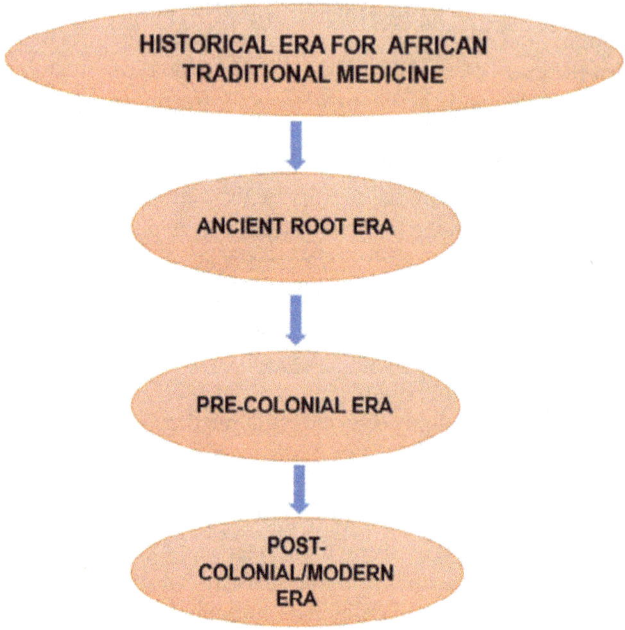

Fig. 12.1 Historical era in emergence of ATM

the traditional practitioners after Mozambique independence in 1975 but the advent of colonization, Christianity belief coupled with involvement of doctor and health practitioner completely shut down the use of ATM in treatment of disease. By twentieth century, which depicts the modern or post -colonial era, ATM resurfaced as part of modern health care system after being revived, owing to high cost, hectic in supplier and inaccessibility to people in rural areas (Abdullahi 2011; Chaitanya et al. 2022).

Ever since this time, ATM has continued to be relevant in human healthcare system in both rural and urban areas. Mostly importantly, ATM has provided a distinct means for treatment different from alternative and complementary medicine, and has helped in drug discovery, all the while preserving the African heritage, traditional knowledge and community identity (White 2015).

12.3 Concept of Health and longevity in African Cultures

The concept of health and longevity in Africa encompasses the totality of well-being of Africans physically, spiritually, emotionally, and mentally. Healthy life and increasing life expectancy were known for people living in African region, evidence of which was record by World Health Organization (WHO) in year 2022 that African healthy life expectancy increased on average of 10 years per person between 2000 and 2019 and this is recorded rise in greater than those observed in other region of the world (United Nations 2022).

The main contributing factor is the incorporation of ATM into the modern health care system. Aside from these, traditional practices, modern healthcare, and socio-economic factors also contribute to this improvement in the health and longevity of Africans (United Nations 2022). Although health is defined as the absence of any disease or impairment (Ajima and Ubana 2018). As for Africans, health is not just about the absence of diseases or the normal functioning of bodily organs, but a state of well-being in both mental, physical, spiritual, and emotional aspects of an individual or their family and community. These various states of health and well-being remain the holistic approaches used as measures to determine the state of health among Africans (Fig. 12.2).

The holistic physical approach to health entails a healthy lifestyle in terms of a healthy diet or nutrition, regular exercise, and adequate rest. This also included avoidance of harmful behavior, incorporation of healthy habits to mental health and reduced risk exposure, and enhancement of physical function of the body part (Capio et al. 2014; Australian National University 2025). Africans were ranked higher in their physical well-being with an index rate of 62.4 in a global study conducted by Cigna, which is more than the global average of 61.4 (Cigna 2021). Although the physical health and well-being of Africans are still challenged by stress, infectious disease burden, and high prevalence of non-communicable diseases (Ramirez 2024).

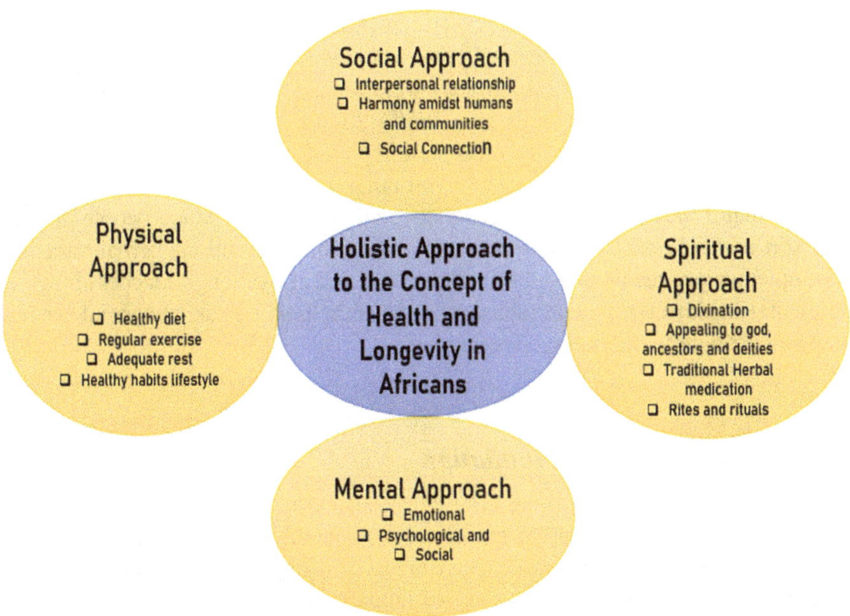

Fig. 12.2 Approaches to the concept of health and longevity in Africa

Africans so much believed that the spiritual well-being is also a core integral aspect of human health. Africans strongly believed there is a strong connection between healthy living and ancestors, a god or God. Some believed a sacrifice needed to be made to the ancestors and deities for spiritual protection to stay healthy. Other religions believed there is a supremacy or supernatural forces that greatly influence human health. Many depend on these ancestors or spiritual beings for healing (Ajima and Ubana 2018). However, many of the traditional practitioners or diviners use ATM, such as herbal medicine or decoction, alongside incantation or rituals as instructed by the gods or the supreme being to do healing from various diseases (Gabriel 2022).

The social health well is tied to the relationship and interaction with the environment and community. Africans believed there must be a balance or harmony between humans, nature, and community. The social well-being had a much greater impact on mental health. African social health is greatly challenged by health literacy, poverty, poor sanitation, limited access to basic resources (such as clean water, good sanitation, healthcare facilities, and a nutritious diet), and gender inequality (Jimoh et al. 2023). The contributory effect of these factors has greatly affected the social well-being and health of Africa, and these have led to reduced social connection, interpersonal relations, and longevity in many African countries (Salami et al. 2019; Bergeron-Boucher et al. 2024).

The emotional, psychological, and social well-being of an individual is a collective measure of human mental health, and this is an integral aspect of overall human health. This greatly affects the thinking, reasoning, and creativity abilities. The mental health in Africa is believed to be socially constructed, greatly influenced by social, cultural, and religious factors. Even some Africans believed that mental illness is caused by external forces. The mental health problem is increasing yearly in Africa, as about 17.9 million Africans were recorded in 2015 to be living with disability due to mental health issues (WHO 2019). The prevalence and risk of mental illness and psychiatric disorders were reported to be higher in urban regions than in rural regions. Depression, anxiety, trauma, and schizophrenia were reported to be the major mental illnesses (Jörns-Presentati et al. 2021). Many people in rural areas sort for traditional healer for a pharmacological approach via herbal remedies as complementary medicine (Anjorin and Yusuf 2022). Although inaccessibility to the mental health care system, abuse of substance, poverty, and financial burden greatly contribute to the outrageous number.

12.3.1 Philosophical Foundation

The philosophical foundation of the concept of health and longevity of African culture is deeply rooted in holism and Ubuntu philosophy. From a holistic view, Africans believed health and longevity of humans solely depend on the interconnection between the social, spiritual, physical, and environmental aspects of life (Bayuo 2025). African philosophy of health also extends further to interlink with other aspects of life, such as daily, family, community, and individual lifestyles. The Ubuntu philosophy of health promotes general well-being or community health rather than individual health. The health state of a person depends on the person, our co-existence, and the relationship, with the hope that the ill health and imbalanced social and metabolic state of one would affect the entire community. This is why their communities' healthcare system is designed for both body, mind, non-body, and spiritual elements, and this is why healing with traditional medicine, rituals, and rites is common amidst the communities (Bolden 2014; Imafidon 2022).

The philosophical foundation of health and longevity of Africans also incorporates ontological, epistemological, and ethical theories of human existence, as well as critical reflections on lived experiences (Imafidon 2022). From the traditional African worldview, the health and longevity philosophical foundation is also tied around or requires integrations of religion, spirituality, and ancestral references. Likewise, appealing to gods or a supreme or ancestral being is also essential for healthy living and longevity (Tosam 2021). This philosophical belief was deeply intertwined and used strongly in ATM. As ATM practitioners do put these above philosophies into consideration when diagnosing and treating illness, as against the common practices in modern biomedical sciences. They strongly believed that diseases and illnesses were caused both attributed to impair in the relationship between human, natural, and supernatural beings. However, the ATM practitioners used a wide range of herbs, animal parts, and other natural source material in the preparation of decoction, powdered or chopped for treatment of ailments and for ritual and rites for appeasing the supreme being (Ezekwesilli-Ofili and Ogbonna 2022; Gabriel 2022).

12.3.2 Indigenous Concept of Wellness and Aging

Africans view good health and long-life expectancy as essential and valuable tools to human survival. As defined by WHO, wellness is a good approach to living (HIFASS 2024). It is an optimum state of health, either for an individual or a community. Wellness is an indication of an active growth process. Wellness in health has a strong interconnection with aging, which is an inevitable stage in life when there is a decline in physiological functioning for survival and growth. Well-being in terms of physical, emotional, intellectual, and social, coupled with good nutrition and healthy habits, improves the quality of life and thus delays the aging process, i.e., extending the individual or community lifespan (WHO 2024). As applicable to the global nations, the concept of wellness and aging amidst the African indigenous community is holistic and multidimensional or complex, as it is an integration of political, economic, intellectual, physical, spiritual, social, environmental, ideological, and emotional factors. All these interplays bring about wellness and aging well. Any imbalance amidst these or impairment in these factors leads to disability, disease conditions, or illness (Dudgeon et al. 2017; Quigley et al. 2022). Africa as a non-Western community, extending to wellness, includes traditions, religious beliefs, community and family support, age-related changes adaptation, and harmony amidst all.

Both globally and in Africans people, addressing challenges of life with ease, adapting to age-related changes with positivity with resilience, maintaining social connection with people, community and their culture, positive attitude to life and attainment of holistic health and well-being were also recognized among Africa as fundamental factors that promote aging well (Quigley et al. 2022). Pedigree of the knowledge enables Africans to live longer, however, healthy aging among Africans is still impaired by factors like cultural differences, historical trauma, persistence of infectious disease, chronic health issues, poverty and limited health facilities, and a poor health care system. All these need to be greatly addressed to improve health and wellness in aging.

12.4 Role of ATM in Health and Longevity

The pursuit of longevity has intensified interest in understanding the biological mechanisms of aging, which, despite longstanding observation, remains complex and multifaceted. Aging is increasingly defined as the accumulation of life's consequences such as molecular and cellular damage that result in functional decline, chronic illness, and ultimately death (Moqri et al. 2023). In the context of longevity, this process is recognized not only as inevitable but also as a modifiable trajectory influenced by genetics, environment, and lifestyle.

The role of ATM in management of human health and diseases cannot be over emphasized and this is the reason why its usage is gaining much attention worldwide. ATM are effective for treating both acute and chronic diseases with little side effect being natural. Generally, ATM helps to promote overall health and wellness, thus enhancing longevity (Okaiyeto and Oguntibeju 2021) (Fig. 12.3).

12.4.1 Management of Communicable and Non-communicable Diseases

Reports from many clinical trials have shown that ATM is potent in the management of communicable and non-communicable, including incurable or terminal diseases as claimed by allopathic medicine. Among these diseases include infertility, diabetes, dementia, hepatitis B, Alzheimer's disease, sickle cell anemia, HIV/AIDs, and cancer (Gaobotse et al. 2023). At times, there is no strong claim that they cure these diseases, they are adjudged as potent in the suppression of pain and symptoms associated with these diseases (Oladele et al. 2020a, b; Sifuna 2022). Likewise, ATM has also been claimed to cure some common 25 diseases in an East African country named Kenya. Among these include blood pressure, dental problems,

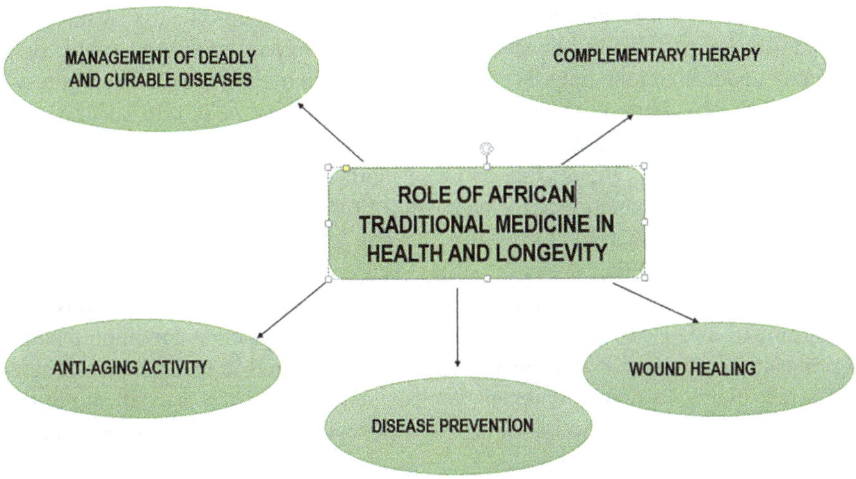

Fig. 12.3 Role of ATM in health and longevity

backache, typhoid, malaria, diarrhea, chest problems, diabetes, skin rashes, stomachache, tetanus, pneumonia, arthritis, tuberculosis, and heartburn (Mabona and Van Vuuren 2013; Oladele et al. 2020a; Wanjohi et al. 2020). A critical, yet often underemphasized dimension of longevity is reproductive health. Fertility problems particularly those arising from age-related decline in reproductive function can significantly undermine both individual and generational longevity. In women, for instance, the natural depletion of ovarian reserve and hormonal shifts with age can reduce fertility, increase the risk of pregnancy complications, and accelerate systemic aging through endocrine changes. Similarly, in men, aging is associated with reduced sperm quality, DNA fragmentation, and increased risk of offspring with genetic disorders. These fertility issues not only affect reproductive potential but may also reflect a broader age-associated pathologies such as oxidative stress, inflammation, and mitochondrial dysfunction. Moreover, recent studies suggest that interventions promoting reproductive health such as antioxidant therapies, lifestyle modifications, and targeted treatments could have systemic anti-aging benefits, thereby contributing to enhanced health span and longevity. Thus, preserving fertility is not merely a matter of reproductive success, it is integrally tied to the broader biological processes that govern aging and longevity.

12.4.1.1 African Medicinal Plants for Managing Fertility Problems That Undermine Longevity

Polycystic ovarian syndrome (PCOS) is an endocrine disorder characterized by hormonal imbalance, leading to irregular menstrual cycles, hyperandrogenism (manifesting as hirsutism and acne), and ovulatory dysfunction, which collectively impair fertility. While PCOS is often recognized as a reproductive disorder, it extends far beyond fertility concerns and poses significant long-term health risks that may undermine longevity. Women with PCOS frequently exhibit insulin resistance and obesity, both of which are critical contributors to the development of type 2 diabetes mellitus and cardiovascular diseases (Royal College of Obstetricians and Gynaecologists 2020), the leading cause of morbidity and mortality in women. In addition, chronic anovulation associated with PCOS results in prolonged unopposed estrogen exposure, thereby increasing the risk of endometrial hyperplasia and endometrial cancer (Frandsen et al. 2024). The combination of metabolic, cardiovascular, and oncologic risks with subtle inflammation and linked to PCOS positions it not only as a cause of infertility but also as a chronic systemic condition that can significantly compromise a woman's long-term health and life expectancy. Therefore, early diagnosis and comprehensive management of PCOS are critical not just for restoring fertility but also for mitigating the broader health consequences that can adversely affect longevity.

Similarly, endometriosis, a chronic, estrogen-dependent condition affecting approximately 10% of women of reproductive age globally is defined by the presence of functionally active endometrial glands and stroma outside the uterine cavity, most commonly involving the ovaries, pelvic peritoneum, and rectovaginal septum. Clinically, it presents with chronic pelvic pain, dysmenorrhea, dyspareunia, dysuria, dyschezia, and is strongly associated with subfertility or infertility. Symptom severity tends to correlate with disease progression, particularly in cases of deep infiltrating endometriosis (Maulenkul et al. 2024). Beyond its immediate reproductive implications, endometriosis is increasingly recognized as a systemic inflammatory disorder with potential to compromise long-term health. Chronic inflammation and oxidative stress associated with the disease have been linked to an increased risk of cardiovascular disease, epithelial ovarian cancer (Farland and Harris 2020), and autoimmune conditions. These systemic effects, combined with the psychological burden of chronic pain and infertility, and the side effects of prolonged hormonal or surgical interventions, may collectively reduce overall health span and potentially undermine longevity. Importantly, as a fertility related disorder, endometriosis exemplifies how reproductive dysfunction can serve as a marker of broader systemic disease. This underscores the need for a multidisciplinary management approach that not only addresses fertility goals but also mitigates long-term health risks and promotes overall well-being.

In the same vein, premature Ovarian Insufficiency (POI), formerly known as premature ovarian failure, is the loss of normal ovarian function before age 40. It is marked by menstrual irregularities, elevated gonadotropins—particularly FSH and low estrogen levels. Affected women often present with infertility, vasomotor symptoms, urogenital atrophy, and psychological distress due to early estrogen deficiency (Wasserman et al. 2025). Beyond reproductive impairment, POI has significant systemic implications. The early loss of estrogen increases the risk of cardiovascular disease, osteoporosis, cognitive decline, mood disorders, and metabolic dysfunction, all of which may compromise long-term health and reduce life expectancy (Sánchez-García et al. 2025; Ishizuka 2021). These effects highlight POI as not only a fertility concern but a chronic endocrine disorder requiring comprehensive care. Early diagnosis, hormone replacement therapy, and long-term monitoring are essential to mitigate complications and improve quality of life and longevity in affected women.

Male factor infertility is responsible for approximately 40–50% of infertility cases (Kumar and Singh 2015), it is increasingly recognized not just as a reproductive issue but as a clinical indicator of broader health risks (Choy and Eisenberg 2020). Men with impaired semen parameters such as low sperm count, poor motility, or abnormal morphology often present with underlying systemic conditions that may negatively impact longevity (Goldman 2014). Emerging evidence links male infertility to elevated risks of cardiovascular disease (Chen et al. 2022), metabolic syndrome, type 2 diabetes, and certain malignancies, particularly testicular and prostate cancers (Panahi and Vaismoradi 2021). Additionally, studies have reported higher mortality rates among infertile men, independent of age and lifestyle factors (Del Giudice et al. 2021). These associations are thought to be mediated by mechanisms such as hormonal dysregulation (e.g., low testosterone), chronic inflammation, oxidative stress, environmental exposures, and shared genetic vulnerabilities. The psychological burden of infertility including anxiety, depression, and reduced self-esteem can further contribute to long-term health decline, as these factors drive men to seek remedy with aphrodisiacs which usually have grave consequences in the long run. Taken together, male infertility serves as a sentinel marker for underlying systemic dysfunction and increased mortality risk. Recognizing this connection highlights the need for early intervention, holistic evaluation, and multidisciplinary care not only to improve reproductive outcomes but also to support overall health and longevity.

Obesity-related infertility, a dual threat to reproductive and metabolic health in both males and females. Obesity-related infertility represents a well-established risk factor for reproductive dysfunction in both men and women (Uddandrao et al. 2024). In women, excess adiposity disrupts the hypothalamic-pituitary-gonadal (HPG) axis, impairs folliculogenesis, and alters sex hormone profiles, resulting in anovulation and subfertility (Zheng et al. 2024). In men, obesity is linked to hypogonadism, reduced testosterone levels, impaired spermatogenesis, diminished sperm quality, and erectile dysfunction primarily due to increased estrogen levels, oxidative stress, and chronic low-grade inflammation. Adipose tissue also promotes the peripheral conversion of androgens to estrogens, further exacerbating endocrine disruption (Service et al. 2023). While it impacts reproductive function, obesity is a major contributor to chronic diseases such as cardiovascular disease, type 2 diabetes, dyslipidemia, atherosclerosis, certain cancers, and cognitive decline all of which are associated with diminished quality of life and significantly shortened life expectancy (Antouri et al. 2024). The endocrine and metabolic dysregulation inherent in obesity characterized by insulin resistance, systemic inflammation, and adipokine imbalance not only compromises fertility but also accelerates age-related morbidity. Therefore, obesity-related infertility should be recognized as a manifestation of broader metabolic dysfunction that, if left unaddressed, may undermine both reproductive potential and overall longevity.

The following are a few African medicinal plants useful in the treatment of fertility-related cases in both males and females:

Newbouldia laevis
Commonly known as boundary tree or "Ewe akoko" in Yoruba language (Nigeria), it is a medicinal plant native to West Africa, particularly Nigeria, Ghana, and Cameroon. It is traditionally utilized in the management of female reproductive disorders. Ethnopharmacological reports highlight its use in treating menstrual irregularities, ovulatory dysfunction, and infertility (Ogunlakin and Sonibare 2020). Its therapeutic potential is attributed to its phytoestrogenic activity, along with potent anti-inflammatory and antioxidant properties, which may help modulate hormonal balance, reduce oxidative stress, and support uterine function (Josiane et al. 2022). Given its traditional applications and pharmacological properties, *Newbouldia laevis* holds promise as a natural therapeutic agent for addressing female reproductive health issues. However, further clinical studies are necessary to fully elucidate its efficacy and safety profiles.

Fadogia agrestis
This is a shrub native to West and Central Africa including Nigeria, Sudan, and Chad has been traditionally utilized for its aphrodisiac properties and in the management of male reproductive health issues like erectile dysfunction (Yakubu e al. 2005). Phytochemical analyses have identified bioactive constituents in *Fadogia agrestis*, notably alkaloids and saponins, which are believed to contribute to its pharmacological effects. Experimental studies in male rats have demonstrated that aqueous extracts of *Fadogia agrestis* can significantly enhance sexual behavior parameters, such as mount and intromission frequencies, and prolong ejaculatory latency (Gotep et al. 2021). Furthermore, the plant's antioxidant properties may play a role in mitigating oxidative stress, which is known to impair reproductive function. A study recommended the use of aqueous extract of *Fadogia agrestis* stem in managing erectile dysfunction (ED), it was reported that the extracted restored the NO/cGMP pathway and ED-associated key enzymes in the penile and testicular tissues of male rats via antioxidant means (Ogunro and Yakubu 2023). While these findings are promising, current evidence is primarily based on animal studies. Therefore, further clinical research is necessary to validate the efficacy and safety of *Fadogia agrestis* in humans.

Vernonia amygdalina
This plant is a multifaceted agent in reproductive health commonly known as bitter leaf, is a widely distributed medicinal plant across Sub-Saharan Africa. Traditionally, it has been employed to address various reproductive health challenges, including ovulatory dysfunction and infertility. Recent scientific investigations have begun to elucidate the mechanisms underlying its therapeutic effects. Phytochemical analyses have revealed that *Vernonia amygdalina (V. amygdalina)* is rich in bioactive compounds such as flavonoids, saponins, alkaloids, and phenolic acids, which contribute to its antioxidant, anti-inflammatory, and hypoglycemic properties. These attributes are particularly relevant in the context of reproductive health,

where oxidative stress and metabolic imbalances can impair fertility. In female reproductive health, studies have demonstrated that *Vernonia amygdalina* extracts can ameliorate symptoms associated with polycystic ovary syndrome (PCOS). Administration of the plant extract in animal models has been shown to restore hormonal balance, reduce oxidative stress markers, and improve ovarian morphology, suggesting its potential in managing PCOS-related infertility (Adebodun et al. 2024). Regarding male fertility, research indicates that dietary inclusion of *V. amygdalina* leaf meal can enhance sperm quality parameters. In male rabbits, supplementation with the plant leaf meal resulted in increased live sperm count and improved sperm morphology, without adverse effects on testosterone levels or testicular indices (Adeyemi et al. 2022). These findings suggest that *V. amygdalina* may support male reproductive health by improving spermatogenesis and protecting against oxidative damage. Furthermore, the plant's antioxidant properties have been implicated in protecting reproductive cells from oxidative stress, a known factor in infertility. By scavenging free radicals and enhancing the body's endogenous antioxidant defenses, *V. amygdalina* may help maintain the integrity of sperm and oocytes, thereby supporting overall reproductive function.

Other Plants

Other potent and promising African medicinal plants indicated for fertility treatment include:

Mondia whitei (White's ginger/Mulondo): It enhances libido and overall reproductive health for both sexes through its antioxidant and androgenic effects. It is common to Uganda, Kenya, Nigeria, Cameroon.

Hibiscus sabdariffa (Roselle/"Zobo"): It supports hormonal balance, may help regulate menstrual cycles through its antioxidant, anti-inflammatory, and antihypertensive effects. It is common to West Africa.

Asparagus africanus: It is traditionally used for treating female infertility and regulating menstruation due to its phytoestrogenic and uterine tonic properties. It is common to East and West Africa.

Cassia occidentalis (*Coffee senna*/Sennu): It improves fertility and treats amenorrhea by its hormonal modulation and antioxidant effect. It is common to Nigeria, Senegal, Ghana.

Zingiber officinale (Ginger): It is used to enhance libido, sperm motility, and hormonal regulation due to its antioxidant, anti-inflammatory activities and ability to improve blood flow. It is common across Africa, although not indigenous.

Tribulus terrestris: It is used to treat erectile dysfunction and low libido due to its ability to enhance luteinizing hormone and testosterone production. It is common to many parts of Africa.

Gongronema latifolium: It supports male fertility by boosting sperm count and quality. It is also used to manage diabetes and oxidative stress due to its antioxidant and anti-inflammatory properties. It is found across Africa.

12.4.2 Disease Prevention

ATM provides herbal remedies or dietary intervention as well as lifestyle practice which help prevention of disease (Oladele et al. 2021, 2024). Many herbal remedies contained bioactive compounds which help to boost the immune system to reduce the risk of infection; antioxidants which help to guide against oxidative stress and chronic disease emergence and antimicrobial agents to combat pathogens and treat infections. The impact of such was reported during the COVID19 era as many medicinal plants with antiviral, antimicrobial, anti-inflammatory, cyto-protective and immune-modulatory properties were used alone or in combination with public health convention drug and recommendation to combat this pandemic disease (Christian et al. 2020; Oladele et al. 2020; Abubakar et al. 2022; Mutombo et al. 2023).

12.4.3 Complementary Therapy

ATM serves as a potent therapy for the health care system to be used alongside Western or modern medicine for the treatment of diseases, especially for people with limited access to conventional therapy due to financial constraints (Oladele et al. 2021; Ezekwesilli-Ofili and Ogbonna 2022; Okaiyeto and Oguntibeju 2021). For example, *R. vomitoria* and *S. microphylla* are used as herbal therapies for the management of hypertension in Nigeria and HIV in South Africa respectively (Ezekwesilli-Ofili and Ogbonna 2022).

12.4.4 Anti-aging Activity

The anti-aging potential of some African medicinal plants makes them a drug candidate for longevity, many of the African medicinal plants contain polyphenols and antioxidant phytochemicals which scavenge free radical's attacks (Kapoor et al. 2009). Likewise, some contain bioactive compounds which exhibit anti-collagenase and anti-elastase activities as well as other phytoactive compounds that can prevent skin damage. Classical examples of such plants include *H. madagascariensis* and *P. aurantiacum* found in Cameroon (Manjia et al. 2019; Ajayi et al. 2024).

12.4.5 Wound Healing

Many African medicinal plant organs such as stems, bark, leaves, and roots were used traditionally in preparing remedies for wound healing (Ajayi et al. 2015;

Oladele et al. 2023). Many of these are prepared in the form of gel, infusions, ointments, decoctions, pastes, lotions, and poultices for the treatment of wounds (Mulugeta 2016; Mrynal 2022). A report of Ethnobotanical Systematic Review shows over 200 species of plant identified across South Africa identified to exist wound healing potential. Among these include plants belonging to the family of Solanaceae, Asphodelaceae, and predominantly helichrysum species of the family of Asteraceae (Fisher et al. 2025). Their wound-healing properties have been attributed to their antimicrobial anti-inflammatory and antioxidant activities (Rolnik and Olas 2021).

12.5 States of Matter and Sources of African Medicinal Formulations

African communities are richly blessed with natural resources, and many of these natural resources are used in the formulation of folk medicine for diagnosing and treating diseases. In health practices, 4000 medicinal plants have been reported to be medicinally and pharmacologically beneficial for treating ailments in African communities (Chaitanya et al. 2022). Herbal remedy as an integral part of the physical element of ATM is obtained from nature and formulated in different forms for man's consumption. The choice of the type of natural resources used depends on the medicinal properties of such natural sources, which are prepared in different forms ranging from solids to liquids such as powder, decoction, infusion, snuffs, gruels, and many more, for easier consumption by humans (Josephine-Ozioma and Antoinette 2019). The dependence of Africans on ATM for sustaining health, diagnosing, and treating diseases is high; it is no wonder that the WHO reported that over 80% of Africans depend on ATM for health care. In this section, more discussion about the sources and matter state of ATM will be done.

12.5.1 States of Matter of ATM

ATM, upon preparation, exist in different forms depending on the nature of the material used in preparation, the method of preparation, the patient's nature, and the user's interest. In Africa, ATM exists primarily in four states:

Solid

Powder, tablet, and capsule are various solid forms in which some ATM exists. Local medicines produced or packaged in this form have a long shelf life and are easy to convey over long distances without deterioration. Many of these are packaged into designed and labeled paper, nylon, or bottles and sold commercially in both rural and urban areas. These various forms of solid include:

1. **Powder**: Roots, dried leaves, stem bark, burnt animal and its parts, and minerals are usually prepared in powder form as herbal medicine. These plant parts are usually air-dried and crushed into powder, while in the case of animal products, they are also air-dried or burnt up and then crushed into powder for end users for the treatment of ailments and diseases. For example, the stem bark and leaves of African plants *Acacia senegal* and *Artemisia Herba-alba* are dried and powdered as a tea for the treatment of diseases such as diabetes, infections, fever, gonorrhea, and ailments (Mahomoodally 2013). Dried and powdered scent leave is used in the treatment of diarrhea and as an additive for soup preparation.
2. **Tablets and capsules**: The advancement in pharmaceutical technology has enabled the packaging of African medicinal plant extracts as dietary supplements in the form of tablets and capsules. Packing in this for seems beneficial as it aids accessibility, standardization and quality control. Many of these are available and sold commercial for treatment of many diseases and aliment. A typical is Rooibos extract obtained from *Aspalathus linearis* which is now formulated in tablet form for management of diabetes and stomach cramps (Jourbert and de-Beer 2011).

Liquid: Decoction, Infusion, Syrups

Some ATM are formulated and packaged in liquid form. Preparation in liquid form is done to ensure efficient delivery and absorption. Water and alcohol were the most common solvents used as a medium for extraction (Ezekwesilli-Ofili and Okaka 2019). Although corn starch, water, and lime juice were other solvent mediums used by some African communities. The choice of solvent depends on the nature and bioactive composition of the materials used. Many of these are packaged in plastic containers, glass bottles, and nylon for sale or dispensed directly to the users upon preparation at the shop of traditional practitioners. Albeit, calabash and clay pots are used for dispensing to the end users in the rural communities.

Liquid herbal medicines exist in different forms depending on the method of preparation. These forms include:

1. **Decoctions**: This is prepared by boiling medicinal plant hard parts, such as stem-bark, seed, rhizome, and root in dried or fresh forms for a certain period and further filtered to obtain the extract (Son et al. 2023). For longer storage, additives like potash may be added as preservative (Luo et al. 2012). This method is the best choice for extracting tonic or therapeutically bioactive compounds from plants that can't give a high yield of their bioactive component at lower temperatures, but such components are not destroyed at slightly higher temperatures. Examples of the plant used for decoction include the bark of Willow (*Salix alba*) and oark (*Quercus robur*) trees, the root of ginger (*Zingiber officinale*) and Dandelion (*Taraxacum officinale*) plants (Grass Roots Remedies 2025).
2. **Infusion**: In this, cold or hot water treatment of the herbs to access the bioactive component. This method can also be applied to crude drugs or extracts which has already been prepared or extracted in solid form. The crude drug in powder is macerated in cold, hot, or warm water and prepared as tea or macerated in hot watery pap as a common practice by some communities before usage. The mixture is then allowed to steep for a while or overnight before usage. To prevent

spoilage, additives like honey can be added to increase the shelf life. The method is mostly useful for extracting aromatic compounds. Examples of plants commonly used in the preparation of infusion include *Melissa officinalis* (Lemon Balm), *Mentha piperita* (Peppermint), *Matricaria recutita* (Chamomile), etc. (Grass Roots Remedies 2025).

3. **Syrups**: This form is suitable for preparing herbal preparations for children and those who find bitter-tasting herbs unpleasant. The method employed here involves mixing the concentrated herbal extract obtained via infusion or decoction methods with either sweeteners like honey, sugar, or vegetable glycerin (Korem et al. 2020). Additions of this sweetener, such as honey, further preserve and increase its acceptability. Herbal dispensation in this form enables children to take phytoactive aromatic compounds. Commonly prepared herd in syrups form includes Elderberry (*Sambucus nigra*) and elecampane (*Inula helanium*) used for treating respiratory infections due to antiviral problems (WebMD 2024).

Semi-solid

Most ATM applied topically are usually made in semi-solid forms, and many of these are widely used for the treatment of skin diseases or used in cosmetic formulation (Sakthivel et al. 2022). This method is adopted to prevent systemic side effects and toxicity of such herbs on the skin or body system. Herbal preparations in this form do contain one or more therapeutic ingredient(s) which are uniformly dispersed in in suitable base and excipients. The whole plant or its parts, or microbes or extract from the whole plants or their parts obtained via methods of infusions, decoctions, wicking, etc., and further processed by concentration or fractionation, distillation, or combination of any of these are used. They are packaged in small plastic containers for sale. The semisolid herb-based formulation occurs in different forms, among which are:

1. **Ointments:** These are greasy upon preparation. Hydrophobic and anhydrous bases like beeswax, petrolatum, wool fat, hard and soft paraffin. They are mostly used for healing superficial wounds. Vetas Ghana is a popular ointment produced in Ghana. The ointment is prepared using an extract of the stem bark of a plant named *Salix alba,* and it is used popularly for the treatment of musculoskeletal pain (Verma et al. 2022). Other example includes Arnica ointment (obtained from the plant *Arnica Montana*), chamomile ointment (made from the plant *Matricaria recutita*), and marigold ointment (made from the extract of marigold blossoms).
2. **Paste:** Herbal toothpaste is a classic example of this form of semi-solid herbal medicine. It usually contains 50% of the powder of the plant extract dispersed in the base. A classic example of such is Dabur herbal toothpaste, which contains mainly Basil, Meswak, and Neem plant extract dispersed in a base containing calcium carbonate as an abrasive agent, carrageenan as a thickening agent, sodium lauryl sulfate as a surfactant agent, and many more (Sakthivel et al. 2022).
3. **Cream**: Herbal creams are usually prepared as emulsions to retain a viscous nature. Oil and water are used to form this emulsion. This is used to disperse the beneficial plant extract. Additives like antimicrobial agents are also preservatives (Pranaya et al. 2025). A classic example is turmeric and Aloe vera cream, which

are used to treat skin conditions like burns, itching, wounds, etc., due to their antioxidant, anti-inflammatory, and antibacterial activities. Other examples include Papaya Tulsi, neem, and turmeric body cream, which are sold commercially.

4. **Jellies and Gel:** Mainly used for antiseptic and lubricating purposes (Sahu et al. 2024). Jellies are non-greasy, semi-soft-solid herbal formulations. Usually translucent or transparent in appearance. Most herbal jellies contain one or two plant extracts mixed with honey as sweeteners, spices as flavoring agents, vinegar, corn starch as a binding agent, oils, and other additives (Chamoli and Tangri 2022). Common plant extract composition includes Aloe vera, ginger, papaya, rosemary, etc. Movit and Alana herbal Jellies are classical examples of Jellies made in Africa. As for Gel, it is made by the combination of one or more plant extracts mixed with gelling agents such as Carbopol and Oil, like olive oil as a solvent and preservatives. Examples of plant extracts which has been reported to be used in gel formulation include turmeric, garlic, and aloe vera (Ipsita et al. 2021, Sahu et al. 2024).
5. **Poultices:** It is usually a moist and soft mass when touched. Made from herbs and clay, and mostly applied on the skin directly or indirectly on a cloth to generate heat and to drain abscesses. It is used traditionally to treat skin problems like infection, pain, and inflammation, or to relax muscles and to draw toxins, such as insect bites, from affected skin regions (Bunpean and Singweratham 2025; Jeet 2025). Roots of blackthorn (*Acacia mellifera*), leaves of honey flower (*Melianthus comosus*), leaves of Caster bean (*Malvina parviflora*), and leaves and fruits of Mahogany (*Trichilia emetic*) are a few among the plant preparations as a poultice in African herbal medicine for wound healing in South Africa.

Gaseous

Some herbal medicines are taken as inhalant, smoke (smudge) or vapor. Most commonly, the steam inhalant which are used to treats cold and nasal congestion, respiratory infections. Steaming help to extract the volatile compounds. Upon taken as warms moisture, it helps to open the nasal airways. Classical examples of herbs taken as inhalant includes thyme basil, rosemary and cinnamon. Some are prepared as snuffs e.g. *Entada rheedii*, *Moringa oleifera,* which are made by powdering dried plant parts most especially leaves and then inhaled through nostril (Josephine-Ozioma and Antoinette 2019). In South Africa, plant herbs like *Sceletium tortuosum*, and *Helichrysum petiolare/odoratissium* are burned and their smoke are inhaled to treatment of respiratory problem and as psychoactive compounds (Nyazema et al. 2024).

12.5.2 Natural Sources of Medicinal Ingredients

The medicinal ingredients used for the preparation of traditional medicine were obtained naturally from different sources. Among these include medicinal plants,

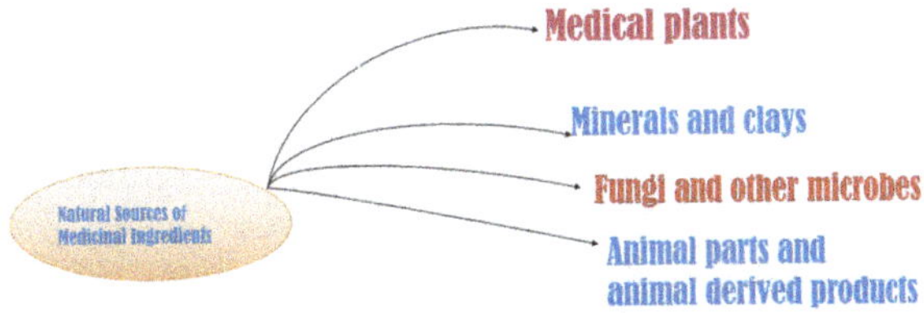

Fig. 12.4 Natural sources of medicinal ingredients

minerals and clays, animal-derived products, and microbial sources (Josephine and Antoinette 2019). The choice of ingredients used depends on resources available, location, knowledge about the pharmacological functions of the plants, nature of the disease, and treatment methods adopted. In Africa, the most sorted ingredient in the preparation of herbal formulation is medicinal plants; as documented, 80% of Africans as well relies on medicinal plants for health care (Mahomoodally 2013; Africa Health Research Institute 2020). Ease of accessibility, low cost, free availability, and low toxicity attributed to its usage make medical plant usage more acceptable and popular amongst Africans (Fig. 12.4).

Some of these medicinal ingredients were available at the local market. Some were displayed at the shop for the public to see, and they were sold commercially at low cost for the public user and traditional practitioners. In some of these local shops, traditional medicine is cooked or prepared in different forms such as syrups, decoctions, ointments, teas, distillates, snuff, fractions, etc.

12.5.2.1 Medical Plants

Medicinal plants serve as a core ingredient in the preparation of plant-based herbal remedies as part of ATM. Right from ancient times, Plants have been valuable sources of material for drug discovery as medicinal plants contain therapeutics and pharmacological substances as secondary metabolites, which are effective for combating diseases. Saponins, alkaloids, flavonoids, terpenoids, tannins, and polyphenols were part of these therapeutic compounds (Chintada and Golla 2025). Thus, making them serve as a source of phyto-pharmaceutical drugs.

Different plant parts are used as ingredients for herbal formulation, which include:

1. **Roots**: This plant part is usually woody, fleshy, or woody. Decoction, tinctures, solvent, or cold or hot extraction were common methods used in processing this part for herbal formulation. Classical examples of plants whose root is used in herbal medicine are *Siphonochilus aethiopicus* (African Ginger), *Zingiber officinale* (ginger), and *Vitellaria paradoxa* (Shea Butter). These were reported to exhibit anti-inflammatory and antimicrobial properties; thus, they were used in Africa for the management of rheumatism and skin diseases (Bulia 2023).
2. **Bulbs**: This plant part is mostly found under the ground with numerous fleshy scales as an extension, as seen in *Allium sativa* (garlic) and *Allium fistulosum*. They are used as a local remedy for the management of diabetes, cancer, and

infectious diseases due to their Antiviral, immune modulating, glucose-lowering, hypertension, and cancer-suppressing potentials (Papu 2014; Tesfaye and Mengesha 2015; Alam et al. 2016).

3. **Rhizomes**: This plant looks like an underground stem, usually fleshy or woody, and can grow horizontally to produce shoots. *Zingiber officinale* (ginger) and *Curcuma longa* (turmeric) are classical examples of African medical plants whose rhizomes are used as ingredients for herbal remedies and formulations. e.g., Zingiber officinale (ginger), which is used for respiratory problems; *Imperata cylindrica* (spear grass) for potency in men and *Curcuma longa* (turmeric), Antioxidant, anti-inflammatory, and anticancer and anti-asthmatic activities were reported for ginger; hence they are used in ATM for treatment of pain, malaria, infection, and respiratory problems (Adebayo et al. 2021).
4. **Leaves**: This is the most commonly used part of the plant for herbal and ATM due to its soft and succulent nature. It is often squeezed or boiled in water or blended, or cooked as a soup in most herbal formulations. Commonly known plants whose leave is widely used in the preparation of herbal medicine are *Aloe barbadensis* (Aloe vera) and *Moringa oleifera* (Oda and Erena 2017). O*leifera Moringa* leave is used traditionally as a sacred remedy for the management of insomnia, joint pain, stomach ulcers, insomnia, and malaria (El Bilali et al. 2024).
5. **Tubers**: Some herbal formulations do involve the use of tubers, the use of this part of the plant in herbal formulation or preparation is not always common. For the most part, used, the tuber is fleshy and swollen in structure. Examples of medical plants whose tubers in herbal medicine are *Gloriosa superba* and *Dioscorea dumetorum*, which are reported to exhibit anti-diabetic and anticancer potentials (Josephine and Antoinette 2019). *Gloriosa superba* is also a Nigerian and African heritage plant used in the traditional management of antipyretic and gonorrhea (Abd El-Ghani 2016).
6. **Gum, resin, nectar, sap and exudate**: These are fluids that ooze out of plant tissues or roots. Some plants secrete any of these to cover up wounded areas or to discourage insect invasion and animals from grazing them. In some medicinal plant such as *Acacia senegal* (Gum Arabic) and *Strerculia urens* (Karaya gum). For example, Gum Arabic is used traditionally for the treatment of infections such as gonorrhea, diarrhea, bronchitis, etc., and is well applied to bind pills and as aromatherapy in Senegal (Jain et al. 2012; Mahomoodally 2013).
7. **Fruits and seeds**: The fruit and seeds of medical plants are another plant part used in the preparation of ATM. These parts are known to contain bioactive compounds and essential oils. A classic example of such is the moringa oleifera seed, which is boiled and prepared as a tea or added to soup as a common practice in the West African region. The seed is used locally for the management of digestive diseases, skin infections, infertility, venereal diseases, and heart disease, etc. (Agoyi et al. 2017).
8. **Flower and stem barks:** The flower, which is the most colorful part of the plant, is also used in the preparation of traditional medicine. A classic example is the flower of the plant *Artemisia afra* and *Azadirachta indica* (Neem). The flower of *A. afra* is yellow and it is used locally for the treatment of fever, diabetes,

rheumatism, and cough (Okoye et al. 2014). The use of stem bark is a common practice in ATM. In most local markets, the air-dried ones are hand and sold commercially in the local market. A classic example is *Mangifera indica* stem-bark, which is used locally for fever, diarrhea, and liver problems (Adeneye et al. 2015).

12.5.2.2 Minerals and Clays

Minerals and clays are natural resources that are also used as medicinal ingredients in the preparation of ATM. The use of minerals as part of traditional medicine is a common practice among South Africans. The minerals are used and sold along with herbs in a traditional medicinal store, and they are labeled with local names. A few of these minerals are shown in the table below (Table 12.1).

Umanyazini is used majorly in South Africa, as part of the ingredients for preparing traditional medicine for the management of diverse ailments and diseases such as infection, sexually transmitted diseases, swelling, pains, aches, and kidney and bladder problems (Street et al. 2013). Ndonya is used for preservation in traditional medicine processes and antimicrobial agents. Isibabuli is used for the treatment of skin infections, acne, and rashes as an ointment in Nigeria.

12.5.2.3 Animal Parts and Animal-Derived Products

Aside from medicinal plants popularly used as ingredients in ATM, animal parts and their derived products are used as ingredients for traditional Medicine. This practice is common among African countries like South Africa, Nigeria, and Tanzania. Many of these have been documented in the literature (Vats and Thomas 2015; Ajagun et al. 2017; Salome et al. 2018; Nieman et al. 2019). These include both domestic and wild animals. They fall within the group of animals classified as mammals, insects, arachnida, fish, oligochaeta, reptiles, chilopoda, aves, etc. They are used as component materials for the preparation of traditional medicine for the treatment of a wide range of diseases and mental disorders and non-medical uses such as rituals and rites, charms, protection, command of longevity, favor, and authority; wealth and riches acquisition, fetches power and magico-medical uses (Salome et al. 2018;

Table 12.1 Some minerals used as ingredients for ATM

Minerals	Local name (in the South African language)	Scientific name
$CuSO_4$	Cwathile	Copper tetraoxosulphate VI
KMnO4	Umanyazini	Potassium tetraoxomanganate VI
NH_4Cl	Itshelomsimpha	Ammonium chloride
S_4	Isibabuli	Sulphur
NaCl	Ndonya	Sodium chloride
$CaSO_4$	Umkhandu omhloph	Calcium tetraoxosulphate VI

Street et al. (2013), Street and Cele (2013)

Nieman et al. 2019). The animal-derived products include egg and egg shells, animal fats, feces, urine, milk, and honey. Some of the animal parts and their derived product with their application in traditional medicine are shown in the table below (Table 12.2).

With advancements in technology and pharmaceutical methods, the bioactive therapeutic compounds present in these animal products can be isolated and validated for drug discovery, development, and formulation processes. Thus, further enabling validation of the ethnomedicinal use of these animals' parts and their derived products.

Table 12.2 Selected animal, animal parts, and their derived products with their medicinal traditional uses and methods of application

Class	Animal English/ Scientific name	Animal part or derived product	Traditional uses	Method of application
Reptiles	Rock python/ *Python sebae*	Oil and Skin	Spinal cord injury and back pain	Skin is dry, powder, mix with oil, and apply while the oil is applied directly
	Royal python/ *Python regius*	Excreta (feces)	Inflammation and pain-related problems	Dry powder and mix with water
	Black cobra/*Naja nigricollis*	Venom	Rheumatoid arthritis	Apply directly to the body part
		Oil	Rheumatoid arthritis, hypertension, and eczema	
	Nile Crocodile/ *Crocodylus niloticus*	Skin	Gastric problems and tuberculosis	Dry, burn into ashes, and mix with honey or herbs
	Cobra/*Naja siamensis*	Head	Commanding power and protection	Dry, powder, and mix with herbs
		Skin	fractured bone and skin burn	Dry, powder, and mix with water
Fish	Octopus/*Octopus vulgaris*	Tail	Urinary problem	Dry, burn into ashes, and swallow
	Nile Perch/*Lates niloticus*	Gills	Abnormal cramp problem	Dry, blind into powder, and mix with water
Aves	African ducks/ *Anas indica*	Animal fats	Angina and Pneumonia	Melt and rob on the chest
	Sacred Ibis/ *Threskiornis aethiopicus*	Fresh blood	Rheumatoid arthritis	Intake of fresh blood
	Tawny Eagle/ *Aquila rapax*	Feathers	Angina	Inhalation of the smoke obtained from burning
Mollusca	Snail/*Achatina fulica*	Shell	Leg pain and inflammation-related problems	Dry, burn, powder, and honey or water
	Snail/*Achatina*	Blood	Ulcer or gastrointestinal problems and hypertension	Drinking of the fresh blood

(continued)

Table 12.2 (continued)

Class	Animal English/ Scientific name	Animal part or derived product	Traditional uses	Method of application
Insect	Earthworm/ *Lumbricus terrestris*	Whole animal	Impotency and infidelity problems in men	Dry, grind, and turn into a paste with hot water or mix with honey
	Millipede/ *Trigoniulus corallines* & *Scutigera coleoptrata*	Whole animal	Skin dandruff and diseases	Dry, burn into ashes, and mix with water
	Spider (Araneus spp)	Cob web	Wound bleeding	Directly apply to the wound
	Honey bee (*Apis mellifera*)	Honey	Skin burns, coughs, arthritis, hemorrhoids, hair loss, and snake bites	Rob on the body parts
Mammals	Lion/*Panthera leo*	Fats/fatty skin	Ear pus problem	Rub on the ears
	Rock Hyrax/ *Procavia capensis*	Urine	Syphilis	Mix the soil that contains hyrax urine with water, filter, and drink
	Brown Rat/*Rattus norvegicus*	Whole body	Protection against thieves	Dry and process further by traditional practitioners
	African Elephant/ *Loxodonta Africana*	Skin	Liver hepatitis	Dry, burn, powder into ashes, and mix with water
	Spotted Hyena/ *Crocuta Crocuta*	Excreta and Skin	Tuberculosis	Intake of the dry meat
	Hippopotamus/ *Hippopotamu amphibious*	Blood	Increased CD4 cells to boost immunity in HIV patients	Dry and leak
	Black Rhinoceros/ *Diceros bicornis*	Horn	Tuberculosis, gastro-intestinal, and asthmatic problems	Dry, grind, and mix with hot water

Vats and Thomas (2015), Salome et al. (2018), Nieman et al. (2019)

12.5.2.4 Microbial Sources

Mushrooms as macrofungi remain the most notable microbial source ingredient used in the preparation of ATMs. The use of mushrooms is a common practice among traditional practitioners in Nigeria and Tanzania (Oyetayo 2011; El Enshasy et al. 2013). A larger percentage of these mushrooms are wild, and the methods and processes subjected to for preparation ATM are usually disclosed by Traditional practitioners (Oyetayo 2011). Historically, before and after civilization in many African countries and the Middle East, some edible mushrooms and truffles (hypogeous macrofungi) are used in the preparation of food for royal families by many

Table 12.3 Selected macrofungi (mushrooms) and indicated diseases/conditions in African folk medicine

Mushroom	African country	Diseases/medical conditions use to managed
Auricularia auricular	Nigeria	Hemoptysis and hemorrhoids
Ganoderma lucidum		Arthritis
Daldinia concentrica		skin disease, whooping cough, stomach ulcer and disturbances
Calvatia cyathiformis		Barreness and leucorrhea
Lentinus squarullosus		heart problem and mumps
Pleurotus tuber-regium		Constipation, cold, stomach ache, headache and fever
Termitomyces microcarpus		Gonorrhea
Parkia biglobosa	Burkina Faso	Malarial-induced respiratory distress problems
Toadstool mushroom	Uganda	Ease delivery of child
Termitomyces titanicus	Tanzania	Constipation, stomach disturb, ulcer and pain
Puffball mushrooms		Wound healing

cultures (Tibuhwa 2012; El Enshasy et al. 2013). For example, the Yoruba tribes in Nigeria used in preparation of Egusi (melon) soup. This is because of the high nutritional value of mushrooms.

Mushrooms usually grow on decaying logs of wood; it has neither flowers nor fruits nor leaves nor buds, and when grown on the top of wood or earth, they are visible to the naked eye. Truffles also grow underground as well via a mutual nutritional relationship with the root of the woody plant. Their harvesting requires expert and technical skills.

Diverse species of mushrooms have been identified in African countries such as Nigeria, Tanzania and Burkina Faso, Libya, Tunisia, South Africa, Cameroon, Zambia, Zimbabwe etc. (El Enshasy et al. 2013). *Parkia, biglobosa, Agaricus, Termitomyces, Pleurotus, Boletus, Lentinus, Ganoderma, Cantharellus, Macrolepiota, Coriolopsis, Calvatia, Lenzites, Trametes, Pycnoporus, and Geastrum* were among species identified in Nigeria, Tanzania and Burkina Faso, which were reported to be used as traditional fork medicine (Oyetayo 2011; El Enshasy et al. 2013).

Examples of mushrooms used in some African countries and the diseases they manage as folk medicine are shown in the table below (Table 12.3).

12.6 Conclusions

ATM formulations encompass a vast array of health-promoting remedies, including those targeting stress relief, mood stabilization, reproductive and maternal health, aphrodisiac enhancement, and postpartum recovery. These holistic interventions are often delivered via oral, topical, inhalational, or even spiritual routes, highlighting the complexity and depth of Indigenous healing systems. Despite their widespread use and cultural value, many of these formulations remain underexplored scientifically, especially in terms of active compound characterization, standardization, and clinical efficacy.

Nanotechnology presents a promising frontier for improving the bioavailability, delivery, and stability of traditional formulations. Encapsulation of herbal extracts in nanoemulsions or biopolymer-based carriers could revolutionize their therapeutic potential while making administration more convenient and acceptable to modern users. Additionally, fostering equitable collaborations between local healers, academic researchers, and pharmaceutical industries is crucial for ethical bioprospecting, knowledge preservation, and innovation. Protecting indigenous intellectual property through culturally sensitive frameworks and enforceable patent systems will also be vital to preventing exploitation and cultural appropriation.

12.7 Future Directions

To fully harness the therapeutic potential of ATM for health and longevity, future research must adopt a transdisciplinary framework that integrates ethnomedicine, phytochemistry, pharmacology, and biotechnology. This includes advancing quality control protocols and dosage standardization, particularly for polyherbal and multifunctional formulations whose preparation methods, such as decoction, fermentation, or maceration directly influence therapeutic outcomes. Institutions like the African Organization for Standardization (ARSO) and its Regional Economic Committees (RECs) must intensify efforts to harmonize regulatory standards and facilitate formal recognition of these remedies across Africa and internationally.

Ultimately, the future of African medicinal formulations lies in bridging indigenous knowledge with modern science. By validating, refining, and responsibly commercializing these therapies, Africa can contribute unique, culturally rich, and scientifically grounded solutions to global health challenges in aging, chronic disease management, and wellness.

References

Abd El-Ghani MM (2016) Traditional medicinal plants in Nigeria: an overview. Agric Biol J N Am 7(5):220–247

Abdullahi AA (2011) Trends and challenges of traditional medicine in Africa. Afr J Tradit Complement Altern Med 8(5S):115–123

Abubakar IB, Kankara SS, Malami I, Danjuma JB, Muhammad YZ, Yahaya H, Singh D, Usman UJ, Ukwuani-Kwaja AN, Muhammad A, Ahmed SJ, Folami SO, Falana MB, Nurudeen QO (2022) Traditional medicinal plants used for treating emerging and re-emerging viral diseases in northern Nigeria. Eur J Integr Med 49:102094. https://doi.org/10.1016/j.eujim.2021.102094

Adebayo SA, Amoo SO, Mokgehle SN, Aremu AO (2021) Ethnomedicinal uses, biological activities, phytochemistry and conservation of African ginger (Siphonochilus aethiopicus): a commercially important and endangered medicinal plant. J Ethnopharmacol 266:113459. https://doi.org/10.1016/j.jep.2020.113459

Adebodun GO, Dikne A, Egbeke A, Arikawe H, Bello B, Ekwobi K, Ogunsola AO (2024) Ameliorative potentials of Vernonia amygdalina in high-fat diet and letrozole-induced polycystic ovary syndrome in female Wistar rats. Trop J Nat Prod Res 8(4):6995–6999

Adeneye AA, Awodele O, Aiyeola SA, Benebo AS (2015) Modulatory potentials of the aqueous stem bark extract of Mangifera indica on carbon tetrachloride-induced hepatotoxicity in rats. J Tradit Complement Med 5:106–115

Adeyemi AA, Oloyede C, Adedotun A (2022) Effect of Vernonia amygdalina leaf meal on the reproductive indices of male rabbits. Arch Zootech 25(2):63–74. https://doi.org/10.2478/azibna-2022-0014

Africa Health Research Institute (2020) Marking African Traditional Medicine Day 2020. https://www.ahri.org/marking-african-traditional-medicine-day-2020/

Agoyi EE, Okou FAY, Assogbadjo EA, Sinsin B (2017) Medicinal uses of Moringa oleifera in southern Benin (West Africa). Acta Hortic 1158:303–308

Ajagun EJ, Anyaku CE, Afolayan MP (2017) A survey of the traditional medical and non-medical uses of animals species and parts of the indigenous people of Ogbomoso, Oyo State. Int J Herb Med 5(3):26–32

Ajayi EIO, Popoola G, Ojediran E (2015) Wound healing potential of Nauclea latifolia and Manihot esculenta leaf extracts in type 1diabetic rats. Afr J Tradit Complement Altern Med 13(1):1–5. https://doi.org/10.21010/ajtcam.v13i1.1

Ajayi EIO, Oladele JO, Babatunde IO (2024) Carotenoids and their formulation supplements in Alzheimer's disease. NeuroLife 2(1):1–24. https://lifeneuro.de/article-1-107-en.html

Ajima OM, Ubana EU (2018) The concept of health and wholeness in traditional African religion and social medicine. Arts Social Sci J 9:388. https://doi.org/10.4172/2151-6200.1000388

Alam MK, Hoq MO, Uddin MS (2016) Medicinal plant Allium sativum—a review. J Med Plant Stud 4(6):72–79

Anjorin O, Yusuf HW (2022) Impact of traditional healers in the provision of mental health services in Nigeria. Ann Med Surg (Lond) 24(82):104755

Antouri Z, Mezouaghi A, Djilali S, Zeb A, Khan I, Omer ASA (2024) The impact of obesity on chronic diseases: type 2 diabetes, heart disease, and high blood pressure. Appl Math Sci Eng 32 (1)

Australian National University (2025) Physical Wellbeing https://www.anu.edu.au/physical-wellbeing

Baily D (2022) Traditional Healing vs. Western Medicine in the African Continent. Afro News, 19 March 2021, afro.com/traditional-healing-vs-western-medicine-in-the-african-continent/. Accessed 6 Oct 2022

Bayuo J (2025) Indigenous African philosophy as a paradigm for health and social care research: a philosophical and methodological discussion. Nurs Inq 32(2):e70002. https://doi.org/10.1111/nin.70002

Bergeron-Boucher M, Callaway J, Strozza C et al (2024) Inequalities in lifespan and mortality risk in the US, 2015–2019: a cross-sectional analysis of subpopulations by social determinants of health. BMJ Open 14:e079534. https://doi.org/10.1136/bmjopen-2023-079534

Bolden R (2014) Ubuntu. In: Coghlan D, Brydon-Miller M (eds) Encyclopaedia of action research. Sage, pp 799–801

Bulia F (2023) African herbal Remedies: traditional healing plants and their modern uses. https://naturalpoland.com/en/artykuly/products-from-africa/african-herbal-remedies-traditional-healing-plants-and-their-modern-uses/

Bunpean A, Singweratham N (2025) Analysis of herbal poultices for knee pain relief in hospitals in the Office of the Permanent Secretary, Ministry of Public Health, Thailand: Tropical Journal of

Natural Product Research. Trop J Nat Product Res (TJNPR) 9(2):776–781. https://doi.org/10.26538/tjnpr/v9i2.46

Capio CM, Sit CHP, Abernethy B (2014) Physical well-being. In: Michalos AC (ed) Encyclopedia of quality of life and Well-being research. Springer, Dordrecht. https://doi.org/10.1007/978-94-007-0753-5_2166

Chaitanya, V.N.L., Gebremariam Baye, M., Saad Ali, H., and Belayneh Usamo, F. (2022). Traditional African medicine. IntechOpen. https://doi.org/10.5772/intechopen.96576

Chamoli M, Tangri SK (2022) Formulation and development of pediatric herbal jelly from papaya seeds for anthelmintic infection. Int Res J Plant Sci 13(5):1–13

Chen PC, Chen YJ, Yang CC, Lin TT, Huang CC, Chung CH, Sun CA, Chien WC (2022) Male infertility increases the risk of cardiovascular diseases: a Nationwide population-based cohort study in Taiwan. World J Men's Health 40(3):490–500

Chintada V, Golla N (2025) Exploring the therapeutic potential of bioactive compounds from plant sources. In: Devi J (ed) Biotechnological intervention in production of bioactive compounds. Sustainable landscape planning and natural resources management. Springer, Cham. https://doi.org/10.1007/978-3-031-76859-0_15

Choy JT, Eisenberg ML (2020) Comprehensive men's health and male infertility. Trans Androl Urol 9(Suppl 2):S239

Christian E, Maduka TOD, EbereEnyoh C (2020) Potential plants for treatment and management of COVID-19 in Nigeria. International Journal of Medicinal Plants and Natural Products (IJMPNP) 6(3):1–12. https://doi.org/10.20431/2454-7999.0603001

Cigna (2021) Africa ranks higher in physical and social Well-being compared to global average despite higher stress levels, finds Cigna's 360° well-being study. Africa Updates. https://www.africaupdates.com/News/News.aspx?NewsId=13598

Del Giudice F, Kasman AM, Chen T, De Berardinis E, Busetto GM, Sciarra A, Ferro M, Lucarelli G, Belladelli F, Salonia A, Eisenberg ML (2021) The association between mortality and male infertility: systematic review and meta-analysis. Urology 154:148–157

Dudgeon P, Bray A, D'Costa B, Walker R (2017) Decolonising psychology: validating social and emotional wellbeing. Aust Psychol 52:316–25. https://doi.org/10.1111/ap.12294

Ebu V, Anoh R, Offiong R et al (2021) Survey of medicinal plants used in the treatment of "Ailments of utmost native importance" in cross river state, Nigeria. Open J Forestry 11:330–339

El Bilali H, Dan Guimbo I, Nanema RK, Falalou H, Kiebre Z, Rokka VM, Tietiambou SRF, Nanema J, Dambo L, Grazioli F, Naino Jika AK, Gonnella M, Acasto F (2024) Research on Moringa (Moringa oleifera Lam.) in Africa. Plants (Basel) 13(12):1613. https://doi.org/10.3390/plants13121613

El Enshasy H, Elsayed EA, Aziz R, Wadaan MA (2013) Mushrooms and truffles: historical biofactories for complementary medicine in Africa and in the middle east. Evid Based Complement Alternat Med 2013:620451. https://doi.org/10.1155/2013/620451

Ezekwesilli-Ofili OJ, Ogbonna OA (2022) Africa traditional and complementary therapies. InterOpen. https://doi.org/10.5772/intechopen.103060

Ezekwesilli-Ofili JO, Okaka ANC (2019) Herbal medicines in African traditional medicine. IntechOpen https://doi.org/10.5772/intechopen.80348

Farland LV, Harris HR (2020) Long-term health consequences of endometriosis – pathways and mediation by treatment. Curr Obstet Gynecol Rep 9(3):79–88

Fisher NRF, Africa C, Klaasen J, Fisher R (2025) South African Medicinal plants traditionally used for wound treatment: an ethnobotanical systematic review. Plants (Basel) 14(5):818. https://doi.org/10.3390/plants14050818. PMID: 40094805; PMCID: PMC11901878

Frandsen CLB, Gottschau M, Nøhr B, Viuff JH, Maltesen T, Kjær SK, Jensen A, Svendsen PF (2024) Polycystic ovary syndrome and endometrial cancer risk: results from a nationwide cohort study. Am J Epidemiol 193(10):1399–1406

Gabriel I (2022) African traditional religion and medicine in contemporary times. Int J Curr Res Hum 26. https://doi.org/10.4314/ijcrh.v26i1.15. ISSN:0855-9740

Gaobotse G, Venkataraman S, Brown PD, Masisi K, Kwape TE, Nkwe DO, Rantong G, Makhzoum A (2023) The use of African medicinal plants in cancer management. Front Pharmacol 14: 1122388. https://doi.org/10.3389/fphar.2023.1122388

Goldman B (2014) Male infertility linked to mortality in study. In: Standford medicine. https://www.med.stanford.edu/news/all-news/2014/05/male-infertility-linked-to-mortality-in-study?

Gotep JG, Sabo SY, Makama SM, Uguru O, Okwuasaba FK (2021) Aphrodisiac activity of ethanol extract and fractions of Fadogia cienkowskii Shweinf. Rubiaceae roots in albino rats. J Med Plant Res 15(2):86–95

Grass Roots Remedies (2025) Infusion and decoctions. https://www.grassrootsremedies.co.uk

HIFASS (2024) What is health and wellness. Health Initiative for society and stability in Africa. https://hifass-hfi.org/what-is-health-and-welness

Ishizuka B (2021) Current understanding of the etiology, symptomatology, and treatment options in premature ovarian insufficiency (POI). Front Endocrinol 12:626924

Imafidon E (2022) Beyond continental and African philosophies of personhood, healthcare and difference. Nurs Philos 23(3):e12393. https://doi.org/10.1111/nup.12393

Ipsita S, Akshya M, Gardia D, Khamari A (2021) Herbal gel formulation and evaluation using locally available plants. Int J Pharmacognosy Pharm Sci 3(1):32–34. https://doi.org/10.33545/27067009.2021.v3.i1a.80. ISSN 1876-3820, https://doi.org/10.1016/j.eujim.2019.05.011

Jain R, Sharma P, Bhagchandani T, Jain SC (2012) Phytochemical investigation and antimicrobial activity of Acacia Senegal root heartwood. J Pharm Res 5:4934–4938

Jeet S (2025) Master Africa Herbal Poultice preparation like a pro. https://www.newsbytesaapp.com/news/lifestyle/master-africa-Herba-poultice-preparation-like-a-pro/story

Jimoh A, Aminu K, Muchabveyo B (2023) Social determinants of health in Africa. In: Sooryamoorthy R, Khalema NE (eds) The Oxford handbook of the sociology of Africa. Academic, Oxford. https://doi.org/10.1093/oxfordhb/9780197608494.013.12

Jörns-Presentati A, Napp AK, Dessauvagie AS, Stein DJ, Jonker D, Breet E, Charles W, Swart RL, Lahti M, Suliman S, Jansen R, van den Heuvel LL, Seedat S, Groen G (2021) The prevalence of mental health problems in sub-Saharan adolescents: a systematic review. PLoS One 16(5): e0251689. https://doi.org/10.1371/journal.pone.0251689

Josephine-Ozioma EO, Antoinette NCO (2019) Herbal medicines in african traditional medicine. IntechOpen. https://doi.org/10.5772/intechopen.80348

Josiane BNB, Ayina CN, Bagbila A-P, Etaga NB, Nzekuie QS, Ouedraogo MS, Mandengue SH, Minkande J (2022) Newbouldia's laevis biological effects against metabolic syndrome components: a review. World J Ad Res Rev 16(02):121–138

Jourbert E, de-Beer D (2011) Rooibos (Aspalathus linearis) beyong the farm gate. From herbal tea to potential phytopharmaceutical. S Afr J Bot 77(4):869–886

Kapoor VK, Dureja J, Chadha R (2009) Herbals in the control of ageing. Drug Discov Today 14(19–20):992–998. https://doi.org/10.1016/j.drudis.2009.06.014

Korem RA, Sharone RB, Rohini P, Sahaja G, Shylaja SS (2020) Formulation and evaluation of anti diabetic herbal syrup. Res J Pharmacogn Phytochem 12(3):141–144. https://doi.org/10.5958/0975-4385.2020.00023.0

Kumar N, Singh AK (2015) Trends of male factor infertility, an important cause of infertility: a review of literature. J Human Reproduc Sci 8(4):191

Luo H, Li Q, Flower A, Lewith G, Liu J (2012) Comparison of effectiveness and safety between granules and decoction of Chinese herbal medicine: a systematic review of randomized clinical trials. J Ethnopharmacol 140(3):555–567. https://doi.org/10.1016/j.jep.2012.01.031

Mabona U, Van Vuuren SF (2013) Southern African medicinal plants used to treat skin diseases. S Afr J Bot 87:175–193. https://doi.org/10.1016/j.sajb.2013.04.002. ISSN 0254-6299

Mahomoodally MF (2013) Traditional medicines in Africa: an appraisal of ten potent african medicinal plants. Evid Based Complement Alternat Med 2013(2013):617459. https://doi.org/10.1155/2013/617459

Manjia JN, Njayou FN, Apeksha J, Upadhyay K, Shirsath K, Devkar RV, Moundipa PF (2019) The anti-aging potential of medicinal plants in Cameroon – Harungana madagascariensis Lam. and Psorospermum aurantiacum Engl. prevent in vitro ultraviolet B light-induced skin damage. Eur J Integr Med 29:100925. https://doi.org/10.1016/j.eujim.2019.05.011

Maulenkul T, Kuandyk A, Makhadiyeva D, Dautova A, Terzic M, Ohibayeva A, Moldaliyev I, Ayazbekov A, Maimakov T, Saruarov Y, Foster F, Sarria-santamera A (2024) Understanding the impact of endometriosis on women's life: an integrative review of systematic reviews. BMC Womens Health 24:524

Moqri M, Herzog C, Poganik JR, Biomarkers of Aging Consortium, Justice J, Belsky DW, Higgins-Chen A, Moskalev A, Fuellen G, Cohen AA, Bautmans I, Widschwendter M, Ding J, Fleming A, Mannick J, Han JJ, Zhavoronkov A, Barzilai N, Kaeberlein M, Cummings S, Kennedy BK, Ferrucci L, Horvath S, Verdin E, Maier AB, Michael OS, Sebastiano V, Gladyshey VN (2023) Biomarkers of aging for the identification and evaluation of longevity interventions. Cell 186(18):3758–3775

Mrynal C (2022) Formulation and development of pediatric herbal jelly from papaya seeds for anthelmintic infection. IRJPS 13:031

Mulugeta K (2016) Knowledge of medicinal plants used in and around Fincha'a Town, Western Ethiopia. J Pharmacogn Phytochem 5(6):110–114

Mutombo PN, Kasilo OMJ, James PB, Wardle J, Kunle O, Katerere D, Wambebe C, Matsabisa MG, Rahmatullah M, Nikiema JB, Mukankubito I, Sheridan R, Sanogo R, Nissapatorn V, Sivakorn C, Tripathy S, Goyal R, Dhobi M (2023) Experiences and challenges of African traditional medicine: lessons from COVID-19 pandemic. BMJ Glob Health 8(8):e010813. https://doi.org/10.1136/bmjgh-2022-010813

Nieman WA, Leslie AJ, Wilkinson A (2019) Traditional medicinal animal use by Xhosa and Sotho communities in the Western Cape Province, South Africa. J Ethnobiol Ethnomed 15:34. https://doi.org/10.1186/s13002-019-0311-6

Nyazema NZ, Chanyandura JT, Egan B (2024) The use and potential abuse of psychoactive plants in southern Africa. An overview of evidence and future potential. Front Pharmacol 15:1269247. https://doi.org/10.3389/fphar.2024.1269247

Oda BK, Erena BA (2017) Aloes of Ethiopia: a review on uses and importance of aloes in Ethiopia. Int J Plant Biol Res 5(1):1059

Ogunlakin A, Sonibare M (2020) Ethnobotanical survey of medicinal plants used as remedy for female infertility and menstrual disorder in southwestern Nigeria. Niger J Pharm Res 15(2):205–217

Ogunro OB, Yakubu MT (2023) Fadogia agrestis (Schweinf. Ex Hiern) stem extract restores selected biomolecules of erectile dysfunction in the testicular and penile tissues of paroxetine-treated Wistar rats. Reprod Sci 30(2):690–700

Okaiyeto K, Oguntibeju OO (2021) African herbal medicines: adverse effects and cytotoxic potentials with different therapeutic applications. Int J Environ Res Public Health 18(11): 5988. https://doi.org/10.3390/ijerph18115988

Okoye TC, Uzo PF, Onyeto CA, Okereke EK (2014) Safe African medicinal plant for clinical studies. In: Toxicological survey of African medicinal plants, pp 535–555. ISBN 9780128000182

Oladele JO, Ajayi EIO, Oyeleke OM, Oladele OT, Olowookere BD, Adeniyi BM, Oyewole OI, Oladiji AT (2020a) A systematic review on COVID-19 pandemic with special emphasis on curative potentials of medicinal plants. Heliyon 6:1–17

Oladele JO, Oyeleke OM, Oladele OT, Olaniyan MD (2020b) Neuroprotective mechanism of Vernonia amygdalina in a rat model of neurodegenerative diseases. Toxicol Rep 7:1223–1232

Oladele JO, Oyeleke OM, Awosanya OO, Olowookere BD, Oladele OT (2021) Fluted pumpkin (Telfaira occidentalis) protects against phenyl hydrazine-induced anaemia and toxicities in rats. Adv Tradit Med. https://doi.org/10.1007/s13596-020-00499-7

Oladele JO, Oyeleke MO, Adewole ST, Ademiluyi OA, Olowookere BD, Oladele OT (2023) Ethanol extract of Curcuma longa protects against carbon tetrachloride-induced oxidative stress renal and hepatic damage in rats. J Toxicol 18:32–41

Oladele OT, Oladele JO, Ajayi EIO, Alabi KE, Oyeleke OM, Atolagbe OS, Olowookere BD, Bamigboye MO (2024) Bioactive composition and protective properties of Talium triangulare in dextran sodium sulphate-induced ulcerative colitis in rats. Pharmacol Res-Modern Chinese Med 10:100344

Oyetayo VO (2011) Medicinal uses of mushrooms in Nigeria: towards full and sustainable exploitation. Afr J Tradit Complement Altern Med 8(3). https://doi.org/10.4314/ajtcam.v8i3.65289

Panahi MH, Vaismoradi M (2021) A systematic review and meta-analysis of male infertility and the subsequent risk of cancer. Front Oncol 11:696702

Papu S (2014) Medicinal values of garlic (Allium sativum L) in human life: an overview. Greener Journal of Agricultural Sciences 4(6):265–280

Pranaya P, Tadikonda RR, Guthpala M, Rekulapally NR, Cheerla SB (2025) A comprehensive review on herbal creams. J Pharmacogn Phytochem 14(3):115–119. https://doi.org/10.22271/phyto.2025.v14.i3b.15353

Quigley R, Russell SG, Larkins S, Taylor S, Sagigi B, Strivens E, Redman-MacLaren M (2022) Aging well for indigenous peoples: a scoping review. Front Public Health 10:780898. https://doi.org/10.3389/fpubh.2022.780898

Ramirez L (2024) Health challenges in African countries: a focus on physical, mental and emotional well-being. AccedaCRIS:130–140. https://accedacris.ulpgc.es

Rolnik A, Olas B (2021) The plants of the Asteraceae family as agents in the protection of human health. Int J Mol Sci 22:3009. https://doi.org/10.3390/ijms22063009

Royal College of Obstetricians and Gynaecologists (2020) Assisted vaginal birth: green-top guideline no. 26. RCOG, London

Sahu GK, Bisai A, Singh V, Kumar N, Tandi DY, Sharma H (2024) Formulation and evaluation of herbal antifungal cream. Acta Sci Pharm Sci 8(6):34–40

Sakthivel M, Arikaran S, Balajeevarshan KNR, Deepak A, Karthikeyan R, Preethi C (2022) Review about herbal based semisolid dosage forms. Res J Pharm Biol Chem Sci 13(5):11

Salami D, Shaaban AN, Oliveira Martins MDR (2019) Africa rising: a narrative for life expectancy gains? Evidence from a health production function. Ann Glob Health 85(1):63. https://doi.org/10.5334/aogh.2307. PMID: 31050394; PMCID: PMC6646944

Salome KT, Danmalam UH, Ayeni EA (2018) Survey of zoological materials used in traditional medicine in Sabon Gari and Zaria Local Government Areas, Kaduna State, Nigeria. J Complement Med Res 8(1):32–39. https://doi.org/10.5455/jcmr.20180329091359

Sánchez-García M, León-Wu K, López-Juárez N, Ramírez-Rentería C, Espinosa-Cárdenas E, Sosa-Eroza E, García-Sáenz MR (2025) Metabolic changes in patients with premature ovarian insufficiency: adipose tissue focus—a narrative review. Meta 15(4):242

Service CA, Puri D, Al Azzawi S, Hsieh T, Patel DP (2023) The impact of obesity and metabolic health on male fertility: a systematic review. Fertil Steril 120(6):1098–1111

Sifuna N (2022) African traditional medicine: its potential, Limitations and Challenges. J Healthcare 5(1):141–150

Son MJ, Kim S, Kim YE, Kim BY, Yeum CS, Lee JC, Cha JH, Kang S, Chang CH, Son S (2023) Clinical effectiveness of decoction form of herbal medicine in primary care treatment of allergic rhinitis: a retrospective cohort study. Integr Med Res 12(3):100973. https://doi.org/10.1016/j.imr.2023.100973

Street RA, Cele MP (2013) Commonly used metal and crystalline salts in South African traditional medicine. J Ethnopharmacol 148(1):329–331. https://doi.org/10.1016/j.jep.2013.03.054

Street RA, Kabera GM, Connolly C (2013) Ethnopharmacological use of potassium permanganate in South African traditional medicine. SAMJ 108(3):187–189

Tesfaye A, Mengesha W (2015) Traditional uses phytochemistry and pharmacological properties of garlic (Allium Sativum) and its biological active compounds. Int J Scientific Res Sci Eng Technol 1:142–148

Tibuhwa DD (2012) Folk taxonomy and use of mushrooms in communities around Ngorongoro and Serengeti National Park, Tanzania. J Ethnobiol Ethnomed 8:36–45. https://doi.org/10.1186/1746-4269-8-36

Tosam MJ (2021) Healthcare and spirituality: a traditional African perspective. Ann Studi Relig 22: 255–277. https://doi.org/10.14598/Annali_studi_relig_22202120

Uddandrao VV, Brahma Naidu P, Chandrasekaran P, Saravanan G (2024) Pathophysiology of obesity-related infertility and its prevention and treatment by potential phytotherapeutics. Int J Obes 48(2):147–165

United Nations (2022) Healthy life expectancy in Africa grows by nearly 10 years. https://news.un.org/en/story/2022/08/1123973

Vats R, Thomas S (2015) A study on use of animals as traditional medicine by Sukuma Tribe of Busega District in North-Western Tanzania. J Ethnobiol Ethnomed 11:38. https://doi.org/10.1186/s13002-015-0001-y

Verma R, Jain V, Rawat H, Kushwaha V, Gandhi Y, Mishra SK, Kumar V, Tandon S, Charde V, Jagtap C, Babu G, Singh A, Singh R, Sharma BS, Srikanth N (2022) First-order kinetic study-based comparative shelf-life assessment of Raw Vetas (Salix alba L. stem bark) and Vetas Ghana (a semisolid form of Salix alba L. stem bark aqueous extract). J Drug Res Ayurvedic Sci 7(2):97–104. https://doi.org/10.4103/jdras.jdras_10_22

Wanjohi BK, Sudoi V, Njenga EW et al (2020) An ethnobotanical study of traditional knowledge and uses of medicinal wild plants among the marakwet community of Kenya. Evid-Based Complement Altern Med J 2020:3208634

Wasserman H, Kalkwarf HJ, Altaye M, Yolton K, Gordon CM (2025) Effect of transdermal estrogen therapy on bone and neurobehavioral health in youth with premature ovarian insufficiency: a case-control study. J Pediatr Adolesc Gynecol

WebMD (2024) Health Benefit of Elderberry syrup. https://www.webmd.com/diet/health-benefits-elderberry-syrup

White P (2015) The concept of diseases and health care in African traditional religion in Ghana. HTS Teol Stud 71(3):01–07. https://doi.org/10.4102/HTS.V71I3.2762

WHO (2019) Global Health Estimates 2016: Burden of disease by cause, age, sex, by country and by region, 2000–2016 http://www.who.int/healthinfo/global_burden_disease/estimates/en/index1.html

WHO (2024) Ageing and health. https://www.who.int/news-room/fact-sheets/detail/ageing-and-health

WHO (2025) Traditional medicine. https://www.afro.who.int/health-topics/traditional-medicine

Yakubu MT, Akanji MA, Oladiji AT (2005) Aphrodisiac potentials of the aqueous extract of Fadogia agrestis (Schweinf. Ex Hiern) stem in male albino rats. Asian J Androl 7(4):399–404. https://doi.org/10.1111/j.1745-7262.2005.00052.x

Zheng L, Yang L, Guo Z, Yao N, Zhang S, Pu P (2024) Obesity and its impact on female reproductive health: unraveling the connections. Front Endocrinol 14:1326546

Zuma T, Wight D, Rochat T et al (2016) The role of traditional health practitioners in rural KwaZulu-Natal, South Africa: generic or mode specific? BMC Complement Altern Med 16: 304. https://doi.org/10.1186/s12906-016-1293-8

Chapter 13
Greek Medicinal Phytochemicals with Anti-Ageing Properties

Marianna Kapetanou, Evripidis Giannimpas, and Efstathios S. Gonos

Abstract Ageing is a complex, multifactorial biological process driven by a network of molecular and cellular mechanisms collectively known as the twelve hallmarks of ageing: genomic instability, telomere attrition, epigenetic alterations, loss of proteostasis, deregulated nutrient sensing, mitochondrial dysfunction, cellular senescence, stem cell exhaustion, altered intercellular communication, disabled macro-autophagy, chronic Inflammation (Inflammageing) and dysbiosis. These interconnected hallmarks contribute to progressive physiological decline and increased susceptibility to chronic diseases. Greece's rich botanical heritage offers a unique reservoir of natural compounds with demonstrated activity against these core mechanisms. This chapter explores the potential of Greek phytochemicals—including oleuropein, chlorogenic acid, α-terpineol, silymarin, crocin, and secoisolariciresinol diglucoside (SDG)—to modulate these ageing pathways. Through mechanistic insights from *in vitro, in vivo*, and emerging clinical studies, we highlight how these compounds stabilize DNA, protect telomeres, regulate nutrient-sensing pathways, promote proteostasis and mitochondrial integrity, reduce

M. Kapetanou
Institute of Chemical Biology, National Hellenic Research Foundation, Athens, Greece

Hellenic Pasteur Institute, Athens, Greece
e-mail: mkapetanou@eie.gr

E. Giannimpas
Institute of Chemical Biology, National Hellenic Research Foundation, Athens, Greece

Hellenic Pasteur Institute, Athens, Greece

School of Medicine, Department of Health Sciences, Aristotle University of Thessaloniki, Thessaloniki, Greece
e-mail: egiannim@auth.gr

E. S. Gonos (✉)
Institute of Chemical Biology, National Hellenic Research Foundation, Athens, Greece
e-mail: sgonos@eie.gr

S. I. S. Rattan, C. Gezer (eds.), *Supplements and Drugs for Healthy Longevity*, Healthy Ageing and Longevity 22, https://doi.org/10.1007/978-3-032-11608-6_13

senescence and inflammation, and support stem cell function. Integrating traditional ethnopharmacological knowledge with modern geroscience, this chapter positions Greek natural compounds as multi-targeted, culturally relevant tools for delaying ageing and extending healthspan. By bridging phytochemistry, systems biology, and translational research, we propose evidence-based strategies for leveraging Greece's phytodiversity in the development of holistic anti-ageing interventions.

Keywords Epigenetics · Inflammaging · Phytochemicals · Proteostasis · Senescence · Stem cells · Telomeres

13.1 Introduction

Ageing is a multifactorial biological process marked by a gradual decline in physiological integrity, increased vulnerability to chronic diseases, and eventual organismal death. At the cellular and molecular level, this process is driven by interconnected mechanisms known as the hallmarks of ageing: genomic instability, telomere attrition, epigenetic alterations, loss of proteostasis, deregulated nutrient sensing, mitochondrial dysfunction, cellular senescence, stem cell exhaustion, altered intercellular communication, disabled macroautophagy, chronic Inflammation (Inflammageing) and dysbiosis (López-OtíN et al. 2023). These hallmarks are both independent and interrelated, forming a dynamic network of decline that ultimately disrupts tissue homeostasis and regenerative capacity.

Rather than treating age-related diseases in isolation, the emerging field of geroscience emphasizes interventions that target the biological foundations of ageing itself. By doing so, researchers aim to delay multiple chronic conditions simultaneously and extend healthspan—the period of life spent in good health (Campisi et al. 2019). Within this framework, increasing attention is being given to the potential of natural compounds to modulate ageing pathways. Epidemiological studies of long-lived populations, including Mediterranean centenarians, suggest that daily consumption of plant-derived bioactives contributes to reduced inflammation, lower oxidative stress, and preserved metabolic health throughout life (Couteur and Barzilai 2022; Caruso and Puca 2021).

Greece, situated at the confluence of Europe, Asia, and Africa, is a biodiversity hotspot, home to more than 5800 plant species, including 1200 endemics (Axiotis et al. 2018). Traditional Greek medicine, rooted in ancient systems established by Theophrastus, Hippocrates, and Dioscorides, continues to influence modern phytotherapy. In regions like Crete and Ikaria—known for their high longevity—wild greens, herbal infusions, and bitter plant extracts form part of the everyday diet and are closely associated with improved health outcomes (Axiotis et al. 2018). Ethnopharmacological studies confirm that many of these species are rich in flavonoids, phenolic acids, terpenes, and lignans—compounds known to regulate inflammation, oxidative stress, and proteostasis (Tsioutsiou et al. 2022).

Modern geroscientific tools now make it possible to rigorously study these traditional remedies. Notably, several Greek-derived compounds—such as oleuropein, chlorogenic acid, α-terpineol, and secoisolariciresinol diglucoside (SDG)—demonstrate activity across multiple ageing hallmarks. While these compounds are not exclusive to Greece, their plant sources—such as *Olea europaea*, *Cynara scolymus*, *Pistacia lentiscus*, and *Linum usitatissimum*—hold a long-standing role in Greek traditional medicine and diet, supporting their selection as representative examples of bioactive agents rooted in the region's natural pharmacopoeia.

This chapter aims to integrate traditional ethnobotanical knowledge of Greek medicinal plants with contemporary geroscience by exploring the molecular pathways through which selected natural compounds modulate the hallmarks of ageing. By examining evidence from *in vitro*, *in vivo*, and emerging clinical studies, we propose that these bioactives represent promising candidates for multi-targeted anti-ageing strategies, combining cultural heritage with scientific innovation.

13.2 Compounds and Extracts from Greek Flora Targeting Ageing Hallmarks

13.2.1 Genomic Instability

Genomic instability—marked by DNA strand breaks, mutations, and chromosomal aberrations—is a central hallmark of ageing. Among Greek phytochemicals, secoisolariciresinol diglucoside (SDG), a lignan abundant in *Linum usitatissimum* (flaxseed), shows strong genome-protective effects. In vitro, SDG pretreatment (50 μM) significantly reduced γ-H2AX foci and comet tail lengths in murine lung fibroblasts and endothelial cells exposed to ionizing radiation, indicating reduced DNA double-strand breaks and enhanced post-damage survival. This effect is mediated by activation of the Keap1–Nrf2 axis, which upregulates cytoprotective enzymes like heme oxygenase 1 (HO-1), NAD(P)H quinone dehydrogenase 1 (NQO1), and glutathione S-transferase mu 1 (GSTM1), bolstering the cellular antioxidant response (Velalopoulou et al. 2015). Rodent studies confirm these findings *in vivo*: flaxseed lignan-rich diets (35% SDG) administered before thoracic irradiation reduced lung fibrosis, inflammation, and DNA damage, while preserving radiosensitivity in tumor tissues (Christofidou-Solomidou et al. 2012). SDG thus functions as both a radioprotectant and a potential adjunct to cancer therapy. Pharmacokinetic studies demonstrate that oral SDG is well absorbed, with systemic conversion to enterodiol and enterolactone peaking at 12–36 hours, ensuring bioavailability (Setchell et al. 2014). Clinically, long-term supplementation reduces oxidative DNA adducts and enhances redox homeostasis. Altogether, SDG offers a potent, bioavailable strategy for preserving genomic integrity and counteracting age-related DNA damage, supporting its inclusion in anti-ageing interventions.

13.2.2 Telomere Attrition

Telomere shortening is a well-established hallmark of ageing, representing the progressive erosion of repetitive DNA sequences at chromosome ends during successive cell divisions. Once telomeres reach a critically short length, they can no longer effectively shield chromosomes from degradation or fusion, activating DNA damage responses (DDR) and triggering replicative senescence or apoptosis (Barnes et al. 2019). Natural compounds from the Greek pharmacopeia, such as α-terpineol and alpha-linolenic acid (ALA), have demonstrated potential to preserve telomere integrity through telomerase-independent mechanisms.

α-Terpineol, a monoterpenoid found in *Pistacia lentiscus* (Chios mastic), enhances the expression of shelterin proteins TRF1 and TRF2, key components that safeguard telomeres by binding directly to double-stranded telomeric DNA. In human dermal fibroblasts, α-terpineol treatment increased TRF1/TRF2 levels, reduced telomeric oxidative damage, and maintained telomere length across passages—delaying senescence without altering telomerase activity (Kapetanou et al. 2024). This mechanism avoids the oncogenic risk associated with telomerase reactivation, which is re-expressed in ~85% of human cancers (Shay and Wright 2005).

Alpha-linolenic acid (ALA), an omega-3 fatty acid from flaxseed, protects telomeres through antioxidant and anti-inflammatory actions. Clinical studies show higher omega-3 levels correlate with slower telomere shortening and modest telomerase activation, supporting ALA's role in preserving genomic stability and promoting healthy ageing (Farzaneh-Far et al. 2010; Kiecolt-Glaser et al. 2013).

Together, these findings position α-terpineol and ALA as promising telomere-protective agents. By enhancing shelterin complex stability and modulating systemic oxidative stress, these compounds delay telomere shortening without resorting to telomerase-based strategies that may carry long-term carcinogenic potential. Their integration into phytotherapeutic formulations offers a targeted, mechanistically grounded approach to preserving telomere integrity and extending cellular healthspan.

13.2.3 Epigenetic Alterations

Ageing is marked by epigenetic drift—progressive changes in DNA methylation and histone modification patterns—that disrupt chromatin organization and transcriptional homeostasis (La Torre et al. 2023). Among Greek-derived compounds, silymarin, a flavonolignan extract from *Silybum marianum*, and its major constituent silibinin, exhibit potent epigenetic-modulating properties by targeting histone deacetylases (HDACs), key enzymes involved in age-related transcriptional repression (Vargas-Mendoza et al. 2020).

Silymarin promotes histone H3 and H4 hyperacetylation, leading to a less compressed chromatin state and enhanced expression of genes governing DNA repair, oxidative stress responses, and metabolic balance (Singh et al. 2016). Silibinin refines this effect by selectively inhibiting HDAC1 and HDAC2, enzymes often elevated in aged and senescent cells, thereby restoring acetylation marks such as H3K9 and H4K16 crucial for genomic integrity and mitochondrial function (Zhou et al. 2016). Unlike synthetic HDAC inhibitors, silibinin shows dose-dependent and tissue-selective activity, avoiding broad cytotoxicity. Its epigenetic benefits are complemented by reduced senescence-associated secretory phenotype (SASP) expression, improved autophagy, and enhanced antioxidant defenses, indicating a broader impact on multiple ageing hallmarks including inflammation and proteostasis (Peleg et al. 2016).

13.2.4 Loss of Proteostasis and Disabled Macroautophagy

Proteostasis—the maintenance of protein homeostasis through balanced synthesis, folding, trafficking, and degradation—is fundamental for cellular function and longevity. A key component of this system is macroautophagy, a lysosome-dependent pathway responsible for clearing damaged proteins and organelles. Ageing disrupts both proteostasis and macroautophagy, leading to the accumulation of oxidized or misfolded proteins, which are implicated in neurodegenerative diseases and systemic decline (Kapetanou et al. 2022; López-OtíN et al. 2023).

18β-Glycyrrhetinic Acid (18β-GA), derived from *Glycyrrhiza glabra*, is a potent modulator of proteostasis. In human fibroblasts, it enhances the expression and activity of proteasome subunits (β5, α7, Rpt6), promoting the clearance of oxidized proteins via Nrf2-mediated upregulation of HO-1 and NQO1 (Kapeta et al. 2010). These effects translate in vivo, where *C. elegans* treated with 18β-GA exhibit increased proteasome activity, reduced aggregation, and extended lifespan. In Alzheimer's-like models, the compound decreased amyloid burden and delayed paralysis through SKN-1 activation—the worm orthologue of Nrf2 (Papaevgeniou et al. 2016). In mice, co-administration of 18β-GA with omega-3 fatty acids improved cognitive and motor functions, highlighting translational potential (Mladenovic Djordjevic et al. 2021).

Beyond proteasome activation, several Greek-derived phytochemicals also restore proteostasis by modulating macroautophagy, a critical lysosomal pathway responsible for degrading damaged proteins and organelles. Silibinin, the primary flavonolignan from *Silybum marianum*, enhances autophagic flux by promoting ROS-induced LC3-II formation and activating the Nrf2 pathway, while simultaneously suppressing NF-κB to mitigate inflammatory signaling and proteotoxic burden (Jiang et al. 2015). Similarly, dicaffeoylquinic acids (DCQAs) extracted from *Cynara scolymus* and *Helichrysum italicum* have been shown to induce autophagy in microglial cells. These compounds elevate LC3-II levels and downregulate JAK2/STAT3 signaling, effectively reducing the accumulation of damaged proteins and alleviating neuroinflammation (Park et al. 2022).

Adding to this proteoprotective arsenal, verbenone, a monoterpenoid found in *Rosmarinus officinalis*, has demonstrated the ability to lower protein carbonylation and oxidative stress, pointing to a role in preserving neural protein quality and redox homeostasis (Song et al. 2024). Likewise, pinocarveol, present in *Eucalyptus globulus* essential oils, exerts cytoprotective antioxidant effects by attenuating protein oxidation and enhancing cellular antioxidant defenses, thereby indirectly supporting the maintenance of proteome integrity (Ancuceanu et al. 2024).

Together, these phytochemicals offer a systems-level approach to preserving proteostasis by activating the proteasome, enhancing autophagy, and strengthening antioxidant networks—hallmarks of a promising anti-ageing strategy with high translational relevance.

13.2.5 Deregulated Nutrient Sensing

Ageing is closely tied to imbalances in nutrient-sensing pathways, particularly AMPK, SIRT1, and insulin/IGF-1 signaling (López-OtíN et al. 2013). These networks regulate energy metabolism, oxidative stress responses, and mitochondrial function. SIRT1, a NAD^+-dependent deacetylase, enhances genome stability and longevity by regulating transcription factors such as FOXO3 and PGC-1α, while AMPK acts as a cellular energy sensor that activates catabolic pathways during energy stress (Guan et al. 2025). Chlorogenic acid (CGA), a phenolic compound abundant in Greek plants like *Cynara scolymus*, modulates these key pathways. CGA activates AMPK, increases NAD^+ levels, and subsequently stimulates SIRT1, promoting mitochondrial biogenesis and antioxidant gene expression (Zhou et al. 2016). In endothelial cells, CGA restores AMPK/SIRT1 signaling after oxidative stress, improving mitochondrial integrity and cell survival. In *C. elegans*, CGA extends lifespan by 20% via DAF-16/FOXO and SKN-1/Nrf2, aligning with conserved insulin/IGF-1 nutrient sensing but acting independently of SIR-2.1 (Zheng et al. 2017). Clinically, green coffee extract rich in CGA (400 mg/day) has improved metabolic profiles in humans with metabolic syndrome, reducing glucose, lipids, and liver enzymes in randomized trials (Khalili-Moghadam et al. 2023). CGA thus mimics calorie restriction by reprogramming metabolic signaling, offering a promising strategy to restore homeostasis and delay ageing.

13.2.6 Mitochondrial Dysfunction

Mitochondrial dysfunction—a decline in energy production, redox balance, and mitochondrial quality control—is a central hallmark of ageing, contributing to oxidative stress, inflammation, and cell death. Crocin, a carotenoid glycoside from *Crocus sativus* (saffron), exerts powerful mitochondrial-protective effects across diverse cellular models. In human endothelial cells under hyperglycemic stress, crocin restores mitochondrial membrane potential, enhances mitochondrial mass, and increases antioxidant enzymes like GPx1, thereby reducing ROS and preserving endothelial function (Li et al. 2021). In neuronal cells, it attenuates glutamate-induced toxicity by stabilizing mitochondria, reducing calcium overload, and modulating apoptotic regulators Bax, Bcl-xL, and caspase-3 (Wang et al. 2019). Crocin activates the Akt/mTOR pathway, enhances ATP production, and protects against mitochondrial collapse in Parkinson's models such as MPP^+-exposed PC12 cells (Zhang et al. 2015). It also exerts cardioprotective effects by modulating TLR4/PTEN/Akt/mTOR and NF-κB signaling, supporting mitochondrial enzyme function and reducing apoptosis (Zununi Vahed et al. 2024). Together, these actions position crocin as a promising natural agent for sustaining mitochondrial health and preventing age-related cellular degeneration.

13.2.7 Cellular Senescence

Cellular senescence is a key hallmark of ageing, characterized by a stable arrest of the cell cycle in response to stressors such as DNA damage, oxidative insult, telomere attrition, or oncogenic activation. While initially protective—preventing the proliferation of damaged cells—senescent cells persist with age and actively contribute to tissue deterioration through the secretion of bioactive molecules collectively termed the senescence-associated secretory phenotype (SASP). This includes pro-inflammatory cytokines (e.g., IL-6, IL-8), matrix metalloproteinases, and growth factors that promote chronic inflammation, tissue remodeling, and paracrine induction of senescence in neighboring cells. As such, targeting senescent cells has become a focal point in anti-ageing therapeutics (López-OtíN et al. 2013).

Two principal strategies have emerged: senolytics, which selectively eliminate senescent cells, and senomorphics, which suppress the SASP while sparing the cell. Quercetin, a flavonoid present in many fruits and vegetables, functions as a senolytic by inducing apoptosis in senescent cells without harming proliferating counterparts. In senescent adipocytes and fibroblasts, quercetin reduces SA-β-Gal activity and oxidative stress markers, while activating AMPK and suppressing NF-κB, leading to SIRT1 upregulation and mTOR inhibition—thereby promoting autophagy and SASP suppression (Robbins et al. 2021). Fisetin, another dietary flavonoid, exhibits potent senolytic activity across various cell types, including human endothelial cells and dermal fibroblasts. In vivo, fisetin administration reduces senescent cell burden, lowers systemic inflammation, and rejuvenates skin architecture by increasing collagen density and reducing MMP activity (Yousefzadeh et al. 2018). In contrast, resveratrol acts as a senomorphic, attenuating the harmful SASP via SIRT1 activation and NF-κB inhibition in vertebrate ageing models such as the annual fish Nothobranchius guentheri (Liu et al. 2018).

Together, these compounds highlight a dual approach to targeting senescence: eliminating harmful cells with senolytics and modulating the SASP with senomorphics. Their combined use holds promise for preserving tissue homeostasis, delaying age-related disease onset, and extending healthspan.

13.2.8 Stem Cell Exhaustion

Stem cell exhaustion is a hallmark of ageing characterized by the gradual loss of regenerative potential, impaired self-renewal, and a bias toward dysfunctional differentiation (Oh et al. 2014). This decline compromises tissue repair and is central to the onset of degenerative conditions. Among phytochemicals derived from Greek flora, oleuropein, a polyphenol found abundantly in Olea europaea (olive leaves), has demonstrated robust activity in counteracting stem cell exhaustion, particularly within mesenchymal stem cells (MSCs). A critical target of oleuropein is connexin43 (Cx43), a gap junction protein that influences MSC fate decisions. In aged MSCs, Cx43 is overexpressed and correlates with a shift toward adipogenesis at the expense of osteogenic and chondrogenic lineages. This imbalance contributes to bone loss and marrow fat accumulation—hallmarks of stem cell ageing. Oleuropein suppresses Cx43 expression and promoter activity, thereby restoring the balance toward bone- and cartilage-forming lineages, while reducing adipogenic differentiation (Varela-EiríN et al. 2020). In osteoarthritic models, oleuropein also reduces cellular senescence markers in chondrocytes, synoviocytes, and osteoblasts, while enhancing the synthesis of cartilage ECM proteins such as collagen type II (Col2A1) and aggrecan. These effects are mediated through the downregulation of the pro-senescent transcription factor Twist-1 and inhibition of NF-κB signaling, resulting in decreased expression of IL-1β, IL-6, COX-2, and MMP-3 (Varela-EiríN et al. 2020). I. Collectively, these findings support oleuropein as a dual-action compound that promotes regenerative lineage commitment and attenuates inflammatory senescence, offering a compelling strategy for delaying stem cell exhaustion and preserving tissue homeostasis during ageing.

13.2.9 Altered Intercellular Communication and Chronic Inflammation (Inflammageing)

Two interconnected hallmarks of ageing are the disruption of intercellular communication and the presence of chronic low-grade inflammation, commonly referred to as "inflammageing". These processes involve the persistent release of pro-inflammatory cytokines, ROS, and matrix-degrading enzymes, which undermine tissue homeostasis, impair regeneration, and contribute to the onset of age-related diseases. Several natural compounds derived from Greek flora show promise in restoring balance to this disrupted signaling environment. For example, secoisolariciresinol diglucoside *(SDG)*, a lignan from *Linum* species, has been shown to suppress neuroinflammatory cytokines such as IL-6 and TNF-α, while mitigating β-amyloid aggregation. These actions support glial-neuronal communication and help preserve synaptic function during ageing, offering neuroprotective benefits in degenerative models (Rom et al. 2018). Similarly, chlorogenic acid (CGA), abundant in plants like *Cynara scolymus*, plays a protective role in skin ageing by maintaining extracellular matrix integrity. It achieves this by downregulating MMPs and pro-inflammatory mediators, thereby preserving the functional dialogue between fibroblasts and keratinocytes (Nguyen et al. 2024). Beyond anti-inflammatory effects, compounds like 18β-glycyrrhetinic acid (18β-GA) directly modulate gap junctions by regulating connexin43 expression, thus stabilizing intercellular calcium signaling and preventing the spread of apoptotic signals under oxidative stress (Mahmoud and Al Dera 2015). Flavonolignans such as silibinin and silymarin further reinforce intercellular homeostasis by inhibiting NF-κB and iNOS pathways, reducing the systemic inflammatory load across tissues (Zhu et al. 2025; Surai 2015). Compounds like dicaffeoylquinic acid (Jang et al. 2022) and verbenone (González-Velasco et al. 2022) add to this synergistic action by stabilizing cell–cell communication in joint and immune tissues, respectively, through MAPK inhibition and cytokine suppression. Together, these bioactives exemplify how Greek natural compounds can restore signaling fidelity, reduce inflammageing, and preserve tissue integrity throughout ageing.

13.2.10 Dysbiosis

Accumulating evidence shows that several Greek-derived phytochemicals also counteract dysbiosis, a recognized hallmark of ageing. **Oleuropein** from *Olea europaea* functions as a prebiotic—enhancing *Lactobacillus* and *Bifidobacterium* growth —and restores gut microbial balance in ulcerative colitis models, reducing inflammation and strengthening barrier function (Zang et al. 2025). Chlorogenic acid undergoes microbial metabolism and, together with EGCG, has been shown to improve age-related dysbiosis and intestinal integrity in murine models (Pereira et al. 2024). These findings align with broader evidence that plant polyphenols act as "duplibiotics," selectively inhibiting pathogens while nurturing health-associated taxa such as *Akkermansia*, *Bifidobacteria*, and *Lactobacilli* (Rodríguez-Daza et al. 2021). By modulating the gut microbiome, these phytochemicals help maintain mucosal homeostasis and combat chronic low-grade inflammation—key features of ageing.

Taken together, the natural compounds derived from Greek flora demonstrate robust, multi-targeted activity across the spectrum of recognized ageing hallmarks as summarized in Table 13.1. From preserving genomic stability and telomere integrity to enhancing proteostasis, mitigating cellular senescence, and restoring intercellular communication, these phytochemicals offer an integrative approach to slowing age-related decline. Importantly, their actions converge on conserved regulatory pathways—such as AMPK, SIRT1, FOXO3, NF-κB, and the proteasome—that orchestrate cellular resilience, redox balance, and regenerative capacity. Beyond *in vitro* and model organism studies, several of these compounds have also shown translational promise in human trials, further supporting their relevance as nutraceutical or therapeutic agents. As a whole, these findings underscore the untapped potential of Greece's botanical biodiversity in contributing to the development of evidence-based interventions that not only delay the onset of age-related pathologies but also promote systemic homeostasis and healthspan extension.

Table 13.1 Greek botanical compounds targeting hallmarks of ageing

Compound	Botanical source from the Greek flora	Targeted hallmarks of ageing	Chemical structure
Oleuropein	Olive leaf (*Olea europaea*)	Stem cell exhaustion	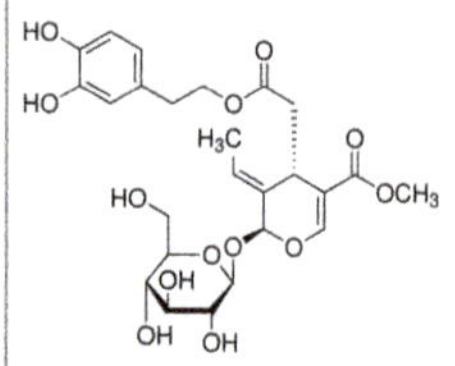
Chlorogenic Acid (CGA)	*Artichoke (Cynara scolymus)*	Nutrient sensing, Intercellular communication	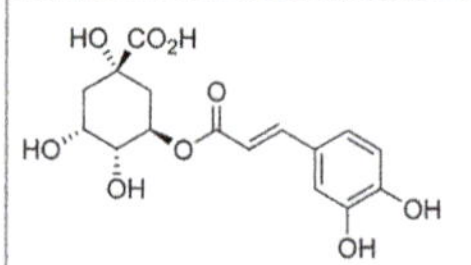
α-Terpineol	Mastic tree resin (*Pistachia lentiscus*), Pine oil (*Pinus* spp.)	Telomere attrition, Genomic instability	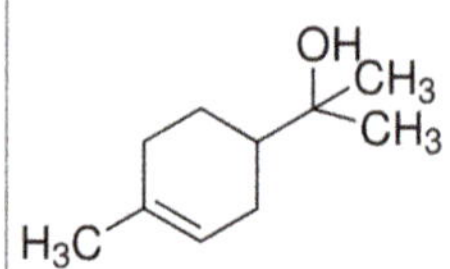
Silymarin/Silibinin	Milk thistle (*Silybum marianum*)	Epigenetic alterations, Proteostasis, Intercellular communication	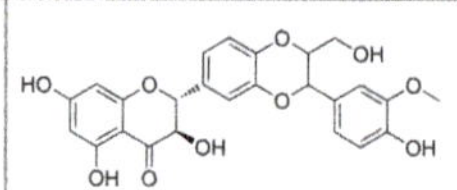
Crocin	Saffron (*Crocus sativus*)	Mitochondrial dysfunction	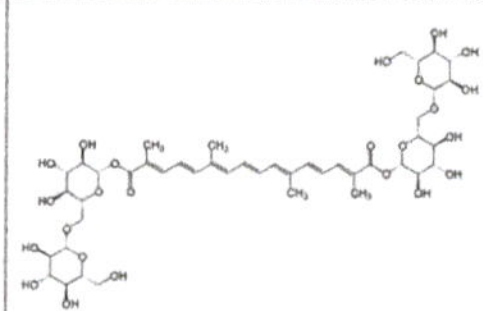
Secoisolariciresinol diglucoside (SDG)	Flaxseed (*Linum usitatissimum*)	Genomic instability, Intercellular communication	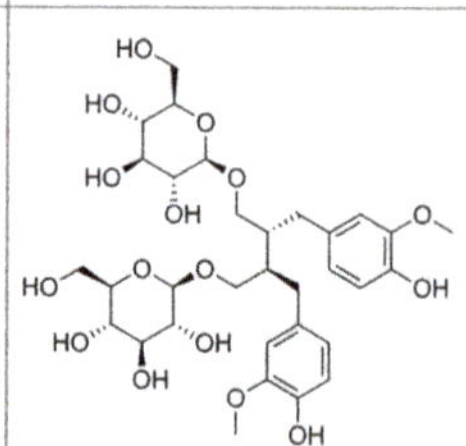
18α/β-Glycyrrhetinic Acid	Licorice root (*Glycyrrhiza glabra*)	Proteostasis, Intercellular communication	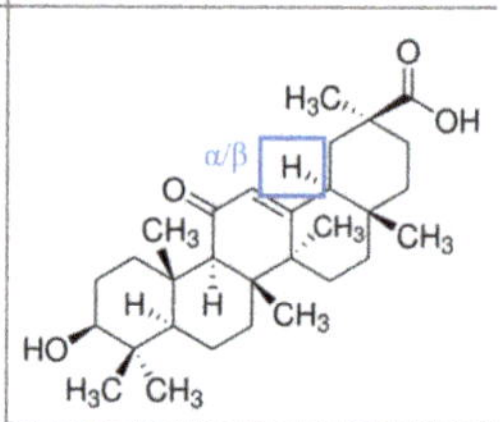

(continued)

Table 13.1 (continued)

Compound	Botanical source from the Greek flora	Targeted hallmarks of ageing	Chemical structure
Fisetin	Strawberry (Fragaria × ananassa), apple (*Malus domestica, Malus pumila*), onion (*Allium cepa*)	Cellular senescence	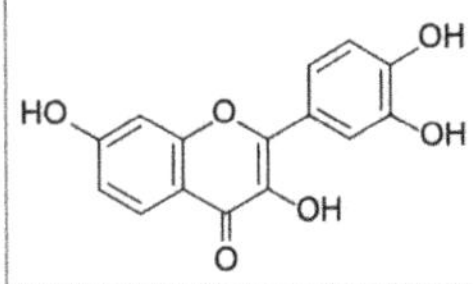
Quercetin	Onion (*Allium cepa*), apple, capers (*Capparis spinosa*)	Cellular senescence	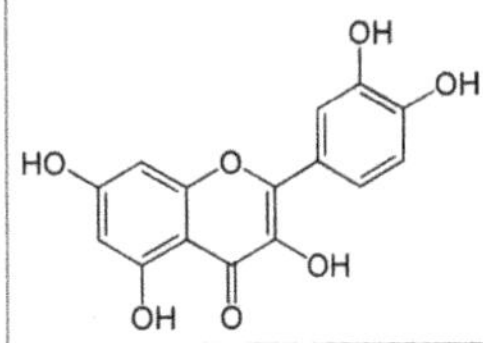
Linolenic Acid	Flaxseed (*Linum usitatissimum*),	Telomere attrition	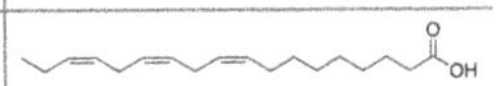
Pinocarveol	Mastic tree resin (*Pistachia lentiscus*), Eucalyptus (*Eucalyptus globulus)*	Proteostasis	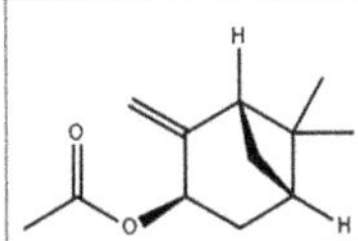
Verbenone	*Rosemary (Rosmarinus officinalis), Thyme (Thymus vulgaris),* Mastic tree resin (*Pistachia lentiscus*)	Proteostasis, Intercellular communication	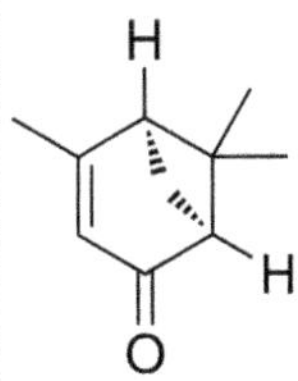
Dicaffeoylquinic acid	*Artichoke (Cynara scolymus), Immortelle (Helichrysum italicum)*	Intercellular communication	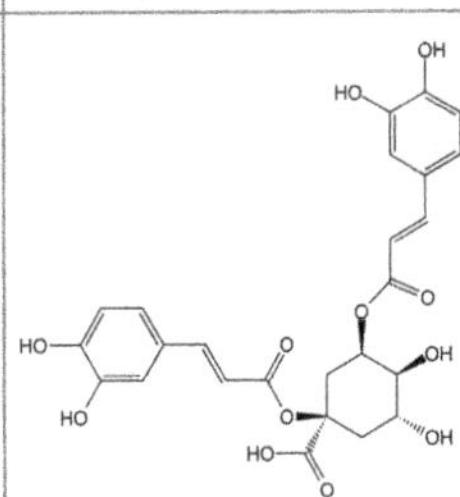
Resveratrol	*Grape (Vitis vinifera)*	Senescence	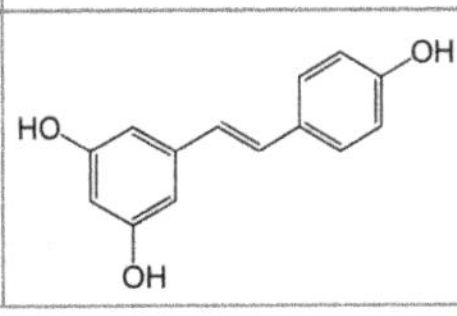

Summary of key natural compounds derived from Greek flora, highlighting their botanical sources, targeted hallmarks of ageing, and chemical structures. The compounds presented have been shown to modulate molecular pathways involved in genomic instability, telomere attrition, epigenetic alterations, loss of proteostasis, deregulated nutrient sensing, mitochondrial dysfunction, cellular senescence, stem cell exhaustion, altered intercellular communication, disabled macroautophagy, chronic Inflammation (Inflammageing) and dysbiosis

13.3 Clinical Translation of Greek Phytochemicals for Healthy Ageing

13.3.1 Safety, Bioavailability, and Pharmacokinetics

Translating the anti-ageing potential of Greek phytochemicals into clinically relevant applications depends on thorough evaluation of their safety, pharmacokinetics, and formulation. While many of these natural compounds exhibit promising bioactivity, several face pharmacokinetic challenges—such as low solubility, poor intestinal absorption, and rapid metabolism—that limit their systemic efficacy.

For instance, oleuropein, the principal phenolic in *Olea europaea*, has GRAS (Generally Recognized as Safe) status and is rapidly metabolized to hydroxytyrosol, which retains biological activity in neuroprotection and inflammation control (Frumuzachi et al. 2024). Silymarin and silibinin, derived from *Silybum marianum*, also demonstrate a favorable safety profile but suffer from low water solubility. Phytosome formulations have markedly improved their absorption (Ranjan and Gautam 2023). Similarly, chlorogenic acid (CGA) from *Cynara scolymus* undergoes rapid hydrolysis, necessitating nanovesicular or self-emulsifying systems to improve bioavailability (Nguyen et al. 2024).

Crocin, a neuroprotective carotenoid from *Crocus sativus*, is hydrolyzed into crocetin, a more absorbable metabolite, but still exhibits low oral bioavailability (Xiao et al. 2024). SDG, a lignan in flaxseed, is well tolerated, though its biological activity depends on microbiome-driven conversion to enterodiol and enterolactone (Zhang et al. 2025).

The senolytic doses for fisetin and quercetin are higher than dietary intake but no adverse effects were reported in clinical trials (Romashkan et al. 2021). Nanoformulations are under study to improve the low bioavailability of fisetin (Szymczak and Cielecka-Piontek 2023).

To overcome these limitations, advanced delivery strategies—such as phytosomes, cyclodextrin complexes, nanoparticles, and bioenhancers like piperine—are being employed to improve absorption, protect compounds from degradation, and extend systemic exposure. These innovations are crucial for transforming traditional phytochemicals into standardized, bioavailable therapeutic agents.

13.3.2 Clinical Evidence and Real-Life Applications

Robust clinical and preclinical data support the health-promoting effects of Greek phytochemicals across metabolic, cognitive, hepatic, dermal, and inflammatory domains. A flagship study evaluated a synergistic nutraceutical blend containing extracts from Linum usitatissimum (SDG), *Silybum marianum* (silymarin), *Cynara scolymus* (chlorogenic acid), and *Pistacia lentiscus* (mastic resin). *In vitro*, this formulation enhanced proteasome activity and proteostasis, delayed senescence, and stabilized telomere length and DNA methylation. A randomized, placebo-controlled trial mirrored these effects, demonstrating reduced oxidized plasma proteins, especially among women aged 45–55 with higher BMI, and correlating inversely with 20S proteasome levels (Athanasopoulou et al. 2022).

Other key single-compound studies further substantiate this bioactivity. Oleuropein supplementation improved blood pressure, lipid profiles, and cognitive function in hypertensive and elderly individuals (Frumuzachi et al. 2024). Silymarin reduced liver enzyme levels and hepatic steatosis in NAFLD patients, and showed benefits in topical applications for UV-induced skin damage (Chantarojanasiri 2023). Chlorogenic acid lowered postprandial glucose and inflammatory cytokines in overweight adults (Kim et al. 2025), while crocin rivaled fluoxetine in alleviating depression and enhanced cognition in elderly subjects (Khaksarian et al. 2019).

SDG supplementation (600 mg/day for 12 weeks) was shown to reduce IL-6 and TNF-α, improve antioxidant capacity, and beneficially modulate gut microbiota in elderly individuals (Nattagh-Eshtivani et al. 2025). These findings underscore the translation al value of Greek phytochemicals in real-life settings and advocate for their integration into dietary and therapeutic regimens targeting ageing hallmarks (Fig. 13.1).

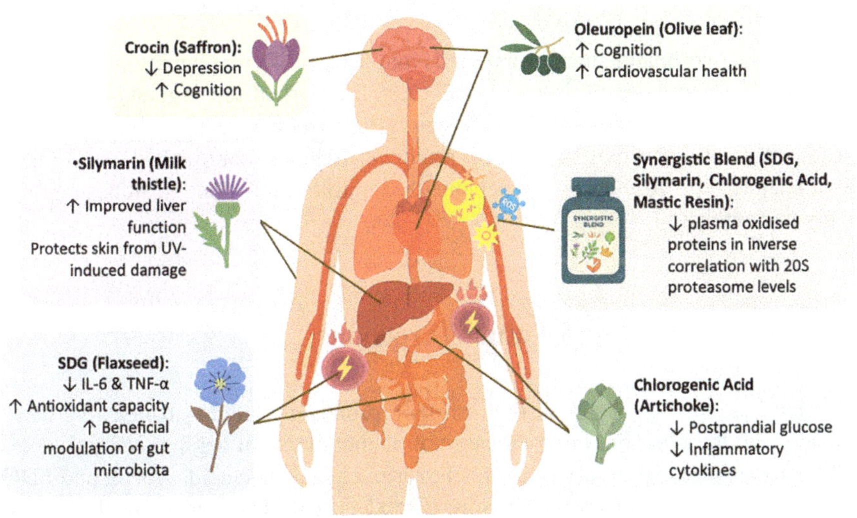

Fig. 13.1 Clinical effects of key Greek-derived phytochemicals. Graphical summary of well-characterized Greek phytochemicals and their documented clinical benefits. Illustrated compounds have been shown to exert significant cognitive, cardiovascular, hepatic, metabolic, dermal, antioxidant and anti-inflammatory effects in human studies, supporting their translational potential in ageing-related interventions

13.3.3 Evidence-Based Recommendations

To guide implementation, Greek phytochemicals can be stratified using a graded evidence system inspired by GRADE. Compounds supported by at least one human clinical trial and consistent preclinical data are classified as Level A, while those with strong preclinical but limited clinical data are Level B. Preliminary or ethnobotanical compounds are Level C, and under-studied candidates are Level D.

- **Level A**: Oleuropein, silymarin, and chlorogenic acid—supported by human trials and mechanistic insights—are recommended for nutraceutical and cosmeceutical use.
- **Level B**: Crocin, SDG, fisetin, and quercetin demonstrate promising effects but require formulation refinement and larger trials.
- **Level C-D**: Compounds like α-terpineol, linolenic acid, and pinocarveol have shown preclinical efficacy in senescence, telomere integrity, and proteostasis but remain in exploratory phases.

Practical application should consider evidence grade, target pathway, and formulation. For example, fisetin and quercetin are effective senolytics in intermittent protocols, while CGA and oleuropein suit long-term dietary use. Topical agents like silymarin, crocin, and Revised to align with the updated classification of the 12 hallmarks of ageing are well-suited for skin applications, especially when delivered through nanocarriers.

Integrating complementary compounds that act on different ageing hallmarks—e.g., combining quercetin for senescence, CGA for nutrient sensing, and oleuropein for stem cell modulation—may yield synergistic benefits. Ultimately, translating these botanicals into evidence-based protocols requires coordinated research in phytochemistry, geroscience, and personalized nutrition to extend healthspan effectively and safely.

13.4 Conclusions and Future Directions

The rich biodiversity of Greek flora offers a promising source of bioactive compounds with the potential to modulate multiple hallmarks of aeging. Natural molecules such as oleuropein, α-terpineol, silymarin, chlorogenic acid, crocin, and SDG have demonstrated efficacy in preclinical models and emerging clinical settings. These compounds target critical ageing mechanisms, including genomic stability, telomere maintenance, proteostasis, cellular senescence, nutrient sensing, mitochondrial function, and chronic inflammation.

Many of these phytochemicals act on interconnected pathways, offering multitargeted benefits. For example, chlorogenic acid modulates both AMPK/SIRT1 signaling and inflammatory cascades, while SDG exerts estrogenic and antioxidant effects that support DNA protection and intercellular communication. Their use within Mediterranean dietary patterns and cosmeceutical formulations underscores both safety and cultural integration.

However, several gaps hinder full clinical translation. Most evidence stems from in vitro and animal studies, with limited human trials directly assessing ageing biomarkers such as epigenetic clocks, senescence markers, or functional healthspan. Pharmacokinetic and bioavailability data remain insufficient, especially in older adults. The combinatorial potential of these compounds also requires deeper investigation.

Future priorities include:

- Conducting large-scale, randomized clinical trials with validated ageing biomarkers.
- Advancing delivery technologies (e.g., phytosomes, liposomes, nanoparticles) to enhance bioavailability.
- Applying multi-omics and systems biology approaches to elucidate mechanisms and personalize interventions.
- Exploring synergistic combinations with gerotherapeutics (e.g., metformin, senolytics).

Greek phytochemicals are well-suited for development into nutraceuticals, functional foods, and longevity-targeted formulations. Strategic collaborations among academic, clinical, and industry partners—both local and international—will be essential to accelerate innovation and ensure regulatory and scientific rigor.

In summary, rigorously validated Greek natural compounds can offer sustainable, accessible, and culturally resonant solutions for promoting healthy human ageing.

References

Ancuceanu R, Anghel AI, Hovaneţ MV, Ciobanu A-M, Lascu BE, Dinu M (2024) Antioxidant activity of essential oils from Pinaceae species. Antioxidants 13:286

Athanasopoulou S, Kapetanou M, Magouritsas MG, Mougkolia N, Taouxidou P, Papacharalambous M, Sakellaridis F, Gonos E (2022) Antioxidant and antiaging properties of a novel synergistic nutraceutical complex: readouts from an in Cellulo study and an in vivo prospective, randomized trial. *Antioxidants* 11:468

Axiotis E, Halabalaki M, Skaltsounis LA (2018) An ethnobotanical study of medicinal plants in the Greek Islands of North Aegean region. Front Pharmacol 9

Barnes RP, Fouquerel E, Opresko PL (2019) The impact of oxidative DNA damage and stress on telomere homeostasis. Mech Ageing Dev 177:37–45

Campisi J, Kapahi P, Lithgow GJ, Melov S, Newman JC, Verdin E (2019) From discoveries in ageing research to therapeutics for healthy ageing. Nature 571:183–192

Caruso C, Puca AA (2021) Special issue "Centenarians—A model to study the molecular basis of lifespan and Healthspan". Int J Mol Sci 22:2044

Chantarojanasiri T (2023) Silymarin treatment and reduction of liver enzyme levels in non-alcoholic fatty liver disease: a case report. Drugs Context 12

Christofidou-Solomidou M, Tyagi S, Pietrofesa R, Dukes F, Arguiri E, Turowski J, Grieshaber PA, Solomides CC, Cengel KA (2012) Radioprotective role in lung of the flaxseed lignan complex enriched in the phenolic secoisolariciresinol diglucoside (SDG). Radiat Res 178:568–580

Couteur DGL, Barzilai N (2022) New horizons in life extension, healthspan extension and exceptional longevity. Age Ageing 51

Farzaneh-Far R, Lin J, Epel ES, Harris WS, Blackburn EH, Whooley MA (2010) Association of marine omega-3 fatty acid levels with telomeric aging in patients with coronary heart disease. JAMA 303:250–257

Frumuzachi O, Gavrilaş LI, Vodnar DC, Rohn S, Mocan A (2024) Systemic health effects of oleuropein and hydroxytyrosol supplementation: A systematic review of randomized controlled trials. Antioxidants (Basel) 13

González-Velasco HE, Pérez-Gutiérrez MS, Alonso-Castro ÁJ, Zapata-Morales JR, Niño-Moreno PDC, Campos-Xolalpa N, González-Chávez MM (2022) Anti-inflammatory and antinociceptive activities of the essential oil of Tagetes parryi A. Gray (Asteraceae) and Verbenone. Molecules 27

Guan G, Chen Y, Dong Y (2025) Unraveling the AMPK-SIRT1-FOXO pathway: the in-depth analysis and breakthrough prospects of oxidative stress-induced diseases. Antioxidants (Basel) 14

Jang G, Lee SA, Hong JH, Park BR, Kim DK, Kim CS (2022) Chondroprotective effects of 4,5-Dicaffeoylquinic acid in osteoarthritis through Nf-κB signaling inhibition. Antioxidants (Basel) 11

Jiang K, Wang W, Jin X, Wang Z, Ji Z, Meng G (2015) Silibinin, a natural flavonoid, induces autophagy via ROS-dependent mitochondrial dysfunction and loss of ATP involving BNIP3 in human MCF7 breast cancer cells. Oncol Rep 33:2711–2718

Kapeta S, Chondrogianni N, Gonos ES (2010) Nuclear erythroid factor 2-mediated proteasome activation delays senescence in human fibroblasts. J Biol Chem 285:8171–8184

Kapetanou M, Athanasopoulou S, Gonos ES (2022) Transcriptional regulatory networks of the proteasome in mammalian systems. IUBMB Life 74:41–52

Kapetanou M, Athanasopoulou S, Goutas A, Makatsori D, Trachana V, Gonos E (2024) α-Terpineol induces shelterin components TRF1 and TRF2 to mitigate senescence and telomere integrity loss via A telomerase-independent pathway. Antioxidants (Basel) 13

Khaksarian M, Behzadifar M, Behzadifar M, Alipour M, Jahanpanah F, Re TS, Firenzuoli F, Zerbetto R, Bragazzi NL (2019) The efficacy of Crocus sativus (Saffron) versus placebo and fluoxetine in treating depression: a systematic review and meta-analysis. Psychol Res Behav Manag 12:297–305

Khalili-Moghadam S, Hedayati M, Golzarand M, Mirmiran P (2023) Effects of green coffee aqueous extract supplementation on glycemic indices, lipid profile, CRP, and malondialdehyde in patients with type 2 diabetes: a randomized, double-blind, placebo-controlled trial. Front Nutr 10

Kiecolt-Glaser JK, Epel ES, Belury MA, Andridge R, Lin J, Glaser R, Malarkey WB, Hwang BS, Blackburn E (2013) Omega-3 fatty acids, oxidative stress, and leukocyte telomere length: A randomized controlled trial. Brain Behav Immun 28:16–24

Kim H, Kim SR, Jung UJ (2025) Coffee and its major polyphenols in the prevention and management of Type 2 diabetes: A comprehensive review. Int J Mol Sci 26:5544

La Torre A, Lo Vecchio F, Greco A (2023) Epigenetic mechanisms of aging and aging-associated diseases. Cells 12

Li X, Liu Y, Cao A, Li C, Wang L, Wu Q, Li X, Lv X, Zhu J, Chun H, Laba C, Du X, Zhang Y, Yang H (2021) Crocin improves endothelial mitochondrial dysfunction via GPx1/ROS/KCa3.1 signal Axis in diabetes. Front Cell Dev Biol 9:651434

Liu S, Zheng Z, Ji S, Liu T, Hou Y, Li S, Li G (2018) Resveratrol reduces senescence-associated secretory phenotype by SIRT1/NF-κB pathway in gut of the annual fish Nothobranchius guentheri. Fish Shellfish Immunol 80:473–479

López-OtíN C, Blasco MA, Partridge L, Serrano M, Kroemer G (2013) The hallmarks of aging. Cell 153:1194–1217

López-OtíN C, Blasco MA, Partridge L, Serrano M, Kroemer G (2023) Hallmarks of aging: an expanding universe. Cell 186:243–278

Mahmoud AM, Al Dera HS (2015) 18β-Glycyrrhetinic acid exerts protective effects against cyclophosphamide-induced hepatotoxicity: potential role of PPARγ and Nrf2 upregulation. Genes Nutr 10:41

Mladenovic Djordjevic AN, Kapetanou M, Loncarevic-Vasiljkovic N, Todorovic S, Athanasopoulou S, Jovic M, Prvulovic M, Taoufik E, Matsas R, Kanazir S, Gonos ES (2021) Pharmacological intervention in a transgenic mouse model improves Alzheimer's-associated pathological phenotype: involvement of proteasome activation. Free Radic Biol Med 162:88–103

Nattagh-Eshtivani E, Barghchi H, Hatami A, Barati M, Heydari A, Abdi Moghaddam Z, Vahedi Fard M, Mohammadhasani K, Miri H, Talebi S, Pahlavani N (2025) The effects of flaxseed supplementation on metabolic syndrome risk factors; A comprehensive review. Clin Nutr Open Sci 60:262–285

Nguyen V, Taine EG, Meng D, Cui T, Tan W (2024) Chlorogenic acid: A systematic review on the biological functions, mechanistic actions, and therapeutic potentials. Nutrients 16

Oh J, Lee YD, Wagers AJ (2014) Stem cell aging: mechanisms, regulators and therapeutic opportunities. Nat Med 20:870–880

Papaevgeniou N, Sakellari M, Jha S, Tavernarakis N, Holmberg CI, Gonos ES, Chondrogianni N (2016) 18α-Glycyrrhetinic acid proteasome activator decelerates aging and Alzheimer's disease progression in Caenorhabditis elegans and neuronal cultures. Antioxid Redox Signal 25:855–869

Park J, Kim Y, Lee C, Kim YT (2022) 3,5-Dicaffeoylquinic acid attenuates microglial activation-mediated inflammatory pain by enhancing autophagy through the suppression of MCP3/JAK2/STAT3 signaling. Biomed Pharmacother 153:113549

Peleg S, Feller C, Ladurner AG, Imhof A (2016) The metabolic impact on histone acetylation and transcription in ageing. Trends Biochem Sci 41(8):700–711. https://doi.org/10.1016/j.tibs.2016.05.008

Pereira QC, Fortunato IM, Oliveira FD, Alvarez MC, Santos TW, Ribeiro ML (2024) Polyphenolic compounds: orchestrating intestinal microbiota harmony during aging. Nutrients [Online] 16

Ranjan S, Gautam A (2023) Pharmaceutical prospects of Silymarin for the treatment of neurological patients: an updated insight. Front Neurosci 17:1159806

Robbins PD, Jurk D, Khosla S, Kirkland JL, Lebrasseur NK, Miller JD, Passos JF, Pignolo RJ, Tchkonia T, Niedernhofer LJ (2021) Senolytic drugs: reducing senescent cell viability to extend health span. Annu Rev Pharmacol Toxicol 61:779–803

Rodríguez-Daza MC, Pulido-Mateos EC, Lupien-Meilleur J, Guyonnet D, Desjardins Y, Roy D (2021) Polyphenol-mediated gut microbiota modulation: toward prebiotics and further. Front Nutr 8

Rom S, Zuluaga-Ramirez V, Reichenbach NL, Erickson MA, Winfield M, Gajghate S, Christofidou-Solomidou M, Jordan-Sciutto KL, Persidsky Y (2018) Secoisolariciresinol diglucoside is a blood-brain barrier protective and anti-inflammatory agent: implications for neuroinflammation. J Neuroinflammation 15:25

Romashkan S, Chang H, Hadley EC (2021) National Institute on Aging workshop: repurposing drugs or dietary supplements for their Senolytic or Senomorphic effects: considerations for clinical trials. J Gerontol A Biol Sci Med Sci 76:1144–1152

Setchell KD, Brown NM, Zimmer-Nechemias L, Wolfe B, Jha P, Heubi JE (2014) Metabolism of secoisolariciresinol-diglycoside the dietary precursor to the intestinally derived lignan enterolactone in humans. Food Funct 5:491–501

Shay JW, Wright WE (2005) Senescence and immortalization: role of telomeres and telomerase. Carcinogenesis 26:867–874

Singh T, Prasad R, Katiyar SK (2016) Therapeutic intervention of silymarin on the migration of non-small cell lung cancer cells is associated with the axis of multiple molecular targets including class 1 HDACs, ZEB1 expression, and restoration of miR-203 and E-cadherin expression. Am J Cancer Res 6:1287–1301

Song HY, Jin S, Lee S, Jalin AMA, Roh KH, Kim WK (2024) The therapeutic effects of SP-8356, a Verbenone derivative, with multimodal Cytoprotective mechanisms in an ischemic stroke rat model. Int J Mol Sci 25

Surai PF (2015) Silymarin as a natural antioxidant: an overview of the current evidence and perspectives. Antioxidants (Basel) 4:204–247

Szymczak J, Cielecka-Piontek J (2023) Fisetin-in search of better bioavailability-from macro to nano modifications: A review. Int J Mol Sci 24

Tsioutsiou EE, Amountzias V, Vontzalidou A, Dina E, Stevanović ZD, Cheilari A, Aligiannis N (2022) Medicinal plants used traditionally for skin related problems in the South Balkan and East Mediterranean region—A review. Front Pharmacol 13

Varela-EirÍN M, Carpintero-Fernández P, Sánchez-Temprano A, Varela-Vázquez A, Paíno CL, Casado-Díaz A, Calañas-Continente A, Mato V, Fonseca E, Kandouz M, Blanco A, Caeiro JR, MayáN MD (2020) Senolytic activity of small molecular polyphenols from olive restores chondrocyte redifferentiation and promotes a pro-regenerative environment in osteoarthritis. Aging (Albany NY) 12:15882–15905

Vargas-Mendoza N, Morales-González Á, Morales-Martínez M, Soriano-UrsúA MA, Delgado-Olivares L, Sandoval-Gallegos EM, Madrigal-Bujaidar E, Álvarez-González I, Madrigal-SantilláN E, Morales-Gonzalez JA (2020) Flavolignans from Silymarin as Nrf2 bioactivators and their therapeutic applications. Biomedicine 8:122

Velalopoulou A, Tyagi S, Pietrofesa R, Arguiri E, Christofidou-Solomidou M (2015) The flaxseed-derived Lignan phenolic Secoisolariciresinol Diglucoside (SDG) protects non-malignant lung cells from radiation damage. Int J Mol Sci 17:7

Wang C, Cai X, Hu W, Li Z, Kong F, Chen X, Wang D (2019) Investigation of the neuroprotective effects of crocin via antioxidant activities in HT22 cells and in mice with Alzheimer's disease. Int J Mol Med 43:956–966

Xiao F, Song Y, Wang G, Aa J (2024) Intestinal metabolism of Crocin and a pharmacokinetics and pharmacodynamics study in the chronic social defeat stress mouse model. Pharmaceuticals (Basel) 17

Yousefzadeh MJ, Zhu Y, Mcgowan SJ, Angelini L, Fuhrmann-Stroissnigg H, Xu M, Ling YY, Melos KI, Pirtskhalava T, Inman CL, Mcguckian C, Wade EA, Kato JI, Grassi D, Wentworth M, Burd CE, Arriaga EA, Ladiges WL, Tchkonia T, Kirkland JL, Robbins PD, Niedernhofer LJ (2018) Fisetin is a senotherapeutic that extends health and lifespan. EBioMedicine 36:18–28

Zang R, Zhou R, Li Y, Liu Z, Wu H, Lu L, Xu H (2025) Oleuropein regulates bile acid metabolism via modulating the gut microbiota, thereby alleviating DSS-induced ulcerative colitis in mice. Foods [Online] 14

Zhang GF, Zhang Y, Zhao G (2015) Crocin protects PC12 cells against MPP(+)-induced injury through inhibition of mitochondrial dysfunction and ER stress. Neurochem Int 89:101–110

Zhang Y, Liu X, Zheng Z, Huang H, Wang Y, Wu S, Shu Y, Yang Y, Zhong Y, Liao P, Wang Y, Pan Z (2025) Network pharmacology uncovers that secoisolariciresinol diglucoside ameliorate premature ovarian insufficiency via PI3K/Akt pathway. Sci Rep 15:1493

Zheng SQ, Huang XB, Xing TK, Ding AJ, Wu GS, Luo HR (2017) Chlorogenic acid extends the lifespan of Caenorhabditis elegans via insulin/IGF-1 signaling pathway. J Gerontol A Biol Sci Med Sci 72:464–472

Zhou Y, Ruan Z, Zhou L, Shu X, Sun X, Mi S, Yang Y, Yin Y (2016) Chlorogenic acid ameliorates endotoxin-induced liver injury by promoting mitochondrial oxidative phosphorylation. Biochem Biophys Res Commun 469:1083–1089

Zhu K, Ni H, Hafeez E, Hu Y, Hu F, Du D, Chen D (2025) Effects of Silibinin on delaying aging in Drosophila melanogaster. Antioxidants (Basel) 14

Zununi Vahed S, Zuluaga Tamayo M, Rodriguez-Ruiz V, Thibaudeau O, Aboulhassanzadeh S, Abdolalizadeh J, Meddahi-Pellé A, Gueguen V, Barzegari A, Pavon-Djavid G (2024) Functional mechanisms of dietary Crocin protection in cardiovascular models under oxidative stress. Pharmaceutics 16

Chapter 14
Mediterranean Diet-Based Supplements for Healthy Longevity

Ceren Gezer

Abstract The Mediterranean diet (MedDiet) is a plant-based diet characterised by a high consumption of fruits, vegetables, nuts, olive oil, and unrefined grains, which are major sources of polyphenols. Among these, hydroxybenzoic acids are the most abundant phenolic acids, while flavanols, flavonols, flavanones, and anthocyanins are the most prevalent flavonoids in the MedDiet. Polyphenols found in the MedDiet have been shown to influence mitochondrial dysfunction, deregulated nutrient sensing, autophagy, cellular senescence, and dysbiosis through hormesis in cellular and animal models. According to epidemiological studies, dietary intake of total polyphenols is nearly equivalent to that obtained through commercially available polyphenol supplements. However, factors such as the food matrix and bioavailability must be considered within a hormetic model when suggesting intake levels for promoting healthy longevity. While combining dietary and supplemental intake may be a viable alternative, more clinical trials including those incorporating nutrigenetic data are needed to establish evidence-based daily intake recommendations for polyphenols. Longitudinal epidemiological and clinical studies on polyphenol intake and longevity could provide valuable insights into optimal amounts and patterns of phytochemical consumption for supplement development. Until stronger evidence supports supplementation, consuming polyphenols through a healthy, balanced diet remains the safest and most effective approach.

Keywords Mediterranean diet · Polyphenols · Supplements · Ageing · Longevity · Hormesis

---

C. Gezer (✉)
Department of Nutrition and Dietetics, Faculty of Health Sciences, Eastern Mediterranean University, Famagusta, Cyprus
e-mail: ceren.gezer@emu.edu.tr

S. I. S. Rattan, C. Gezer (eds.), *Supplements and Drugs for Healthy Longevity*, Healthy Ageing and Longevity 22, https://doi.org/10.1007/978-3-032-11608-6_14

14.1 Introduction

Mediterranean Diet (MedDiet) is a dietary pattern associated with the lifestyle and culture of Mediterranean countries. Three different meta-analysis of prospective studies indicated that MedDiet is associated with decreased risk of all-cause mortality (Bonaccio et al. 2018; Eleftheriou et al. 2018; Soltani et al. 2019). Another meta-analysis also suggested that MedDiet is associated with prolonged survival in individuals with a history of cardiovascular diseases (Tang et al. 2021). A meta-analysis included randomized controlled trials, prospective cohort, case–cohort, case–control studies, and showed that MedDiet is associated with decreased cancer mortality risk, and all-cause mortality among cancer survivors (Morze et al. 2021). Therefore, MedDiet has been widely touted as an ideal diet for healthy ageing and longevity (Guasch-Ferré and Willett 2021; Mazza et al. 2021; Andreo-López et al. 2023).

MedDiet is characterized by the low consumption of meat/meat-products, moderate consumption of dairy products, fish and red wine, and high consumption of fruits, vegetables, nuts, olive oil and non-refined cereals. Therefore, MedDiet is rich in oleic acid and omega-3 fatty acids, polysaccharides and polyphenols (Serra-Majem et al. 2020). The European Prospective Investigation into Cancer and Nutrition (EPIC) study indicated that phenolic acids were the most prevalent, followed by flavonoids, in terms of total polyphenol intake. Furthermore, coffee was identified as the primary dietary source of phenolic acids across all European countries. In contrast, the main sources of flavonoids differed by region: fruits were the primary contributors in Mediterranean countries, whereas tea was the major source in non-Mediterranean countries. Additionally, vegetable oils were the predominant sources of lignans in Mediterranean countries (Zamora-Ros et al. 2016).

Another Italian cohort study demonstrated similar findings, showing that phenolic acids are the primary contributors to total polyphenol intake, followed by flavonoids. Among phenolic acids, hydroxybenzoic acids were the most abundant, while among flavonoids, flavanols, flavonols, flavanones, and anthocyanins were the most prevalent (Godos et al. 2017a). Accordingly, the major dietary sources of polyphenols in the MedDiet include olives; fruits—such as cherries for anthocyanins and citrus fruits for flavanones; vegetables—such as artichokes for flavones and spinach and beans for flavonols; and coffee (Godos et al. 2017a). The study demonstrated that increased polyphenol intake is related to higher adherence to the MedDiet (Kapolou et al. 2021). Moreover, another study demonstrated that among polyphenol subgroups, only lignans, anthocyanins, and flavanones showed a linear positive correlation with high adherence to the MedDiet. At the individual compound level, apigenin, hesperetin, naringenin, lariciresinol, matairesinol, pinoresinol, secoisolariciresinol, and ferulic acid were also positively associated with greater adherence (Godos et al. 2017b). MedDiet as a dietary pattern or its individual food components have potential beneficial effects on the ageing process (Alì et al. 2021; Grande de França et al. 2023). Polyphenols in the MedDiet can modulate various hallmarks of ageing, especially chronic inflammation, loss of

Table 14.1 Commonly consumed polyphenols of MedDiet

Main classes	Subclasses	Food sources
Phenolic acids	**Hydroxybenzoic acids**	
	Gallic acid	Tea, oak bark, gallnuts, apples, strawberries, pineapples, bananas, blackberry, lemon, grape
	Ellagic acid	Walnut, chestnut, pomegranate and berries
	Hydroxycinnamic acids	
	Chlorogenic acid	Coffee, eggplant, carrot, artichoke, apple, pear, peach, plum, cherry, tomato and potato
	Ferulic acid	Bananas, beetroot, broccoli, cabbage, citrus fruits, coffee, eggplant, grain bran, spinach and whole grains
Flavonoids	**Flavonols**	
	Quercetin	Onion, cabbage, green pepper, tomato, coriander, okra, grape, cranberry, asparagus, caper, apple,buckwheat
	Kaempferol	Tea, beans, broccoli, apple, and strawberries
	Myricetin	Orange, grape, spinach, cauliflower, carrot, blueberry, tea, red wine
	Flavanones	
	Naringenin, naringin, hesperetin, hesperidin	Citrus fruits such as grapefruit, pumelo, mandarin, lemon
	Flavones	
	Apigenin	Parsley, celery, artichoke, oregano and chamomile tea
	Luteolin	Parsley, thyme, green pepper, celery, carrot, artichoke, apple
	Anthocyanins	
	Delphinidin, cyanidin, pelargonidin, malvidin, peonidin, petunidin	Blueberry, cranberry, pomegranate, red grapes, black soybean, purple corn, red cabbage, raspberry are
Lignans	Secoisolariciresinol	Linseed and oat
Others	**Phenylethanoids**	
	Tyrosol, hydroxy tyrosol	Olive, olive oil
	Secoiridoids	
	Oleoropein	Olive, olive oil

proteostasis and cellular senescence (Meccariello and D'Angelo 2021; Pereira et al. 2023). Therefore, polyphenols may improve cognitive function and gut barrier integrity, impair photo-ageing, age-related muscle dysfunction, cardiovascular and metabolic disorders (Liu et al. 2024).

In this chapter, more focus is given on MedDiet's commonly consumed polyphenols such as phenolic acids, flavonoids, lignans and tyrosols to understand whether supplementation of these compounds exerts healthy longevity effects (Table 14.1).

14.2 Polyphenols in Plant-Based Foods

Polyphenols are classified to four major subgroups as phenolic acids, flavonoids, lignans, stilbenes and others (Truzzi et al. 2021). Various polyphenols can regulate metabolism and expand longevity due to targeting mitophagy, senescence pathways and circadian rhythms (Shen et al. 2022). Besides, telomere shortening, nuclear factor kappa B (NF-κB), insulin and insulin like growth factor (IGF)-1, adenosine monophosphate-activated protein kinase (AMPK), mammalian target of rapamycin (mTOR), and mitochondrial dysfunction are pathways initiate age related disorders such as neurodegenerative diseases, cardiovascular diseases, other metabolic diseases and cancer (Muzammil et al. 2021). In an updated review it has been suggested that various physical, cellular, or molecular biomarkers of dietary phytochemicals are most commonly studied in worm, rats and mice animal models (Chen et al. 2024; Park et al. 2024). The scarcity of human studies may stem from difficulties in measuring metabolites and biomarkers, integrating *in vitro* biological effects, and accounting for the matrix effects of polyphenols (Bertelli et al. 2021). Also, bioavailability which depends on the process of absorption and metabolism of polyphenols is important. Polyphenols are generally reacted with glycoside enzymes in mouth, then hydroxylated in small intestine or directly move to colon to metabolized by gut microbiota, then metabolites transfer to liver for conjugation. In addition, during all these processes solubility and permeability of the polyphenols related with the health effects in the body. According to biopharmaceutical classification system, polyphenols generally have low solubility and high permeability, which limits their potential health effects (Truzzi et al. 2021).

14.2.1 Phenolic Acids

14.2.1.1 Hydroxybenzoic Acids

Gallic acid is a well-known benzoic acid derivative found in tea, oak bark, gallnuts, apples, strawberries, pineapples, bananas, blackberries, lemons, and grapes. It exerts antioxidant and anti-inflammatory effects, thereby helping to prevent chronic diseases such as cancer, diabetes, and cardiovascular diseases, and potentially promoting longevity (Hadidi et al. 2024). Gallic acid has been shown to enhance antioxidant and anti-inflammatory responses and reduce cellular senescence in embryonic fibroblasts and pancreatic islet cells of rats, suggesting its potential to modulate diabetes and promote healthy ageing (Rahimifard et al. 2020). Additionally, gallic acid has been identified as a potential geroprotector in human stem cells (Shan et al. 2022). It may also influence gut microbiota composition, as demonstrated in atherosclerotic $ApoE^{-/-}$ male mice (Clark et al. 2022). Furthermore, gallic acid exhibits neuroprotective effects by restoring the levels of key antioxidant enzymes-catalase

(CAT), glutathione (GSH), and superoxide dismutase (SOD)-in aluminum chloride-induced Alzheimer's disease in Wistar rats (Ogunlade et al. 2022).

Ellagic acid is particularly found in walnut, chestnut, pomegranate and berries (Cannataro et al. 2021). Ellagic acid has been shown to extend lifespan by increasing resilience to heat shock, starvation, and oxidative stress in *Drosophila melanogaster (D. melanogaster)* (Kharat et al. 2020). Ellagic acid has been shown to reduce the risk of age-related diseases and promote longevity through its anti-proliferative, antioxidant, and anti-inflammatory activities (Deepika and Maurya 2022). These effects are mediated not only by ellagic acid itself via upregulation of nuclear factor erythroid 2–related factor 2 (Nrf2) and downregulation of nuclear factor kappa B (NF-κB) but also by its gut microbiota-derived fermentation metabolites, urolithins (Xian et al. 2023). Urolithins have been suggested to modulate the gut microbiota, thereby enhancing intestinal barrier function and exerting beneficial effects on metabolic systems (Leng et al. 2025). In addition, ellagic acid activates AMP-activated protein kinase (AMPK) and increases mitochondrial biogenesis in muscle cells (Li et al. 2022). A randomized controlled trial also demonstrated that supplementation with 180 mg/day of ellagic acid significantly reduced insulin resistance in individuals with diabetes, potentially through the activation of sirtuin 1 (SIRT1) (Ghadimi et al. 2021).

14.2.1.2 Hydroxycinnamic Acids

Chlorogenic acid and ferulic acid are most commonly known hydroxycinnamic acids (Li et al. 2023). Chlorogenic acid is mostly found in coffee, vegetables and fruits such as eggplant, carrot, artichoke, apple, pear, peach, plum, cherry, tomato and potato (Lu et al. 2020). Chlorogenic acid increase lifespan of yeast by alleviating oxidative damage via triggering oxidative stress response (Tungmunnithum et al. 2020). Chlorogenic acid combat oxidative stress and expand lifespan through skinhead (SKN)-1/Nrf2 and DAF-16/forkhead box O (FOXO) in *Caenorhabditis elegans (C. elegans)* (Siswanto et al. 2022). Chlorogenic acid hinders H_2O_2 induced senescence in human umbilical vein endothelial cells via Nrf2/heme oxygenase (HO)-1 pathway as well as diminish vascular senescence in a dose-dependent manner (20 and 40 mg/kg/day) due to increase in SIRT1 and endothelial nitric oxide synthase in C57/BL6 mice (Hada et al. 2020). Chlorogenic acid not only show anti-inflammatory effects via NF-kB inhibition and anti-oxidant effects via Nrf2 activation but also lipid and glucose metabolism modulation by AMPK and neuromodulation (Nguyen et al. 2024).

Ferulic acid is mostly found in bananas, beetroot, broccoli, cabbage, citrus fruits, coffee, eggplant, grain bran, spinach and whole grains (Kumar et al. 2024). Ferulic acid demonstrate neuroprotective effect due to enhance autophagy in stable GFP-RFP-LC3 U87 cells and PC-12 cells and *C. elegance* (Long et al. 2022). Ferulic acid supplementation extends lifespan and increase stress tolerance through mitochondrial and insulin/IGF-1 signaling pathways in *C. elegans* (Li et al. 2021). Ferulic acid enhances longevity and stress tolerance via insulin/IGF-1 signaling pathway integrated DAF-16 transcription factor in *C. elegance* (Yang et al. 2022). Ferulic acid may also extend lifespan by promoting SIRT1 and AMPK (DInicolantonio et al. 2022). Intraperitoneal injection of ferulic acid (50 mg/kg) suppresses total body irradiation-mediated bone marrow damage, stem cell senescence, and enhancing antioxidant response in C57BL/6 mice (Wagle et al. 2021). In addition, ferulic acid increases mitochondrial biogenesis and antioxidant capacity via Nrf1 and Nrf2 in mice skeletal muscle (Huang et al. 2024b).

14.2.2 Flavonoids

Flavanoids classified to six major subgroups as flavones, flavonols, flavanones, isoflavones, flavanols and anthocyanins. It has been suggested that flavonoids can show senolytic and senomorphic effects to increase longevity (Fan et al. 2022). A cross-sectional analysis of National Health and Nutrition Examination Survey data it has been shown that increased intake of dietary flavonoids especially anthocyanidins, flavones and flavonols may have a positive effect on the reproductive life span of women and on reproductive health (Liu et al. 2025). Mostly intaken flavonoids are flavanols, flavonols, flavanones and anthocyanins in MedDiet (Godos et al. 2017a).

14.2.2.1 Flavonols

Quercetin, kaempferol, and myricetin are grouped as flavonols and food sources rich from flavonols are fruits and vegetables such as broccoli, strawberry, leak, onion, kale (Li et al. 2023). It has been indicated that quercetin can prolonged lifespan through NF-kB inhibition, modulation of phosphatidylinositol-3 kinase (PI3K)/Akt/mTOR and p38 mitogen-activated protein kinase (MAPK) pathways and induction of SOD, CAT enzyme transcriptions via Nrf2 (Sorrenti et al. 2023). A meta-analysis of randomized controlled human trials quercetin supplementation decreased C-reactive protein levels in individuals who have chronic diseases, and decreased interleukin (IL)-6 levels in females, individuals who have chronic diseases and who take >500 mg/day quercetin supplementation. It has been underlined in this meta-analysis that there are not any clinical or epidemiologic studies have reported any compound related adverse effects of quercetin supplementation 3–1000 mg/day for up to 12 weeks (Ou et al. 2020).

Kaempferol can modulate endoplasmic reticulum stress and insulin pathways in intestinal stem cells in *D. melanogaster,* hence improving stress tolerance and enhancing longevity (Zhao et al. 2024). A study conducted with C57BL/6N mice indicated that kaempferol (25 and 100 mg/kg) increased anti-oxidant capacity and physical performance and *in vitro* analysis in C2C12 cells indicated that kaempferol modulate glucose and protein metabolism and mitochondrial biogenesis through PI3K/Akt and p38/MAPK signaling pathways (Ji et al. 2024). Kaempferol shows antioxidant and anti-inflammatory effects in hydrogen peroxide-induced ARPE-19 retinal pigment epithelium cells via upregulation of SIRT1, downregulation of polyADP-ribose polymerase-1 in a dose-dependent manner (Al Sabaani 2020). Kaempferol can manage age-related chronic diseases such as neurodegenerative, cardiovascular and metabolic diseases show due to anti-oxidant and anti-inflammatory effects. Even though kaempferol has low toxicity and the toxicity dose range is wide in animal and human studies, the dose management for this aim is not clear yet (Hussain et al. 2024). A review on kaempferol and quercetin-rich herbs supplementation suggested the inhibition of amyloid (Aβ) plaque formation and tau hyperphosphorylation, the suppression of oxidative stress, and regulation of PI3K/Akt pathways, based on pre-clinical study results (Alexander et al. 2023). It has been found that 50 mg/day kaempferol aglycone supplementation for four weeks is safe for healthy adults in a randomised, placebo-controlled study (Akiyama et al. 2023).

Myricetin against Aβ42 oligomer-induced toxicity through decreasing tau phosphorylation, reactive oxygen species generation, lipid peroxidation, and deoxyribonucleic acid (DNA) oxidation in neuronal SH-SY5Y cells (Wang et al. 2023b). Dihydromyricetin inhibits DNA methyltransferase-1 dose dependently in human skin cells and may rejuvenate aged human skin (Falckenhayn et al. 2023).

14.2.2.2 Flavanones

Naringenin, naringin, hesperetin and hesperidin are grouped as flavanones (Najmanová et al. 2020). Flavanons are particularly found in citrus fruits such as grapefruit, pumelo, mandarin, lemon (Kumar et al. 2023). Naringenin extends lifespan in *C. elegans* via MAPK, as well as delayed brain ageing via SIRT1 activation in C57BL/6 N mice (100 mg/kg/day) (Ge et al. 2021; Piragine et al. 2024). Naringenin initiates SIRT1 activity and inhibits senescence in H9c2 cells, which is a cellular model for myocardial ageing. Also, 100 mg/kg/day naringin supplementation reduces oxidative stress, inflammatory cytokines TNF-α and IL-6 in C57BL/6 J mice (Testai et al. 2020). Naringin impedes H_2O_2-induced oxidative damage in cells and extends *D. melanogaster* lifespan through insulin signaling pathway (Du et al. 2024). Naringin increases stress tolerance and prolongs lifespan in *C. elegans* models of Alzhemier's disease and Parkinson disease via FOXO/DAF-16 (Zhu et al. 2020). Naringin also attenuates glucose-induced ageing due to a decrease of fat accumulation and initiates autophagy in *C. elegans* (Guo et al. 2023). Naringin demonstrate a hormetic effect in nucleus pulpous cells, cardiac cells, stem cells such as periodontal, endothelial commonly via Nrf2 transcription factor (Calabrese et al. 2023). A study conducted with human aortic endothelial cells and ApoE$^{-/-}$ C57BL/6 mice indicated that naringenin (200 mg/kg/day) mitigate vascular senescence via upregulation of SIRT1 (Wang et al. 2023a). A single-ascending-dose randomized crossover study conducted with 18 individuals demonstrated that 150 mg, 300 mg, 600 mg and 900 mg naringenin supplementation was safe and suggested that 300 mg twice a day may produce a physiological effect (Rebello et al. 2020).

Hesperidin exhibits anti-oxidant and anti-inflammatory effects via initiating AMPK, as well as take roles on macro molecules metabolism hence may indicate expanding longevity by preventing or treating age-related diseases (Ahmad et al. 2023). Three citrus fruits species (tangerine, orange, and grapefruit) peels supplementation prolonged lifespan of *D. melanogaster* (Oboh et al. 2021). Hesperetin prolongs health span due to increasing Cisd2 expression in C57BL/6 background CISD2 reporter transgenic old mice (Yeh et al. 2022). A review includes naringenin and hesperidin suggested that human studies conducted with between 200–500 mg supplementation of naringenin and hesperidin are safe and show vascular and neuroprotection (Cannataro et al. 2021).

14.2.2.3 Flavones

Apigenin and luteolin are grouped as flavones. Flavons most abundantly found in parsley, celery, oregano, rosemary, green olive, pumpkin, watermelon, bell pepper, honey, fava beans, chickpea (Kumar et al. 2023; Singh et al. 2024). Apigenin extend lifespan by phosphatase and tensin homolog (PTEN)/Akt signaling pathway and antioxidant response via reverse D-galactose inhibited CAT, glutathione S-transferase transcription in *D. melanogaster* (Oyebode et al. 2020; Sun et al. 2025). Apigenin prolong lifespan by insulin/IGF-1 and AMPK/NRF-2/FOXO signaling pathways and show mitohormesis effect in *C. elegans* (Elkhedir et al. 2022; Cheng et al. 2023; Kramer and Johnson 2024). In C57BL/6N mice retina, a solid dispersion of apigenin (40 and 60 mg/kg) increased the expression of antioxidant enzymes via Nrf2 pathway and upregulated autophagy dose dependently (Zhang et al. 2020). A study conducted with C57BL/6N mice indicated that apigenin supplementation (0.5 mg/ml drinking water) increased nitric oxide bioavailability and normalize reactive oxygen species and antioxidant expression hence restored carotid artery endothelium-dependent dilation and alleviate age-associated aortic stiffening in old mice (Clayton et al. 2021). Another study conducted with C57BL/6N mice showed that apigenin supplementation (0.5 mg/ml drinking water) increased transcripts which decrease with ageing process in brain while apigenin treatment *in vitro* decreased inflammation and cellular senescence markers in human astrocytes (Cavalier et al. 2024). In addition, apigenin enhances dose dependently osteoblast differentiation potential and reduction of oxidative stress and senescent cell burden in human bone marrow stromal stem cells (Ali et al. 2024). Apigenin increased antioxidant enzyme expression via increasing Nrf2 transcription factor in PIG3V cell (Zhang et al. 2020). The review on apigenin display that it can induce antioxidant response and mitochondrial biogenesis through apigenin can activate the peroxisome proliferator-activated receptor gamma coactivator (PGC)1-α/Nrf2 signaling pathway thus improved endogenous antioxidant adaptation induced by exercise in healthy individuals (Poblete-Aro et al. 2023).

Luteolin which is commonly found in tomato, capsicum, carrot and celery shows especially anti-inflammatory, anti-oxidant and neuroprotective effects due to NF-κB, toll-like receptor, MAPK, Wnt/β-catenin, PI3K/Akt, AMPK/mTOR, and Nrf-2 according to cell culture and animal models (Zhu et al. 2024). Luteolin impedes H_2O_2-induced cellular senescence through SIRT1 and p53 in HEI-OC1 (Zhu et al. 2021). Also, intraperitoneal injection of dimethyl sulfoxide 80 mg/kg/day, dissolved in 1% luteolin alleviate brain ageing via attenuation of oxidative stress, mitochondrial dysfunction, neuro-inflammation, and D-galactose induced senescence in Wistar rats brain due to activation of SIRT1 (Younis et al. 2024). Moreover, luteolin can modulate mitochondria, endoplasmic reticulum, golgi apparatus, and lysosomes functions during in postovulatory oocyte ageing via upregulation of SIRT1 (Xing et al. 2023). In addition, a polyphenol-rich natural extract in which luteolin is the most concentrated flavonoid inhibited age-induced inflammation, fibrosis and senescence C57BL/6 N mice (Zumerle et al. 2024).

14.2.2.4 Anthocyanins

Anthocyanins commonly found in blueberry, cranberry, pomegranate, red grapes, black soybean, purple corn, red cabbage, raspberry are delphinidin, cyanidin, pelargonidin, malvidin, peonidin, petunidin (Bendokas et al. 2020; Kumar et al. 2023). Cyanidin-3-glycoside prolong lifespan of C. elegans due to increase heat, ultra violate a and oxidative stress tolerance (Doringo et al. 2023; Nas and Medina 2023). Anthocyanins can prevent reactive oxygen species formation and apoptosis of mitochondria (Bendokas et al. 2020). In addition, it has been indicated that anthocyanins can modulate gut microbiota thus provide prevention from age-related diseases (Hair et al. 2021). Moreover, a meta-analysis of randomized controlled trials suggested that anthocyanin rich supplementation affects cognitive process and neuronal activities positively of healthy middle-aged and older adults (Feng et al. 2023). It has been proposed that anthocyanins avoid liver diseases and promote liver longevity via anti-inflammatory and anti-oxidant mechanisms (Mohammed and Khan 2022).

14.2.3 Lignans

Secoisolariciresinol is a lignan molecule which is most commonly found in linseed and oat (Li et al. 2023). It has been demonstrated that secoisolariciresinol diglucoside can increase mitochondrial DNA copy number and delay telomere shortening thus raising ovarian reserve by suppressing oxidative stress in female C57BL/6 J mice (He et al. 2022). Also, secoisolariciresinol diglucoside impede cerebral Aβ deposition, neuroinflammatory responses and modulate gut microbiota in female Alzheimer's disease mice (Jia et al. 2024).

Sesamin is a lignan found in sesame seed and oil. Sesamin demonstrated anti-inflammatory effects via MAPK and NF-κB pathways activation, hence preventing inflammatory metabolic diseases (Majdalawieh et al. 2022). In addition, sesamin activated Nrf2/Cnc in *D. melanogaster* neuronal cells (Le and Inoue 2021). Sesamin and sesamol showed neuroprotective effects on H_2O_2-induced SH-SY5Y cells via activation of SIRT1-SIRT3-FOXO3a signaling pathway (Ruankham et al. 2021). Also, sesamol may increase *C. elegans* lifespan under heat stress, ultraviolet irradiation stress and oxidative stress (Huang et al. 2024a). Sesamin showed anti-inflammatory, antioxidative, and antiapoptotic effects due to activation of Nrf2/HO-1/NAD(P)H quinone dehydrogenase (NQO)1 signaling pathways in intestinal ischemia/reperfusion injured Sprague-Dawley rats (Wang et al. 2021).

14.2.4 Phenylethanoids and Secoiridoids

Tyrosol and hydroxy tyrosol are phenylethanoids and oleuropein is a secoiridoid composed of an ester bond between elenolic acid and hydroxytyrosol found in extra virgin olive oil (Finicelli et al. 2021; Gervasi and Pojero 2024). The absorption of olive oil phenolics is estimated to be 60–95%, but there is a general lack of data regarding the bioavailability and safety of oleoropein and hyrdoxytyrosol in the elderly (Pojero et al. 2023). Oleuropein improve stress resiliance and promote longevity through insulin/IGF-1 and SKN-1/Nrf2 signaling in *C. elegans* (Feng et al. 2021). Hydroxytyrosol and oleuropein show antioxidant, anti-inflammatory, and neuroprotective activities. SIRT1, AMPK, FOXO, PGC1-α, NF-κB and Nrf-2 signaling activation are likely the key mechanisms of hydroxytyrosol and oleuropein to modulate oxidative stress, inflammation, autophagy and mitochondrial biogenesis (Serreli and Deiana 2020; Bucciantini et al. 2021; Micheli et al. 2023). In addition, oleuropein and hydroxytyrosol may attenuate oxidative stress, promote autophagy and facilitate clearance of amyloid-beta aggregates, mitigate mitochondrial damage and dysfunction, thereby reducing neuroinflammation and improving cognitive function and brain health (Riquelme and Pando 2025). European Food Safety Authority has approved the health claim that olive oil polyphenols help protect blood lipids from oxidative damage when 5 mg/day of hydroxytyrosol and derivatives such as oleuropein complex and tyrosol in olive oil consumed (EFSA 2011). A recent systematic review of randomized controlled trials suggested that hydroxytyrosol and oleuropein supplementation (5 mg/day hydroxytyrosol to 990 ml/day olive leaf infusion-320.8 mg oleuropein and 11.9 mg hydroxytyrosol) shows beneficial effects on cardiovascular, bone and cognitive health (Frumuzachi et al. 2024). A recent meta-analysis of 14 randomized controlled trial on cardiometabolic risk factors propose that oleuropein, hydroxytyrosol, and tyrosol intake reduced trygliceride, total cholesterol and insulin levels (Frumuzachi et al. 2025). Therefore, olive oil polyphenols can prolong longevity by preventing chronic diseases such as cardiovascular diseases.

These pathways activations by polyphenols as explained above are cellular stress response pathways. Most of the polyphenols are identified as nutritional hormetins by triggering one or more of the stress response pathways such as autophagy, sirtuins, inflammatory- and oxidative-responses (Rattan 2022). Among these plant-based polyphenols included in MedDiet, ferulic acid, quercetin, luteolin, and naringin are the widely known polyphenols, which indicate their hormetic mode of action (Calabrese et al. 2021a, b, 2023, 2024). Polyphenols have been explained individually mostly in cellular models, but it is important to underline that it is important to consider the synergistic and antagonistic effects of these components' combinations for the stress-induced hormesis for healthy longevity (Rattan 2022).

Almost all polyphenols except quercetin, resveratrol and curcumin commonly found in herbal extracts are sold commercially (Bertelli et al. 2021). Therefore, it has been suggested that a combination of different dietary supplements which affect different pathways, or a combination of dietary supplements with dietary patterns, or a combination of different dietary patterns with personalised dietary strategies can be more effective for healthy ageing and longevity of humans (Duan et al. 2022). Furthermore, it has been suggested that these plant-based polyphenols act as hormetins due to their presence in low doses within the MedDiet, consequently, their long-term, moderate activation of stress response pathways may promote healthy longevity (Martucci et al. 2017).

14.3 Conclusions

Dietary polyphenols play a critical role in healthy ageing and longevity through a biphasic hormetic dose-response mechanism, and MedDiet contains a wide array of these phenolic compounds. MedDiet polyphenols have the potential to modulate mitochondrial dysfunction, deregulated nutrient sensing, autophagy, cellular senescence, and dysbiosis. However, so far these effects are essentially reported on cellular and animal study models. To clearly understand their effects, bioavailability, metabolism, interaction with other components and other biomarkers on human beings need to be clarified either by dietary or supplemental intake.

In addition, understanding the interactions of polyphenols is important, as they not only interact with each other but also with other components, and they are also converted to other metabolites by the gut microbiota. Current doses of individual polyphenol supplements on the market are close to the total daily dietary polyphenol intake. Dietary pattern analysis with respect to polyphenols is important while considering genetic characteristics of people in order to decide if there is a need for polyphenol supplementation.

As the majority of current research is preclinical, further clinical trials are necessary to establish evidence-based recommendations regarding the sufficient daily intake suggestions for polyphenols for healthy longevity. Longitudinal epidemiologic and clinical studies on polyphenol intake and longevity can shed light on the amounts and patterns of phytochemicals to design supplements.

Therefore, until stronger evidence supports supplementation, obtaining polyphenols through a healthy and balanced diet, such as the MedDiet, remains the safest and most effective approach. Polyphenols from the MedDiet are regarded as particularly advantageous in promoting longevity, since characterized by the regular consumption of polyphenol-rich foods in moderate quantities and a high degree of polyphenol diversity, improved bioavailability—enhanced by the presence of healthy fats such as those in olive oil and nuts that facilitate the absorption of fat-soluble compounds—and synergistic interactions with other essential nutrients, including omega-3 fatty acids, dietary fiber, vitamins, and minerals.

References

Ahmad A et al (2023) A comprehensive review of the therapeutic potential of citrus bioflavonoid hesperidin against lifestyle-related disorders. Cogent Food Agric. Informa Healthcare. https://doi.org/10.1080/23311932.2023.2226427

Akiyama M et al (2023) A randomized, placebo-controlled trial evaluating the safety of excessive administration of kaempferol aglycone. Food Sci Nutr 11(9):5427–5437. https://doi.org/10.1002/fsn3.3499

Al Sabaani N (2020) Kaempferol protects against hydrogen peroxide-induced retinal pigment epithelium cell inflammation and apoptosis by activation of SIRT1 and inhibition of PARP1. J Ocul Pharmacol Ther 36(7). https://doi.org/10.1089/jop.2019.0151

Alexander C, Parsaee A, Vasefi M (2023) Polyherbal and multimodal treatments: Kaempferol- and Quercetin-Rich herbs alleviate symptoms of Alzheimer's disease. Biology. Multidisciplinary Digital Publishing Institute (MDPI). https://doi.org/10.3390/biology12111453

Alì S et al (2021) Healthy ageing and Mediterranean diet: a focus on hormetic phytochemicals. Mech Ageing Dev 200. https://doi.org/10.1016/j.mad.2021.111592

Ali D et al (2024) Apigenin and Rutaecarpine reduce the burden of cellular senescence in bone marrow stromal stem cells. Front Endocrinol 15. https://doi.org/10.3389/fendo.2024.1360054

Andreo-López MC et al (2023) Influence of the Mediterranean diet on healthy aging. Int J Mol Sci. https://doi.org/10.3390/ijms24054491

Bendokas V et al (2020) Anthocyanins: from plant pigments to health benefits at mitochondrial level. Crit Rev Food Sci Nutr. Bellwether Publishing, Ltd:3352–3365. https://doi.org/10.1080/10408398.2019.1687421

Bertelli A et al (2021) Polyphenols: from theory to practice. Foods. MDPI. https://doi.org/10.3390/foods10112595

Bonaccio M et al (2018) Mediterranean diet and mortality in the elderly: a prospective cohort study and a meta-analysis. Br J Nutr 120(8). https://doi.org/10.1017/S0007114518002179

Bucciantini M et al (2021) Olive polyphenols: antioxidant and anti-inflammatory properties. Antioxidants 10(7). https://doi.org/10.3390/antiox10071044

Calabrese EJ et al (2021a) Luteolin and hormesis. Mech Ageing Dev 199. https://doi.org/10.1016/j.mad.2021.111559

Calabrese EJ, Agathokleous E, Calabrese V (2021b) Ferulic acid and hormesis: biomedical and environmental implications. Mech Ageing Dev 198. https://doi.org/10.1016/j.mad.2021.111544

Calabrese, E.J. *et al.* (2023) 'Naringin commonly acts via hormesis', *Sci Total Environ* Elsevier BV https://doi.org/10.1016/j.scitotenv.2023.164728

Calabrese EJ et al (2024) Quercetin induces its chemoprotective effects via hormesis. Food Chem Toxicol. https://doi.org/10.1016/j.fct.2023.114419

Cannataro R et al (2021) Polyphenols in the Mediterranean diet: from dietary sources to microrna modulation. Antioxidants. MDPI:1–24. https://doi.org/10.3390/antiox10020328

Cavalier AN et al (2024) Protective effects of apigenin on the brain transcriptome with aging. Mech Ageing Dev 217. https://doi.org/10.1016/j.mad.2023.111889

Chen JC, Wang R, Wei CC (2024) Anti-aging effects of dietary phytochemicals: from Caenorhabditis elegans, Drosophila melanogaster, rodents to clinical studies. Crit Rev Food Sci Nutr. Taylor and Francis Ltd:5958–5983. https://doi.org/10.1080/10408398.2022.2160961

Cheng Y et al (2023) Transient inhibition of mitochondrial function by chrysin and apigenin prolong longevity via mitohormesis in C. Elegans. Free Radic Biol Med 203:24–33. https://doi.org/10.1016/j.freeradbiomed.2023.03.264

Clark M et al (2022) Gallic acid ameliorates atherosclerosis and vascular senescence and remodels the microbiome in a sex-dependent manner in ApoE−/− mice. J Nutr Biochem 110. https://doi.org/10.1016/j.jnutbio.2022.109132

Clayton ZS et al (2021) Apigenin restores endothelial function by ameliorating oxidative stress, reverses aortic stiffening, and mitigates vascular inflammation with aging. Am J Physiol Heart Circ Physiol 321(1):H185–H196. https://doi.org/10.1152/ajpheart.00118.2021

DInicolantonio JJ et al (2022) Ferulic acid and berberine, via Sirt1 and AMPK, may act as cell cleansing promoters of healthy longevity. Open Heart. BMJ Publishing Group. https://doi.org/10.1136/openhrt-2021-001801

Doringo JAM et al (2023) Cyanidin-3-glucoside enhances longevity and heat stress resilience in Drosophila melanogaster. Biomed Biotechnol Res J 7(4):537–544. https://doi.org/10.4103/bbrj.bbrj_194_23

Du X et al (2024) Naringin ameliorates H2O2-induced oxidative damage in cells and prolongs the lifespan of female Drosophila melanogaster via the insulin signaling pathway. Food Sci Hum Wellness 13(3):1231–1245. https://doi.org/10.26599/FSHW.2022.9250103

Duan H et al (2022) Dietary strategies with anti-aging potential: dietary patterns and supplements. Food Res Int. Elsevier Ltd. https://doi.org/10.1016/j.foodres.2022.111501

Eleftheriou D et al (2018) Mediterranean diet and its components in relation to all-cause mortality: meta-analysis. Br J Nutr 120(10). https://doi.org/10.1017/S0007114518002593

Elkhedir AE et al (2022) Apigenin glycosides from green pepper enhance longevity and stress resistance in Caenorhabditis elegans. Nutr Res 102:23–34. https://doi.org/10.1016/j.nutres.2022.02.003

Falckenhayn C et al (2023) Identification of dihydromyricetin as a natural DNA methylation inhibitor with rejuvenating activity in human skin. Front Aging 4. https://doi.org/10.3389/fragi.2023.1258184

Fan X et al (2022) Flavonoids—natural gifts to promote health and longevity. Int J Mol Sci. MDPI. https://doi.org/10.3390/ijms23042176

Feng S et al (2021) Oleuropein enhances stress resistance and extends lifespan via insulin/igf-1 and skn-1/nrf2 signaling pathway in caenorhabditis elegans. Antioxidants 10(11). https://doi.org/10.3390/antiox10111697

Feng RC et al (2023) Effects of anthocyanin-rich supplementation on cognition of the cognitively healthy middle-aged and older adults: a systematic review and meta-analysis of randomized controlled trials. Nutr Rev. Oxford University Press:287–303. https://doi.org/10.1093/nutrit/nuac055

Finicelli M et al (2021) Polyphenols, the healthy brand of olive oil: insights and perspectives. Nutrients. MDPI. https://doi.org/10.3390/nu13113831

Frumuzachi O et al (2024) Systemic health effects of oleuropein and hydroxytyrosol supplementation: a systematic review of randomized controlled trials. Antioxidants 13(9):1040. https://doi.org/10.3390/antiox13091040

Frumuzachi O et al (2025) The impact of oleuropein, hydroxytyrosol, and tyrosol on cardiometabolic risk factors: a meta-analysis of randomized controlled trials. Crit Rev Food Sci Nutr [Preprint]. https://doi.org/10.1080/10408398.2025.2453090

Ge Y et al (2021) Naringenin prolongs lifespan and delays aging mediated by IIS and MAPK in: Caenorhabditis elegans. Food Funct 12(23). https://doi.org/10.1039/d1fo02472h

Gervasi F, Pojero F (2024) Use of oleuropein and hydroxytyrosol for cancer prevention and treatment: considerations about how bioavailability and metabolism impact their adoption in clinical routine. Biomedicines. Multidisciplinary Digital Publishing Institute (MDPI). https://doi.org/10.3390/biomedicines12030502

Ghadimi M et al (2021) Decreased insulin resistance in diabetic patients by influencing Sirtuin1 and Fetuin-a following supplementation with ellagic acid: a randomized controlled trial. Diabetol Metab Syndr 13(1). https://doi.org/10.1186/s13098-021-00633-8

Godos J, Marventano S et al (2017a) Dietary sources of polyphenols in the Mediterranean healthy Eating, Aging and Lifestyle (MEAL) study cohort. Int J Food Sci Nutr 68(6):750–756. https://doi.org/10.1080/09637486.2017.1285870

Godos J, Rapisarda G et al (2017b) Association between polyphenol intake and adherence to the Mediterranean diet in Sicily, southern Italy. NFS J 8:1–7. https://doi.org/10.1016/j.nfs.2017.06.001

Grande de França NA et al (2023) The role of dietary strategies in the modulation of hallmarks of aging. Ageing Res Rev. https://doi.org/10.1016/j.arr.2023.101908

Guasch-Ferré M, Willett WC (2021) The Mediterranean diet and health: a comprehensive overview. J Intern Med. https://doi.org/10.1111/joim.13333

Guo P et al (2023) Naringin alleviates glucose-induced aging by reducing fat accumulation and promoting autophagy in Caenorhabditis elegans. Nutrients 15(4). https://doi.org/10.3390/nu15040907

Hada Y et al (2020) The protective effect of chlorogenic acid on vascular senescence via the Nrf2/HO-1 pathway. Int J Mol Sci 21(12):1–17. https://doi.org/10.3390/ijms21124527

Hadidi M et al (2024) The potential health benefits of Gallic acid: therapeutic and food applications. Antioxidants. Multidisciplinary Digital Publishing Institute (MDPI). https://doi.org/10.3390/antiox13081001

Hair R, Sakaki JR, Chun OK (2021) Anthocyanins, microbiome and health benefits in aging. Molecules. MDPI AG. https://doi.org/10.3390/molecules26030537

He XL et al (2022) Secoisolariciresinol Diglucoside improves ovarian reserve in aging mouse by inhibiting oxidative stress. Front Mol Biosci 8. https://doi.org/10.3389/fmolb.2021.806412

Huang J et al (2024a) Antioxidant capacity of sesamol in Caenorhabditis elegans model system. Food Prod Process Nutr 6(1). https://doi.org/10.1186/s43014-023-00186-5

Huang T et al (2024b) Dietary Ferulic acid increases endurance capacity by promoting skeletal muscle oxidative phenotype, mitochondrial function, and antioxidant capacity. Mol Nutr Food Res 68(4). https://doi.org/10.1002/mnfr.202200719

Hussain MS et al (2024) Kaempferol: paving the path for advanced treatments in aging-related diseases. Exp Gerontol. Elsevier Inc. https://doi.org/10.1016/j.exger.2024.112389

Ji X et al (2024) Kaempferol improves exercise performance by regulating glucose uptake, mitochondrial biogenesis, and protein synthesis via PI3K/AKT and MAPK signaling pathways. Foods 13(7). https://doi.org/10.3390/foods13071068

Jia M et al (2024) Secoisolariciresinol diglucoside attenuates neuroinflammation and cognitive impairment in female Alzheimer's disease mice via modulating gut microbiota metabolism and GPER/CREB/BDNF pathway. J Neuroinflammation 21(1). https://doi.org/10.1186/s12974-024-03195-4

Kapolou A et al (2021) Association of mean daily polyphenols intake with Mediterranean diet adherence and anthropometric indices in healthy greek adults: a retrospective study. Appl Sci (Switzerland) 11(10). https://doi.org/10.3390/app11104664

Kharat P et al (2020) Ellagic acid prolongs the lifespan of Drosophila melanogaster. GeroScience 42(1):271–285. https://doi.org/10.1007/s11357-019-00135-6

Kramer DJ, Johnson AA (2024) Apigenin: a natural molecule at the intersection of sleep and aging. Front Nutr. Frontiers Media SA. https://doi.org/10.3389/fnut.2024.1359176

Kumar A et al (2023) Major phytochemicals: recent advances in health benefits and extraction method. Molecules. MDPI. https://doi.org/10.3390/molecules28020887

Kumar M et al (2024) Ferulic acid: extraction, estimation, bioactivity and applications for human health and food. J Sci Food Agric [Preprint]. John Wiley and Sons Ltd. https://doi.org/10.1002/jsfa.13931

Le TD, Inoue YH (2021) Sesamin activates nrf2/cnc-dependent transcription in the absence of oxidative stress in drosophila adult brains. Antioxidants 10(6). https://doi.org/10.3390/antiox10060924

Leng P, Wang Y, Xie M (2025) Ellagic acid and gut microbiota: interactions, and implications for health. Food Sci Nutr. John Wiley and Sons Inc. https://doi.org/10.1002/fsn3.70133

Li H et al (2021) Ferulic acid supplementation increases lifespan and stress resistance via insulin/igf-1 signaling pathway in C. elegans. Int J Mol Sci 22(8). https://doi.org/10.3390/ijms22084279

Li H et al (2022) Ellagic acid enhances muscle endurance by affecting the muscle fiber type, mitochondrial biogenesis and function. Food Func 13(3). https://doi.org/10.1039/d1fo02318g

Li W et al (2023) Research progress on classification, sources and functions of dietary polyphenols for prevention and treatment of chronic diseases. J Future Foods. Beijing Academy of Food Sciences:289–305. https://doi.org/10.1016/j.jfutfo.2023.03.001

Liu Y et al (2024) Dietary polyphenols as anti-aging agents: targeting the hallmarks of aging. Nutrients 16(19). https://doi.org/10.3390/nu16193305

Liu L et al (2025) Association of dietary flavonoid intake with reproductive lifespan: a cross-sectional study. Eur J Nutr 64(3). https://doi.org/10.1007/s00394-025-03641-w

Long T et al (2022) Ferulic acid exerts neuroprotective effects via autophagy induction in C. elegans and cellular models of Parkinson's disease. Oxidative Med Cell Longev 2022. https://doi.org/10.1155/2022/3723567

Lu H et al (2020) Chlorogenic acid: a comprehensive review of the dietary sources, processing effects, bioavailability, beneficial properties, mechanisms of action, and future directions. Compr Rev Food Sci Food Saf 19(6):3130–3158. https://doi.org/10.1111/1541-4337.12620

Majdalawieh AF et al (2022) Immunomodulatory and anti-inflammatory effects of sesamin: mechanisms of action and future directions. Crit Rev Food Sci Nutr. Taylor and Francis Ltd:5081–5112. https://doi.org/10.1080/10408398.2021.1881438

Martucci M et al (2017) Mediterranean diet and inflammaging within the hormesis paradigm. Nutr Rev 75(6). https://doi.org/10.1093/nutrit/nux013

Deepika and Maurya, P.K. (2022) 'Ellagic acid: insight into its protective effects in age-associated disorders', *3 Biotech* Springer Science and Business Media Deutschland GmbH https://doi.org/10.1007/s13205-022-03409-7

Mazza E et al (2021) Mediterranean diet in healthy aging. J Nutr Health Aging 25(9). https://doi.org/10.1007/s12603-021-1675-6

Meccariello R, D'Angelo S (2021) Impact of polyphenolic-food on longevity: an elixir of life. An overview. Antioxidants. MDPI. https://doi.org/10.3390/antiox10040507

Micheli L et al (2023) Role of hydroxytyrosol and oleuropein in the prevention of aging and related disorders: focus on neurodegeneration, skeletal muscle dysfunction and gut microbiota. Nutrients. MDPI. https://doi.org/10.3390/nu15071767

Mohammed HA, Khan RA (2022) Anthocyanins: traditional uses, structural and functional variations, approaches to increase yields and products' quality, hepatoprotection, liver longevity, and commercial products. Int J Mol Sci. MDPI. https://doi.org/10.3390/ijms23042149

Morze J et al (2021) An updated systematic review and meta-analysis on adherence to Mediterranean diet and risk of cancer. Eur J Nutr 60(3). https://doi.org/10.1007/s00394-020-02346-6

Muzammil A et al (2021) Anti-aging natural compounds and their role in the regulation of metabolic pathways leading to longevity. Mini-Rev Med Chem 21(18). https://doi.org/10.2174/1389557521666210405162006

Najmanová I et al (2020) The pharmacokinetics of flavanones. Crit Rev Food Sci Nutr. https://doi.org/10.1080/10408398.2019.1679085

Nas JSB, Medina PMB (2023) CYANIDIN-3-glucoside promotes longevity and tolerance against UVA and oxidative stress in CAENORHABDITIS ELEGANS. Explor Anim Med Res 13(1):105–110. https://doi.org/10.52635/eamr/13.1.105-110

Nguyen V et al (2024) Chlorogenic acid: a systematic review on the biological functions, mechanistic actions, and therapeutic potentials. Nutrients. https://doi.org/10.3390/nu16070924

Oboh G et al (2021) Effect of citrus peels-supplemented diet on longevity, memory index, redox status, cholinergic and monoaminergic enzymes in Drosophila melanogaster model. J Food Biochem 45(2). https://doi.org/10.1111/jfbc.13616

Ogunlade B, Adelakun SA, Agie JA (2022) Nutritional supplementation of gallic acid ameliorates Alzheimer-type hippocampal neurodegeneration and cognitive impairment induced by aluminum chloride exposure in adult Wistar rats. Drug Chem Toxicol 45(2):651–662. https://doi.org/10.1080/01480545.2020.1754849

Ou Q et al (2020) Impact of quercetin on systemic levels of inflammation: a meta-analysis of randomised controlled human trials. Int J Food Sci Nutr 71(2):152–163. https://doi.org/10.1080/09637486.2019.1627515

Oyebode OT et al (2020) Apigenin ameliorates D-galactose-induced lifespan shortening effects via antioxidative activity and inhibition of mitochondrial-dependent apoptosis in Drosophila melanogaster. J Funct Foods 69. https://doi.org/10.1016/j.jff.2020.103957

Park SH et al (2024) Scientific evidence of foods that improve the lifespan and healthspan of different organisms. Nutr Res Rev 37(1):169–178. https://doi.org/10.1017/S0954422423000136

Pereira QC et al (2023) The molecular mechanism of polyphenols in the regulation of ageing hallmarks. Int J Mol Sci. https://doi.org/10.3390/ijms24065508

Piragine E et al (2024) The citrus flavanone naringenin prolongs the lifespan in C. elegans and slows signs of brain aging in mice. Exp Gerontol 194. https://doi.org/10.1016/j.exger.2024.112495

Poblete-Aro C et al (2023) The potential role of apigenin in the enhancement of the antioxidant adaptation to exercise by bioactive dietary compounds. Rev Chil Nutr. Sociedad Chilena de Nutricion Bromatologia y Toxilogica:691–699. https://doi.org/10.4067/s0717-75182023000600691

Pojero F et al (2023) Effects of oleuropein and hydroxytyrosol on inflammatory mediators: consequences on Inflammaging. Int J Mol Sci. MDPI. https://doi.org/10.3390/ijms24010380

Rahimifard M et al (2020) Therapeutic effects of Gallic acid in regulating senescence and diabetes; an in vitro study. Molecules 25(24). https://doi.org/10.3390/MOLECULES25245875

Rattan SIS (2022) Physiological hormesis and hormetins in biogerontology. Curr Opin Toxicol. https://doi.org/10.1016/j.cotox.2022.01.001

Rebello CJ et al (2020) Safety and pharmacokinetics of naringenin: a randomized, controlled, single-ascending-dose clinical trial. Diabetes Obes Metab 22(1):91–98. https://doi.org/10.1111/dom.13868

Riquelme CV, Pando ME (2025) Oleuropein: a narrative review on its role in neurodegenerative diseases. Aging Adv 2(2):75–83. https://doi.org/10.4103/AGINGADV.AGINGADV-D-25-00002

Ruankham W et al (2021) Sesamin and sesamol attenuate H2O2-induced oxidative stress on human neuronal cells via the SIRT1-SIRT3-FOXO3a signaling pathway. Nutr Neurosci 24(2):90–101. https://doi.org/10.1080/1028415X.2019.1596613

Scientific Opinion on the substantiation of health claims related to polyphenols in olive and protection of LDL particles from oxidative damage (ID 1333, 1638, 1639, 1696, 2865), maintenance of normal blood HDL cholesterol concentrations (ID 1639), mainte (2011) EFSA J 9(4):2033. https://doi.org/10.2903/j.efsa.2011.2033

Serra-Majem L et al (2020) Updating the mediterranean diet pyramid towards sustainability: focus on environmental concerns. Int J Environ Res Public Health 17(23). https://doi.org/10.3390/ijerph17238758

Serreli G, Deiana M (2020) Extra virgin olive oil polyphenols: modulation of cellular pathways related to oxidant species and inflammation in aging. Cells. Multidisciplinary Digital Publishing Institute (MDPI). https://doi.org/10.3390/cells9020478

Shan, H. *et al.* (2022) ', pp. 532–539. https://doi.org/10.1007/s13238-021-00872-5

Shen J et al (2022) Dietary phytochemicals that can extend longevity by regulation of metabolism. Plant Foods Hum Nutr. Springer:12–19. https://doi.org/10.1007/s11130-021-00946-z

Singh A et al (2024) A comprehensive review of apigenin a dietary flavonoid: biological sources, nutraceutical prospects, chemistry and pharmacological insights and health benefits. Crit Rev Food Sci Nutr [Preprint]. Taylor and Francis Ltd. https://doi.org/10.1080/10408398.2024.2390550

Siswanto FM et al (2022) Chlorogenic acid activates Nrf2/SKN-1 and prolongs the lifespan of Caenorhabditis elegans via the Akt-FOXO3/DAF16a-DDB1 pathway and activation of DAF16f. J Gerontol A Biol Sci Med Sci 77(8):1503–1516. https://doi.org/10.1093/gerona/glac062

Soltani S et al (2019) Adherence to the Mediterranean diet in relation to all-cause mortality: a systematic review and dose-response meta-analysis of prospective cohort studies. Adv Nutr. https://doi.org/10.1093/advances/nmz041

Sorrenti V et al (2023) Cell survival, death, and proliferation in senescent and cancer cells: the role of (poly)phenols. Adv Nutr. https://doi.org/10.1016/j.advnut.2023.05.014

Sun Z, Li L, Zhang L (2025) Apigenin enhancing oxidative resistance and proteostasis to extend lifespan via PTEN-mediated AKT signalling pathway. Biochim Biophys Acta Mol basis Dis 1871(3). https://doi.org/10.1016/j.bbadis.2025.167670

Tang C et al (2021) Mediterranean diet and mortality in people with cardiovascular disease: a meta-analysis of prospective cohort studies. Nutrients. https://doi.org/10.3390/nu13082623

Testai L et al (2020) The citrus flavonoid naringenin protects the myocardium from ageing-dependent dysfunction: potential role of SIRT1. Oxidative Med Cell Longev 2020. https://doi.org/10.1155/2020/4650207

Truzzi F et al (2021) An overview on dietary polyphenols and their biopharmaceutical classification system (Bcs). Int J Mol Sci. MDPI. https://doi.org/10.3390/ijms22115514

Tungmunnithum D et al (2020) Almond skin extracts and chlorogenic acid delay chronological aging and enhanced oxidative stress response in yeast. Life 10(6). https://doi.org/10.3390/life10060080

Wagle S et al (2021) Supplemental ferulic acid inhibits total body irradiation-mediated bone marrow damage, bone mass loss, stem cell senescence, and hematopoietic defect in mice by enhancing antioxidant defense systems. Antioxidants 10(8). https://doi.org/10.3390/antiox10081209

Wang Y et al (2021) Sesamin protects against and ameliorates rat intestinal ischemia/reperfusion injury with involvement of activating Nrf2/HO-1/NQO1 signaling pathway. Oxidative Med Cell Longev 2021. https://doi.org/10.1155/2021/5147069

Wang J et al (2023a) Naringenin ameliorates vascular senescence and atherosclerosis involving SIRT1 activation. J Pharm Pharmacol 75(8):1021–1033. https://doi.org/10.1093/jpp/rgad053

Wang L et al (2023b) Myricetin protected against Aβ oligomer-induced synaptic impairment, mitochondrial function and oxidative stress in SH-SY5Y cells via ERK1/2/GSK-3β pathways. https://doi.org/10.1101/2023.01.12.523781

Xian W et al (2023) Ameliorative effect of Ellagic acid on aging in rats with the potential mechanism relying on the gut microbiota and urolithin A-producing ability. J Agric Food Chem 71(19):7396–7407. https://doi.org/10.1021/acs.jafc.3c00960

Xing X et al (2023) Luteolin regulates the distribution and function of organelles by controlling SIRT1 activity during postovulatory oocyte aging. Front Nutr 10. https://doi.org/10.3389/fnut.2023.1192758

Yang W et al (2022) DAF-16 is involved in colonic metabolites of ferulic acid-promoted longevity and stress resistance of Caenorhabditis elegans. J Sci Food Agric 102(15):7017–7029. https://doi.org/10.1002/jsfa.12063

Yeh CH et al (2022) Hesperetin promotes longevity and delays aging via activation of Cisd2 in naturally aged mice. J Biomed Sci 29(1). https://doi.org/10.1186/s12929-022-00838-7

Younis RL et al (2024) Luteolin mitigates D-galactose-induced brain ageing in rats: SIRT1-mediated neuroprotection. Neurochem Res 49(10):2803–2820. https://doi.org/10.1007/s11064-024-04203-y

Zamora-Ros R et al (2016) Dietary polyphenol intake in europe: the european prospective investigation into cancer and nutrition (EPIC) study. Eur J Nutr 55(4):1359–1375. https://doi.org/10.1007/s00394-015-0950-x

Zhang B et al (2020) Apigenin protects human melanocytes against oxidative damage by activation of the Nrf2 pathway. Cell Stress Chaperones 25(2):277–285. https://doi.org/10.1007/s12192-020-01071-7

Zhao L et al (2024) Kaempferol enhances intestinal repair and inhibits the hyperproliferation of aging intestinal stem cells in Drosophila. Front Cell Dev Biol 12. https://doi.org/10.3389/fcell.2024.1491740

Zhu Q et al (2020) A dihydroflavonoid naringin extends the lifespan of C. elegans and delays the progression of aging-related diseases in PD/AD models via DAF-16. Oxidative Med Cell Longev 2020. https://doi.org/10.1155/2020/6069354

Zhu RZ et al (2021) Luteolin inhibits H2O2-induced cellular senescence via modulation of SIRT1 and p53. Korean J Physiol Pharmacol 25(4):297–305. https://doi.org/10.4196/KJPP.2021.25.4.297

Zhu M et al (2024) Luteolin: a promising multifunctional natural flavonoid for human diseases. Phytother Res John Wiley and Sons Ltd:3417–3443. https://doi.org/10.1002/ptr.8217

Zumerle S et al (2024) Targeting senescence induced by age or chemotherapy with a polyphenol-rich natural extract improves longevity and healthspan in mice. Nat Aging [Preprint]. https://doi.org/10.1038/s43587-024-00663-7

Chapter 15
Geroprotective Effects of Plant Extracts

Alexey Moskalev

Abstract The search for natural compounds that can extend lifespan and improve health during aging has led to significant interest in plant extracts as potential geroprotectors. This comprehensive assessment examines the evidence for geroprotective effects of various plant extracts, focusing on their ability to extend lifespan and healthspan in model organisms. An analysis of the major classes of bioactive compounds in plants—primarily polyphenols and terpenoids—reveals diverse mechanisms of action in modulating aging processes. Key pathways influenced by these botanical preparations include antioxidant defenses, stress resistance, metabolic regulation, and inflammation. Findings from many studies demonstrate lifespan extension in model organisms like *Caenorhabditis elegans*, *Drosophila melanogaster*, and mice through administration of extracts from diverse plant sources. The documented research suggests multiple mechanisms by which these extracts exert their effects, including activation of evolutionarily conserved longevity and hormetic pathways like DAF-16/FOXO, SKN-1/Nrf2, and AMPK. Finally, the most promising candidates for further development are identified, along with future research directions including the need for more extensive mammalian studies and eventual clinical trials.

Keywords Geroprotectors · Plant extracts · Longevity pathways · Polyphenols · Model organisms

15.1 Introduction

Aging is characterized by progressive physiological decline and increased vulnerability to diseases. The concept of geroprotection—interventions that slow aging processes and extend healthy lifespan—has gained traction as a potential approach to address multiple age-related pathologies simultaneously. Geroprotectors are

A. Moskalev (✉)
Institute of Biology of Aging and Healthy Longevity Medicine, Petrovsky Russian Research Center for Surgery, Moscow, Russia

S. I. S. Rattan, C. Gezer (eds.), *Supplements and Drugs for Healthy Longevity*, Healthy Ageing and Longevity 22, https://doi.org/10.1007/978-3-032-11608-6_15

compounds or interventions that can slow aging, increase lifespan, and improve healthspan (Proshkina et al. 2024). Unlike conventional nutraceuticals that may address specific health conditions, geroprotectors target fundamental aging mechanisms to broadly improve health during aging. Nutraceuticals typically focus on specific symptoms or disorders, whereas geroprotective agents aim to delay the underlying aging process that contributes to multiple diseases. This is important as the global aging population continues to increase, driving the significance of healthy longevity medicine in modern society (Golubev et al. 2024).

Plant extracts represent a rich source of potential geroprotectors due to their complex mixtures of bioactive compounds (Plyusnin et al. 2023). Among these, polyphenols and terpenoids present in the extracts stand out as major classes of the molecules with demonstrated geroprotective properties. Polyphenols are a diverse group of plant-derived compounds characterized by multiple phenol structural units, including flavonoids, phenolic acids, and stilbenes (Proshkina et al. 2024). These compounds have gained attention for their potent anti-inflammatory and antioxidant properties and ability to modulate various signaling pathways related to aging. A systematic assessment identified 62 polyphenols that have been shown to increase lifespan and improve biomarkers of aging in animal models (Proshkina et al. 2024). Among the evaluated phytochemicals, chlorogenic acid, quercetin, epicatechin, genistein, resveratrol, and curcumin were identified as compounds with the highest geroprotective potential, presenting excellent research and therapeutic opportunities (Proshkina et al. 2024).

Terpenoids, on the other hand, constitute the largest group of plant secondary metabolites but have been less frequently associated with geroprotective properties compared to polyphenols (Proshkina et al. 2020). However, emerging evidence suggests that terpenoids can effectively influence aging mechanisms and age-related diseases. The biological activities of terpenoids conform to both primary criteria of geroprotectors (lifespan extension, improved aging biomarkers, low toxicity) and secondary criteria (evolutionarily conserved mechanisms, prevention of age-related diseases) (Proshkina et al. 2020). Such findings indicate that terpenoids are an underestimated source of potential geroprotectors that can effectively influence the mechanisms of aging and age-related diseases.

The desire to enhance both lifespan and healthspan has led researchers to explore plants traditionally used in various ethnic cuisines and medication systems. From *Withania somnifera* (Ashwagandha) in Ayurvedic medicine to *Abies sibirica* (Siberian fir) in Russian traditional medicine, a wealth of plant extracts is now being systematically studied for their potential geroprotective effects (Kumar et al. 2013; Lipatova et al. 2021). The current investigation aims to comprehensively examine the evidence for geroprotective effects of various plant extracts, their underlying mechanisms, and their potential as interventions to promote healthy aging.

15.2 Main Mechanisms and Targets of Plant Extract Geroprotective Effects

Plant extracts exert their geroprotective effects through multiple mechanisms targeting key pathways involved in aging. One of the primary mechanisms is through their antioxidant properties. Oxidative stress, characterized by an imbalance between reactive oxygen species (ROS) production and antioxidant defenses, is considered a major contributor to aging (Proshkina et al. 2024). *Berberis vulgaris* L. extract, for example, exhibits antioxidant activity both *in vitro* (using red blood cell tests) and *in vivo* in *Drosophila melanogaster (D. melanogaster)*, increasing lifespan by up to 6% while promoting healthspan through improved stress resistance, locomotor activity, and intestinal barrier integrity (Golubev et al. 2024). Similarly, honeysuckle (*Lonicera pallasi*) extract and its component cyanidin-3-O-glucoside (C3G) demonstrate antioxidant and membrane-protective activities in various *in vitro* models while extending the median and maximum lifespan of *D. melanogaster* by up to 8% (Golubev et al. 2022).

The antioxidant effects of plant extracts often translate to enhanced stress resistance in model organisms. *Sorbus aucuparia* (mountain ash) berry extract, abundant in rutin and cyanidin-3-rutinoside, shows pronounced antioxidant activity *in vitro* and increases resistance to starvation in *D. melanogaster* males while reducing resistance to hyperthermia (Platonova et al. 2023). Wheat bran extracts with increased anthocyanin content demonstrated both radical scavenging activity *in vitro* and the ability to extend the lifespan of female *D. melanogaster*, indicating geroprotective properties (Mikhailova et al. 2023). Notably, anthocyanin-containing wheat bran extract (+AWBE) showed higher radical scavenging activity and membrane protective activity compared to anthocyanin-free wheat bran extract, highlighting the importance of these compounds in mediating antioxidant effects (Mikhailova et al. 2023).

Beyond direct antioxidant activities, plant extracts often target evolutionarily conserved signaling pathways that regulate lifespan across species. The insulin/insulin-like growth factor-1 (IGF-1) signaling pathway, with the transcription factor DAF-16/FOXO playing a central role, is frequently modulated by plant extracts. *Jujubae Fructus* extract prolongs lifespan in *C. elegans* through a mechanism dependent on DAF-16 and its downstream target SOD-3, promoting nuclear translocation of DAF-16 in transgenic worms (Zhang et al. 2024b). Similarly, *Anoectochilus roxburghii* extract extends lifespan while mitigating ROS and lipofuscin accumulation and boosting resistance to ultraviolet radiation (UV) and heat stress in *C. elegans* (Xu et al. 2024). The extract modulates the expression of key healthy aging genes including *daf-16*, *sod-3*, and *gst-4*, facilitating nuclear translocation of DAF-16 (Xu et al. 2024). Importantly, the extract fails to extend lifespan in *daf-16*-deficient *Caenorhabditis elegans (C. elegans)*, indicating dependency on the DAF-16/FOXO signaling pathway (Xu et al. 2024).

The SKN-1/Nrf2 pathway, critical for cellular defense against oxidative stress and xenobiotics and one of the major hermetic mechanism (Rattan 2022), is another common target. Moringa seed extract rich in isothiocyanates activates the SKN-1/Nrf2 pathway in *C. elegans*, upregulating several glutathione S-transferases involved in detoxification (Farias-Pereira et al. 2024). The extract and its active component MIC-1 upregulate skn-1 expression and induce SKN-1 nuclear translocation (Farias-Pereira et al. 2024). *Punica granatum* (pomegranate) leaf extract significantly enhances stress resistance and extends lifespan of *C. elegans* through coordinated regulation of stress response, metabolic homeostasis, and longevity signaling pathways (Todorova et al. 2024). The molecular mechanisms highlight the essential roles of HLH-30/TFEB, in conjunction with DAF-16/FOXO and SKN-1/NRF2, as mediators of the extract's effect on healthspan (Todorova et al. 2024).

AMP-activated protein kinase (AMPK) signaling represents another key pathway influenced by plant extracts. *Aronia melanocarpa* extract extends both lifespan and healthspan of *C. elegans* via the AMPK signaling pathway, specifically through mitogen-activated protein kinase 1 (PMK-1) (Zhang et al. 2024a). Indian almond (*Terminalia catappa*) leaf extract's effect on prolonging lifespan by inhibiting lipid and ROS accumulation under high glucose conditions operates through *atgl-1*, *daf-16*, and *skn-1* being regulated by *aak-1* (AMPK) (Kim et al. 2023). The effectiveness of the extract was reduced in aak-1/AMPK-deficient mutants and completely abolished under high glucose conditions, confirming the central role of AMPK signaling (Kim et al. 2023).

Autophagy induction represents a crucial mechanism by which plant extracts may promote longevity. The Siberian fir terpenes extract Abisil activates autophagy in human fibroblasts while reversing rotenone-induced elevation of ROS levels (Lipatova et al. 2021). Treatment with Abisil results in predominant downregulation of genes involved in cell cycle progression, oxidative phosphorylation, and fatty acid biosynthesis (Lipatova et al. 2021). Similarly, procyanidin A1 from peanut skin extends lifespans and alleviates cellular senescence by enhancing oxidative stress resistance and inducing autophagy (Wei et al. 2024). Treatment with an autophagy inhibitor abolished the antioxidative stress and senescence-alleviating effects of procyanidin A1, confirming autophagy's central role (Wei et al. 2024).

Many plant extracts influence metabolic processes that affect aging, including lipid metabolism, glucose homeostasis, and mitochondrial function. *Cyclocodon lancifolius* fruit extract promotes longevity and reduces oxidative stress in *C. elegans* (Liu et al. 2024). Metabolomic analysis revealed that purine metabolism is a key regulatory pathway for this extract to exert healthy aging effects (Liu et al. 2024). Similarly, grape seed procyanidins have neuroprotective effects against amyloid-β-induced toxicity in *C. elegans* (Gonzalez-Manzano et al. 2024). The extract reduces proteasomal activity, potentially leading to a reduction in misfolded protein accumulation, and modulates the expression of key genes involved in autophagy and proteostasis (Gonzalez-Manzano et al. 2024). These findings underscore the multifaceted ways in which plant extracts can modulate metabolic processes to promote longevity.

Chronic inflammation, a hallmark of aging, is another target of plant extracts with geroprotective properties. A standardized extract of *Salvia haenkei* extends lifespan and healthspan of naturally aged mice (Zumerle et al. 2024). HK treatment inhibits age-induced inflammation, fibrosis, and senescence markers across several tissues, increases muscle strength, and improves fur thickness compared to age-matched controls (Zumerle et al. 2024). Mechanistically, luteolin—the most concentrated flavonoid in HK—was identified as a senomorphic compound that disrupts the p16-CDK6 interaction (Zumerle et al. 2024). *Hylotelephium erythrostictum* extract ameliorates intestinal injury in *D. melanogaster* by scavenging ROS and inhibiting multiple signaling pathways (Kim et al. 2024). The extract significantly prolongs the lifespan of flies fed toxic compounds, reduces cell mortality, maintains intestinal integrity and gut acid-base homeostasis, and eliminates dextran sodium sulphate-induced ROS accumulation (Kim et al. 2024).

Thus, plant extracts exert geroprotective effects through multiple, often complementary mechanisms. They directly scavenge free radicals, enhance endogenous antioxidant systems, modulate conserved longevity pathways (DAF-16/FOXO, SKN-1/Nrf2, AMPK), induce autophagy, regulate metabolism, and reduce inflammation. The multiplicity of mechanisms likely contributes to the robust geroprotective effects observed in various model organisms and suggests potential for synergistic combinations of plant extracts or their constituents in future interventions.

A critical concept underlying the geroprotective effects of many plant extracts is hormesis—the phenomenon whereby mild stressors that would be harmful at high doses become beneficial at low doses (Rattan 2022). Plant-derived compounds that elicit such hormetic responses are termed hormetins, distinguishing them from classical antioxidants that directly neutralize free radicals.

The hormetic mechanism explains why many plant extracts, rather than simply acting as direct antioxidants, actually stimulate the body's endogenous stress response systems. This mild stress triggers adaptive responses that ultimately enhance cellular resilience and longevity. For instance, the activation of the SKN-1/Nrf2 pathway by isothiocyanate-rich Moringa seed extract (Farias-Pereira et al. 2024) represents a classic hormetic response—the extract induces a mild oxidative challenge that upregulates antioxidant defense systems.

Similarly, the autophagy induction observed with Siberian fir terpenes extract Abisil (Lipatova et al. 2021) exemplifies hormesis in action. Rather than preventing cellular stress, the extract appears to impose a controlled stress that activates protective mechanisms like autophagy, ultimately leading to improved cellular function and longevity.

The hormetic nature of plant extracts also explains why optimal effects are often achieved at specific dosage ranges, with diminished or even harmful effects at higher concentrations. This dose-response relationship, characteristic of hormesis, has been observed in multiple studies, such as the differential effects of various Abisil concentrations (Lipatova et al. 2021).

The hormetin concept fundamentally reframes our understanding of how plant extracts promote longevity. Rather than viewing these compounds merely as antioxidants that prevent damage, we can appreciate them as sophisticated modulators of cellular stress responses that enhance the organism's intrinsic capacity for self-repair and adaptation. This paradigm shift from damage prevention to adaptive enhancement represents a more nuanced and mechanistically accurate understanding of how phytochemicals promote healthy aging.

15.3 Geroprotective Effects of Different Plant Extracts

A diverse array of plant extracts has demonstrated geroprotective potential in model organisms. Berry extracts are particularly rich in polyphenols, especially anthocyanins, and have shown significant lifespan-extending effects. Honeysuckle (*Lonicera pallasii*) extract, which contains abundant anthocyanins, extends the median and maximum lifespan of *D. melanogaster* by up to 8% (Golubev et al. 2022). The extract also increases stress resistance by up to 10% and improves intestinal barrier integrity by 4% (Golubev et al. 2022). These effects are associated with increased expression of the pro-longevity gene *Hif1* and decreased expression of the anti-longevity gene *Keap1* (Golubev et al. 2022). Similarly, cloudberry extract, which has a relatively large carotenoid content, increases the median lifespan and the age of 90% mortality of female *D. melanogaster* by 11–19% and decreases the rate of aging by up to 13% (Lashmanova et al. 2018). However, the effects on males are less pronounced and, in some cases, even negative (Lashmanova et al. 2018), highlighting the importance of considering sex-specific responses to interventions.

Aronia melanocarpa (chokeberry) extract extends the lifespan of *C. elegans* by 22.2% after high-dose treatment (Zhang et al. 2024a). The extract improves behavior and physiological functions by increasing pharyngeal pumping rate, decreasing lipofuscin accumulation and ROS levels, enhancing oxidative stress resistance, and increasing the activities of antioxidant enzymes (Zhang et al. 2024a). Transcriptome analysis identifies pmk-1 (mitogen-activated protein kinase 1) as the gene with the largest fold change after intervention (Zhang et al. 2024a). Bilberry extract has also been evaluated in clinical settings, where it demonstrates significant healthy aging, brightening, and antioxidant efficacy in human skin (Nobile et al. 2024). After 84 days of supplementation, wrinkle depth decreased by 10.6%, skin smoothness improved by 7.9%, skin firmness improved by 13.3%, and skin antioxidant capacity increased by 20.8% (Nobile et al. 2024).

Ziziphus jujuba (jujube) extract and its derivatives have been extensively studied for their geroprotective properties. The fruit extract significantly extends the lifespan of *C. elegans* in a dose-dependent manner (Zhang et al. 2024b), prolongs the reproductive period, and increases normal activity during aging (Zhang et al. 2024b). Polysaccharides from jujube increase body length and width, head thrashes, body bends, and brood size of nematodes (Han et al. 2024). They also reduce ROS levels, increase activities of antioxidant enzymes, and upregulate the expression of *daf-16*, *skn-1*, *sod-3*, and *gcs-1* in nematodes (Han et al. 2024).

However, sometimes extracts may have contradictory effects, as demonstrated by a study showing that *Artemisia argyi* extract, while increasing mean lifespan of both female and male flies by 23.74% and 22.30% respectively, also decreased resistance to oxidative stress in males (Yang et al. 2024).

Tea and its derivatives have received significant attention for their potential geroprotective properties. Green tea extract extends lifespan in *C. elegans*, with the methanol fraction showing significantly stronger effects than other fractions (Ke et al. 2024). Correlation analysis between mass spectroscopic data and healthy aging activity suggests that ester-type catechins (ETCs) are the major healthy aging components, including common ETCs, phenylpropanoid-substituted ester-type catechins, cinnamoylated catechins, ester-type flavoalkaloids, and cinnamoylated flavoalkaloids (Ke et al. 2024). Cinnamoylated flavoalkaloids at 200 μM show the strongest healthy aging effects, with up to 73% lifespan extension (Ke et al. 2024). White tea aqueous extract inhibits high-fat diet-induced senescence in *Drosophila*, significantly extending the lifespan of flies on a high-fat diet and partially restoring their climbing ability (Huang et al. 2024). This is accompanied by increased activities of antioxidant enzymes and decreased lipid hydroperoxide levels (Huang et al. 2024). Black tea extracts enhance stress-induced sleep in *C. elegans* to resist UV damage (Chen et al. 2024). Feeding black tea extract causes worms to enter a superior stress-induced sleep state, potentially boosting organismal recovery (Chen et al. 2024).

Xianhu tea water extract exhibits superior DPPH radical-scavenging and ABTS inhibition capacities, with the greatest reductions in ROS, malondialdehyde (MDA), and lipofuscin levels, and highest upregulation of superoxide dismutase (SOD) and catalase (CAT) activities among several tea varieties (Zhang et al. 2025). The lifespan was 23.50% and 21.07% significantly longer than nematodes in the control group (Zhang et al. 2025). Moreover, Xianhu tea modulates 13 aging-related genes, which strongly correlate with 23 identified phenolic compounds (Zhang et al. 2025). These results underscore the potential of various tea extracts as geroprotectors, with particularly strong effects observed for specific components like cinnamoylated flavoalkaloids.

Various herbs and spices have demonstrated remarkable geroprotective properties. White and red ginseng aqueous extracts extend lifespan and improve stress resistance in *C. elegans* (Hu et al. 2025), with red ginseng exhibiting superior healthy aging effects compared to white ginseng at the same concentration, potentially due to its higher content of reducing sugars, polyphenols, melanoidins, total saponins, and especially the conversion of ginsenosides (Hu et al. 2025). Ginseng extract also attenuates injury from ultraviolet irradiation in female *D. melanogaster* through the autophagy signaling pathway (She et al. 2024). Timut pepper extract (a spice native to Nepal) extends *C. elegans* lifespan at different maintenance temperatures and increases the proportion of active nematodes in their early adulthood (Jongsma et al. 2024). The extract enhances speed and distance moved as the nematodes age and slows the age-dependent decline of collagen expression, ensuring extracellular matrix homeostasis (Jongsma et al. 2024).

Rhodiola rosea, widely used in Traditional Chinese Medicine, has demonstrated remarkable geroprotective effects across multiple model organisms. In *Drosophila melanogaster*, *Rhodolia rosea* extract extends lifespan in both sexes through a mechanism independent of dietary restriction and not dependent on the SIR2, insulin/IGF-1 signaling (IIS), or target of rapamycin (TOR) pathways that are typically associated with lifespan extension (Schriner et al. 2013) Interestingly, *Rhodolia rosea* produces sex-specific responses in *D. melanogaster*, where female flies show down-regulation of glycolytic genes and dSir2 with decreased NADH levels, while males show no protection against heat stress, no effect on the major heat shock protein HSP70, and down-regulation of the mitochondrial HSP22 (Schriner et al. 2013). In *C. elegans*, *Rhodiola* extract remarkably extends lifespan by 37.1% at high doses (480 μg/mL), and enhances stress resistance against heat shock, UV, and paraquat-induced oxidative stress (Jiang et al. 2021). This occurs while reducing levels of ROS and MDA and increasing activities of SOD and CAT (Jiang et al. 2021). Mechanistically, *Rhodiola* extract upregulates the expression of stress resistance genes (*sod-3*, *gst-4*, *daf-16*, *skn-1*), downregulates *daf-2* and *age-1*, and accelerates the translocation of DAF-16 and SKN-1 into the nucleus (Jiang et al. 2021). The extension of lifespan triggered by *Rhodiola* extract is reversed in daf-16 (mu86) and skn-1(zu169) mutants, confirming these genes are involved in *Rhodiola*-regulated longevity (Jiang et al. 2021). In silkworm, *Bombyx mori*, *Rhodolia rosea* extract extends lifespan without affecting daily food intake, body weight, or fecundity, while enhancing stress resistance against heat stress (37 °C) and starvation (Chen et al. 2016). At the molecular level, the extract increases the activity of major antioxidant enzymes including glutathione S-transferase and CAT, alters the content of glutathione and MDA, and increases the expression of BmFoxO, a downstream regulator of insulin/IGF-1 signaling pathway (Chen et al. 2016).

Pomegranate (*Punica granatum* L.), known for its high polyphenol content, has demonstrated remarkable geroprotective effects across multiple model organisms. In *D. melanogaster*, supplementation with 10% (v/v) pomegranate juice extended the lifespan of male and female flies by 18% and 8% respectively. When male and female flies were reared together, there was a 19% increase in longevity of pomegranate juice-fed flies compared to controls. Notably, pomegranate juice outperformed resveratrol in several health and lifespan parameters. Beyond mere lifespan extension, pomegranate supplementation conferred a two-fold enhancement in fecundity, improved resistance to oxidative stress (H_2O_2 and paraquat-induced), enhanced resistance to *Candida albicans* infection, and maintained physical activity over an extended period as assessed by climbing assay (Balasubramani et al. 2014). In *C. elegans*, similar geroprotective effects were observed, with pomegranate juice increasing lifespan by a remarkable 56% in wild-type worms and 18% in *daf-16* mutants, suggesting partial dependence on the *daf-16* pathway. Pomegranate extract also extended lifespan by 28% and 10% in wild-type and *daf-16* mutant worms, respectively. Interestingly, while pomegranate juice significantly reduced intestinal fat deposition in *C. elegans* by 68% in wild-type and 33% in *daf-16* mutants, pomegranate extract surprisingly increased intestinal fat deposition. Ellagic acid, a component of pomegranate, extended lifespan by 11% while increasing intestinal fat deposition. These divergent effects on fat metabolism despite consistent lifespan extension suggest complex and potentially multiple mechanisms of action for pomegranate constituents (Zheng et al. 2017). The significant conservation of aging pathways between these model organisms and humans suggests potential application of pomegranate for healthy aging in human populations.

Schisandra chinensis (SC), a traditional herb widely used in Asian countries, has demonstrated significant effects on muscle strength, muscle mass, and metabolic health in various models of aging. In postmenopausal women, supplementation with 1000 mg of SC extract daily for 12 weeks significantly increased quadriceps muscle strength and decreased lactate levels at rest (Park et al. 2020). Similarly, a randomized, double-blind, placebo-controlled trial in adults over 50 years showed that 1 g of SC extract daily for 12 weeks, combined with low-intensity exercise, significantly enhanced knee extensor strength compared to placebo, although no differences were observed in muscle mass or inflammatory markers (Cho et al. 2021). In animal models, SC has shown more extensive effects. In ovariectomized rats (a model for aging), ethanol-hexane extract of SC exhibited anti-inflammatory and antioxidant properties while improving muscle regeneration by promoting mitochondrial biogenesis and autophagy when combined with exercise. This extract also demonstrated osteogenic effects, modulating both bone formation and bone resorption (Kim et al. 2019). In aged mice (16–20 months), SC Fructus ethanol extract supplementation for 4 months prevented age-related decreases in insulin sensitivity, reduced body weight gain and fat mass while increasing lean mass, and enhanced grip strength. Histological analysis revealed increased myofiber size and slow-twitch red muscle. At the molecular level, SC increased expression of proteins related to muscle protein synthesis (phospho-Erk1 and phospho-S6K1) and genes related to myogenesis (*MyoD*, *Myf5*, *MRF4*, *myogenin*), while decreasing expression of genes related to muscle degradation (*atrogin-1*, *MuRF-1*, and *myostatin*) (Choi et al. 2019). These findings collectively suggest that SC extract may be a promising intervention for age-related muscle wasting and metabolic decline.

Traditional medicinal herbs offer a particularly rich source of geroprotective compounds. Ashwagandha (*Withania somnifera*) root extract extends lifespan in *C. elegans*, particularly in mutants for the human nicotinic acetylcholine receptor, nAchR, α7 equivalent (acr-16), which show around 20% lifespan extension (Kumar et al. 2013). In mice, ashwagandha ethanol extract attenuates sarcopenia-related muscle atrophy, reduces epidermal and abdominal fat while significantly increasing muscle mass (Ko et al. 2024). The extract balances muscle protein synthesis/degradation, promotes fiber differentiation, and enhances mitochondrial biogenesis through the IGF-1/Akt/mTOR pathway (Ko et al. 2024). Withaferin A, a compound found in ashwagandha, prolongs the median lifespan of male *D. melanogaster* by 7.7–9.6% and the maximum lifespan by 11.1% (Koval et al. 2021). The compound also improves intestinal barrier permeability in older flies and affects the expression of genes involved in antioxidant defense, DNA damage recognition, heat shock proteins, and repair of double-strand breaks (Koval et al. 2021).

Maral root (*Rhaponticum carthamoides*) extract and its main constituent 20-hydroxyecdysone significantly enhance stress resilience in *C. elegans* (Todorova et al. 2025). The extract not only extends lifespan but also improves neuromuscular functions and cognitive abilities (Todorova et al. 2025). Similarly, *Curcumae Rhizoma* extract improves resistance to UV, thermal, oxidative, and pathogen-induced stress in *C. elegans*, extending lifespan in a dose-dependent manner (Jing et al. 2025). Mechanistically, it reduces ROS, increases SOD activity, and enhances UV resistance via the insulin/IGF-1 pathway and DAF-16 translocation (Jing et al. 2025).

Nelumbo nucifera (lotus) leaf extract ameliorates amyloid beta-induced toxicity in *C. elegans* models of Alzheimer's disease by activating autophagy (Wen et al. 2024). RNAi-mediated inhibition of autophagy-related genes negates the protective effects of the extract against amyloid beta-induced paralysis and aggregate accumulation (Wen et al. 2024). *Dendrobium nobile* alcohol extract extends the lifespan of *C. elegans* via hsf-1 and daf-16 (Li et al. 2024). The extract slows lipofuscin accumulation, decreases ROS levels, elevates SOD activity, enhances oxidative and heat stress resistance, and protects dopamine neurons from 6-hydroxydopamine-induced neurodegeneration (Li et al. 2024). It also reduces Aβ-induced neurotoxicity, upregulates the mRNA expression of DAF-16 and HSF-1, promotes the nuclear localization of DAF-16, and enhances the fluorescence intensity of HSP-16.2 (Li et al. 2024).

Several studies have demonstrated the geroprotective potential of terpenoid-rich extracts. The Siberian fir terpenes extract Abisil has geroprotective, anti-tumorigenic, and anti-angiogenic effects (Lipatova et al. 2021). Treatment with Abisil results in activation of autophagy, reversal of rotenone-induced elevation of ROS levels, and several-fold decrease of mitochondrial potential. Lower doses of Abisil (25 μg/ml) show better oxidative effects than high doses (50 or 125 μg/ml) (Lipatova et al. 2021). A clinical study of Siberian fir terpenes demonstrated that supplementation in middle-aged patients of both sexes reduces the biological age reflecting the condition of the arteries by 2.5 years compared to baseline levels (Maganova et al. 2023). The study showed a statistically significant decrease in the minimum thickness of the intima-media complex (by 45%) and a significant decrease in pulse wave velocity (by 10%) compared to initial measurements (Maganova et al. 2023).

Interestingly, some plant extracts demonstrate significant geroprotective effects despite the persistence of other conditions typically associated with accelerated aging. A polyphenol-rich plant extract improves plasma lipid levels and endotoxemia in obese mice, prevents macrophage recruitment to adipose tissues, reduces adipose accumulation of cholesterol and cholesterol oxides, and extends median lifespan—all despite the persistence of obesity (Aires et al. 2019). The extract drives the normalization of high-fat/high-sucrose-mediated functional enrichment of genes associated with immunity and inflammation (Aires et al. 2019). This suggests the existence of a "healthy adiposity" trait and highlights the potential for plant extracts to promote healthy aging even in the context of persistent metabolic challenges.

Several other unique plant extracts have demonstrated geroprotective potential. Whale meat extract, which contains many imidazole dipeptides (especially the unique component balenine), prolongs the lifespan of *C. elegans* mainly through sir-2.1, daf-2, and daf-16, and the myo-3, unc-54, unc-22, and fhod-1 genes involved in locomotor function (Shibato et al. 2024). The expression of the antioxidant enzymes sod-2 and sod-3 is also increased (Shibato et al. 2024). Sake lees extract obtained using a continuous phase-transition extraction method is abundant in bioactives and contains novel components, thus exerting prominent healthy aging effects *in vivo* (Zou et al. 2025). The extract not only promotes a 19.32% increase in the lifespan of *C. elegans* by enhancing oxidative stress and neuroprotective effects but also ameliorates age-related phenotypes like motoricity and age pigment (Zou et al. 2025).

The flavonoid corylin from *Psoralea corylifolia* extends the replicative lifespan in yeast by targeting the Gtr1 protein (Wang et al. 2022). In human umbilical vein endothelial cells, RNA sequencing data shows that corylin ameliorates cellular senescence (Wang et al. 2022). In an *in vivo* mammalian model, corylin extends lifespan in mice fed a high-fat diet (Wang et al. 2022). This study effectively connects findings across multiple model systems, from yeast to mammalian cells to whole organisms, providing a comprehensive view of corylin's geroprotective potential.

The carotenoid fucoxanthin exhibits geroprotective activity in *D. melanogaster* and *C. elegans* (Guvatova et al. 2020). In human fibroblasts, fucoxanthin at a concentration of 5 μM causes the most pronounced antioxidant effect in late passage cells (Guvatova et al. 2020). Transcriptomic data show increased expression levels of genes related to the Nrf2/ARE pathway (Guvatova et al. 2020). According to the analysis of enriched KEGG pathways, fucoxanthin alters cellular processes like ribosome biogenesis, lipid metabolism, and cell cycle regulation including some age-related pathways such as Wnt, JAK-STAT, and FoxO signaling pathways (Guvatova et al. 2020).

Plant extracts containing mixtures of polyphenols can have synergistic effects on lifespan extension. Rosemary and ginger extracts produce robust lifespan extension when used in combination in *Drosophila* (Aparicio et al. 2025). The healthy aging effects are linked to improved intestinal barrier function in aged flies and are not associated with reduced food intake (Aparicio et al. 2025). In terms of cellular hallmarks of aging, the plant extracts, including rosemary and ginger, lead to AMPK activation and improved markers of autophagy and proteostasis in aged flies (Aparicio et al. 2025). Furthermore, feeding the combination of rosemary and ginger enhances cognitive function in aged flies (Aparicio et al. 2025). Interestingly, several observations reveal effects from the combination that are more pronounced or entirely absent when either extract is administered alone (Aparicio et al. 2025).

Dietary interventions supplemented with plant extracts can also modulate lifespan. A study on the effects of dietary restriction combined with polyphenols showed that polyphenol compounds may potentiate intermittent feeding (IF)-induced longevity by minimizing specific components of IF-induced cell stress (Aires et al. 2012). While IF overall promotes longevity, some aspects of IF-induced stress may paradoxically lessen this effect (Aires et al. 2012). Even seemingly harmful plant extracts can have surprising effects on lifespan. *Lepista nuda* extract from the fruiting body and mycelium prepared using water, 75% ethanol, and petroleum ether significantly reduces lifespan in *D. melanogaster* (Huang et al. 2025). The extract decreases the content of 20E, JH, and Vg, as well as the transcripts of Halloween, ECR, JHAMT, and Yolk1 (Huang et al. 2025). This serves as a reminder of the complex and sometimes contradictory effects that plant extracts can have on longevity and reproductive fitness.

Vicatia thibetica de Boiss root extract extends the lifespan of *C. elegans* and improves healthspan (Liu et al. 2021). The extract counteracts senescence-associated biomarkers in doxorubicin-induced premature aging in adult mice (Liu et al. 2021). Interestingly, the extract promotes the expression of collagen in *C. elegans* and mammalian cell systems, which might be one of the essential factors to exert the healthy aging effects (Liu et al. 2021). Lifespan extension by the extract is dependent on DAF-16 and HSF-1 via oxidative stress resistance and heat stress resistance (Liu et al. 2021).

Summing up, a diverse array of plant extracts has demonstrated geroprotective effects in various model organisms. Berry extracts, tea derivatives, traditional medicinal herbs, and spices have shown particularly promising results, with some extracts extending lifespan by over 70% in model organisms. The effects are often dependent on conserved longevity pathways, particularly DAF-16/FOXO and SKN-1/Nrf2, and many extracts improve not just lifespan but also various aspects of healthspan, including stress resistance, mobility, and cognitive function. Interestingly, some extracts show sex-specific effects or differential responses depending on genetic background, highlighting the importance of personalized approaches to geroprotection.

15.4 Conclusions

This review has comprehensively examined the geroprotective effects of a diverse range of plant extracts across multiple model organisms. The evidence collectively demonstrates that plant extracts can effectively modulate aging processes and extend both lifespan and healthspan through multiple complementary mechanisms. Among the reviewed extracts, several demonstrate particularly promising geroprotective potential based on their consistent effects across different model systems and their ability to target evolutionarily conserved longevity pathways.

Green tea extract and its derivatives showed remarkable lifespan extension in *C. elegans*, with some components extending lifespan by up to 73% (Ke et al. 2024). This powerful effect, particularly associated with cinnamoylated flavoalkaloids, underscores the potential of refined plant extracted components as potent geroprotectors. Berries rich in anthocyanins, such as honeysuckle, bilberry, and *Aronia melanocarpa,* consistently demonstrated life-extending effects across different model organisms (Golubev et al. 2022; Nobile et al. 2024; Zhang et al. 2024a). Most notably, Aronia extract extended lifespan by over 22% in *C. elegans* while improving multiple aspects of healthspan (Zhang et al. 2024a).

Extracts from adaptogenic herbs like Ashwagandha and Maral root (*Rhaponticum carthamoides*) showed broad physiological benefits beyond simple lifespan extension, improving stress resilience, muscle function, and cognitive performance

(Ko et al. 2024; Kumar et al. 2013; Todorova et al. 2025). These effects translate across multiple model systems, from nematodes to mice, suggesting conservation of the underlying mechanisms. The combined effects of rosemary and ginger extracts on *D. melanogaster* highlight the potential for synergistic combinations of plant extracts to produce more robust geroprotective effects than individual extracts alone (Aparicio et al. 2025).

The clinical study of Siberian fir terpenes demonstrating a reduction in biological age by 2.5 years as measured by arterial indices in middle-aged patients provides promising evidence for the translation of findings from model organisms to humans (Maganova et al. 2023). Similarly, the demonstrated benefits of bilberry extract on human skin aging parameters (Nobile et al. 2024) suggest direct applications of plant extracts for promoting healthy aging in humans.

The mechanisms by which these extracts exert their effects are diverse but often converge on evolutionarily conserved longevity pathways. Many botanical preparations activate the DAF-16/FOXO and SKN-1/Nrf2 pathways, enhancing stress resistance and antioxidant defenses (Farias-Pereira et al. 2024; Xu et al. 2024; Zhang et al. 2024b). Other extracts modulate metabolic processes through AMPK activation (Kim et al. 2023; Zhang et al. 2024a), reduce inflammation (Kim et al. 2023; Zumerle et al. 2024), or promote autophagy (Lipatova et al. 2021; Wei et al. 2024; Wen et al. 2024). The multiplicity of mechanisms likely contributes to the robust geroprotective effects observed and suggests potential for tailored interventions targeting specific aspects of the aging process.

15.5 Challenges

Despite the promising results in model organisms, several challenges and future directions remain for the field, as follows:

First, there is a need for more studies in mammalian models to validate findings from invertebrate studies. While some extracts have shown efficacy in mice (Liu et al. 2021; Wang et al. 2022; Zumerle et al. 2024), most studies have been conducted in *C. elegans* or *D. melanogaster*. The translation of findings from these simpler organisms to mammals cannot be assumed and requires dedicated research efforts.

Second, standardization of extract composition and dosage remains a significant challenge for translation to clinical applications. The complex mixture of bioactive compounds in plant extracts can vary significantly depending on plant species, growing conditions, and extraction methods. Several investigations have highlighted dose-dependent effects, with some extracts showing hormetic responses where lower doses have better effects than higher doses (Lipatova et al. 2021). Identifying optimal dosing regimens and understanding the interplay between different components of complex extracts will be crucial for developing effective interventions.

Third, more comprehensive clinical evaluations are needed to assess the safety and efficacy of promising plant extracts in humans. While some research has demonstrated benefits in human subjects (Cho et al. 2021; Maganova et al. 2023; Nobile et al. 2024; Park et al. 2020), more extensive clinical trials with larger sample sizes, longer durations, and more comprehensive assessment of aging biomarkers will be essential to establish plant extracts as legitimate geroprotectors for human use.

Future research should focus on identifying the specific bioactive compounds responsible for the geroprotective effects of promising extracts and determining their optimal dosages and combinations. The synergistic effects observed with combinations like rosemary and ginger (Aparicio et al. 2025) suggest that intelligently designed mixtures of plant extracts or their components may yield more potent geroprotective interventions than single extracts. Additionally, long-term safety evaluations are essential, as the extended use of bioactive compounds may have unforeseen consequences.

The sex-specific and strain-specific responses observed in certain investigations (Banse et al. 2024; Lashmanova et al. 2018; Platonova et al. 2023) highlight the importance of considering individual genetic backgrounds in the development of geroprotective interventions. The concept of personalized geroprotection, where interventions are tailored to an individual's genetic makeup, sex, and metabolic status, represents an exciting frontier for future research.

In summary, plant extracts represent a rich and largely untapped source of potential geroprotectors with diverse mechanisms of action. The evidence gathered from lots of studies across multiple model organisms provides a strong foundation for the further development of plant-derived interventions to promote healthy aging. As understanding of aging mechanisms continues to advance, the targeted use of plant-derived geroprotectors, either alone or in synergistic combinations, may become an increasingly important approach to extending healthy lifespan in humans.

Funding The study was supported by the state assignment of Petrovsky National Research Center of Surgery (FURG-2025-0059, Project Reg. No. 125090910140-8).

References

Aires DJ et al (2012) Potentiation of dietary restriction-induced lifespan extension by polyphenols. Biochim Biophys Acta 1822:522–526. https://doi.org/10.1016/j.bbadis.2012.01.005

Aires V et al (2019) Healthy adiposity and extended lifespan in obese mice fed a diet supplemented with a polyphenol-rich plant extract. Sci Rep 9:9134. https://doi.org/10.1038/s41598-019-45600-6

Aparicio R, Salazar AM, Schmid ET, Khanbabaei A, Rajgopal A, Randolph RK, Walker DW (2025) The impact of rosemary and ginger extracts on aging and healthspan in Drosophila. Aging Dis. https://doi.org/10.14336/AD.2024.1558

Balasubramani SP, Mohan J, Chatterjee A, Patnaik E, Kukkupuni SK, Nongthomba U, Venkatasubramanian P (2014) Pomegranate juice enhances healthy lifespan in Drosophila melanogaster: an exploratory study. Front Public Health 2:245. https://doi.org/10.3389/fpubh.2014.00245

Banse SA et al (2024) Antioxidants green tea extract and nordihydroguaiaretic acid confer species and strain-specific lifespan and health effects in Caenorhabditis nematodes. Geroscience 46: 2239–2251. https://doi.org/10.1007/s11357-023-00978-0

Chen C et al (2016) Rhodiola rosea extends lifespan and improves stress tolerance in silkworm, Bombyx mori. Biogerontology 17:373–381. https://doi.org/10.1007/s10522-015-9622-8

Chen Y, Ouyang J, Tang X, Tong J, Liu H, Liu Z, Gong Y (2024) Black tea extracts enhance stress-induced sleep of Caenorhabditis elegans to resist UV damage. Food Res Int 196:115025. https://doi.org/10.1016/j.foodres.2024.115025

Cho YH, Lee SY, Lee CH, Park JH, So YS (2021) Effect of Schisandra chinensis Baillon extracts and regular low-intensity exercise on muscle strength and mass in older adults: a randomized, double-blind, placebo-controlled trial. Am J Clin Nutr 113:1440–1446. https://doi.org/10.1093/ajcn/nqaa447

Choi H, Seo E, Yeon M, Kim MS, Hur HJ, Oh BC, Jun HS (2019) Anti-aging effects of Schisandrae chinensis Fructus extract: improvement of insulin sensitivity and muscle function in aged mice. Evid Based Complement Alternat Med 2019:5642149. https://doi.org/10.1155/2019/5642149

Farias-Pereira R, Camayoc P, Raskin I (2024) Isothiocyanate-rich Moringa seed extract activates SKN-1/Nrf2 pathway in Caenorhabditis elegans. Int J Mol Sci 25. https://doi.org/10.3390/ijms252010917

Golubev D et al (2022) Honeysuckle extract (Lonicera pallasii L.) exerts antioxidant properties and extends the lifespan and healthspan of Drosophila melanogaster. Biogerontology 23:215–235. https://doi.org/10.1007/s10522-022-09954-1

Golubev D et al (2024) Berberis vulgaris L. extract supplementation exerts regulatory effects on the lifespan and healthspan of Drosophila through its antioxidant activity depending on the sex. Biogerontology 25:507–528. https://doi.org/10.1007/s10522-023-10083-6

Gonzalez-Manzano S, Ayuda-Duran B, Martin-Sanz R, Garzon-Garcia L, Santos-Buelga C, Gonzalez-Paramas AM (2024) Exploring the neuroprotective effects of grape seed Procyanidins on amyloid-beta-induced toxicity in Caenorhabditis elegans. Foods 13. https://doi.org/10.3390/foods13233865

Guvatova Z et al (2020) Protective effects of carotenoid fucoxanthin in fibroblasts cellular senescence. Mech Ageing Dev 189:111260. https://doi.org/10.1016/j.mad.2020.111260

Han S, Hu F, Ji X, Liu Y, Zhang S, Wang Z, Qiao K (2024) Polysaccharides from Ziziphus jujuba prolong lifespan and attenuate oxidative stress in Caenorhabditis elegans via DAF-16 and SKN-1. Int J Biol Macromol 282:137482. https://doi.org/10.1016/j.ijbiomac.2024.137482

Hu M et al (2025) Red ginseng aqueous extract superior to white ginseng exhibited anti-aging property through IIS signaling pathway in Caenorhabditis elegans. Chem Biodivers: e202500604. https://doi.org/10.1002/cbdv.202500604

Huang Y et al (2024) White tea aqueous extract: A potential anti-aging agent against high-fat diet-induced senescence in Drosophila melanogaster. Foods 13. https://doi.org/10.3390/foods13244034

Huang Y, Li J, Wang D, Ye Z, Wang L, Liu M (2025) Diets supplemented with different solvents extracts of Lepista nuda alters longevity and fecundity, and the expression of related genes in Drosophila melanogaster (Diptera: Drosophilidae). J Econ Entomol 118:403–415. https://doi.org/10.1093/jee/toae241

Jiang S, Deng N, Zheng B, Li T, Liu RH (2021) Rhodiola extract promotes longevity and stress resistance of Caenorhabditis elegans via DAF-16 and SKN-1. Food Funct 12:4471–4483. https://doi.org/10.1039/d0fo02974b

Jing L et al (2025) Unlocking the potential of Curcumae Rhizoma aqueous extract in stress resistance and extending lifespan in Caenorhabditis elegans. Molecules 30. https://doi.org/10.3390/molecules30081668

Jongsma E, Grigolon G, Baumann J, Weinkove D, Ewald CY, Wandrey F, Grothe T (2024) Timut pepper extract slows age-dependent decline of mobility and collagen loss and promotes longevity. Nutrients 16 https://doi.org/10.3390/nu16132122

Ke JP et al (2024) Unraveling anti-aging mystery of green tea in C. elegans: chemical truth and multiple mechanisms. Food Chem 460:140510. https://doi.org/10.1016/j.foodchem.2024.140510

Kim JS, Takanche JS, Kim JE, Jeong SH, Han SH, Yi HK (2019) Schisandra chinensis extract ameliorates age-related muscle wasting and bone loss in ovariectomized rats. Phytother Res 33:1865–1877. https://doi.org/10.1002/ptr.6375

Kim Y, Lee SB, Cho M, Choe S, Jang M (2023) Indian almond (Terminalia catappa Linn.) leaf extract extends lifespan by improving lipid metabolism and antioxidant activity dependent on AMPK signaling pathway in Caenorhabditis elegans under high-glucose-diet conditions. Antioxidants (Basel) 13. https://doi.org/10.3390/antiox13010014

Kim H et al (2024) Extracts ofHylotelephiumerythrostictum (miq.) H. Ohba ameliorate intestinal injury by scavenging ROS and inhibiting multiple signaling pathways in Drosophila. BMC Complement Med Ther 24:397. https://doi.org/10.1186/s12906-024-04686-w

Ko JS et al (2024) Ashwagandha ethanol extract attenuates Sarcopenia-related muscle atrophy in aged mice. Nutrients 16. https://doi.org/10.3390/nu16010157

Koval L, Zemskaya N, Aliper A, Zhavoronkov A, Moskalev A (2021) Evaluation of the geroprotective effects of withaferin A in Drosophila melanogaster. Aging (Albany NY) 13:1817–1841. https://doi.org/10.18632/aging.202572

Kumar R, Gupta K, Saharia K, Pradhan D, Subramaniam JR (2013) Withania somnifera root extract extends lifespan of Caenorhabditis elegans. Ann Neurosci 20:13–16. https://doi.org/10.5214/ans.0972.7531.200106

Lashmanova EA, Kuzivanova OA, Dymova OV, Moskalev AA (2018) The effects of cloudberry extract on Drosophila melanogaster lifespan and stress resistance. Adv Gerontol 31:958–965

Li L et al (2024) Dendrobium Nobile alcohol extract extends the lifespan of Caenorhabditis elegans via hsf-1 and daf-16. Molecules 29. https://doi.org/10.3390/molecules29040908

Lipatova A et al (2021) Effects of Siberian fir terpenes extract Abisil on antioxidant activity, autophagy, transcriptome and proteome of human fibroblasts. Aging (Albany NY) 13:20050–20080. https://doi.org/10.18632/aging.203448

Liu W et al (2021) Antiaging effects of Vicatia thibetica de Boiss root extract on Caenorhabditis elegans and Doxorubicin-induced premature aging in adult mice. Oxid Med Cell Longev 2021:9942090. https://doi.org/10.1155/2021/9942090

Liu Y, Zhang X, Wang Y, Wang J, Wei H, Zhang C, Zhang Q (2024) Cyclocodon lancifolius fruit prolongs the lifespan of Caenorhabditis elegans via antioxidation and regulation of purine metabolism. Food Funct 15:3353–3364. https://doi.org/10.1039/d3fo02931j

Maganova F, Voevoda M, Popov V, Moskalev A (2023) A prospective randomized comparative placebo-controlled double-blind study in two groups to assess the effect of the use of biologically active additives with Siberian fir terpenes for the biological age of a person. Front Pharmacol 14:1150504. https://doi.org/10.3389/fphar.2023.1150504

Mikhailova DV et al (2023) Antioxidant properties and geroprotective potential of wheat bran extracts with increased content of Anthocyanins. Antioxidants (Basel) 12. https://doi.org/10.3390/antiox12112010

Nobile V, Dudonne S, Kern C, Roveda G, Garcia C (2024) Antiaging, brightening, and antioxidant efficacy of fermented bilberry extract (Vaccinium myrtillus): a randomized, double-blind, Placebo-Controlled Trial. Nutrients 16. https://doi.org/10.3390/nu16142203

Park J, Han S, Park H (2020) Effect of Schisandra Chinensis extract supplementation on quadriceps muscle strength and fatigue in adult women: a randomized, double-blind, Placebo-Controlled Trial. Int J Environ Res Public Health 17. https://doi.org/10.3390/ijerph17072475

Platonova EY, Golubev DA, Zemskaya NV, Shevchenko OG, Patov SA, Shaposhnikov MV, Moskalev AA (2023) The antioxidant and geroprotective properties of an extract of mountain ash (Sorbus aucuparia L.) fruits. Mol Biol (Mosk) 57:979–994

Plyusnin SN et al (2023) Sources of potential geroprotectors in the flora of the European northeast. J Herb Med 41:100717. https://doi.org/10.1016/j.hermed.2023.100717

Proshkina E et al (2020) Terpenoids as potential geroprotectors. Antioxidants (Basel) 9. https://doi.org/10.3390/antiox9060529

Proshkina E et al (2024) Polyphenols as Potential Geroprotectors. Antioxid Redox Signal 40:564–593. https://doi.org/10.1089/ars.2023.0247

Rattan SIS (2022) Physiological hormesis and hormetins in biogerontology. Curr Opin Toxicol 29: 19–24. https://doi.org/10.1016/j.cotox.2022.01.001

Schriner SE et al (2013) Extension of Drosophila lifespan by Rhodiola rosea through a mechanism independent from dietary restriction. PLoS One 8:e63886. https://doi.org/10.1371/journal.pone.0063886

She J et al (2024) Ginseng extract attenuates the injury from ultraviolet irradiation for female Drosophila melanogaster through the autophagy signaling pathway. J Med Food 27:348–358. https://doi.org/10.1089/jmf.2023.K.0195

Shibato J, Takenoya F, Kimura A, Yamashita M, Rakwal R, Shioda S (2024) Lifespan extension and motor function improvement effects of whale meat extract in Caenorhabditis elegans. Int J Mol Sci 25. https://doi.org/10.3390/ijms252312833

Todorova MN, Savova MS, Mihaylova LV, Georgiev MI (2024) Punica granatum L. leaf extract enhances stress tolerance and promotes healthy longevity through HLH-30/TFEB, DAF16/FOXO, and SKN1/NRF2 crosstalk in Caenorhabditis elegans. Phytomedicine 134:155971. https://doi.org/10.1016/j.phymed.2024.155971

Todorova V, Todorova MN, Savova MS, Ivanov K, Georgiev MI, Ivanova S (2025) Maral root extract and its main constituent 20-Hydroxyecdysone enhance stress resilience in Caenorhabditis elegans. Int J Mol Sci 26. https://doi.org/10.3390/ijms26083739

Wang TH et al (2022) The flavonoid corylin exhibits lifespan extension properties in mouse. Nat Commun 13:1238. https://doi.org/10.1038/s41467-022-28908-2

Wei H, Zhang B, Shi S, Wang Z, Zhang F, Fang J (2024) Selenol Switch Assay for Selenoprotein Derivatization. Chembiochem 25:e202400594. https://doi.org/10.1002/cbic.202400594

Wen YP et al (2024) Exploring the therapeutic potential of Nelumbo nucifera leaf extract against amyloid-beta-induced toxicity in the Caenorhabditis elegans model of Alzheimer's disease. Front Pharmacol 15:1408031. https://doi.org/10.3389/fphar.2024.1408031

Xu P, Wang J, Wang J, Hu X, Wang W, Lu S, Sheng Y (2024) Anoectochilus roxburghii extract extends the lifespan of Caenorhabditis elegans through activating the daf-16/FoxO pathway. Antioxidants (Basel) 13. https://doi.org/10.3390/antiox13080945

Yang Y, Cao Y, Zhang J, Fan L, Huang Y, Tan TC, Ho LH (2024) Artemisia argyi extract exerts antioxidant properties and extends the lifespan of Drosophila melanogaster. J Sci Food Agric 104:3926–3935. https://doi.org/10.1002/jsfa.13273

Zhang H et al (2024a) Aronia melanocarpa extract extends the lifespan and health-span of Caenorhabditis elegans via mitogen-activated protein kinase 1. Food Funct 15:11020–11035. https://doi.org/10.1039/d4fo02479f

Zhang Z et al (2024b) Jujubae Fructus extract prolongs lifespan and improves stress tolerance in Caenorhabditis elegans dependent on DAF-16/SOD-3. Sci Rep 14:13713. https://doi.org/10.1038/s41598-024-64045-0

Zhang G, Jiang W, Hu Q, Luo J, Peng X (2025) Characterization and anti-aging potency of phenolic compounds in Xianhu tea extracts. Foods 14. https://doi.org/10.3390/foods14050737

Zheng J et al (2017) Pomegranate juice and extract extended lifespan and reduced intestinal fat deposition in Caenorhabditis elegans. Int J Vitam Nutr Res 87:149–158. https://doi.org/10.1024/0300-9831/a000570

Zou X et al (2025) Sake lees extract obtained using a novel continuous phase-transition extraction method: evaluation of its bioactive composition, anti-aging efficacy and mechanism. Food Funct. https://doi.org/10.1039/d5fo00291e

Zumerle S et al (2024) Targeting senescence induced by age or chemotherapy with a polyphenol-rich natural extract improves longevity and healthspan in mice. Nat Aging 4:1231–1248. https://doi.org/10.1038/s43587-024-00663-7

Part III
Drugs for Health and Longevity

Chapter 16
Metformin as a Senotherapeutic Agent

Şeydanur Turgut, Erdem Atasever, Gülnur Andican, and Ufuk Çakatay

Abstract Metformin, a first-line antidiabetic agent, has emerged as a promising senotherapeutic agent capable of modulating various senescent-related signaling pathways implicated in aging and age-related diseases. Beyond its glucose-lowering effects, metformin targets several metabolic features of aging, including impaired nutrient sensing, mitochondrial dysfunction, genomic instability, epigenetic alterations, and chronic inflammation as a repurposed conventional drug. Mechanistically, metformin activates AMP-activated protein kinase (AMPK), thereby inhibiting mammalian target of rapamycin (mTOR) signaling, enhancing autophagy, and promoting metabolic resilience. In parallel, it exerts AMPK-independent effects, including inhibiting mitochondrial reactive oxygen species (ROS) formation, suppressing advanced glycation end-products formation, and activating DNA repair mechanisms through the ATM kinase. Metformin also influences the epigenetic landscape by modulating DNA methyltransferases, histone modifiers, and microRNAs. Moreover, it attenuates the detrimental effects of senescence-associated secretory phenotype by inhibiting NF-κB and activating Nrf2-related redox responses. These wide-spectrum effects of metformin make it a potent modulator of cellular senescence and a candidate for delaying the onset of age-related diseases. This chapter provides the signaling mechanisms underlying metformin's senotherapeutic effects and discusses its potential role in promoting healthy aging and extending healthspan.

Keywords Metformin · Drug repurposing · Age-related diseases · Senescence · Senotherapeutic

---

Ş. Turgut · E. Atasever · G. Andican · U. Çakatay (✉)
Cerrahpaşa Faculty of Medicine, Department of Medical Biochemistry, İstanbul University-Cerrahpaşa, İstanbul, Türkiye
e-mail: cakatay@iuc.edu.tr

S. I. S. Rattan, C. Gezer (eds.), *Supplements and Drugs for Healthy Longevity*, Healthy Ageing and Longevity 22, https://doi.org/10.1007/978-3-032-11608-6_16

Abbreviations

AD	Alzheimer's disease
AGE	Advanced glycation end-products
AMPK	AMP-activated protein kinase
ARDs	Age-related diseases
ATM kinase	Ataxia telangiectasia mutated kinase
BMAL1	Basic helix-loop-helix ARNT-like 1
CKD	Chronic kidney disease
CRY1	Cryptochrome circadian regulator 1
DMAPs	Damage-associated molecular patterns
DNMTs	DNA methyltransferases
FOXO	Forkhead box O
HDACs	Histone deacetylases
IIS	Insulin/IGF-1 signaling
mTOR	Mammalian target of rapamycin
Nrf2	Nuclear factor erythroid 2-related factor 2
OCT1	Organic cation transporter 1
PD	Parkinson's disease
PGC-1α	Peroxisome proliferator-activated receptor-γ (PPAR-γ) coactivator-1α
ROS	Reactive oxygen species
SASP	Senescence-associated secretory phenotype
TNF-α	Tumor necrosis factor alpha
VEGF	Vascular endothelial growth factor

16.1 Introduction

Numerous chronic diseases, particularly those associated with metabolic syndrome and the cardiovascular system, such as type II diabetes, atherosclerosis, hypertension, myocardial infarction, and stroke, become increasingly likely with aging. The progressive accumulation of cellular and molecular damage over time contributes to the functional decline associated with aging and the development of age-related diseases (ARDs) (Turgut and Andican 2025). A growing body of evidence supports the concept that biological aging is not only modifiable but is also potentially targetable through pharmacological interventions aimed at its underlying hallmarks. Geropharmacology is becoming an increasingly complex topic as life expectancy rises. Senotherapeutics are promising pharmacological agents that can impede or decrease the rate of aging-related degenerative processes and extend lifespans by repairing damage or modulating stress resistance. Ents that can impede or decrease the rate of aging-related degenerative processes and extend lifespans by repairing damage or modulating stress resistance (Atasever et al. 2025).

Among the most intensively studied conventional drugs is metformin, a biguanide initially developed for treating type 2 diabetes mellitus and now recognized for its extensive senotherapeutic effects on health and lifespan in humans and various model organisms (Kulkarni et al. 2020). Metformin has garnered considerable attention as a senotherapeutic agent targeting the fundamental mechanisms of cellular senescence (Atasever et al. 2025). Its senotherapeutic potential arises from its ability to modulate numerous intracellular signaling pathways that intersect with several hallmarks of aging, including deregulated nutrient sensing, mitochondrial dysfunction, genomic instability, telomere attrition, epigenetic alterations, stem cell exhaustion, and chronic inflammation (Kulkarni et al. 2020). Upon cellular uptake via the organic cation transporter 1 (OCT1), metformin accumulates in the mitochondrial matrix and inhibits respiratory chain complex I. This inhibition decreases the cellular ATP/AMP ratio, leading to the activation of AMP-activated protein kinase (AMPK)—a crucial energy-sensing enzyme that coordinates metabolic stress responses and cellular repair programs (Barzilai et al. 2016; Rena et al. 2017). AMPK activation triggers a series of downstream signaling events, including inhibition of the insulin/IGF-1 signaling (IIS) and mammalian target of rapamycin (mTOR) pathways, enhancement of autophagy, and maintenance of proteostasis, all of which alleviate aging-related metabolic stress (Saxton and Sabatini 2017; Kulkarni et al. 2020). Beyond its AMPK-mediated action, metformin exhibits AMPK-independent actions, notably through direct suppression of mitochondrial ROS production and glycation, thereby reducing oxidative stress and the accumulation of advanced glycation end-products (AGEs) (Cameron et al. 2016) (Fig. 16.1). Metformin activates DNA damage response kinases such as ataxia telangiectasia mutated kinase (ATM kinase) and enhances AMPK independent of genome surveillance and repair (Algire et al. 2010). Metformin's senoprotective effects are further supported by its epigenetically mediated influences on important regulators, such as DNA methyltransferases (DNMTs), histone deacetylases (HDACs), and microRNA processing enzymes like DICER1. These regulators maintain transcriptional fidelity and stability (Zhang et al. 2025). MicroRNAs are a family of naturally occurring non-coding RNAs ranging from 18 to 24 nucleotides. MicroRNAs can regulate post-translational gene expression by binding to the target mRNAs' 3′-untranslated region (3′-UTR). Due to its elevated expression in several aged tissues and cells, microRNA-34a has been proposed as a potential senescence biomarker. Metformin activates the microRNA-34a/SIRT1 pathway to regulate cellular senescence and proliferation (Yan et al. 2023). Importantly, metformin modulates inflammatory responses and cellular senescence by activating nuclear factor erythroid 2-related factor 2 (Nrf2), a master regulator of the antioxidant response. This leads to the upregulation of detoxifying enzymes such as glutathione peroxidase 7, which protect against oxidative damage and stem cell exhaustion (Martin-Montalvo et al. 2013). Concurrently, metformin inhibits NF-κB signaling, thereby suppressing the senescence-associated secretory phenotype (SASP), characterized by pro-inflammatory cytokines such as IL-6, tumour necrosis factor alpha (TNF-α), and vascular endothelial growth factor (VEGF) (Moiseeva et al. 2013).

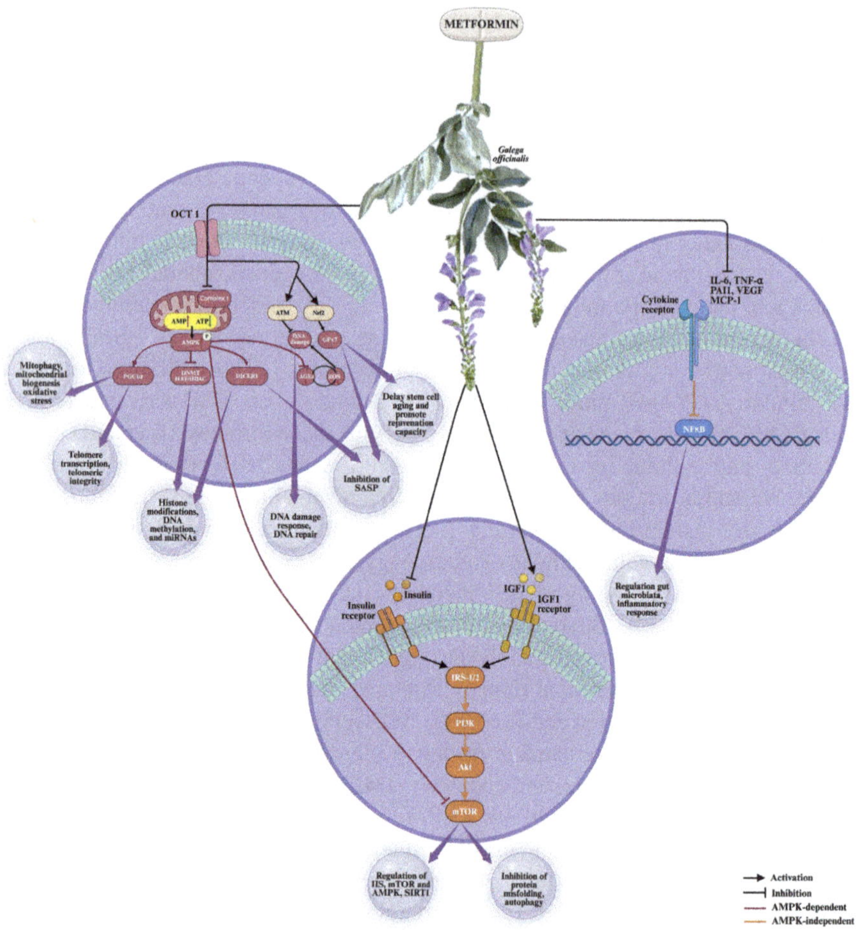

Fig. 16.1 Several signaling pathways of metformin targeting senescence: senescent forms of mitotic and postmitotic cells are characterized by altered cell morphology and broad group signaling activities. Metformin exhibits its senotherapeutic effects on senescent cells through various AMPK-dependent and AMPK-independent signaling mechanisms in different subcellular compartments

Together, these AMPK-dependent and -independent pathways establish metformin as a multi-target modulator of aging mechanisms, capable of delaying cellular senescence, enhancing cellular resilience, and attenuating chronic inflammation (Fig. 16.1). Its systemic effects on nutrient sensing, mitochondrial function, genome stability, epigenetic regulation, and inflammatory modulation underscore its potential as a geroprotective agent that may also extend healthspan and delay the onset of ARDs such as cardiovascular disease, neurodegeneration, cancer, osteoporosis, and chronic kidney disease (Evans et al. 2005; Patil et al. 2014; Chen et al. 2022; de Vries et al. 2023).

Identifying possible senotherapeutic targets, using cutting-edge data analysis technologies, integrating existing drug expertise, and gaining mechanistic insights are key aspects of drug repurposing. In recent years, interest has grown regarding its potential to prevent or treat a variety of ARDs. The therapeutic benefit of metformin, one of the most popular repurposed medications for senotheraphy, is evaluated in this chapter.

16.2 Senotherapeutic Signaling Targets of Metformin

As illustrated in Fig. 16.1, metformin modulates numerous senescence-related signaling pathways that intersect with the hallmarks of aging, exerting both AMPK-dependent and AMPK-independent actions. Upon entry into cells via the OCT1, metformin accumulates in mitochondria, inhibiting respiratory complex I, resulting in a lower ATP/AMP ratio and consequent activation of AMPK. This energy-sensing kinase regulates a metabolic shift that favors catabolic processes, mitochondrial remodeling, and cellular rejuvenation (Steinberg and Carling 2019).

AMPK activation downregulates IIS by inhibiting IRS-1/2, PI3K, and Akt, ultimately suppressing the mTOR pathway. This downregulation reduces anabolic growth signals, enhances autophagic flux, and improves proteostasis, thereby alleviating two fundamental hallmarks of aging: deregulated nutrient signaling and loss of proteostasis (Saxton and Sabatini 2017; Kulkarni et al. 2020). In parallel, AMPK stimulates peroxisome proliferator-activated receptor-γ (PPAR-γ) coactivator-1α (PGC-1α), promoting transcriptional activity for mitochondrial biogenesis, mitophagy, and metabolic flexibility—core mechanisms that mitigate age-related mitochondrial dysfunction (Wang et al. 2019). Further, AMPK facilitates nuclear responses that preserve genome integrity and function. It induces telomeric repeat-containing RNAs expression to support telomere maintenance and modulates epigenetic regulators such as DNMTs, HDACs, and DICER1, contributing to transcriptional fidelity and miRNA processing (Martin-Montalvo et al. 2013; Steinberg and Carling 2019). These actions mitigate telomere attrition and epigenetic drift, two hallmarks tightly linked to cellular senescence and organismal aging.

Beyond AMPK, metformin influences senescence-related pathways through AMPK-independent mechanisms. Metformin reduces reactive oxygen species (ROS) production and the accumulation of AGEs by inhibiting mitochondrial complex I, thereby alleviating oxidative stress and inflammation (Cameron et al. 2016). Additionally, metformin activates the DNA damage sensor ATM, independent of AMPK, to enhance DNA repair pathways, thus preserving genomic integrity (Algire et al. 2010). Metformin also modulates critical regulators of the inflammatory response and cellular senescence. It activates Nrf2, a master transcription factor for antioxidant defense, resulting in upregulation of antioxidant enzymes and other cytoprotective enzymes that protect against oxidative damage and delay stem cell exhaustion (Martin-Montalvo et al. 2013). Concurrent inhibition of NF-κB leads to repression of the paracrine effects of SASP, including pro-inflammatory cytokines such as IL-1, IL-6, TNF-α, and ultimately reduces chronic inflammation and promotes tissue regeneration (Moiseeva et al. 2013).

All these signaling pathways collectively reveal that metformin acts on multiple interconnected metabolic events, including nutrient sensing, mitochondrial health, telomere function, epigenetic stability, DNA repair, stem cell maintenance, and inflammatory regulation. The senotherapeutic, signal-modulating role of metformin is closely related to these broad metabolic events, giving it great potential to prolong life expectancy and delay or prevent the onset of multiple ARDs.

16.3 Metformin as a Multitarget Senotherapeutic Agent

Cellular senescence, once considered a protective mechanism against malignant transformation, is now recognized as a central driver of embryonic development, wound healing, age-associated metabolic decline, chronic inflammation, and ARDs (Atasever et al. 2025). Senescent cells, characterized by permanent cell cycle arrest and acquisition of a pro-inflammatory SASP, gradually accumulate with the aging process and contribute to the pathogenesis of numerous ARDs, including cardiovascular disease, neurodegeneration, osteoarthritis, and cancer. Given its ability to modulate a spectrum of aging hallmarks and attenuate senescent phenotypes, metformin has gained recognition as a multi-target senotherapeutic agent (Fig. 16.2).

Metformin exerts its senotherapeutic effects via both senostatic (suppressing senescent phenotypes and SASP) and senomorphic (modifying paracrine functions of existing SASP) mechanisms (Atasever et al. 2025; Cebe and Kızılyel 2025). Its senomodulatory actions span several signaling pathways implicated in the aging process. A core mechanism involves the inhibition of mitochondrial complex I, leading to altered cellular energy metabolism and activation of AMPK, a master regulator of metabolic homeostasis and stress response (Barzilai et al. 2016; Rena et al. 2017). Through AMPK activation, metformin downregulates the IIS and mTOR pathways, promoting autophagy, suppressing cellular hypertrophy, and enhancing proteostasis. These modulatory effects mitigate the accumulation of damaged proteins and organelles—a hallmark of senescent cells (Kulkarni et al. 2020). Concurrently, AMPK promotes mitochondrial biogenesis and mitophagy through PGC-1α and Forkhead box O (FOXOs) transcription factors, thereby restoring mitochondrial function and reducing ROS, which are both drivers and consequences of cellular senescence (Wang et al. 2019). In addition to metabolic reprogramming, metformin modulates inflammatory and immune responses. It suppresses the IKK/NF-κB signaling axis, resulting in downregulation of SASP contents such as IL-6, TNF-α, and MCP-1 (Moiseeva et al. 2013). These effects are essential for limiting the paracrine propagation of secondary senescence and the chronic, low-grade inflammation observed in aging tissues ("inflammaging"). Furthermore, metformin enhances Nrf2-mediated antioxidant responses and DNA damage repair pathways—including ATM activation—preserving genomic stability and

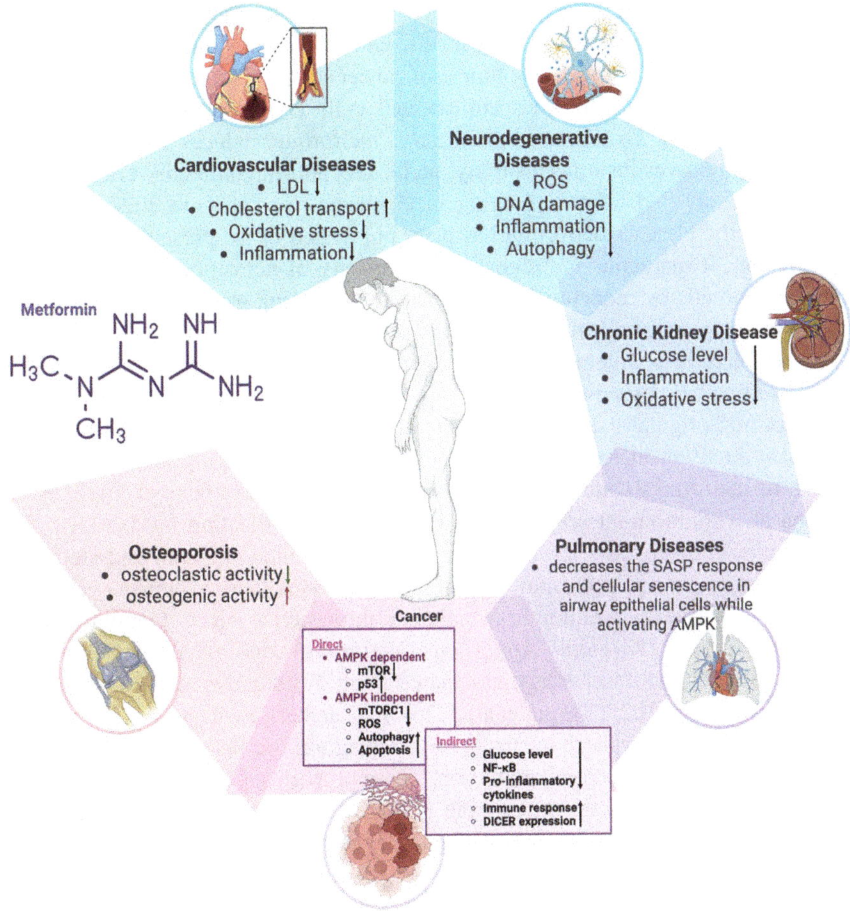

Fig. 16.2 Senescence cell accumulation affects different organ systems at different times and rates; this phenotypic variation is named “mosaic aging.” A wide array of individual and organ-specific senescent phenotypes is responsible for developing ARDs. Targeting specific SASP factors with metformin may provide new therapeutic opportunities to mitigate the detrimental effects of senescent cells that accumulate in aging organ systems

reducing the pro-inflammatory burden of DNA damage-associated molecular patterns (DAMPs) (Algire et al. 2010; Martin-Montalvo et al. 2013).

Importantly, metformin’s pleiotropic activity translates into therapeutic benefits across a range of ARDs. In cardiovascular disease, it improves endothelial function, reduces arterial stiffness, and lowers systemic inflammation through mechanisms involving eNOS activation and oxidative stress suppression (Cebe and Kızılyel 2025). In neurodegenerative diseases, metformin has shown efficacy in reducing amyloid-β deposition, tau pathology, and neuroinflammation via mTOR inhibition and autophagy enhancement, suggesting its role in cognitive protection (Patil et al. 2014).

In oncology, metformin exerts anticancer effects by modulating IIS and mTOR pathways, reducing systemic insulin levels, and improving immune surveillance. Moreover, it impairs the metabolic fitness of cancer stem cells and interferes with the pro-tumorigenic effects of senescent stromal cells (Pollak 2010). In metabolic disorders like type 2 diabetes and obesity, metformin reduces hepatic glucose production, improves insulin sensitivity, and exerts anti-inflammatory effects—outcomes intimately tied to its anti-aging mechanisms (Lv and Guo 2020). Beyond disease-specific effects, metformin has been implicated in the preservation of stem cell function. It maintains the regenerative capacity of hematopoietic and mesenchymal stem cells by reducing ROS formation, enhancing mitochondrial turnover, and modulating autophagy pathways (Marycz et al. 2016; Cheng et al. 2022). These properties directly counteract stem cell exhaustion, a key hallmark of aging that underlies impaired tissue repair.

Metformin, originally developed as a treatment for type 2 diabetes, has gathered significant attention in recent years for its potential use in aging and ARDs. The breadth of metformin's molecular targets and physiological effects exemplifies its function as a multi-target senotherapeutic agent. By modulating nutrient sensing, mitochondrial homeostasis, proteostasis, DNA repair, inflammation, and stem cell biology, metformin addresses multiple aging mechanisms simultaneously—a critical feature for effective intervention in complex, multi-systemic ARDs. Ongoing efforts such as the TAME (*Targeting Aging with Metformin*) trial will further clarify its potential to delay the onset of age-associated morbidities and extend healthspan in humans (Chen et al. 2025; Turgut et al. 2025). Researchers are increasingly focusing on its effects on cellular mechanisms involved in aging, such as cellular senescence, mitochondrial function, and inflammation. These properties position metformin as a potential senotherapeutic agent, capable of mitigating or preventing the onset of ARDs such as cardiovascular disease, neurodegenerative disorders, cancer, osteoporosis, and chronic kidney disease.

A key mechanism by which metformin exerts its effects is the activation of AMPK. AMPK is a central regulator of cellular energy balance, activating processes that promote cell survival, autophagy, and metabolic homeostasis. By activating AMPK, metformin encourages the removal of damaged cellular components, reduces oxidative stress, and induces autophagy, a process that clears dysfunctional proteins and organelles. These effects help prevent cellular senescence, a condition that becomes more prevalent with age and contributes to the pathogenesis of multiple chronic diseases.

Mitotic cells may also respond to intrinsic or extrinsic damage with cell death, but this generally is not problematic in mitotic tissues. In contrast, however, damage or repeated cell division can cause mitotic cells to express a senescent phenotype, which may compromise the function and integrity of the surrounding tissue. It is a fact that nondividing post-mitotic cells can also transform into their senescent forms. The cellular senescence load, especially in postmitotic cells such as neurons, myocytes, cardiomyocytes, adipocytes, and osteocytes, gradually increases with aging, leading to chronic inflammation and tissue dysfunction—key drivers of ARDS. Aging causes changes in several molecular, cellular, and physiological features of circadian rhythms. Emerging evidence suggests that altered circadian rhythm plays a pivotal role in cellular senescence. Metformin additionally exhibits tissue-specific impacts on metabolic rhythms and the circadian clock. Tissue-specific impacts upon metabolic rhythms and the circadian clock (Barnea et al. 2012).

The poor oral absorption and gastrointestinal adverse effects significantly limited the clinical application of metformin. Researchers have attempted to design and synthesize prodrugs of metformin with higher lipophilicity to improve permeability and passive transcellular absorption. Metformin has a very good safety profile. Usually minor, side effects include nausea, upset stomach, or diarrhea. Serious negative effects are uncommon. These include severe allergic reactions and lactic acidosis, a condition in which the blood contains an accumulation of lactic acid. In recent years, interest has grown regarding its potential to prevent or treat a variety of ARDs. Reducing hepatic steatosis, inhibiting the growth and metastasis of various types of cancers, improving lipid profile, reducing endothelial oxidative stress, ameliorating renal fibrosis, modulating the immune system, improving blood flow, angiogenesis, and protection in animal models of ischemia are some of the positive effects that can be achieved (Atasever et al. 2025). Enhancing cognitive function in AD, boosting nerve repair. Here, we summarize the most recent data from clinical investigations and discuss the current understanding of the mechanisms of metformin in ARDs.

16.3.1 Cardiovascular Diseases

Vascular aging describes a process mediated by structural and functional changes in the vascular wall. It is characterized by endothelial dysfunction, which promotes the release of vasoactive substances that increase vascular smooth muscle tone, reduce arterial distensibility, and impair the ability of arteries to adapt to changes in blood pressure (Cebe and Kızılyel 2025). Age-associated declines in cardiac function lead to cardiovascular diseases among the elderly population. Cardiovascular diseases, including atherosclerosis and heart failure, are closely linked to vascular aging. As individuals age, endothelial function declines, and vascular stiffness increases, contributing to the development of hypertension, atherosclerosis, and other cardiovascular diseases (Bu et al. 2022). Senescence, a chronic inflammatory condition, is directly linked to cardiovascular disorders and poses significant management challenges when combined with age-related comorbidities. Atherosclerotic cardiovascular disease is commonly thought to be primarily caused by diabetes. Metformin treatment improves functional and structural changes that occur in aging arteries and alleviates SASP-related detrimental effects (Tai et al. 2022). Metformin also inhibits the activity of the transcription factor NF-kB. It suppresses IkB and IKKα/β phosphorylation and prevents NF-kB from translocating into the cell nucleus, all occurring without AMPK activation. Moreover, metformin inhibits the production of pro-inflammatory cytokines, which are primarily regulated by NF-kB (Cebe and Kızılyel 2025).

Preclinical studies have suggested that metformin reduces arterial stiffness and improves vascular function by modulating the endothelial nitric oxide synthase (eNOS) pathway, which is crucial for maintaining vascular health (Majithiya and Balaraman 2006). Additionally, metformin's inhibition of the mTOR pathway, which regulates cellular growth and metabolism, may help counteract age-related endothelial dysfunction (You et al. 2020). Its ability to improve mitochondrial function may further support heart health by enhancing ATP production in cardiac cells, potentially protecting against heart failure, a common age-associated condition. In this context, metformin has shown potential in improving endothelial function, alleviating senescence-related detrimental events, and contributing to cardiovascular protection.

16.3.2 Neurodegenerative Diseases

Degenerative brain disorders like Alzheimer's Disease (AD) and Parkinson's Disease (PD) may be mitigated by the senotherapeutic properties of metformin. Metformin can rapidly cross the blood-brain barrier and accumulate in different brain regions. Neurodegenerative diseases such as AD and PD are driven by cellular dysfunction, oxidative stress, and chronic inflammation. In the hippocampus, metformin promotes neurogenesis and prevents pathological neuronal apoptosis, hence protecting neuronal loss, the direct source of cognitive deficits in AD. The risk of developing AD appears to be positively correlated with type 2 diabetes, according to emerging data. It has been revealed that impaired glucose metabolism and insulin signaling, two important processes in type 2 diabetes, play a role in the development and progression of AD. Due to the strong correlation, researchers have even suggested the title of "*Type-3-Diabetes*" for AD, suggesting that anti-diabetic medications may be able to prevent or treat AD (Liao et al. 2021). Mitochondrial dysfunction is central to the pathogenesis of these conditions, and metformin's ability to promote mitochondrial biogenesis offers a novel therapeutic approach (de Marañón et al. 2022). By enhancing mitochondrial function, metformin could potentially mitigate the energy deficits that neurons experience in senescence and neurodegeneration. Another important factor that mitigates AD pathogenesis is metformin's antioxidant properties. Administering metformin suppresses the rise in malondialdehyde levels and superoxide dismutase activity.

In Alzheimer's disease, for instance, metformin has been shown to reduce the accumulation of amyloid-beta plaques, a hallmark of the disease. It is believed that metformin's activation of AMPK leads to a reduction in the activity of glycogen synthase kinase 3-beta (GSK3β), an enzyme involved in tau hyperphosphorylation and amyloid precursor protein processing. This reduction in tau and amyloid-beta accumulation may help preserve cognitive function. Moreover, metformin's ability to reduce neuroinflammation, a critical factor in both AD and PD, could further support its neuroprotective potential (Rotermund et al. 2018). Metformin can reduce Aβ synthesis and, consequently, Aβ toxicity via regulating the AMPK/mTOR/S6K/BACE1 signaling pathway. Nuclear factor-kappa B (NF-κB) is a well-known pro-inflammatory transcription factor involved in abnormal neuroinflammation in AD. The NF-κB pathway, thought to be the primary signaling system implicated in AD neuroinflammation, can also be inhibited by AMPK activation. Additionally, metformin can block the AMPK-dependent loss of mitochondrial function caused by Aβ and enhance activation of the caspase cascade, which is implicated in apoptosis-mediated neurodegeneration (Liao et al. 2021).

Parkinson's disease (PD) is one of the second most common degenerative brain diseases, next to AD. Repurposing effective and approved drugs like metformin could be an effective strategy for managing PD. In Parkinson's disease and other neurodegenerative brain disorders, the gene parkin protects against mitochondrial toxicity. Metformin has neuroprotective benefits by increasing the expression of the Parkin gene. Additionally, leucine-rich repeat kinase 2 and Parkinson's disease kinase 1 are two genes that metformin regulates and are implicated in Parkinson's disease. Metformin has a neuroprotective effect through controlling aging, oxidative stress, neurogenesis, mitochondrial failure, autophagy, α-synuclein aggregation, and the expression of genes that protect neurons (Liao et al. 2021).

16.3.3 Chronic Kidney Disease

Metformin may decrease the course of chronic kidney disease (CKD) and the reduction in kidney function over time because it targets many pathways that cause chronic kidney damage, according to emerging research (El-Damanawi et al. 2024). Another age-related disease that significantly impacts the elderly population is CKD. Kidney dysfunction is often exacerbated by factors such as diabetes and hypertension, which are more common in elderly individuals. The role of metformin in CKD is still under investigation, but early studies suggest that metformin's anti-inflammatory and antioxidant effects may help protect kidney function (El-Damanawi et al. 2024). Metformin protects against renal tubular epithelial cell damage caused by higher glucose-induced fibrosis (Liang et al. 2022).

In animal models of CKD, metformin has been shown to reduce oxidative stress and inflammation in the kidneys, which are key drivers of kidney damage (Pan et al. 2020). Additionally, by reducing the accumulation of senescent cells in renal tissue, metformin may help slow the progression of CKD. However, due to the risk of lactic acidosis in individuals with renal insufficiency, the use of metformin in CKD patients must be carefully monitored, particularly in those with advanced kidney disease (Pan et al. 2020).

16.3.4 Pulmonary Diseases

It is well known that cellular senescence plays a part in diseases associated with age. The accumulation of DNA damage brought on by oncogene activity triggers the DNA damage response (DDR) via ATM and ATR, which in turn activates the tumor suppressor p53. Cellular senescence is induced, and the transition from G1 to S phase is further inhibited by activated p53, which also increases the transcription of the cyclin-dependent kinase inhibitor p21. The function of alveolar epithelial cells is impaired in a variety of age-related chronic lung diseases, including cancer, bronchopulmonary dysplasia, pulmonary fibrosis, and chronic obstructive pulmonary disease. Senotherapeutics are currently being used extensively in lung disorders linked to aging. In vitro, metformin decreases the SASP response and cellular senescence in airway epithelial cells while activating AMPK. Metformin has been shown in animal experiments to have radiation protection and radiosensitization effects, as well as to protect healthy lung tissue from radiation-induced senescence and lung damage. By increasing autophagy activity, metformin can also protect mouse lung epithelial cells from senescence, offering possibilities for both preventing and treating acute lung injury. Metformin is comparatively well tolerated and can lower mortality in people with chronic obstructive pulmonary disease (Liu et al. 2024).

16.3.5 Cancer

Oncogene-induced senescence, which is cellular senescence brought on by oncogene activation, can also prevent further tumors. Cancer is another age-associated disease where metformin shows promise. The aging process is associated with increased oxidative stress and cellular damage, factors that can promote tumorigenesis. Metformin's ability to activate AMPK and inhibit the mTOR pathway plays a central role in reducing cancer cell proliferation and survival. The mTOR pathway regulates cell growth and metabolism and is frequently dysregulated in cancer cells (Ben Sahra et al. 2011). By inhibiting mTOR, metformin reduces the energy available to cancer cells, making them more vulnerable to cellular stress and death (Hajimohammadebrahim-Ketabforoush et al. 2024).

Clinical studies have indicated that metformin use in diabetic patients is associated with a reduced risk of certain cancers, including breast, colorectal, and prostate cancer (Evans et al. 2005). Furthermore, metformin has been shown to suppress the growth of various cancer cell lines, both in vitro and in vivo, by reducing oxidative stress and promoting tumor cell apoptosis (Hua et al. 2023). Ongoing clinical trials are investigating the potential of metformin as an adjuvant therapy for cancer. In recent years, metformin has also gained attention as a possible senotherapeutic option for colorectal cancer. Compared to individuals without diabetes, those with diabetes are more likely to develop malignancies of the colon, rectum, pancreas, liver, and many other organs. Consequently, as diabetes is thought to be a risk factor for the development of cancer, treating it may decrease the probability of developing cancer (Chen et al. 2020). In animal models, studies have demonstrated that metformin may reduce the development of intestinal polyps and colorectal abnormalities. According to a brief clinical study, low-dose metformin (250 mg/day) decreased colorectal aberrant crypt foci, suggesting that it may help prevent colon cancer (Chen et al. 2020). Investigations revealed that metformin inhibits the mRNA and protein expression of INHBA as its primary target, down-regulating TGF-β/PI3K/Akt signal transductions, which causes an inhibitory effect on colorectal cancer proliferation. Metformin reduces the viability of most colorectal cancer cell lines while having little impact on normal colon cells, suppressing cell proliferation, but does not significantly affect apoptosis (Xiao et al. 2022). Metformin is not only attributed to reducing the risk of colorectal cancer, but also pancreatic cancer, breast cancer, lung cancer, and prostate cancer, as well, by inhibiting their proliferation potential (Xiao et al. 2022).

16.3.6 Osteoporosis

Osteoporosis is a systemic bone metabolic disease that causes the microstructure of bones to break down and the density and quality of bones to diminish, making the bones more prone to fractures. The prevalence of osteoporosis has risen yearly due to changes in lifestyle and population aging. Osteoporosis, characterized by a decrease in bone mineral density and an increased risk of fractures, is a common phenomenon in the elderly. Osteoporosis is currently mostly treated with oral medications that have limited effects and directly affect osteoblasts and osteoclasts' activities. Hormone replacement therapy is not favored because of its limited applicability and significant side effects (Yang et al. 2024). By promoting the formation of new bone while inhibiting bone resorption, metformin may help slow the progression of osteoporosis and improve bone health in elderly individuals. Metformin's effects on the AMPK pathway may also play an important role in enhancing bone metabolism. AMPK activation is related to the upregulation of key bone formation genes and downregulation of osteoclast differentiation factors (Cai et al. 2024).

Metformin has been shown to affect the balance between osteoblasts and osteoclasts' activities, the two primary cell types responsible for bone remodeling. In preclinical studies, metformin has been shown to stimulate osteoblast differentiation and inhibit osteoclast activity. Metformin treatment is significantly associated with a reduction in the risk of osteoporosis, an increase in bone mineral density, and a decrease in bone fractures (Cai et al. 2024).

16.4 Clinical Evidence and Translational Potential

The compelling preclinical data supporting metformin's senotherapeutic properties have catalyzed a growing interest in its clinical translation. As a first-line therapy for type 2 diabetes mellitus with an extensive safety profile established over decades, metformin offers a uniquely positioned candidate for repurposing in age-related diseases and healthspan extension (Barzilai et al. 2016). Beyond glycemic control, epidemiological and interventional studies have consistently associated metformin use with reduced incidence of several ARDs, including cardiovascular disease, neurodegenerative disorders, cancer, and frailty (Campbell et al. 2017; Kulkarni et al. 2020). A growing body of evidence suggests that metformin, a widely used antidiabetic agent, exerts pleiotropic effects beyond glucose regulation, including modulation of aging-related pathways and neurodegeneration. Two landmark clinical trials—MILES (Metformin in Longevity Study) and TAME (Targeting Aging with Metformin)—have been specifically designed to explore metformin's role in decelerating the biological processes of aging and reducing the incidence of ARDs.

16.4.1 The MILES Trial: Molecular Markers of Aging in Response to Metformin

The MILES trial was an NIH-supported, randomized, placebo-controlled pilot study that investigated whether metformin influences aging-associated molecular signatures in humans. The study enrolled healthy, non-diabetic older adults (aged 60–75) and evaluated gene expression changes in muscle and adipose tissue following 6 weeks of metformin administration (1500–2000 mg/day). Key findings demonstrated that metformin altered the expression of multiple genes associated with inflammation, mitochondrial function, and metabolic regulation—all of which are hallmarks of aging (Kulkarni et al. 2020). Notably, metformin suppressed pro-inflammatory gene networks and enhanced mitochondrial oxidative phosphorylation pathways in skeletal muscle, indicating a potential systemic effect on aging-related decline. (Kulkarni et al. 2020). The question of whether metformin is protective in disease-free individuals is still up for debate.

16.4.2 The TAME Trial: A Paradigm Shift in Geroscience

The TAME trial, a proposed multi-center, double-blind, placebo-controlled study, represents a transformative approach in geroscience. Rather than targeting a specific disease, TAME aims to determine whether metformin can delay the onset of multiple age-related chronic conditions, including cardiovascular disease, cancer, cognitive decline, and mortality. Led by Dr. Nir Barzilai and collaborators, TAME plans to recruit approximately 3000 individuals aged 65–80 years and follow them for 5–6 years. The primary outcome is a composite measure of time to the occurrence of major age-related diseases or death. While enrollment is pending funding, the trial is notable for its conceptual innovation: using an FDA-approved drug to target aging as a common driver of multiple pathologies (Barzilai et al. 2016). Observational studies have already suggested longevity benefits. For instance, diabetic patients treated with metformin demonstrate survival rates that are not only higher than those of untreated diabetic patients but, intriguingly, comparable or superior to non-diabetic individuals, even after adjusting for comorbidities (Bannister et al. 2014). These findings imply that metformin's actions extend beyond glycemic modulation and may impact the biological aging process.

In the realm of oncology, retrospective analyses and meta-analyses have linked metformin use to decreased cancer incidence and improved survival outcomes across several malignancies, including breast, colorectal, and prostate cancers (Decensi et al. 2010; Noto et al. 2012). Mechanistically, these effects are attributed to metformin's inhibition of insulin/IGF-1 and mTOR signaling, reduction in circulating glucose and insulin levels, and activation of anti-proliferative pathways (Pollak 2010).

Similarly, studies in neurodegenerative diseases suggest potential cognitive benefits. A longitudinal cohort study found that metformin users exhibited slower cognitive decline compared to non-users, while others have reported improved executive function in older adults with diabetes (Ng et al. 2014; Kuan et al. 2017). These observations align with preclinical findings on metformin's capacity to reduce neuroinflammation, promote autophagy, and attenuate tau hyperphosphorylation and amyloid-beta accumulation (Patil et al. 2014; Wang et al. 2019).

Despite its promise, metformin's clinical repurposing as a gerotherapeutic agent faces key challenges. Heterogeneity in study design, dosage, duration, and patient populations complicates data interpretation. Additionally, questions remain regarding its efficacy in non-diabetic individuals, optimal treatment windows, sex-specific responses, and long-term safety in elderly populations. Addressing these gaps will require robust, biomarker-informed clinical trials incorporating functional, molecular, and physiological endpoints of aging. Nonetheless, the integration of metformin into clinical geroscience represents a pragmatic and evidence-based strategy to decelerate aging and prevent multimorbidity. Its affordability, tolerability, and systemic pleiotropic actions make it an appealing candidate for immediate translational exploration, particularly in the context of preventive medicine and healthspan promotion.

16.5 Metformin in Chronopharmacology and Polypharmacy

The integration of metformin into chronopharmacology and polypharmacy frameworks represents a critical dimension in optimizing its therapeutic efficacy and minimizing adverse outcomes, particularly in aging populations with multimorbidity. Chronopharmacology—concerned with the effects of biological rhythms on drug action—has emerged as a key field in geroscience, given the profound circadian alterations observed during aging.

Metformin has been shown to interact with the circadian clock at multiple levels. It activates AMPK, which in turn modulates the expression and stability of core clock components, including cryptochrome circadian regulator 1 (CRY1) and basic helix-loop-helix ARNT-like 1 (BMAL1) (Um et al. 2011). Furthermore, metformin influences the rhythmicity of peripheral clocks, particularly in liver and muscle tissues, impacting glucose metabolism and mitochondrial function in a time-dependent manner (Barnea et al. 2012). This temporal modulation suggests that metformin's efficacy and tolerability may vary depending on the time of administration, highlighting the potential of chronotherapeutic strategies to enhance outcomes. Indeed, preclinical studies demonstrate that metformin administration during the active phase improves glycemic control and reduces hepatic steatosis more effectively than during the rest phase (Bailey 2017).

From a clinical perspective, few trials have directly evaluated the chronotherapeutic administration of metformin in humans. However, chronotype, sleep quality, and meal timing have all been shown to influence metforminMetformin response and glucose-lowering efficiency (Reutrakul and Van Cauter 2018). Thus, personalized dosing schedules aligned with circadian biology may be especially valuable for elderly patients or those with disrupted sleep-wake cycles. Continuously using more than five medications, whether prescribed or not, is known as polypharmacy (including vitamins, supplements, and herbal products). This indicates excessive polypharmacy when it exceeds ten or more. In nursing home populations, polypharmacy rates can reach 90%; therefore, controlling medications and their side effects is essential to preserving stable health during these patients' long-term care. In the context of polypharmacy, metformin's widespread use in elderly individuals with multiple comorbidities poses challenges and opportunities. Polypharmacy is associated with increased risks of adverse drug reactions, drug-drug interactions, and reduced adherence (Maher et al. 2014). Metformin is generally considered to have a favorable safety profile with minimal risk for hypoglycemia when used as monotherapy; however, its interactions with commonly co-administered drugs warrant consideration. For instance, metformin and proton pump inhibitors or H2 antagonists can exacerbate vitamin B12 deficiency, a concern particularly relevant in the elderly due to age-related malabsorption and the cognitive risks associated with cobalamin deficiency (de Jager et al. 2010). Additionally, co-administration with nephrotoxic agents such as nonsteroidal anti-inflammatory drugs (NSAIDs), contrast media, or certain diuretics can elevate the risk of lactic acidosis through compromised renal clearance (Salpeter et al. 2010).

Emerging data suggest that metformin may also modulate the pharmacokinetics or pharmacodynamics of other senotherapeutic or anti-aging agents. For example, when combined with rapamycin or acarbose, metformin exhibits additive or synergistic effects on lifespan extension and metabolic reprogramming in animal models, although such combinations necessitate close monitoring for adverse interactions (Harrison et al. 2014; Mannick et al. 2014). Furthermore, polypharmacy considerations are vital in frail or cognitively impaired older adults, where drug burden indexes such as the Drug Burden Index or anticholinergic load may influence decisions about metformin initiation or deprescribing. In such populations, metformin's low sedative or anticholinergic effect makes it an attractive agent, although careful renal function monitoring remains imperative (Hilmer and Gnjidic 2009).

16.6 Conclusion

Metformin offers a multi-target approach to postpone age-related degenerative processes and ARDs. It has a favorable safety profile and a strong cost-benefit ratio. New promising findings from studies such as the TAME and MILES trials suggest that it can be considered a foundational active pharmacological agent for interventions related to geropharmacology. The heterogeneity of the senescence phenotype at the organ and tissue level requires an individualized assessment of senoprotective and senotherapeutic metformin dosing in the aging population. Within the known reliable therapeutic range, an individual prophylactic dosage should be given, taking into account the person's existing or future possible ARDs risks. Moreover, to realize its full therapeutic potential, future research must refine our understanding of its chronopharmacology, interactions in the context of polypharmacy, and long-term effects on human health. In conclusion, optimizing metformin therapy within chronopharmacological and polypharmacy frameworks may enhance its efficacy as a senotherapeutic agent. Future clinical trials should integrate circadian biomarkers, drug interaction profiles, and principles of geriatric pharmacology to personalize the use of metformin in aging populations.

Author Declarations

Conflict of Interest The authors have reported no conflicts of interest. All the data reviewed in this chapter are from the public scientific literature.

Declaration of AI-Assisted Technologies in the Writing Process During this work, the authors used QuillBot to improve the language and readability. After using this tool/service, the authors reviewed and edited the content as needed and took full responsibility for the publication.

References

Algire C, Amrein L, Zakikhani M et al (2010) Metformin blocks the stimulative effect of a high-energy diet on colon carcinoma growth in vivo and is associated with reduced expression of fatty acid synthase. Endocr Relat Cancer 17:351–360. https://doi.org/10.1677/ERC-09-0252

Atasever E, Atayik MC, Çakatay U (2025) Senotherapeutics: milestones, innovations, and prospects. Adv Pharmacol. https://doi.org/10.1016/bs.apha.2025.01.021

Bailey CJ (2017) Metformin: historical overview. Diabetologia 60:1566–1576. https://doi.org/10.1007/s00125-017-4318-z

Bannister CA, Holden SE, Jenkins-Jones S et al (2014) Can people with type 2 diabetes live longer than those without? A comparison of mortality in people initiated with metformin or sulphonylurea monotherapy and matched, non-diabetic controls. Diabetes Obes Metab 16: 1165–1173. https://doi.org/10.1111/dom.12354

Barnea M, Haviv L, Gutman R et al (2012) Metformin affects the circadian clock and metabolic rhythms in a tissue-specific manner. Biochim Biophys Acta (BBA) - Mol Basis Dis 1822:1796–1806. https://doi.org/10.1016/j.bbadis.2012.08.005

Barzilai N, Crandall JP, Kritchevsky SB, Espeland MA (2016) Metformin as a tool to target aging. Cell Metab 23:1060–1065. https://doi.org/10.1016/j.cmet.2016.05.011

Ben Sahra I, Regazzetti C, Robert G et al (2011) Metformin, independent of AMPK, induces mTOR inhibition and cell-cycle arrest through REDD1. Cancer Res 71:4366–4372. https://doi.org/10.1158/0008-5472.CAN-10-1769

Bu Y, Peng M, Tang X et al (2022) Protective effects of metformin in various cardiovascular diseases: clinical evidence and AMPK-dependent mechanisms. J Cell Mol Med 26:4886–4903. https://doi.org/10.1111/jcmm.17519

Cai Y, Jun G, Zhuang X (2024) Metformin treatment reduces the incidence of osteoporosis: a two-sample Mendelian randomized study. Osteoporos Int 35:1089–1098. https://doi.org/10.1007/s00198-023-07013-0

Cameron AR, Morrison VL, Levin D et al (2016) Anti-inflammatory effects of metformin irrespective of diabetes status. Circ Res 119:652–665. https://doi.org/10.1161/CIRCRESAHA.116.308445

Campbell JM, Bellman SM, Stephenson MD, Lisy K (2017) Metformin reduces all-cause mortality and diseases of ageing independent of its effect on diabetes control: a systematic review and meta-analysis. Ageing Res Rev 40:31–44. https://doi.org/10.1016/j.arr.2017.08.003

Cebe T, Kızılyel F (2025) Risk of senescence, polypharmacy, and their outcomes in elderly cardiovascular disease patients. Adv Pharmacol 104:351–392. https://doi.org/10.1016/bs.apha.2025.02.011

Chen K, Li Y, Guo Z et al (2020) Metformin: current clinical applications in nondiabetic patients with cancer. Aging 12:3993–4009. https://doi.org/10.18632/aging.102787

Chen M, Fu Y, Wang X et al (2022) Metformin protects lens epithelial cells against senescence in a naturally aged mouse model. Cell Death Discov 8:8. https://doi.org/10.1038/s41420-021-00800-w

Chen Q, Hartman R, Dankiv L et al (2025) Game changer: navigating between challenges and hopes in geropharmacology. Adv Pharmacol. https://doi.org/10.1016/bs.apha.2025.02.007

Cheng F-F, Liu Y-L, Du J, Lin J-T (2022) Metformin's mechanisms in attenuating hallmarks of aging and age-related disease. Aging Dis 13:970–986. https://doi.org/10.14336/AD.2021.1213

de Jager J, Kooy A, Lehert P et al (2010) Long term treatment with metformin in patients with type 2 diabetes and risk of vitamin B-12 deficiency: randomised placebo controlled trial. BMJ 340: c2181. https://doi.org/10.1136/bmj.c2181

de Marañón AM, Díaz-Pozo P, Canet F et al (2022) Metformin modulates mitochondrial function and mitophagy in peripheral blood mononuclear cells from type 2 diabetic patients. Redox Biol 53:102342. https://doi.org/10.1016/j.redox.2022.102342

de Vries TJ, Kleemann AS, Jin J, Schoenmaker T (2023) The differential effect of metformin on osteocytes, osteoblasts, and osteoclasts. Curr Osteoporos Rep 21:743–749. https://doi.org/10.1007/s11914-023-00828-0

Decensi A, Puntoni M, Goodwin P et al (2010) Metformin and cancer risk in diabetic patients: a systematic review and meta-analysis. Cancer Prev Res (Phila) 3:1451–1461. https://doi.org/10.1158/1940-6207.CAPR-10-0157

El-Damanawi R, Stanley IK, Staatz C et al (2024) Metformin for preventing the progression of chronic kidney disease. Cochrane Database Syst Rev 6(6):CD013414. https://doi.org/10.1002/14651858.CD013414.pub2

Evans JMM, Donnelly LA, Emslie-Smith AM et al (2005) Metformin and reduced risk of cancer in diabetic patients. BMJ 330:1304–1305. https://doi.org/10.1136/bmj.38415.708634.F7

Hajimohammadebrahim-Ketabforoush M, Zali A, Shahmohammadi M, Hamidieh AA (2024) Metformin and its potential influence on cell fate decision between apoptosis and senescence in cancer, with a special emphasis on glioblastoma. Front Oncol 14. https://doi.org/10.3389/fonc.2024.1455492

Harrison DE, Strong R, Allison DB et al (2014) Acarbose, 17-α-estradiol, and nordihydroguaiaretic acid extend mouse lifespan preferentially in males. Aging Cell 13:273–282. https://doi.org/10.1111/acel.12170

Hilmer SN, Gnjidic D (2009) The effects of polypharmacy in older adults. Clin Pharmacol Ther 85: 86–88. https://doi.org/10.1038/clpt.2008.224

Hua Y, Zheng Y, Yao Y et al (2023) Metformin and cancer hallmarks: shedding new lights on therapeutic repurposing. J Transl Med 21:403. https://doi.org/10.1186/s12967-023-04263-8

Kuan Y-C, Huang K-W, Lin C-L et al (2017) Effects of metformin exposure on neurodegenerative diseases in elderly patients with type 2 diabetes mellitus. Prog Neuro-Psychopharmacol Biol Psychiatry 79:77–83. https://doi.org/10.1016/j.pnpbp.2017.06.002

Kulkarni AS, Gubbi S, Barzilai N (2020) Benefits of metformin in attenuating the hallmarks of aging. Cell Metab 32:15–30. https://doi.org/10.1016/j.cmet.2020.04.001

Liang D, Li Z, Feng Z et al (2022) Metformin improves the senescence of renal tubular epithelial cells in a high-glucose state through E2F1. Front Pharmacol 13. https://doi.org/10.3389/fphar.2022.926211

Liao W, Xu J, Li B et al (2021) Deciphering the roles of metformin in Alzheimer's disease: a snapshot. Front Pharmacol 12:728315. https://doi.org/10.3389/fphar.2021.728315

Liu Z, Zhang Y, Li D, Fu J (2024) Cellular senescence in chronic lung diseases from newborns to the elderly: An update literature review. Biomed Pharmacother 173:116463. https://doi.org/10.1016/j.biopha.2024.116463

Lv Z, Guo Y (2020) Metformin and its benefits for various diseases. Front Endocrinol (Lausanne) 11. https://doi.org/10.3389/fendo.2020.00191

Maher RL, Hanlon J, Hajjar ER (2014) Clinical consequences of polypharmacy in elderly. Expert Opin Drug Saf 13:57–65. https://doi.org/10.1517/14740338.2013.827660

Majithiya JB, Balaraman R (2006) Metformin reduces blood pressure and restores endothelial function in aorta of streptozotocin-induced diabetic rats. Life Sci 78:2615–2624. https://doi.org/10.1016/j.lfs.2005.10.020

Mannick JB, Del Giudice G, Lattanzi M et al (2014) mTOR inhibition improves immune function in the elderly. Sci Transl Med 6:268ra179. https://doi.org/10.1126/scitranslmed.3009892

Martin-Montalvo A, Mercken EM, Mitchell SJ et al (2013) Metformin improves healthspan and lifespan in mice. Nat Commun 4:2192. https://doi.org/10.1038/ncomms3192

Marycz K, Tomaszewski KA, Kornicka K et al (2016) Metformin decreases reactive oxygen species, enhances osteogenic properties of adipose-derived multipotent mesenchymal stem cells in vitro, and increases bone density in vivo. Oxid Med Cell Longev 2016:9785890. https://doi.org/10.1155/2016/9785890

Moiseeva O, Deschênes-Simard X, St-Germain E et al (2013) Metformin inhibits the senescence-associated secretory phenotype by interfering with IKK/NF-κB activation. Aging Cell 12:489–498. https://doi.org/10.1111/acel.12075

Ng TP, Feng L, Yap KB et al (2014) Long-term metformin usage and cognitive function among older adults with diabetes. J Alzheimers Dis 41:61–68. https://doi.org/10.3233/JAD-131901

Noto H, Goto A, Tsujimoto T, Noda M (2012) Cancer risk in diabetic patients treated with metformin: a systematic review and meta-analysis. PLoS One 7:e33411. https://doi.org/10.1371/journal.pone.0033411

Pan Q, Lu X, Zhao C et al (2020) Metformin: the updated protective property in kidney disease. Aging 12:8742–8759. https://doi.org/10.18632/aging.103095

Patil SP, Jain PD, Ghumatkar PJ et al (2014) Neuroprotective effect of metformin in MPTP-induced Parkinson's disease in mice. Neuroscience 277:747–754. https://doi.org/10.1016/j.neuroscience.2014.07.046

Pollak M (2010) Metformin and other biguanides in oncology: advancing the research agenda. Cancer Prev Res (Phila) 3:1060–1065. https://doi.org/10.1158/1940-6207.CAPR-10-0175

Rena G, Hardie DG, Pearson ER (2017) The mechanisms of action of metformin. Diabetologia 60: 1577–1585. https://doi.org/10.1007/s00125-017-4342-z

Reutrakul S, Van Cauter E (2018) Sleep influences on obesity, insulin resistance, and risk of type 2 diabetes. Metabolism 84:56–66. https://doi.org/10.1016/j.metabol.2018.02.010

Rotermund C, Machetanz G, Fitzgerald JC (2018) The therapeutic potential of metformin in neurodegenerative diseases. Front Endocrinol (Lausanne) 9:400. https://doi.org/10.3389/fendo.2018.00400

Salpeter SR, Greyber E, Pasternak GA, Salpeter EE (2010) Risk of fatal and nonfatal lactic acidosis with metformin use in type 2 diabetes mellitus. Cochrane Database Syst Rev 2010:CD002967. https://doi.org/10.1002/14651858.CD002967.pub4

Saxton RA, Sabatini DM (2017) mTOR signaling in growth, metabolism, and disease. Cell 168: 960–976. https://doi.org/10.1016/j.cell.2017.02.004

Steinberg GR, Carling D (2019) AMP-activated protein kinase: the current landscape for drug development. Nat Rev Drug Discov 18:527–551. https://doi.org/10.1038/s41573-019-0019-2

Tai S, Sun J, Zhou Y et al (2022) Metformin suppresses vascular smooth muscle cell senescence by promoting autophagic flux. J Adv Res 41:205–218. https://doi.org/10.1016/j.jare.2021.12.009

Turgut Ş, Andican G (2025) Senotherapeutic approach to age-related endocrine diseases. Adv Pharmacol 104:475–514. https://doi.org/10.1016/bs.apha.2025.02.009

Turgut Ş, Atasever E, Cebe T et al (2025) Senotherapeutic repurposing of metformin for age-related diseases and their signaling pathways. Mol Biol Rep 52:410. https://doi.org/10.1007/s11033-025-10524-0

Um J-H, Pendergast JS, Springer DA et al (2011) AMPK regulates circadian rhythms in a tissue- and isoform-specific manner. PLoS One 6:e18450. https://doi.org/10.1371/journal.pone.0018450

Wang Y, An H, Liu T et al (2019) Metformin improves mitochondrial respiratory activity through activation of AMPK. Cell Rep 29:1511–1523.e5. https://doi.org/10.1016/j.celrep.2019.09.070

Xiao Q, Xiao J, Liu J et al (2022) Metformin suppresses the growth of colorectal cancer by targeting INHBA to inhibit TGF-β/PI3K/AKT signaling transduction. Cell Death Dis 13:202. https://doi.org/10.1038/s41419-022-04649-4

Yan S, Dong W, Li Z et al (2023) Metformin regulates chondrocyte senescence and proliferation through microRNA-34a/SIRT1 pathway in osteoarthritis. J Orthop Surg Res 18:198. https://doi.org/10.1186/s13018-023-03571-5

Yang K, Wang X, Zhang C et al (2024) Metformin improves HPRT1-targeted purine metabolism and repairs NR4A1-mediated autophagic flux by modulating FoxO1 nucleocytoplasmic shuttling to treat postmenopausal osteoporosis. Cell Death Dis 15:795. https://doi.org/10.1038/s41419-024-07177-5

You G, Long X, Song F et al (2020) Metformin activates the AMPK-mTOR pathway by modulating lncRNA TUG1 to induce autophagy and inhibit atherosclerosis. Drug Des Devel Ther 14: 457–468. https://doi.org/10.2147/DDDT.S233932

Zhang T, Zhou L, Makarczyk MJ et al (2025) The anti-aging mechanism of metformin: from molecular insights to clinical applications. Molecules 30:816. https://doi.org/10.3390/molecules30040816

Chapter 17
Effects of Procaine, Gerovital H3, on Some Chronic Conditions Often Encountered in Older People

Gabriel-Ioan Prada, Daniela Grădinaru, and Ana-Gabriela Prada

Abstract Two medicines, procaine and its derivative Gerovital H3, have a long history of medical use: 120 years and 75 years, respectively. This chapter describes their mechanisms of action, both classical and pleiotropic. Procaine has been used as a local anesthetic and Gerovital H3 has shown several other effects that are briefly described along with some of the most recent studies that support them. The latter, sometimes together with procaine, has demonstrated antioxidant and free radical scavenger action, antiatherogenic effects, antidepressive activity, protection against deleterious effects of ionizing radiations, some positive effects in case of neoplasia, some improvement in mental symptoms. Several studies with negative results will be also presented, all of them with a critical evaluation and analysis. Many areas in the field of chronic conditions in older people are still not entirely understood.

Keywords Ana Aslan · Procaine · Gerovital H3 · Chronic conditions · Older people

17.1 Introduction

Even though the lifespan of humans increased significantly during the last decades, healthy lifespan did not follow the same rhythm as that of our chronological age. Although we live longer, the number of years we spend without diseases associated with advanced age did not increase proportionally. Various authors consider that European population has never been older than it is today and it is estimated that by the year 2050 over 30% of the population of our continent will be beyond the age of 60 years (EUROSTAT 2020) This positive evolution is connected to the social, economic, scientific and medical progress that we witnessed and it could be

G.-I. Prada
University of Medicine and Pharmacy Carol Davila, Bucharest, Romania

National Institute of Gerontology and Geriatrics Ana Aslan, Bucharest, Romania

Academy of Romanian Scientists, Bucharest, Romania

D. Grădinaru · A.-G. Prada (✉)
University of Medicine and Pharmacy Carol Davila, Bucharest, Romania

S. I. S. Rattan, C. Gezer (eds.), *Supplements and Drugs for Healthy Longevity*, Healthy Ageing and Longevity 22, https://doi.org/10.1007/978-3-032-11608-6_17

considered a triumph of civilization if there were no impact of these changes upon society, pensions and health insurance systems. "The problem of aging is not a problem, it is a pessimistic look at humanity's great triumph" (Daia and Stancu 2021).

Present scientific research is increasingly concerned with studying some potential pharmaceutical interventions that could "fight" aging. One of the paths of exploration is based on reevaluation of the existent geroprotective medication, with new methods and modern means. Procaine was introduced in medical practice under the name of "novocaine" and in the 1960s it became the first, and one of the most controversial medicines, dedicated to older people. Procaine being the main ingredient of Gerovital-H3 (GH3), an original medicine created by Ana Aslan and her colleagues between 1946 and 1956 after numerous experiments and clinical research.

Procaine, a molecule synthesized by the German chemist Alfred Einhorn in 1905, represents the prototype of local anesthetics. Besides its main anesthetic effect, procaine proved to demonstrate multiple mechanisms of action and outcomes, such as antioxidant and geroprotective. Ana Aslan intuited and discovered the "eutrophic" therapeutic actions of Procaine during a chronic and intermittent treatment of both animal and human subjects with a prophylactic and curative role. "Eutrophic" or "biotrophic" means beneficial for cellular metabolism and functions in general.

Gerovital-H3 (GH3), in injectable form, has as active ingredient Procaine hydrochloride along with excipients and stabilizers, and it needs to be stressed that its original formula (composition), developed by Ana Aslan and pharmacist Elena Polovrageanu, has remained unchanged to this day. GH3 was and has been used within the National Institute of Gerontology and Geriatrics in Bucharest, Romania, and in other centers throughout the country and also abroad, and this demonstrates its efficacy and safety that have been tested through long-term clinical experience. One variant of GH3 is Aslavital that contains in addition Vitamin B6.

New scientific evidence supporting beneficial effects of Procaine and GH3 continues to support its widespread use for the prevention of phenomena related to the so-called "premature aging", as well as for achieving and active longevity. GH3 is mainly based on procaine and it has the following composition: Procaine hydrochloride solution 2%, benzoic acid 0.12%, potassium metabisulfite 0.10%, disodium phosphate 0.01%, pH 3–4. To ensure a better stability of the procaine solution, pharmacist Elena Polovrageanu decreased its pH to 3.3 (as compared to 4.5) which allows a larger amount of procaine to be maintained in the body for a longer period, before it is hydrolyzed to *para*-aminobenzoic acid (PABA) and diethylaminoethanol (DEAE) (Gradinaru et al. 2017).

17.2 Mechanisms of Action

Procaine is an ester of PABA with DEAE. The sites of action of procaine at the cellular level are multiple, and its pharmacological properties depend on dosage and administration. As an anesthetic, procaine is a sodium channel inhibitor. It is known that by systemic administration in high doses procaine has an anesthetic effect. Moreover, it is known that of all local anesthetics, procaine is the least toxic (Blaustein and Goldman 1996; Noda 1971; Seeman 1972).

In the body, procaine readily crosses the blood–brain barrier and produces stimulatory effects in the central nervous system. Local anesthetic action manifests at high concentrations and is effective on nerve fibers and nerve endings. *In vivo* it has affinity for M2 muscarinic receptors in Rhesus monkeys. It produces short-lived emotional and sensory stimulation in the central nervous system. This provides evidence of cholinergic modulation through blockade of cerebral muscarinic cholinergic receptors. A study by Benson et al. (2004) performed on human subjects showed that, following intravenous administration, there is an increase in paralimbic cerebral blood flow, as determined by positron emission tomography (PET) (Benson et al. 2004). Procaine exhibits vasodilatory and antispasmodic effects—particularly on intestinal smooth muscle—and also induces myocardial depression (Georgescu et al. 2005).

GH3 was proven superior to procaine from a tolerance and efficacy perspective, while being tested by Ana Aslan, but also by doctors from other scientific institutions. Ingredients of GH3 also have a metabolic action, while the association of procaine with benzoic acid and potassium metabisulphytes is beneficial. In this formulation, benzoic acid serves as an antioxidant, while the potassium salt potentiates procaine's activity in the nervous system and the heart. Potassium ions facilitate procaine penetration into the cell (Aslan 1980, 1985; Aslan et al. 1986).

Pharmacokinetic studies have shown that blood levels of procaine are higher following GH3 administration than after procaine hydrochloride administration. Procaine is a biotrophic compound, as its *in vivo* hydrolysis by a pseudocholinesterase (in the blood) or a microsomal carboxylesterase (in various organs and tissues) produces primary and secondary metabolites that serve as precursors in the biosynthesis of molecules essential for cellular metabolism. Pharmacokinetic data, together with various reports in the literature, have enabled the delineation of the metabolic pathways involved in the *in vivo* hydrolysis of procaine. Procaine undergoes stepwise hydrolysis. In the first stage, DEAE and PABA (primary metabolites) are generated, and the second stage leads to the release of ethanolamine from DEAE (Georgescu et al. 2005; Aslan et al. 1985).

Procaine and its metabolites cross the blood–brain barrier, but in the brain, procaine is hydrolyzed slowly due to the low levels of procainase. The cerebrospinal fluid may serve as a reservoir and re-delivery compartment for procaine. The peak of absorptions and the area under the curve for absorption are higher for GH3 compared to its metabolites. Urinary excretion of procaine is pH-dependent. Renal clearance of procaine is of 2.25 in acidic urine and of 0.25 in alkaline urine. On rats, the studies were done according to age, bioavailability of procaine double marked with ^{99}Tc and ^{131}I, conditioned so that GH3, as well as absorption, biodistribution, metabolism and elimination of DEAE with ^{99}Tc. Absorption of GH3 and its metabolites is fast, the radioactive trace being present in all organs, according to their physiologic load and of the organs' circulatory context. In rats, procaine hydrolysis takes place in the liver, and its conditioned half-life as GH3 is 122 min, compared to 36 min for procaine hydrochloride. The elimination of DEAE follows first-order kinetics, with a half-life of 28 h and an elimination period of approximately 72 h (Dobre et al. 1992a, b, c).

Ethanolamine is a precursor in the biosynthesis of membrane phospholipids: phosphatidylethanolamine (cephalin) and phosphatidylcholine (lecithin). This second metabolic step is notable from a pharmacodynamic perspective. When, under normal or pathological conditions, the cellular demand for ethanolamine increases, the cell draws on its membrane phospholipid constituents—particularly phosphatidylserine—which it sacrifices. In these circumstances, GH3, by supplying exogenous ethanolamine, protects membrane integrity. Furthermore, ethanolamine undergoes progressive conversion into acetylcholine, a neurotransmitter that plays a key role in cognitive functions (Barbulescu-Poli 2015).

In turn, the PABA released by procaine's cleavage is incorporated into the structures of folic acid and tetrahydrofolic acid (the "H3" designation highlights procaine's vitamin-like action). PABA chiefly contributes to protecting hepatic metabolic capacity against carbon-tetrachloride toxicity and safeguarding neural tissue from levomepromazine-induced damage (a nootropic effect). Moreover, the long-term memory enhancement and improved emotional adjustment to environmental stress seen with GH3 are largely attributable to PABA. Together, these two primary metabolites underlie the parent compound's stimulatory effects on learning processes, its protection against pentetrazol-induced convulsions, and its increase of resistance to chemically induced hypoxia (Barbulescu-Poli 2015).

Kietzmann and Kaemmerer (1989) have tested the influence of oral administration of procaine hydrochloride on intermediary metabolism in rats and they have found that the ratio between acetyl coenzyme A and coenzyme A was increased in the liver and in the cerebellum. Also, procaine hydrochloride, Gerovital H3, as well as DEAE, have increased in a dose and time-dependent manner, the speed of integration of amino acids in proteins at the hepatic level, while PABA did not exert any effect (Kietzmann and Kaemmerer 1989).

17.3 Antioxidant Effect

Numerous pathological processes—arteriosclerosis, cancer, Alzheimer's disease, Parkinson's disease, diabetes mellitus, and macular degeneration—are associated with oxidative stress and cellular damage. Thus, oxidative stress does not appear to be the cause but rather a pathophysiological factor accompanying these conditions. It is noteworthy that there is no single definition of oxidative stress, nor is the way it contributes to the worsening of clinical symptoms fully explained. Moreover, there is currently no universally accepted strategy for its experimental assessment. Despite all these uncertainties, limiting the intensity of oxidative stress has become the most widespread and employed therapeutic target (Gradinaru 2013; Rusu 2001).

The fundamental concept of oxidative stress theory is easy to understand, which is why it quickly gained widespread popularity. More precisely, almost everyone knows that a chronic excess of free radicals leads to tissue damage and disease. This concept has naturally driven the broad use of various antioxidant vitamins and dietary supplements, which both the general population and patients with conditions linked to oxidative stress unfortunately take, in most cases, without medical prescription. At the same time, the role of reactive oxygen species (ROS) in cellular signaling is systematically ignored, which may help explain one reason adverse effects occur during antioxidant vitamin treatment (Sies 1997; Flora 2009; Gutteridge and Halliwell 2010; Mayne 2003).

The antioxidative action is attributed to procaine, but also to the other components of GH3 and of Aslavital: potassium salts, benzoic acid and glutamic acid (Gradinaru et al. 2017; Rusu and Lupeanu 1989).

Several studies have investigated the effect of procaine and GH3 in the systems that generate free radicals, *in vitro*. The inhibitory effect of these products over the production of the superoxide radical ($O_2^{\cdot -}$) was demonstrated by using the Nishikimi chemical system, which is composed of NADH-phenazine methosulfate (PMS)- nitroblue tetrazolium (NBT). Thus, GH3 and Aslavital in their different concentrations (0.2, 0.4, 0.6, 1 and 2 mM procaine), as well as their ingredients, inhibit in different modalities the generation of the superoxide radical (Gradinaru et al. 2017; Rusu and Lupeanu 1989).

By employing an enzymatic superoxide-generating system, consisting of xanthine and xanthine oxidase, researchers examined how these pharmaceutical preparations act on an endogenous source of free radicals that contribute to in vivo oxidative tissue damage during ischemia followed by reperfusion (Gradinaru et al. 2017; Gradinaru et al. 2009).

Superoxide radical ($O_2^{\cdot -}$) formed during the reaction between xanthine and xanthine-oxidase reduces 2-[4-iodophenyl]-3-[4-nitrophenol]-5-pheniltetrazolium chloride [INT] to formazans. Reduction of INT was recorded in the presence of various concentrations of procaine and Gerovital H3, compared to superoxide dismutase (SOD)—antioxidant enzyme that detoxifies the superoxide under physiological conditions.

In vitro studies carried out on cells isolated from blood (erythrocytes, lymphocytes, thrombocytes), on subcellular organelles isolated from rat tissue homogenates and also animal cells in cultures, have generated a significant number of positive results regarding the effect of procaine and GH3 on lipid peroxidation (Gradinaru et al. 2017).

Fluorescence electron microscopy studies have shown that in rats treatment with GH3 reduces the accumulation of lipofuscin—pigment of aging—in brain, testicles, heart and liver (Aslan et al. 1984a, b; Radaceanu et al. 1991; Revnic 1990).

An important number of studies have demonstrated the *in vivo* beneficial effects of the treatment with procaine and G H3 in laboratory animals, by modulating the expression of some antioxidant enzymes and the inhibition of lipid peroxidation. Thus, it was demonstrated the GH3 has a modulatory effect on the activity of SOD in rat brain. Also, chronic treatment with procaine, GH3 and Aslavital inhibits lipid peroxidation of mitochondrial fragments and rat brain homogenate (Gradinaru et al. 2017; Aslan and Enachescu 1980).

Studies carried out by Lupeanu (1999, 2000) showed that the alteration of mitochondrial function, followed by the accelerated degradation of purines in aging, is accomplished with the additional generation of ROS (the superoxide radical and hydrogen peroxide) at the mitochondrial level, in the cytosol, and peroxisomes. The increased production of ROS in aging can modulate the activity of certain cellular antioxidant protection systems, such as catalase. During aging, catalase activity increases significantly in the liver and heart, but decreases significantly in the kidneys of female Wistar rats. In male rats, liver enzyme activity decreases significantly during the aging process. Procaine and Gerovital H3 induce a significant increase in catalase activity in the liver and kidneys across all age groups, while reducing catalase activity in the heart of adult and aged female rats (Lupeanu 1999, 2000).

NADH oxidation in liver mitochondria, representing electron transport through the entire respiratory chain, decreased significantly in aged rats, concomitantly with the intensified formation of ROS. Treatment with procaine and Gerovital H3 induced stimulation of electron transport in the respiratory chain (NADH oxidation) and a significant reduction in the generation of ROS. This result also demonstrates the antioxidant action of procaine and GH3 *in vivo* (Lupeanu 1990; Lupeanu and Trandaburu 1994–1995).

17.4 Anti-atherogenic Actions

Several clinical studies have noted the anti-atherogenic effect of these geriatric products through their antithrombogenic action, by stimulating the fibrinolysis process, and, not least, through beneficial modifications of lipid metabolism. These effects have been demonstrated through numerous experimental and clinical investigations (Gradinaru et al. 2017; Aslan and Enachescu 1980; Aslan et al. 1984a, b; Rusu et al. 1996).

In an open-label, controlled clinical study conducted on 88 patients with systemic atherosclerosis who underwent 7 months of treatment with GH3 and injectable Aslavital, Rusu et al. (1996) have highlighted the hypolipidemic and antioxidant effect of these products (Rusu et al. 1996).

Also, through *in vitro* determinations performed by incubating native LDL isolated from human plasma in the presence of various concentrations of GH3, Aslavital, and procaine, Gradinaru et al. (2009) demonstrated their inhibitory effect on the formation of conjugated dienes and lipid hydroperoxides, the maximum inhibition being exerted by Aslavital, followed by GH3 and procaine. A significant inhibition of LDL oxidation was observed after 20 and 40 min from the induction of oxidation by copper ions in the presence of Aslavital at all tested concentrations, whereas GH3 produced a significant reduction in conjugated diene formation only at the highest concentration of 1 mM. After 180 min, GH3 and Aslavital inhibited conjugated diene formation only at 1 mM, while lower concentrations of 0.5 and 0.1 mM had no inhibitory effect on LDL oxidation. Procaine at 1 mM significantly inhibited LDL oxidation during the first 60 min (Gradinaru et al. 2017; Gradinaru et al. 2009). By using the same experimental model, the inhibitory effect of the two hydrolysis products of procaine, DEAE and PABA was also tested. An inhibitory action was evidenced only in the case of DEAE. Thus, at a concentration of 5 mM in the reaction medium, Aslavital inhibited lipid peroxidation by 46%, GH3 by 17%, and procaine and DEAE by 23% and 28%, respectively.

Studies have confirmed the "statin-like" action of procaine exerted by regulating the activity of hydroxy-methyl-glutaryl-coenzyme A (HMG-CoA) reductase, involved in cholesterol biosynthesis (Xu et al. 2003). A series of relatively recent clinical evidence confirms the hypolipidemic, statin-like effects of procaine, which regulates the activity of HMG-CoA reductase that is involved in cholesterol metabolism (Xu et al. 2003; Istvan and Deisenhofer 2001). In this circumstance, we could suggest that adding Aslavital or GH3 to hypolipidemic therapy would allow lowering statin doses and thus reduce the risk of adverse reactions.

17.5 Antidepressive Activity

A series of studies have highlighted the antidepressant effect of these products, based on the inhibition of monoamine oxidase activity (IMAO) (Zung et al. 1974; Yau 1974). In 1940, Phillpot demonstrated on liver homogenate the effect of procaine as an inhibitor of the oxidation of tyramine and adrenaline, using high concentrations in the incubation medium (33 mM) (Philpot 1940).

Studies were continued in the years of 1970–1980 by other researchers (Bucci 1972; Hrachovec 1972a, b), showcasing that for procaine and for GH3, the characteristic of IMAO in the brain, liver, and heart in the rat (Bucci 1972; Hrachovec 1972a, b).

Through MacFarlane's studies (MacFarlane and Besbris 1974; MacFarlane 1975), GH3 was included in the category of reversible and competitive inhibitors, while Stroescu (Stroescu et al. 1980; Stroescu 1988), through pharmacodynamic tests, highlights for procaine an IMAO type B (MacFarlane and Besbris 1974; MacFarlane 1975; Stroescu et al. 1980; Stroescu 1988).

Using *in vitro* and *in vivo* experimental models, Rusu et al. (1995) and Borsa et al. (2002) studied the effect of GH3, Aslavital, and procaine on the oxidative deamination of four substrates specific to MAO isoforms A and B, namely serotonin, tryptamine, tyramine, and benzylamine. *In vitro*, GH3 and Aslavital, at concentrations of 1–5 mM, significantly inhibit the oxidative deamination of serotonin and tryptamine, with no effect on benzylamine. Using tyramine as substrate, MAO activity is inhibited only by Aslavital. Procaine inhibits only the oxidative deamination of serotonin. After *in vivo* treatment, a slight inhibition of MAO activity in the brain is observed for all four substrates. Compared with procaine, higher percentages of inhibition are seen with GH3. These results place GH3 and Aslavital in the category of MAO-A inhibitors. The MAO-inhibitory effect was associated with inhibition of lipid peroxidation in rat brain homogenate and mitochondrial fractions, suggesting that MAO activity is inhibited by limiting free-radical reactions (Rusu et al. 1995; Borsa et al. 2002).

Clinically, chronic treatment with GH3 leads to improved mood and increased motivation, the removal of negative energy and irritability, acting as an MAOI without adverse effects (Aslan and Vrabiescu 1986; Balaceanu-Stolnici et al. 1996). From a clinical standpoint, patients exhibit improved mood and a marked reduction in depressive symptoms, driven by a pathophysiological mechanism that raises serotonin levels in synaptic clefts through a modulatory effect on MAO—characterized by mild, reversible, and competitive inhibition.

17.6 Protective and Antioxidant Effects in Case of Exposure to Ionizing Radiations

In human lymphocyte cultures, ionizing radiations (IR) significantly increases DNA damage and causes a decrease in reduced glutathione (GSH) levels and antioxidant enzyme activities (Prasad et al. 2005).

Antioxidant activity of procaine and its pharmaceutical derivative preparation Gerovital H3 has been reported in several *in vitro* şi *in vivo* studies. Chronic treatment with GH3 inhibits lipid peroxidation in rat brain and increases the activity of superoxide dismutase (SOD). Moreover, procaine and GH3 inhibit the generation of superoxide radicals ($O_2^{-\cdot}$) in various *in vitro* systems—enzymatic and non-enzymatic (Rusu and Lupeanu 1989; Gradinaru et al. 2009). In addition, procaine and GH3 induce a significant increase of the activity of catalase in liver and kidney of young, adult and old rats (Lupeanu 1999).

Several cellular processes, such as defense against oxidative stress, telomere shortening, DNA repair, and the inflammatory response, can contribute to the relationship between aging and radiosensitivity (Hernandez et al. 2015). In accordance with the "Free Radical Theory of Aging", also known as the "Oxidative Stress Hypothesis", oxidative stress is caused by the imbalance between the production of ROS and the capacity of the biological system to counteract, with adequate antioxidant defense, the oxidative damage to cell membranes and other structures such as lipids, lipoproteins, proteins, and DNA (Gorbunova et al. 2007; Kregel and Zhang 2007; Sies 2015). IR contributes in a continuous manner to generating ROS, which leads to an increase in cellular lesions and to ageing. The age- and pathology-dependent IR effect was investigated in a group of 202 healthy individuals and in 303 cancer patients; radiosensitivity increased in both study groups with chronological age (Schuster et al. 2018).

The Free Radical Theory of Aging argues that a reduced antioxidant capacity and a decreased capacity to repair damaged DNA are associated with aging (Garinis et al. 2008; Gorbunova et al. 2007; Moskalev et al. 2013; Muralidharan et al. 2017; Harman 1956). Thus, it can be accepted that ionizing radiation may induce more extensive DNA damage in peripheral blood mononuclear cells (PBMCs) from elderly donors than in cells isolated from young donors. As expected, there was a significant donor-age effect on the induction of DNA damage by irradiation. All irradiation doses (2, 4, and 8 Gy) induced a significantly higher number of DNA lesions in the "old" cells compared with the "young" ones.

Endogenous (native) DNA damage is known to be one of the mechanisms responsible for the increased risk of mutagenesis, being involved in cellular survival, vascular aging, cancer, and neurodegenerative diseases (Bautista-Nino et al. 2016; Garm et al. 2013a, b).

Pre-treatment of cells with GH3 before irradiation significantly modified the cellular response to irradiation, partially preventing DNA fragmentation induced by the different doses of X-ray radiation (significant interaction between radiation and GH3 treatment), both in young and in elderly subjects. This protective effect was significant at radiation doses of 4 and 8 Gy. In the experiment carried out on Jurkat lymphoblastic cells, the antioxidant effect of procaine and GH3 was tested by comparison with a natural compound, curcumin, whose membrane-level antioxidant action is well known (Margina et al. 2013, 2015).

The reduced number of endogenous fragmented DNA observed after 24 h of treatment with procaine could be explained by the inhibition of mitochondrial ROS production. Mitochondrial ROS also increases with advancing age, which could explain a more pronounced inhibition of ROS in the elderly (Kuka et al. 2013).

- Through these experiments we provide new evidence regarding the differences that exist between procaine and GH3 in terms of their antioxidant properties.
- GH3 and procaine have exhibited a dose-dependent antioxidant effect; the two products have inhibited the generation of free radicals and lipid peroxidation.
- GH3 has shown *ex vivo* a significant protective effect against DNA lesions produced by X-rays, in the PBMC cells isolated from young and older subjects.
- Procaine significantly reduced the native endogenous lesion of DNA in PBMC cells isolated from older subjects.

17.7 Positive Effects in Case of Neoplasia

In experiments that used the cell line MCF-7, obtained from breast cancer tumors, it was discovered that procaine is a demethylating agent that reduces with 40% the 5-methyl-cytosine from DNA, evaluated by high performance capillary electrophoresis (Lirk et al. 2012). Moreover, procaine can also demethylate the so-called "hypermethylated CpG islands", such as those situated in the promotor region of gene RARβ (*Retinoic acid receptor beta*—associated with initiation/progress of cancer), with the restoration of expression for genes whose expression is reduced or even suppressed (*silenced*). In addition, procaine inhibits the growth of cancer cells and behaves similarly to procainamide, which restores the expression of certain hypermethylated genes in prostate cancer cells and reduces the growth of xenograft tumors (human tumors transplanted into mice). It is important to note that the doses of procaine that achieve significant demethylating and growth-inhibitory effects are like those administered alongside antineoplastic drugs or radiotherapy. The most important finding from these experiments is the protective action of procaine against chemotherapy-related nephrotic and hepatic toxicities. Given its long-standing and safe use as a local anesthetic, with well-known pharmacological properties, procaine can proceed to preclinical and clinical studies for validation in the epigenetic therapy of cancer.

Also, in gastric cancer global measurement of DNA methylation has demonstrated that procaine significantly reduces the global level of methylation of DNA. The analysis of expression and activity of DNA-methyltransferases (DNMTs) has shown that procaine repressed their activity but not the expression of DNMT1/DNM T3A isoforms. Moreover, analyses of RT-qPCR elucidate the fact that procaine leads to CDKN2A and RARβ gene regulation, due to activation of the promotor that acts on them (Li et al. 2018).

Many preclinical studies have shown that local anesthetics have a direct inhibitory effect on tumor activity, including tumor cell survival, proliferation, migration, and invasive activity (Fan 2020, 2021; Liu et al. 2020; Gradinaru et al. 2021).

Gao in 2009 provided the first evidence that procaine is capable of reactivating, in lung cancer cells, Wnt inhibitor factor-1 (WIF-1) "silenced" by promoter hypermethylation (Gao 2009).

So, *in vitro*, procaine and procainamide can inhibit the signaling pathway for canonic Wnt by demethylation of promoter WIF-1 in lung cancer cells type H460 and A549. Procaine *in vivo* antitumor efficacy has been tested mainly using murin model of cancer cells NSCLC10 (*non-small-cell lung carcinoma*). It was observed that in nude mice inoculated with cells A549 or NCIH1975, treatment with procaine (50 mg/kg/zi) significantly attenuated tumor growth, while a higher dose did not produce any change, and this result was in accordance with *in vitro* data that showed that a small dose of procaine (100 nM) inhibits cellular proliferation of the two types of lung cancer cells. In addition, it was noticed that procaine could inhibit the expression of tumor PCNA (a marker of cell proliferation) (Gao 2009).

Bone cancer called osteosarcoma, occurs and develops from immature bone cells that are normally formed in the new bone tissue, generally developing in lower limbs, mainly femur, tibia, around knee joint. Recent studies (Ying et al. 2017) have demonstrated that the inhibitor effect of procaine is achieved by adjusting microARN (miARN), MiR-133b proving to be a tumor suppressor in osteosarcoma. In these studies, MG63 cells have been treated with various concentrations of procaine, and cell viability, apoptosis and expression of miR-133b have been detected by specific methods: MTT, flux cytometry and, respectively, qRT-PCR, then the cells being transfected with the inhibitor miR-133b and treated with 2 μM procaine. The results demonstrated that procaine significantly inhibited cellular viability, promoted apoptosis and the level of expression of miR-133b in a dose-dependent manner. Moreover, during these studies it was noticed that the inhibitor of procaine and of miR-133B dramatically reversed the effects of procaine on cell viability, apoptosis and migration. In addition, procaine significantly decreased the level of p/t-AKT, p/t-ERK and p/t-S6, while the inhibitor of procaine and of miR-133b triggered these effects.

A recent study (Yang et al. 2022) demonstrates that procaine is an anti-metastatic agent in an experimental cellular hepatocarcinoma model, by inhibiting the transition of cancer cells from the epithelial to the mesenchymal phenotype through the activation of markers specific to the two cell types. As a mechanism, procaine inhibits the c-Met receptor (with tyrosine-kinase function) and thus the oncogenesis-specific signaling pathways, and consequently inhibits HGF (hepatocyte growth factor). Another mechanism that highlights the anti-cancer action of procaine is its direct interaction with the DNA molecule. In a spectroscopy and molecular modeling study, Ali (2018) used molecular docking on five different B-DNA structures (extracted from the *Protein Data Bank*) and showed that procaine binds to the adenine-thymine (AT)-rich regions of the five calf thymus B-DNA structures (Rusu et al. 1992). They also discovered that procaine, alone or in combination with doxorubicin, acts as an electron donor to DNA bases.

17.8 Other Effects in Older People

Oettmeier et al. 2019 have reported the clinical use of high-dose procaine hydrochloride infusions with sodium bicarbonate as an additive has been reported for the treatment of various acute illnesses, chronic pain, and inflammatory conditions. Xu in 2016 investigated the effects of daily Gerovital H3 tablet use for 3 months on alleviating mental symptoms and improving health-related quality of life in elderly Chinese subjects. The study was randomized, double-blind with placebo control, and included 100 eligible participants, men and women, aged between 50 and 89 years. Treatment with GH3 showed positive results by supporting mental health and improving general health and well-being, while the cognitive function of the elderly was restored. Mean levels of low mood and anxiety (assessed with the *Self-Rating Depression and Anxiety Scales*) were both decreased, and the prevalence of clinical anxiety diminished (Xu 2016).

17.9 Studies with Negative Results

In a *review* article requested by the *National Institute on Aging* within the *National Institute of Health* (USA), Ostfeld et al. (1977) evaluated the scientific literature regarding the use of procaine in systemic administration in aging and associated chronic diseases, including data published in 285 articles and books that described treatments performed on more than 100,000 patients. Except for a possible antidepressant effect, the authors found no convincing evidence of beneficial effects of procaine or GH3 in the treatment of aging-associated diseases in older patients (cited by Perls 2013). The article published in 1977 by Ostfeld et al. represents a review of several studies on procaine and GH3. They concluded that the preparations would have no effect other than a slight and reversible inhibition of MAO (monoamine oxidase), thereby exerting a mild antidepressant effect.

It is necessary to examine this article carefully. Unfortunately, the authors do not demonstrate objectivity. They consider that the evidence in favor of positive effects is based on studies using a small number of subjects. But the negative studies, which they regard as providing reliable data, also used a small number of subjects, 10 in each group, or 40 or 50. Of course, such a small number of participants in an experiment is beset by errors, especially type II errors—that is, false-negative results. Moreover, the duration of these studies is very short: from a few weeks to 3 months. This fact shows that the investigators who designed the studies did not understand that the effects of procaine and GH3 are not spectacular and do not manifest rapidly, as they may have been accustomed to with other medications. Furthermore, when they investigated the effects on neurocognitive function, they used very heterogeneous groups of patients at various stages of progression. On the other hand, the tests used (which are still available today) do not have very high precision and therefore cannot detect variations that are not sufficiently large. Although the changes in these test values are not large (due to low sensitivity), it is nevertheless possible, from a clinical point of view, to observe a variation (positive or negative). The conclusion is that the articles used in the Ostfeld et al. analysis cannot offer arguments for or against, except for the antidepressant effect. Regarding this last aspect, antidepressant action is very important for several reasons. First, according to the mechanism of action of GH3, it is rather a regulator of serotonin levels in synaptic clefts. Although the clinical consequence is not a major antidepressant effect, the great advantage lies in the absence of significant adverse effects encountered with other types of antidepressant medications. In this context, part of the positive effects observed in neurocognitive disorders may indeed be due to the antidepressant action because neurocognitive impairment is often associated with depressive symptomatology, especially in the early stages of dementia. It should be noted that the prevalence of depression in the elderly is higher than is currently believed, particularly due to the somatization of the clinical manifestations of mood disorders in this population group. Moreover, depression itself is very important because it represents a risk factor for cardiac conditions and neurocognitive disorders. Patients with depression develop these organic conditions more frequently and with greater prevalence. Of course, GH3 cannot be used in major depression of psychotic intensity, but it can be successfully used in mild forms or for treating depressive symptoms, if necessary, in combination with other antidepressants to allow reduction of their doses (followed by a decrease in the risk of adverse reactions).

Thirty years later, Szatmari and Bereckzi in 2008 published in the prestigious journal *Cochrane Database Systematic Reviews* a study in which they evaluated, each of them independently, the efficacy and adverse events of treatment with procaine and GH3, used for improving cognitive function in patients with dementia, who were included in double-blind, randomized trials (clinical test studies) conducted before 1990. The pooled data from the two studies showed an unfavorable effect of procaine treatment in terms of adverse effects caused. In the case of patients with dementia, a single study suggested a negative effect, whereas two trials conducted on healthy elderly subjects suggested a positive effect of procaine on cognitive function. The authors concluded that the evidence regarding the negative effects of procaine and of procaine-based preparations is stronger than the data reporting its benefits in the prevention and/or treatment of dementia or cognitive impairment. Again, this study too suffers from a lack of objectivity. The authors have chosen primarily articles with negative results, all based on experiments using a small number of patients and consequently burdened by a very high risk of type II errors—false-negative results. Moreover, the evaluations rely on tests that do not have very high sensitivity, thus introducing yet another source of type II errors.

Zaric and Pace in 2009 have investigated the *Cochrane Central Register of Controlled Trials* database to study the frequency of transient neurologic symptoms (TNS) (painful pathology that appears immediately after the postoperative period) and of neurological complications after neurological complications after spinal anesthesia with lidocaine compared to other local anesthetics, including procaine. The risk of developing TNS after spinal anesthesia with lidocaine was higher than when bupivacaine, prilocaine, and procaine were used. Here the issue is the direct contact between the nervous system and the anesthetic, a fact that does not occur in the case of therapy with GH3.

Ghafari in 2012 in a prospective, randomized, double-blind clinical trial conducted on 110 patients (aged 20–70 years), evaluated the effect of lidocaine compared to that of procaine on the cognitive deficit manifested after coronary artery surgery. In the group treated with procaine, the total neurocognitive score decreased significantly compared to the preoperative score and compared to the group treated with lidocaine. Again, a study of very low scientific value given the very broad age spectrum: from 20 years to 70 years.

Takenami in 2012, set out to compare the neurotoxicity of procaine, bupivacaine, levobupivacaine, and ropivacaine administered intrathecally in rats. Although all four local anesthetics caused identical neurotoxic lesions initiated in the dorsal root and extending into the posterior horn through axonal degeneration, bupivacaine was the most neurotoxic of them, and neurotoxicity increased proportionally with the volume of anesthetic administered, in the order: procaine > levobupivacaine > ropivacaine. The same observation regarding the mode of administration: GH3 does not come into direct contact with the nervous system, so the risk of neurotoxicity is greatly reduced (if not entirely absent, given the clinical experience from treating over one million patients over the years).

Yilbas in 2018 studied the effect of injected procaine into the joints in the knee cartilage and in the *synovium* of Sprague-Dawley rats. The results showed no significant differences in inflammation (assessed histologically) between the procaine group and the control group treated with physiological saline for any of the treatment periods (after 1, 2, 7, 14, and 21 days). Moreover, no significant differences were detected between the same treatment groups regarding the percentage of apoptotic chondrocytes.

17.10 Discussion and Concluding Remarks

The idea of "youth without aging and life without death" has a long history in mankind. Perhaps few people are aware that this has already been achieved by cancer cells: they are always "young" (do not become mature and this is the reason they cannot fulfil their role in the host tissue) and never die provided adequate nutrients are offered. Of course, nobody wants that. Actually, the achievable goal, for the time being, is to maintain the desired level of functioning for each person as long as possible. This concept could be summarized by the expression "active longevity", in other words maintaining the desired level of independence to advanced age. To date we do not have the necessary means to maintain complete health up to the end of life and almost all older people have chronic conditions that often reduce their capacity of functioning independently. This could offer a more realistic perspective regarding problem-solving when dealing with older people. As it is often said the main objective of medical care is to postpone disability-years as closer as possible to the end of life. In older people often it is more important to address quality of life than complete health. As many physicians who deal with older people know we do treat most of the diseases but cure only a few.

Taking into account previous aspects we might understand better the nature of effects of GH3 and procaine in conditions often encountered in older people. They do not exert a fast or sudden influence on chronic diseases. Sometimes it takes several months to see a clinical meaningful effect. Moreover, we do not always have measurement instruments sensitive enough to show a positive or negative change, even though it can be observed clinically.

Several experiments demonstrated antioxidant capacity of GH3 and procaine. This is very important since free radicals are involved in chronic degenerative diseases that have a significant impact on older people: atherosclerosis, Alzheimer's disease, Parkinson's disease. The free radicals scavenging effect needs a long time to be observed since the elimination of these deleterious products is not sudden but gradual and the repair mechanisms of the body need several months to produce improvements observable clinically. Since the destructions generated by free radicals accumulate over long periods of time it is recommended that the prophylactic therapy with GH3 starts around the age of 40 years. Degenerative diseases usually develop over a long period. In Alzheimer's disease the present aim is to identify a so-called "brain at risk" that most of time occurs around the age of 40 years without symptoms or signs of disease. Moreover, it is well known that atherosclerotic lesions develop early in life, sometimes even around the age of 20 years. This could support the idea that a prophylactic treatment with these free radical scavengers should start even during adulthood.

Another important action of GH3 and procaine is its antidepressive effect. Depression is very important in older people and most of the time underdiagnosed. It is important because many studies in cardiology have shown that patients with depression develop more often and at a younger age myocardial infarction. If they had such an infarction and present untreated depression, they are more prone to develop another myocardial infarction. In addition, depression is a risk factor for Alzheimer's disease since the depressed patients develop more often and at a younger age this degenerative disease. Depression in older people is often overlooked because this category of patients does not express depressive symptoms, but somatic manifestations without any significant changes at physical examination—it is the so-called "somatization of depression". Consequently, depression is often a late diagnosis in patients with advanced age. As it was shown in previous sections, procaine and Gerovital H3 are weak, reversible and competitive inhibitors of MAO—the enzyme that breaks-down serotonin in synaptic cleft. Due to these characteristics, they do not create very high and dangerous levels of serotonin and do not produce the classical "serotoninergic syndrome" observed with the use of other older monoamine-oxidase inhibitors like tranylcypromine. As it is often said many things come at a cost. Gerovital H3 and procaine are weak, reversible and competitive MAO inhibitors and therefore do not exhibit the dangerous manifestations of serotoninergic syndrome but their anti-depressive effect is not very strong and cannot be used in major depression being efficient only in mild forms of disease. Depression being a very frequent and often ignored diagnosis in older people it warrants a preventive approach treating this condition even in milder forms to avoid progression to late more complicated stages. Maybe the antidepressive effect explains, at least in part, why many patients report feeling "younger"—perhaps meaning that their energy level is higher.

Clinically it was observed that a long-term treatment with GH3 has the tendency to normalize the values of cholesterol and triglyceride in mild to moderate forms of dyslipidemia. Initially this was less understood but relative recent studies in early 2000 demonstrated that procaine and GH3 have a statin-like effect inhibiting the enzyme involved in synthesis of cholesterol: hydroxy-methyl-glutaryl Coenzyme A reductase. The inhibition is not as strong as a classical statin but it does not have the well-known adverse reactions of this class of medicine: rhabdomyolysis and increase in liver enzymes. Therefore, GH3 can be used as an adjunct to classic statin therapy in order to reduce the doses of those medicines and to decrease the risk of side effects.

The effects of GH3 have been demonstrated in relation to viability of cells. This raised the problem of a possible stimulation of neoplastic cells. As previously shown several studies highlighted the anti-cancer effects of procaine, GH3, as well as other local anesthetics. Even though this activity is not strong enough to warrant their use as anti-neoplastic agents it proves that they do not stimulate the development of cancers, on the contrary.

An interesting aspect was related to the effects of GH3 on cells that were subjected to the action of ionizing radiations. Recent studies performed by an international team of scientists in Germany, as described through this chapter, have proven a radio-protective activity if the cells have been pre-treated with GH3. This is an important finding and supports the need for a preventive treatment with this medicine.

A number of studies claimed that procaine and GH3 had no significant effect. Unfortunately, these studies use a small number of patients and over a short period of time and also measure the outcome with instruments that are not very sensitive. Some scientists tried to identify effects on cognitive function (dementia, Alzheimer's disease) in a small number of subjects, with a large spectrum of ages, after a few months of treatment. This objective was not feasible since the effects on cognitive function take a lot of time to be noticeable and the group of subjects need to be as homogenous as possible taking into account their small number. Moreover, the available instruments are not sensitive enough in order to identify effects over short periods. Most of these studies have a type II error—could not identify and effect even though the effect could be present.

GH3 can induce an allergic reaction. This is why we recommend a test before starting the therapy: injection of 1 ml medicine subcutaneously on the anterior aspect of forearm. It needs to be evaluated after 24 h. If there will be a red maculo -papule (a reddish raised area of the skin, similar to intra-dermo-reaction to PPD for tuberculosis) with a diameter larger than 20 mm, then the subject is allergic and should not start the treatment. Careful interpretation should be considered because a reddish coloration immediately after injection is normal since Gerovital H3 produces also a small vasodilation. Second test is injection of 1 ml of solution in the brachial triceps. The number of allergic reactions is very small, something around 1 in 7000 cases. We have over 40 years of direct clinical experience with this medicine and have seen only a few cases of allergic reactions, small and self-limited. There is another phenomenon that can occur in a small number of patients during the first doses of injectable treatment: some subjects report mild dizziness after injection, a sensation that disappears spontaneously after 5 or 10 min and after a week of therapy. Therefore, it is advisable that patients who follow an injectable therapy to lie on a couch for about 15 min after injection.

There is one conceptual problem that often leads to a lot of confusion: the difference between chronologic and biologic age. Many try to superimpose these two notions. Actually, they are completely different. Chronologic age is merely a passage of time, it does not represent an "aging", while there is a biologic aging meaning the loss of functional reserve. Perhaps it is not appropriate to speak about "biologic age" since the changes that occur in older people take place at very different chronologic ages. For example, wrinkles, even though are common in older people, do not occur in every person and at the same age. There are very old persons that have only a few wrinkles. The same situation could be seen regarding the greying of the hair: there are very old persons with very little grey hair. Perhaps it is the time to make a change in our perspective concerning "age" and risk factors in general. We all learn in medical school that there are two main categories of risk factors: non-modifiable and modifiable. The most important non-modifiable risk factor is considered "age" (meaning "chronologic age"). But maybe it is not exactly true. If one considers that "age" is the most important non-modifiable risk factor most of all older people should have the same transformations at roughly the same age. But there is a huge variability in the transformations that accompany advanced age, they occur at different ages and in various amounts. Maybe a change in the concept will create several advantages. If we consider that not the "age" (chronologic) is the real risk factor but the "time of exposure to other risk factors", we might have some benefits. First, we remove one risk factor from the non-modifiable category ("chronologic age") and transfer it into the modifiable category ("time of exposure to other risk factors"). This is important since a non-modifiable risk factor is actually a label and we cannot do anything about it (we cannot reverse the chronologic age). The consideration of the time of exposure as the real risk factor will also stimulate research because if we might control all known risk factors and the behavior of a disease will not change, there might be another risk factor (or risk factors) that we should look for. The confusion between age and time of exposure could occur because the longer one lives the longer will be exposed to risk factors and the correlation between chronologic age and diseases will be very strong but it will obscure the real "culprit": time of exposure to other risk factors (other than "age").

In conclusion, one could say that the current therapies, including that with GH3, does not "make people younger", cannot reverse the time and send subjects to a younger age. What treatment with GH3 can do is preventing the deleterious effects produced by the accumulation of free radicals—both primary prevention and secondary prevention, eliminating the excess of these free radicals and giving recovery mechanisms the opportunity to act. It is also a mild anti-depressive medicine, a mild statin-like therapy for dyslipidemia. It does not stimulate development of cancer cells, it protects against the effects of ionizing radiations, but it does not replace classical medicines being mainly an adjunct to classical therapies that, if associated, could reduce the therapeutic dosages and diminish the risk of adverse effects.

References

Ali MS (2018) Comprehensive exploration of the anticancer activities of procaine and its binding with calf thymus DNA: a multi spectroscopic and molecular modelling study. RSC Adv 8(17): 9083–9093

Aslan A (1980) Theoretical bases of procaine therapy (Gerovital H3 and Aslavital) in the prophylaxis of ageing. Rom J Gerontol Geriatr 1:5–15

Aslan A (1985) Specifications regarding the technique and action of GerovitalH3 treatment after 34 years of usage. Rom J Gerontol Geriatr 6(1):3–11

Aslan A, Enachescu G (1980) Researches on the antithrombophylic activity of the biotrophic therapy with GH3 and Aslavital. Rom J Gerontol Geriatr 1:185–204

Aslan A, Vrabiescu A (1986) Gerovital H3- Produs original "Prof. dr. Ana Aslan". Ed. MIC, Bucuresti

Aslan A, Matei C, Radaceanu V, Apostol G (1984a) The effects of the preventive and curative Aslavital treatment on experimental atero-arteriosclerosis in rabbits. I. Effects on the time course of the development of hypercholesterolemia and hypertriglyceridemia. Rom J Gerontol Geriatr 5:171–182

Aslan A, Radaceanu V, Vrabiescu A, Matei-Vladescu C (1984b) The effect of Gerovital H3 and Aslavital on the lipofuscin pigment from old brain, heart and testis, evaluated spectrofluorimetrically. Rom J Gerontol Geriatr 5:147–155

Aslan A, Cofaru S, Strungaru C (1985) Study on Gerovital H3 pharmacokinetics in relation to the administration route. Rom J Gerontol Geriatr 6(4):303–307

Aslan A, Balaceanu-Stolnici C, Manoiu A (1986) A double blind study on the antidepressive effect of Gerovital H3 and Aslavital in the elderly. Rom J Gerontol Geriatr 7:79–88

Balaceanu-Stolnici C, Covic M, Manoiu A, Vrabiescu M (1996) Double blind study concerning the anti-depressive effects and the clinical tolerance of Gerovital H3 without potassium metabisulphite. Rom J Gerontol Geriatr 17:46–62

Barbulescu-Poli T (2015) In Memoriam – Ana Aslan, volum alcătuit sub egida Institutului Naţional de Gerontologie şi Geriatrie "Ana Aslan", Editura Istros- Muzeul Brăilei (1997) și reeditat sub egida INGG „Ana Aslan" şi a Fundaţiei „Ana Aslan", Editura Curtea Veche, București

Bautista-Nino PK, Portilla-Fernandez E, Vaughan DE, Danser AH, Roks AJ (2016) DNA damage: a main determinant of vascular aging. Int J Mol Sci 17(5)

Benson BE, Carson RE, Kiesewetter DO, Herscovitch P, Eckelman WC, Post RM, Ketter TA (2004) A potential cholinergic mechanism of procaine‘s limbic activation. Neuropsychopharmacol 45:30–37

Blaustein MP, Goldman DE (1996) Competitive action of calcium and procaine on lobster axon. A study of the mechanism of action of certain local anesthetics. J Gen Physiol 49(5):1043–1063

Borsa C, Rusu C, Gradinaru D, Ionescu C (2002) Gerovital new formula and Aslavital effect on monoamine oxidase activity in mitochondrial fraction from rat liver and brain. Rom J Gerontol Geriatr 24:7–23

Bucci L (1972) The MAO inhibitory activity of procaine, international symposium of gerontology, Bucharest, Romania; June 26–27, 111

Daia A, Stancu S (2021) Population aging and the strength of the European Union, International Journal of Multidisciplinary and Current Research; Motto: "The problem of aging is not a problem, but a pessimistic view of the great triumph of mankind." Frank W. Notestein. https://www.researchgate.net/publication/350603844_Population_Aging_and_the_Strength_of_the_European_Union

Dobre V, Lungu S, Turcu I, Cucu M (1992a) A study of double labelled procaine with ^{99}mTc and ^{131}Iodine biodistribution under the trade mark of Gerovital H3. Rom J Gerontol Geriatr 13:39–46

Dobre V, Turcu I, Cucu M, Mihai V (1992b) The pharmacokinetics of Gerovital H3 and its metabolites old rats. Rom J Gerontol Geriatr 13:61–74

Dobre V, Turcu I, Cucu M, Mihai V (1992c) Experimental pharmacokinetic studies concerning DEAE labelled absorption, biodistribution, metabolization and elimination. Rom J Gerontol Geriatr 13:75–84

EUROSTAT (2020) Ageing Europe – statistics on population developments. https://ec.europa.eu/eurostat/statistics-explained/SEPDF/cache/80393.pdf. Accessed on July 2025

Fan X (2020) Effects of anesthesia on postoperative recurrence and metastasis of malignant tumors. Cancer Manag Res 12:7619–7633

Fan X (2021) Local anesthetics impair the growth and self-renewal of glioblastoma stem cells by inhibiting ZDHHC15-mediated GP130 palmitoylation. Stem Cell Res Ther 12(1):107

Flora SJ (2009) Structural, chemical and biological aspects of antioxidants for strategies against metal and metalloid exposure. Oxid Med Cell Longev 2(4):191–206

Gao Z (2009) Procaine and procainamide inhibit the Wnt canonical pathway by promoter demethylation of WIF-1 in lung cancer cells. Oncol Rep 22(6):1479–1484

Garinis GA, van der Horst GT, Vijg J, Hoeijmakers JH (2008) DNA damage and ageing: new-age ideas for an age-old problem. Nat Cell Biol 10(11):1241–1247

Garm C, Moreno-Villanueva M, Burkle A, Larsen LA, Bohr VA, Christensen K, Stevnsner T (2013a) Genetic and environmental influence on DNA strand break repair: a twin study. Environ Mol Mutagen 54(6):414–420

Garm C, Moreno-Villanueva M, Burkle A, Petersen I, Bohr VA, Christensen K, Stevnsner T (2013b) Age and gender effects on DNA strand break repair in peripheral blood mononuclear cells. Aging Cell 12(1):58–66

Georgescu M, Rusu C, Revnic F, Răchită M, Lupeanu E, Valuch A, Andrei V, Grădinaru D, Borşa C, Ionescu C, Opriş S (2005) Gerovital H3 şi Aslavital—mecanisme geroprotectoare şi antioxidante la nivel celular şi molecular. Rom J Gerontol Geriatr 26(4):27–45

Ghafari R (2012) Cognitive deficit in first-time coronary artery bypass graft patients: a randomized clinical trial of lidocaine versus procaine hydrochloride. Perfusion 27(4):320–325

Gorbunova V, Seluanov A, Mao Z, Hine C (2007) Changes in DNA repair during aging. Nucleic Acids Res 35(22):7466–7474

Gradinaru D (2013) Biochimia stresului oxidativ în patologia asociată îmbătrânirii. Editura Ars Academica, Bucureşti

Gradinaru D, Margina D, Borsa C (2009) In vitro studies regarding the antioxidant effects of procaine, Gerovital H3 and Aslavital. Rev Roum Chim 54(9):761–766

Gradinaru D, Borşa C, Ionescu C, Unguiranu A, Margină D (2017) Mecanisme biochimice implicate în acțiunea antioxidantă a Gerovitalului H3, Farmacist.ro, Anul XIII; Vol. 1, Nr 174, 16–22

Gradinaru D, Ungurianu A, Margina D, Moreno-Villanueva M, Bürkle A (2021) Procaine-the controversial geroprotector candidate: new insights regarding its molecular and cellular effects. Oxid Med Cell Longev 2021:3617042. https://doi.org/10.1155/2021/3617042. PMID: 34373764; PMCID: PMC8349289

Gutteridge JM, Halliwell B (2010) Antioxidants: molecules, medicines, and myths. Biochem Biophys Res Commun 393(4):561–564

Harman D (1956) Aging: a theory based on free radical and radiation chemistry. J Gerontol 11(3): 298–300

Hernandez L, Terradas M, Camps J, Martin M, Tusell L, Genesca A (2015) Aging and radiation: bad companions. Aging Cell 14(2):153–161

Hrachovec JP (1972a) Inhibitory effect of Gerovital H3 on monoaminoxidase of rat brain, liver and heart. Physiologist 15:3–20

Hrachovec JP (1972b) Inhibitory effect of Gerovital H3 on rat brain monoamine oxidase. Fed Proc 31(2):604

Istvan ES, Deisenhofer J (2001) Structural mechanism for statin inhibition of HMG-CoA reductase. Science 292(5519):1160–1164. https://doi.org/10.1126/science.1059344. PMID: 11349148

Kietzmann M, Kaemmerer K (1989) Effect of procaine and procaine metabolites on coenzyme a and acetyl coenzyme a concentration in various tissues of the rat. Z Alternsforsch 44(4): 211–217

Kregel KC, Zhang HJ (2007) An integrated view of oxidative stress in aging: basic mechanisms, functional effects, and pathological considerations. Am J Physiol Regul Integr Comp Physiol 292(1):R18–R36

Kuka S, Tatarkova Z, Racay P, Lehotsky J, Dobrota D, Kaplan P (2013) Effect of aging on formation of reactive oxygen species by mitochondria of rat heart. Gen Physiol Biophys 32(3):415–420

Li YC, Wang Y, Li DD, Zhang Y, Zhao TC, Li CF (2018) Procaine is a specific DNA methylation inhibitor with anti-tumor effect for human gastric cancer. J Cell Biochem 119(2):2440–2449

Lirk P, Berger R, Hollmann MW, Fiegl H (2012) Lidocaine time- and dose-dependently demethylates deoxyribonucleic acid in breast cancer cell lines in vitro. Br J Anaesth 109(2): 200–207

Liu H, Dilger JP, Lin J (2020) Effects of local anesthetics on cancer cells. Pharmacol Ther 212: 107558

Lupeanu E (1990) NADH-dehydrogenase activity in the mitochondrial fraction of rat liver. Gerovital H3 and Aslavital effect on NADH oxidation. Rom J Gerontol Geriatr 11:207–217

Lupeanu E (1999) Effect of procaine and Gerovital H3 treatment on purine catabolism in aging. Rom J Gerontol Geriatr 21:22–32

Lupeanu E (2000) Influence of procaine and Gerovital H3 treatment on fatty acids oxidation in rat liver of different ages. Rom J Gerontol Geriatr 22:27–34

Lupeanu E, Trandaburu I (1994–1995) Effect of Fe(III)-EDTA complex on NADH oxidation by rat liver, kidney and heart mitochondria related to age. Rom J Gerontol Geriatr 15–16:36–50

MacFarlane MD (1975) Procaine HCl (Gerovital H3): a weak, reversible, fully competitive inhibitor of monoamine oxidase. Fed Proc 34(1):108–110

MacFarlane MD, Besbris H (1974) Procaine (GH3) therapy: mechanism of inhibition of monoamine oxidase. J Am Geriatr Soc 8:365–371

Margina D, Gradinaru D, Manda G, Neagoe I, Ilie M (2013) Membrane effects exerted in vitro by polyphenols – quercetin, epigallocatechin gallate and curcumin – on HUVEC and Jurkat cells, relevant for diabetes mellitus. Food Chem Toxicol 61:86–93

Margina D, Olaru OT, Ilie M, Gradinaru D, Gutu C, Voicu S, Dinischiotu A, Spandidos DA, Tsatsakis AM (2015) Assessment of the potential health benefits of certain total extracts from Vitis vinifera, Aesculus hyppocastanum and Curcuma longa. Exp Ther Med 10(5):1681–1688

Mayne ST (2003) Antioxidant nutrients and chronic disease: use of biomarkers of exposure and oxidative stress status in epidemiologic research. J Nutr 133 Suppl 3:933S–940S

Moskalev A, Shaposhnikov MV, Plyusnina EN, Zhavoronkov A, Budovsky A, Yanai H, Fraifeld VE (2013) The role of DNA damage and repair in aging through the prism of Koch-like criteria. Ageing Res Rev 12(2):661–684

Muralidharan N, Bhat T, Kumari NS (2017) A study on effect of ageing on the levels of Total antioxidant and lipid peroxidation. IJCMR 4(12):8–10

Noda K (1971) The mode of action of procaine hydrochloride, viewed from ^{45}Ca ion association with the membrane sites of the Sartorius muscle fibers. Kurume Med J 18(2):71–81

Oettmeier R, Reuter U, Bonilla LBP (2019) The procaine-base-infusion: 20 years of experience of an alternative use with several therapeutical effects. J Altern Complement Integr Med 2019(5: 061)

Ostfeld A, Smith CM, Stotsky BA (1977) The systemic use of procaine in the treatment of the elderly: a review. J Am Geriatr Soc 25:1–20

Perls T (2013) The reappearance of procaine hydrochloride (Gerovital H3) for antiaging. J Am Geriatr Soc 61(6):1024–1025

Philpot FJ (1940) The inhibition of adrenaline oxidation by local anaesthetics. J Physiol 97(3): 301–307

Prasad NR, Menon VP, Vasudev V, Pugalendi KV (2005) Radioprotective effect of sesamol on gamma-radiation induced DNA damage, lipid peroxidation and antioxidants levels in cultured human lymphocytes. Toxicology 209(3):225–235

Radaceanu V, Matei-Vladescu C, Petrescu A (1991) The effects of Gerovital H3 on the lipofuscin accumulation in neurons of old rat cerebral cortex. Fluorescence and electron microscopy studies. Rom J Gerontol Geriatr 12:211–225

Revnic F (1990) The electron-microscopic aspects of hepatocyte aging in Wistar rats. The influence of GH3 treatment. Rom J Gerontol Geriatr 11:225–321

Rusu C (2001) Free radicals, oxidative stress and the degenerative disease of aging. Rom J Gerontol Ger 23:3–23

Rusu C, Lupeanu E (1989) Inhibitory effect of procaine, Gerovital H3 and Aslavital on the production of superoxide radical. Rom J Gerontol Geriatr 10:117–129

Rusu C, Borsa C, Gradinaru D, Ionescu C (1992) Gerovital H3 effect on the peroxidation potential and superoxide dismutase activity in rat liver, brain and kidney homogenates. Rom J Geront Geriat 13:93–100

Rusu C, Borsa C, Gradinaru D, Ionescu C (1995) Gerovital H3 and procaine effect on monoamine oxidase activity and lipid peroxidation in the mitochondrial fraction from rat brain and liver. Rom J Gerontol Geriatr 16:19–30

Rusu C, Borsa C, Gradinaru D, Ionescu C, Babeanu S (1996) Antioxidant and lipid-lowering effects of the original procaine-based products. Rom J Gerontol Geriatr 18:47–62

Schuster B, Ellmann A, Mayo T, Auer J, Haas M, Hecht M, Fietkau R, Distel LV (2018) Rate of individuals with clearly increased radiosensitivity rise with age both in healthy individuals and in cancer patients. BMC Geriatr 18(1):105

Seeman P (1972) The membrane actions of anesthetics and tranquilizers. Pharmacol Rev 24(4): 583–655

Sies H (1997) Oxidative stress: oxidants and antioxidants. Exp Physiol 82(2):291–295

Sies H (2015) Oxidative stress: a concept in redox biology and medicine. Redox Biol 4:180–183

Stroescu V (1988) The experimental and clinical pharmacology of procaine, Gerovital H3 and Aslavital, proceedings of the National Congress of Gerontology and Geriatrics; Vol I, p 43

Stroescu V, Gane P, Constantinescu I, Vrabiescu A (1980) Biochemical and pharmacodynamic arguments to support procaine IMAO-B type action. Rom J Gerontol Geriatr 1:55–61

Szatmari S, Bereczki D (2008) Procaine treatments for cognition and dementia. Cochrane Database Syst Rev (4):CD005993

Takenami T (2012) Intrathecally administered ropivacaine is less neurotoxic than procaine, bupivacaine, and levobupivacaine in a rat spinal model. Can J Anaesth 59(5):456–465

Xu G (2016) Effects of G.H.3. on mental symptoms and health-related quality of life among older adults: results of a three-month follow-up study in Shanghai, China. Nutr J 15:9

Xu J, Lecanu L, Han Z, Yao Z, Greeson J, Papadopoulos V (2003) Inhibition of adrenal cortical steroid formation by procaine is mediated by reduction of the c-AMP-induced 3-hydroxy-3-methylglutaryl-coenzyme a reductase messenger ribonucleic acid levels. J Pharmacol Exp Ther 307:1148–1157

Yang MH, Mohan CD, Deivasigamani A, Chinnathambi A, Alharbi SA, Rangappa KS, Jung SH, Ko H, Hui KM, Sethi G, Ahn KS (2022) Procaine abrogates the epithelial-mesenchymal transition process through modulating c-met phosphorylation in hepatocellular carcinoma. Cancers (Basel) 14(20):4978

Yau TM (1974) Gerovital H3, monoamine oxidase and brain monoamines, theoretical aspects of aging. Academic Press Inc, p 157

Yilbas A (2018) Procaine and saline have similar effects on articular cartilage and synovium in rat knee. BMC Anesthesiol 18(1):51

Ying B, Huang H, Li H, Song M, Wu S, Ying H (2017) Procaine inhibits proliferation and migration and promotes cell apoptosis in osteosarcoma cells by upregulation of MicroRNA-133b. Oncol Res 25(9):1463–1470

Zaric D, Pace NL (2009) Transient neurologic symptoms (TNS) following spinal anaesthesia with lidocaine versus other local anaesthetics. Cochrane Database Syst Rev 2009(2):CD003006

Zung WW, Gianturco D, Pfeiffer E, Hsioh-Shan Wang, Whanger A, Bridge P, Potkin SG (1974) Pharmacology of depression in the aged. Evaluation of Gerovital H3 as an antidepressant drug. Psyhosomatics 15:127–131

Chapter 18
Medicinal Peptide Drugs and Khavinson Peptides®

S. V. Trofimova and A. V. Trofimov

Abstract This chapter is dedicated to the results of experimental and clinical applications of short peptides developed by the outstanding scientist and founder of the theory of peptide regulation of aging, Professor Vladimir Khavinson. In recent years, the key role of short peptides in maintaining the body's homeostasis has been demonstrated. Long-term experimental studies have shown that Khavinson's peptides®, by regulating gene expression, possess high biological activity. The use of pharmaceuticals and dietary supplements based on Khavinson peptides® increases both the average and maximum lifespan of experimental animals, reduces the incidence of malignant tumors, contributes to the elongation of telomeres, and restores the functions of the immune and endocrine systems. According to a 15-year study, this leads to a statistically significant reduction in mortality among people who regularly received peptides.

Keywords Khavinson peptides® · Short peptides · Gene expression nutrition · Homeostasis

18.1 Introduction

Research into peptides as regulators of basic bodily functions has moved markedly from fundamental studies toward clinical application. Although the first peptide-based medicines appeared in the mid-twentieth century, a distinct pharmaceutical direction focused on peptides has truly taken shape only in recent decades. Interest in

S. V. Trofimova (✉)
Saint Petersburg Institute of Bioregulation and Gerontology After Professor Khavinson V., St. Petersburg, Russia

A. V. Trofimov
Saint Petersburg Institute of Bioregulation and Gerontology After Professor Khavinson V., St. Petersburg, Russia

Department of Innovative Technologies in Medicine and Dentistry, "G. D'Annunzio" University, Chieti-Pescara, Italy

S. I. S. Rattan, C. Gezer (eds.), *Supplements and Drugs for Healthy Longevity*, Healthy Ageing and Longevity 22, https://doi.org/10.1007/978-3-032-11608-6_18

this class of molecules is driven by their natural origin, high biological activity at low doses, and generally favorable safety profile, which together make peptides attractive targets for modern pharmacology (Antony and Vijayan 2021; Caputi et al. 2019; Kang et al. 2020; Khavinson et al. 2002, 2021).

Historically, a significant portion of peptide bioregulators consisted of tissue and organ extracts in which the full set of active components was not completely defined in composition and structure. In parallel, physiologists and pharmacologists became increasingly interested in short regulatory peptides with precisely specified amino-acid sequences and reproducible sources of production. The shift to chemical synthesis and to isolating substances with established structures made it possible to study mechanisms of action more rigorously and to design drug candidates rationally (Dilman et al. 1979; Fiorino et al. 2021; Son et al. 2021; Trofimova 2020; Trofimova et al. 2022).

18.2 Historical Overview

As mentioned above, the first peptide drugs appeared on the European pharmaceutical market in the middle of the last century. However, the number of them was low, and the leaders in the creation of medicinal peptides were Austria, Switzerland and Russia. Two of the most well-known Austrian peptide preparations include: Cerebrolysin® (EVER Neuro Pharma GmbH) and Actovegin® (TAKEDA AUSTRIA, GmbH). Cerebrolysin® is isolated from the brain of pigs and contains low molecular weight biologically active neuropeptides that penetrate the blood-brain barrier and regulate the functions of brain neurons. The drug normalizes metabolism and has an organ-specific effect on brain cells under conditions of hypoxia and ischemia, improves intracellular protein synthesis in the brain during aging. Cerebrolysin® is a nootropic drug and has a normalizing effect in cases of impairment of cognitive functions, improves the processes of memorization (Cui et al. 2019). Actovegin®, in contrast to Cerebrolysin®, is a protein-free polypeptide extract from the blood of calves. Actovegin® stimulates energy processes in cells, increases the uptake of glucose and oxygen by tissues, improves blood circulation and tissue regeneration (Brock et al. 2018). Close in pharmacological action to Actovegin is the drug Solcoseryl® (LEGACY PHARMACEUTICALS SWITZERLAND, GmbH), produced in Switzerland. The drug is a deproteinized dialysate from the blood of calves, contains a wide range of low molecular weight components with a molecular weight of up to 10 kDa (including glycoproteins, nucleosides and nucleotides, amino acids, oligopeptides). Solcoseryl® stimulates tissue regeneration, improves oxygen and glucose transport to cells, increases ATP synthesis by cells, stimulates fibroblast proliferation and collagen synthesis in the vascular endothelium (Ochi et al. 2003).

Examples of Tissue-Derived Peptide Preparations and Their Use

Early peptide-based medicinal forms included standardized extracts, among them preparations derived from bovine prostate and thymus tissues. Such agents were reported to normalize urodynamics and the contractility of the bladder's smooth muscle, as well as to exert immunomodulatory effects manifested by increases in T-lymphocyte counts and functional activity in patients with primary and secondary immunodeficiency (Ochi et al. 2003; Bircan et al. 1990; Bertarelli 1994; Hymos et al. 2020; Trofimova and Khavinson 2021). At the same time, extract-based drugs are limited by variability of composition and difficulties in verifying precise targets of action.

Russian Contribution and Age-Related Aspects
A substantial contribution to deciphering peptide regulatory mechanisms was made by studies conducted at the Military Medical Academy and the St. Petersburg Institute of Bioregulation and Gerontology. These works demonstrated that aging is accompanied by a complex remodeling of the peptidergic system, a decline in the synthesis of peptide-like physiologically active substances, and changes in the intensity of protein metabolism in organs and tissues (Khavinson 2002; Vanyushin and Khavinson 2016; Khavinson et al. 2003a, b, 2011, 2014a, b; Anisimov et al. 2000, 2003; Khavinson and Malinin 2005). These observations provided conceptual grounds for using short peptides to support cellular homeostasis and correct age-associated dysfunctions.

The first peptide bioregulators (Khavinson Peptides®) were created by V. Morozov and V. Khavinson in the early 70 s of the last century. Basis for the creation of these extracts was the concept of the Selye's general adaptation syndrome: during the stage of anxiety after exposure to stress, the functions of the thymus and pineal gland are suppressed. As is known, it is the pineal gland and thymus that have the main regulating effect on the work of the neuroimmunoendocrine system and, as a consequence, the aging processes of the human body. The stage of resistance following the stage of anxiety normally develops on day 4–5 and is characterized by the restoration of these organs. However, with various disorders this stage can slow down, which potentially leads to the emergence of various diseases, including those contributing to the accelerated aging of the body. Considering these fundamental physiological prerequisites, scientists decided to create preparations from the thymus and pineal gland to stimulate the stage of resistance (Selye 1955; Dilman et al. 1979; Khavinson 2002).

The first drug from this group in was Thymalin®, created in 1982 and manufactured by Samson-Med LLC (St. Petersburg, Russia), registration certificate 82/1108/8. Thymalin® is a complex of peptides with a molecular weight of up to 10 kDa, isolated from the thymus of cattle. Long-term experimental and clinical studies have shown that the drug restores impaired immunity, regulates the number and ratio of T- and B-lymphocytes and their subpopulations, stimulates cellular immunity reactions, enhances phagocytosis, stimulates the processes of regeneration and hematopoiesis in the event of their suppression, and also improves the course of processes cell metabolism (Khlystova et al. 2003). The introduction of the drug helps to increase life expectancy and reduce the incidence of tumors in animals (Anisimov et al. 2001; Khavinson 2002). Also, the introduction of Thymalin® leads to the normalization of the immune system function and an increase in the quality of life in people of the older age group. In addition, the results of recent studies have shown that the use of Thymalin in the treatment of patients suffering from COVID-19 contributes to the normalization of cytokine synthesis, has an anti-inflammatory effect, prevents development of disseminated intravascular coagulation, acute respiratory distress syndrome and multiple organ failure (Khavinson et al. 2021).

The next drug was Epithalamin®, which was created in 1990 and is produced by Samson-Med LLC (St. Petersburg, Russia), registration certificate 90/250/6. The drug is isolated from the pineal gland of the brain of cattle and is a complex of peptides with a molecular weight of up to 10 kDa. Long-term experimental and clinical studies have shown that the drug normalizes the functions of the anterior pituitary gland, the concentration of gonadotropic hormones in the blood, the secretion of melatonin, and increases the sensitivity of the hypothalamus to endogenous hormonal influences. Administration of the drug to animals led to an increase in the average life span, a decrease in the incidence of tumors, and the restoration of the cyclic activity of the ovaries in old animals. Epithalamin® increased the body's resistance to stress, normalized carbohydrate metabolism (Anisimov et al. 1994; Khavinson 2002).

As is known, a decrease in the melatonin-producing function of the pineal gland is one of the characteristic manifestations of physiological aging. Disruption of melatonin production increases with accelerated aging and age-dependent pathologies: coronary heart disease, arterial hypertension, Alzheimer's disease, non-insulin-dependent diabetes mellitus. The use of Epithalamin® in elderly patients promoted the restoration of a reduced level of melatonin, improved functions of the immune, endocrine, cardiovascular, antioxidant systems, and brain functions (Khavinson and Morozov 2003).

At the same time, the data of studies on the effectiveness of long-term peptide preparation use of the pineal gland and thymus in elderly patients against the background of age-related diseases, as well as signs of accelerated aging of the body, are of considerable interest. For this purpose, several clinical studies of the geroprotective action of peptide bioregulators have been carried out.

Evaluation of long-term results in these patients with a 15-year follow-up is certainly of interest (Korkushko et al. 2011). Given the importance of the scientific data obtained, we present the original of some reports of a clinical study published in several scientific Russian monographs (Korkushko et al. 2002; Khavinson and Morozov 2001). Thus, as a result of 15-year observation of two groups of elderly patients with accelerated aging of CVS, it was found that long-term use of the peptide geroprotector from the pineal gland reduces the rate of aging of CVS, prevents age-dependent decrease in physical performance, has a normalizing effect on the daily rhythm of melatonin production, carbohydrate and lipid exchange. The geroprotective effect of the peptide preparation of the pineal gland is also evidenced by a significant decrease in mortality in the group of patients who received Epithalamin against the background of basic therapy.

Thus, over the past 40 years, under the leadership of Professor Vladimir Khavinson, scientific research has been conducted to create peptides that stimulate the reserve capabilities of the human body. It has been established that these peptides initially isolated from the organs of young animals are substances of a peptide nature. Subsequently, these peptide bioregulators were synthesized from amino acids. Peptidergic regulation of homeostasis occupies an important place in physiological processes, which lead to the aging of cells, tissues, and organs, consisting in the involution of major regulatory systems—the nervous, the endocrine, and the immune. An important scientific discovery by Professor V. Khavinson was the ability of peptide bioregulators to regulate gene activity, protein synthesis and prolong life expectancy. It has been demonstrated that short peptides (Khavinson Peptides®) play an important role in the transmission of biological information, modulation of transcription, and restoring genetically conditioned alterations occurring with age. The conducted researches proved the short peptides (Khavinson Peptides®) could promote the neuronal differentiation and be a promising tool for the application of peptides as a modulator of neurogenesis in neurodegenerative diseases (Caputi et al. 2019; Khavinson et al. 2021).

18.3 Regulatory Mechanisms

Numerous studies on regulatory peptides in recent decades have led to a re-examination of ideas about regulation mechanisms of physiological functions, maintenance of homeostasis processes and adaptation of the body's functional systems to the environment. Nutrition is one of the most important environmental factors that affect the human body throughout its life. Nutrients, being converted through metabolism, ensure the body's vital activity, affecting the quality and duration of life. Eating disorders always lead to some kind of negative consequences. Therefore, various pathological processes in the human body, as well as aging processes, can be considered from the point of view of disorders in the metabolism of nutrients at the organ, tissue and cellular level of the organization of living matter.

However, with age, in various parts of the digestive tract, there is a steady increase in involutional processes that contribute to the disruption of formation and absorption of short peptides, which ultimately leads to a violation of the adaptation processes and homeostasis of the functional systems of the entire organism.

It should be noted that metabolism of short peptides obtained by enzymatic hydrolysis in the gastrointestinal tract of proteins that enter the body with food and their synthetic analogs seems to be uniform, therefore the general biological role of short peptides in the regulation of homeostasis, regardless of their origin, is further examined. The metabolism of short peptides can be divided into three main stages: absorption into the blood in the gastrointestinal tract, interaction with targets of various organs and tissues, and elimination/resorption in the kidneys.

18.3.1 Metabolism of Short peptides in the Gastrointestinal Tract and Their Absorption into the Bloodstream

Protein molecules that enter the body with food, under the action of enzymes of the gastric (pepsin, rennin, gastrixin) and intestinal (aminopeptidase, enteropeptidase) tract, are split mainly into di- and tripeptides. These short peptides are divided into two groups: resistant to hydrolysis and able to break down into amino acids. Hydrolysis of labile short peptides occurs on the surface of the brush border membranes of enterocytes (membrane digestion) or inside enterocytes using cytosolic peptidases (intracellular digestion) (Timofeeva et al. 2005; Shen et al. 2001).

Peptides resistant to hydrolysis (glycyl-glycine, proline-containing peptides, glycyl-sarcosine, carnosine, etc.) penetrate into the circulatory system in an unchanged form, which is confirmed by the data of various researchers. According to all dipeptides have the same transport mechanism from the gastrointestinal tract into the blood. It was established that the Ala-Gly-Gly (AGG) tripeptide and some dipeptides are not hydrolyzed in the small intestine and can enter the bloodstream unchanged. In addition, it has been shown that dipeptides, in particular the Gly-Pro (GP) peptide, can penetrate from the gastrointestinal tract not only into the bloodstream, but can also be transported even through the blood-brain barrier in an unchanged form. In another study, it was shown that in addition to passive transfer in the alkaline border of enterocytes, there is a system of active transport of di- and tripeptides.

When studying the morpho-physiological characteristics of the transport of the cyclic nonapeptide Cys-Tyr-Phe-Gln-Asn-Cys-Pro-Arg-Gly (*CYFQNCPRG*) through the intestinal epithelium, its partial absorption also occurred while maintaining physiological activity. The localization of the label to this nonapeptide in the cytoplasm of enterocytes and the intercellular space in the basal region of the intestinal epithelium of rats and frogs has been shown using the methods of electron and fluorescence confocal microscopy.

These data are consistent with the results of studies of short peptides developed under the guidance of Prof. V.Kh. Khavinson. Numerous studies have shown that short peptides *KE*, *AEDG*, *KEDA* are not hydrolyzed in the stomach, duodenum, jejunum and ileum, and are only slightly hydrolyzed in the colon and liver. The degree of hydrolysis of these peptides was determined by cleavage by L-amino acid oxidase. In addition, it was shown that the *KEDA* peptide is not subject to hydrolysis in blood plasma, which confirms the assumption about the possibility of transport of peptides in the bloodstream to specific target organs for them (Timofeeva et al. 2005).

The results showing that the peptides *KE*, *AEDG*, *KEDA* are able to regulate the activity of digestive enzymes in old animals are important. It was found that oral administration of the *KEDA* peptide in old rats for 2 weeks increased the activity of sucrase and maltase in the duodenum by 60%, maltase in the jejunum and colon by 3 times, and aminopeptidase M and glycyl-L-leucinepeptidase by 2 times (Timofeeva et al. 2005). The *KE* peptide in a similar experiment increased the activity of maltase (by 1.2 times in the duodenum, by 1.5 times in the jejunum, by 1.3 times in the ileum) and alkaline phosphatase (by 2.2 times in the jejunum) and the ileum and 1.8 times—in the duodenum). It is important to note that the peptides *KEDA* and *KE* promoted an increase in the activity of gastrointestinal enzymes in old animals to values typical for young animals. In addition, oral administration of the *KE* and *AEDG* peptides for 1 month in aged rats improved the absorption of glucose and glycine in the medial region under the action of the *KE* peptide, and in the proximal and distal parts of the small intestine under the influence of the *AEDG* peptide (Khavinson et al. 2002).

18.3.2 Tissue-Specific (Gene-Specific) Interaction of Short peptides with Target Cells in Various Organs

Through the bloodstream, short peptides penetrate into various organs and tissues. Probably, short peptides, in accordance with their structure, have an affinity for certain organs and tissues in which they accumulate and exhibit the greatest biological activity. It was found that short peptides *KE* and *AEDG* have a pronounced tissue-specific effect on the thymus and pineal gland (Khavinson et al. 2011). It is assumed that the interaction of short peptides with target tissues is based on their ability to penetrate the cytoplasmic and nuclear membrane into the nucleus and nucleolus and interact with DNA, epigenetically regulating the expression of genes encoding a number of signaling molecules and protein markers of cell functional activity (differentiation factors, proliferation, apoptosis, transcription) (Khavinson et al. 2011).

It was found that short peptides synthesized on the basis of the Tat-protein (activator of transcription of the viral genome of human immunodeficiency HIV-1) are able to penetrate into the cell. These peptides have been combined into the cell-penetrating peptides (CPP) group. Penetration into the cell through the membrane is most often characteristic of alkaline peptides containing an excess of positively charged amino acid residues in the structure. The advantage of these peptides is that they easily overcome the acidic glycocalyx layer that is adjacent to the cell membrane. For synthetic alkaline and amphiphilic peptides containing several lysine residues in the structure, the ability not only to penetrate into the cell, but also to form complexes with DNA and RNA was noted. It was found that the binding of these peptides to DNA leads to strengthening of its double helix.

The direct interaction of the peptide with the membrane is determined by the electrostatic interaction of positively charged side groups of amino acid residues of arginine and lysine with negative carboxyl groups of phosphatidylserine, exposed on the outer side of the cytoplasmic membrane. For negatively charged (carboxyl) side groups of peptides, the binding sites are positively charged groups of phosphatidylcholine and phosphatidylethanolamine. Thus, pinocytosis may be the main mechanism for the penetration of short peptides through the cytoplasmic membrane.

In addition, it was shown that FITC-labeled short peptides *AEDG*, Glu-Asp-Arg (*EDR*), Lys-Glu-Asp-Gly (*KEDG*) penetrate the nucleus and nucleolus of HeLa cells. HeLa cells were incubated with FITC-labeled peptides for 12 hours. It is known that the nucleus of eukaryotic cells has a system of transport pores (nucleopores) formed by protein complexes nucleoporins. The inner diameter of nucleopores is about 50 nm; therefore, they are permeable to freely diffusing low molecular weight substances with a molecular weight of up to 3.5 kDa, which include short peptides. The results obtained make it possible to consider the possibility of direct interaction of short peptides with DNA. In recent years, the method of molecular modeling has been increasingly used to analyze nanostructures, which include short peptides.

Comparison of the spatial arrangement of functional groups on the surface of the large groove of double-stranded DNA and side groups of regulatory peptides showed that the *AEDG* peptide can bind to the complementary DNA site on the promoter region of the gene, causing local strand separation, and thereby initiating the process of RNA gene transcription by polymerase II.

The ATTTS sequence complementary to the *AEDG* peptide was found in the promoter portions of the Ki67, p53, IL-2, MMP2 and Tram1 genes and telomerase. It was experimentally proved that the addition of the *AEDG* peptide to the culture of human lung fibroblasts induces the expression of the telomerase gene, telomerase activity, and promotes the elongation of telomeres by 2.4 times (Khavinson and Malinin 2005).

Activation of gene expression is accompanied by an increase in the number of cell divisions by 42.5%, which demonstrates overcoming the Hayflick limit of cell division. In addition, in the pineal gland cell culture of young and old animals under the influence of the tetrapeptide, an increase in the synthesis of MMP2 and Ki67 proteins and a decrease in the synthesis of the proapoptotic protein p53 were observed.

Epigenetics postulates the tissue, subcellular, and age specificity of DNA methylation and indicates that the character of DNA methylation in cancer cells is different than in normal cells. Taking these facts into account, it can be assumed that the same biologically active short peptide can bind to DNA depending on the nature of its methylation and will have different effects on gene functions in different tissues (cells), in the nucleus and mitochondria, in young and old cells, in normal and malignant cells. Thus, specific (complementary) peptide-DNA interactions can epigenetically control the genetic functions of a cell, and this mechanism probably played an important role at the earliest stages of life and evolution (Vanyushin and Khavinson 2016).

18.3.3 Elimination and Resorption of CP in the Kidneys

The question of whether short peptides are excreted in the urine or resorbed in the kidneys is still controversial (Shen et al. 2001). According to some researchers, short peptides resistant to hydrolysis by gastrointestinal tract enzymes are eliminated in the urine (Timofeeva et al. 2005). In another work, using confocal microscopy, it was shown that the introduction of a fluorescently labeled polypeptide into the intestines of rats led to its accumulation in the epithelium vesicles of the proximal renal tubule, which indicates the role of the kidneys in the metabolism of not only endogenous, but also exogenous peptides and proteins. These data are consistent with the study, which indicates that the metabolism of peptides with a molecular weight of less than 60 kDa occurs in the proximal renal tubules.

Design and Rationale of the KEDA Tetrapeptide

Based on amino-acid analysis of a hepatic peptide preparation, the synthetic tetrapeptide KEDA (Lys-Glu-Asp-Ala) was designed. The sequence was chosen to reproduce key structural motifs responsible for the regulatory effects of the original mixture, but in a fully defined and reproducible form.

The experimental program assessed KEDA's influence on histogenesis and liver function and compared it to a control peptide—AEDG (Ala-Glu-Asp-Gly), known for retinoprotective activity. A model using the ectoderm of Xenopus laevis gastrula served as a system with pluripotent potential and the capacity for directed differentiation under inductive cues. Short peptides were tested over a wide concentration range (2–200 ng/mL), with morphological and immunohistochemical endpoints (Khavinson et al. 2002, 2013).

The results revealed fundamental differences in inductive properties: under AEDG, neural differentiation was initiated in pluripotent tissue, whereas KEDA triggered mesodermal differentiation. Thus, tissue/lineage specificity of action was confirmed for short peptides already at the level of embryonic rudiments, opening prospects for fine regulation of regenerative processes using minimalist amino-acid sequences (Khavinson et al. 2003a, b).

Further studies have shown that the *KEDA* peptide is practically not hydrolyzed by peptide hydrolases of the small intestine, entering first into the blood plasma, and then into various tissues and organs. This is explained primarily by the fact that the regulatory systems of the intestine carry out the transport of peptides formed during digestion from proteins much faster than the absorption of a mixture of free amino acids, to which the body did not adapt during evolution. The small intestine has di- and tripeptide transporters providing absorption of short peptides (Bai et al. 1995; Shen et al. 2001). It is known that the rate of transport of some dipeptides exceeds the rate of transfer of those amino acids of which they are composed. Peptidases of the brush border of enterocytes break down a significant part (about 40–60%) of short peptides only into di- and tripeptides. Then, from the intestine, short peptides are distributed with blood flow to various organs and tissues, penetrate through the cytoplasmic membrane into the nucleus and nucleolus of cells, where they can bind to DNA and epigenetically regulate gene expression. The DNA-peptide interaction is probably the oldest in evolution. This may explain the high biological activity of short peptides, their complete harmlessness, and their successful use as substances with significant physiological effects.

The safety of *KEDA* (Lys-Glu-Asp-Ala) has been confirmed by the results of a study of its general toxic effect. In the study of acute toxicity, it was found that a single administration of a *KEDA* (Lys-Glu-Asp-Ala) solution to animals at a dose 5000 times higher than the dose recommended for clinical use does not cause toxic reactions. The study of the subacute and chronic toxicity of the tetrapeptide indicated the absence of side effects during its long-term use in doses exceeding the therapeutic one by 100–1000 times. When assessing the general state of animals, morphological and biochemical parameters of peripheral blood, morphological state of internal organs, state of the cardiovascular and respiratory systems, liver and kidney function, pathological changes in the body were not found. Like all peptides developed at the St. Petersburg Institute of Bioregulation and Gerontology *KEDA* peptide regulates gene expression. This tetrapeptide caused the activation of ribosomal genes and decondensation of heterochromatin in chromosomes 1, 9 and 16 in the cells of elderly people (Khavinson et al. 2002; Vanyushin and Khavinson 2016).

Tissue Specificity and Effects in Liver-Tissue Models
KEDA's tissue-specific activity was demonstrated in liver explants from aged rats. As an integral metric, investigators used the explant area index—the ratio of the entire explant area (including the peripheral growth zone) to the initial area of the central-zone fragment. Under KEDA, the index increased by ~15–16% relative to baseline, whereas the control peptide did not produce a significant rise (Khavinson et al. 2011; Khavinson 2004). This finding aligns with the hypothesis that KEDA stimulates reparative processes in hepatic parenchyma.

Hepatoprotective and Metabolic Effects
In a series of subsequent experiments on animal models of acute toxic hepatitis, KEDA exhibited both protective and therapeutic properties: total bilirubin and cholesterol levels decreased, transaminase activity (ALT/AST) normalized, dystrophic changes in the hepatic stroma diminished, and the number of glycogen-containing cells increased. In vitro, the tetrapeptide reduced intestinal glycyl-leucine dipeptidase activity by approximately 50%. After 2 weeks of oral administration to young rats, overall activity of several digestive enzymes decreased, whereas in old animals values tended to return toward control, which may reflect age-dependent differences in regulation of enzyme systems (Khavinson 2004; Timofeeva et al. 2005).

Immunological Shifts
Part of KEDA's hepatoprotective action is likely mediated by immunomodulation. In cellular assays, incubation with the peptide increased the proportion of lymphocytes with CD3+, CD4+, and CD8+ phenotypes and reduced the content of CD22+ cells. In clinically oriented observations, increases in neutrophil phagocytic activity were reported in healthy volunteers and in patients with viral hepatitis A (Khavinson et al. 2014a, b). These data indicate KEDA's potential as a fine regulator of both innate and adaptive immune responses.

The *KEDA* peptide, regardless of the dose, inhibited the activation of complement by the classical pathway in healthy people in in vitro experiments. In patients with viral hepatitis A in the same study, the *KEDA* peptide at a dose of 40 μg/mL activated, and at a dose of 60–160 μg/mL inhibited the stimulation of complement along the classical pathway, without affecting the alternative pathway of the complement system activation (Timofeeva et al. 2005).

The results of further studies showed that the tetrapeptide *KEDA*, in addition to immunomodulatory, has antioxidant activity. In addition, this peptide influenced the rheological properties of blood, slowing down its clotting and inhibiting fibrinolysis. Moreover, the revealed effects were present to a greater extent in the blood plasma of patients with viral hepatitis A (Khavinson et al. 2014a, b).

Thus, the short peptide *KEDA* in *in vitro* and *in vivo* experiments has pronounced immunoprotective, antioxidant and hepatoprotective properties, contributing to the restoration of liver function on the molecular and cellular level. Further studies of the hepatoprotective properties of *KEDA* (Lys-Glu-Asp-Ala) will make it possible to create a safe drug with high hepatoprotective activity on its basis.

18.4 Conclusions

Long-term studies have proved that Khavinson peptides® open up a new direction in the regulation of aging processes. For the first time, Professor Vladimir Khavinson developed and introduced into clinical practice a new group of geroprotectors from 10 medicinal peptide preparations and more than 50 peptide biologically active food additives based on peptides that restore the functions of various organs and tissues of the human body. The scale of his creative ideas, scientific foresight, and organizational potential of Professor Vladimir Khavinson allowed him to formulate the concept of a unified mechanism for peptide regulation of gene expression and protein synthesis in wildlife.

References

Anisimov VN et al (1994) Twenty years of study on effect of pineal peptide preparation: epithalamin in experimental gerontology and oncology. Ann N Y Acad Sci 719:483–493

Anisimov VN et al (2000) Immunomodulatory peptide L-Glu-L-Trp slows down aging and inhibits spontaneous carcinogenesis in rats. Biogerontology 1:55–59

Anisimov VN et al (2001) Effect of synthetic thymic and pineal peptides on biomarkers of ageing, survival and spontaneous tumor incidence in female CBA mice. Mech Ageing Dev 122(1): 41–68

Anisimov VN et al (2003) Effect of epitalon on biomarkers of aging, life span and spontaneous tumor incidence in female swiss-derived SHR mice. Biogerontology 4:193–202

Antony P, Vijayan R (2021) Bioactive peptides as potential nutraceuticals for diabetes therapy: a comprehensive review. Int J Mol Sci 22(16):9059

Bai JP et al (1995) Targeting of peptide and protein drugs to specific sites in the oral route. Crit Rev Ther Drug Carrier Syst:339–371

Bertarelli F (1994) Efficacy of thymostimulin. Lancet 343(8890):184

Bircan K et al (1990) Raveron versus placebo in the conservative treatment of benign prostatic hyperplasia (BPH). Int Urol Nephrol 22(4):345–348

Brock J et al (2018) Update on the role of Actovegin in musculoskeletal medicine: a review of the past 10 years. Clin J Sport Med

Caputi S et al (2019) Effect of short peptides on neuronal differentiation of stem cells. Int J Immunopathol Pharmacol 33:1–12

Cui S et al (2019) Cerebrolysin for vascular dementia. Cochrane Database Syst Rev 2019: CD008900

Dilman VM et al (1979) Increase in lifespan of rats following polypeptide pineal extract treatment. J Exp Pathol 17(9):539–545

Fiorino A et al (2021) Amelogenin-derived peptides in bone regeneration: a systematic review. Int J Mol Sci 22(17):9224

Hymos A et al (2020) Thymic peptides reverse immune exhaustion in patients with reactivated human alphaherpesvirus1 infections. Int J Mol Sci 21:2379

Kang YJ et al (2020) Therapeutic targeting strategies for early-to late-staged Alzheimer's disease. Int J Mol Sci 21(24):9591

Khavinson VK (2002) Peptides and ageing. Neuro Endocrinol Lett:144

Khavinson VK (2004) Tetrapeptide stimulating functional activity of hepatocytes and its therapeutical use. European Patent, p 1325026

Khavinson VKh, Malinin VV (2005) Gerontological aspects of genome peptide regulation. Basel: Karger AG, p 104

Khavinson VKh, Morozov VG (2001) Peptides of pineal gland and thymus in ageing regulation. St. Petersburg, Foliant

Khavinson V, Morozov V (2003) Peptides of pineal gland and thymus prolong human life. Neuroendocrinol Lett 24(3–4):233–240

Khavinson VK et al (2002) Effects of Livagen peptide on chromatin activation in lymphocytes from old people. Bull Exp Biol Med 134(4):389–392

Khavinson VK et al (2003a) Effect of peptides on proliferative activity of retinal and pigmented epithelial cells. Bull Exp Biol Med 135(5):597–599

Khavinson VK et al (2003b) Peptide Epitalon activates chromatin at the old age. Neuroendocrinol Lett 24(5):329–333

Khavinson VK et al (2011) Morphofunctional fundamentals for peptide regulation of aging. Biol Bull Rev 1(4):390–394

Khavinson VK et al (2013) Peptidergic regulation of differentiation of embryonic cells. Cell Technol Biol Med 1:172–175

Khavinson VK et al (2014a) Peptide regulation of gene expression and protein synthesis in bronchial epithelium. Lung 192:781–791

Khavinson VK et al (2014b) Peptide geroprotectors – epigenetic regulators of physiological functions of the body. Monograph. A.I. Herzen Russian State Pedagogical University, St. Petersburg, p 271

Khavinson V et al (2021) EDR peptide: possible mechanism of gene expression and protein synthesis regulation involved in the pathogenesis of Alzheimer's disease. Molecules 26(1) 159:1–16

Khlystova ZS et al (2003) Thymalin in developing respiratory organs of human fetus. Bull Exp Biol Med 135:600–602

Korkushko OV et al (2002) Peptide preparations of thymus and pineal gland in prevention of accelerated aging. Nauka, St.Petersburg

Korkushko OV et al (2011) Peptide geroprotector from the pituitarygland inhibits rapid aging of elderly people: results of 15-year follow-up. Bull Exp Biol Med 151(3):366–369

Ochi M et al (2003) Solcoseryl, a tissue respiration stimulating agent, significantly enhances the effect of capacitively coupled electric field on the promotion of bone formation around dental implants. Clin Oral Implants Res 14(3):294–302

Selye H (1955) Stress and disease. Science 122(3171):625–631

Shen H et al (2001) Developmental expression of PEPT1 and PEPT2 in rat small intense, colon, and kidney. Pediatr Res:789–795

Son G-H et al (2021) The role of antimicrobial peptides in preterm birth. Int J Mol Sci 22(16):8905

Timofeeva MN et al (2005) Effect of peptide livagen on the activity of digestive enzymes of the gastrointestinal tract and non-digestive organs of rats of different ages. Adv Gerontol 16:92–96

Trofimova S (2020) Molecular mechanisms of retina pathology and ways of its correction. Springer, Cham, p 90

Trofimova SV, Khavinson VK (2021) Role of short peptides as an important nutritional element in maintenance of body homeostasis. In: Rattan SIS, Kaur G (eds) Nutrition, food and diet in ageing and longevity. Healthy ageing and longevity, vol 14. Springer, Cham. https://doi.org/10.1007/978-3-030-83017-5_23

Trofimova SV et al (2022) Medicinal peptide drugs: a promising direction in modern pharmacology. EC Clin Med Case Rep 5(3):17–24

Vanyushin BF, Khavinson VK (2016) Short biologically active peptides as epigenetic modulators of gene activity. In: Doerfler W, Biihm P (eds) Epigenetics – a different way of looking at genetics. Epigenetics and human health. Springer, Cham, pp 69–90

Chapter 19
Caloric Restriction Mimetics as Drugs for Healthy Longevity

Rajesh Tamatta, Varsha Pai, and Abhishek Kumar Singh

Abstract Caloric restriction (CR) has been widely recognized as a robust intervention for extending lifespan and delaying the onset of age-related diseases across multiple species. However, long-term adherence to CR in humans poses significant practical and psychological challenges. This has increased interest in calorie restriction mimetics (CRMs), compounds that mimic the beneficial effects of CR without reducing caloric intake. CRMs target key cellular and molecular pathways involved in aging, such as AMP-activated protein kinases, sirtuins, and mammalian targets of rapamycin. By modulating these pathways, CRMs can enhance cellular stress resistance, improve metabolic health, and reduce inflammation hallmarks associated with aging and chronic diseases. Examples of well-studied CRMs include resveratrol, metformin, and spermidine, which show promise in preclinical and clinical studies for promoting healthy longevity. While the potential of CRMs is substantial, translating these findings into safe and effective human therapies requires a deeper understanding of their long-term effects, optimal dosing, and population-specific responses. Additionally, systems biology and personalized medicine advancements may pave the way for tailoring CRM-based interventions for individual aging trajectories. As research continues to evolve, CRMs hold promises for extending lifespan and enhancing health span, thereby contributing to healthier aging and reducing the burden of age-related conditions in the global population.

Keywords Caloric restriction mimetics · Caloric restriction · Brain aging · Longevity · Metabolic disorders

---

R. Tamatta · V. Pai · A. K. Singh (✉)
Department of Biotherapeutics Research, Manipal Academy of Higher Education, Manipal, Karnataka, India
e-mail: abhishek.singh@manipal.edu

S. I. S. Rattan, C. Gezer (eds.), *Supplements and Drugs for Healthy Longevity*, Healthy Ageing and Longevity 22, https://doi.org/10.1007/978-3-032-11608-6_19

Abbreviations

2-DG	2-Deoxy-d-glucose
AD	Alzheimer's disease
CR	Caloric restriction
CRMs	Caloric restriction mimetics
CSF	Cerebrospinal fluid
IF	Intermittent fasting
NDDs	Neurodegenerative disorders
PD	Parkinson's disease
Pts	Pterostilbene
T2DM	Type 2 diabetes mellitus
Uro A	Urolithin A
Uro C	Urolithin C

19.1 Introduction

Aging consists of multiple biological mechanisms that cause alterations in the body's equilibrium. The aging process results from the interplay of different molecular and cellular pathways, including DNA damage, impaired mitochondrial function, loss of protein homeostasis, and the accumulation of senescent cells. The progression of aging includes shifts at both the cellular and molecular levels, influencing general physiological functions (Bhaduri et al. 2023). Dietary intake and daily routines play crucial roles in aging by regulating metabolic processes, cellular repair, and inflammation. Diverse ant-iaging strategies, such as regular exercise, stress management, and medicinal and dietary interventions, have been employed to delay aging and age-related diseases. Dietary interventions include intermittent fasting (IF) and caloric restriction (CR) (Fig. 19.1). Alternate cycles of eating and fasting, such as IF, contribute to longevity via the adaptive stress response, activating autophagy and antioxidants production (Strilbytska et al. 2024). IF has been shown to protect against certain diseases by improving health, reducing oxidative stress, and maintaining blood glucose levels by improving insulin sensitivity (Strilbytska et al. 2024). Despite their numerous benefits, IF and CR have adverse effects, such as altered hormonal levels, poor eating habits, disturbed sleep cycles, and poor mental health (Hepsomali and Groeger 2021). Another setback from IF includes proteins and vitamins malnutrition (Grajower and Horne 2019).

CRMs are pharmacologically active substances that mimic the effect of CR without significantly reducing food intake (Singh et al. 2018; Sharma and Singh 2023; Trisal and Singh 2024, 2025). These CRMs trigger stress-related pathways, increasing stress resistance. Integrating phytochemical-based supplements, dietary supplements, and CRMs represents a promising strategy for promoting healthy aging and longevity. Ongoing research reveals the mechanisms of action and

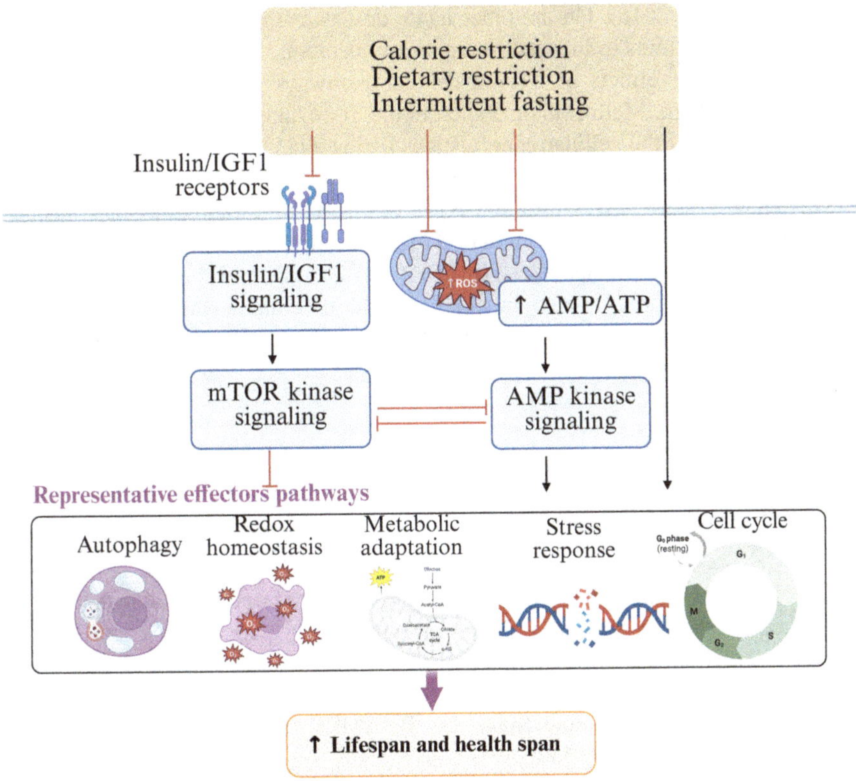

Fig. 19.1 Dietary interventions that promote longevity and health span by modulating metabolic pathways, reducing inflammation, and enhancing cellular resilience

potential therapeutic applications of CRMs. Understanding these interventions and their synergistic effects may pave the way for personalized approaches to increase lifespan and mitigate age-related diseases such as neurodegenerative disorders (NDDs) (Singh et al. 2024a, b; Wang et al. 2024a, b). Lowering caloric intake while avoiding poor nutrition, known as CR, may support a longer lifespan. CRMs have attracted significant research interest because of their remarkable ability to extend both median and maximum lifespans across various species. CRMs mimic the lifespan-extending effects of CR, while also reducing the risk of NDDs. These CRMs can achieve this by modulating the key metabolic pathways, regulating hormonal activity, and thus supporting optimal physiological functions (Singh et al. 2024a, b).

Anti-aging effects are mediated through multiple pathways. CRMs can be effectively structured via upstream and downstream targeting. Upstream targets in CRM focus on early modulation of energy metabolism within signaling pathways. These interventions often work by inhibiting glycolysis or glucose metabolism. For example, 2-deoxyglucose (2DG) and acarbose function as glycolytic inhibitors by mimicking glucose without being metabolized, thereby reducing energy availability

(Ingram and Roth 2015). On the other hand, downstream targets regulate intracellular signaling proteins influenced by upstream metabolic changes. These interventions replicate the effects of CR on cellular pathways without directly altering glucose metabolism. Metformin and spermidine, for example, inhibit mitochondrial complex I, altering the cellular energy status and activating AMP-activated protein kinase (AMPK), a crucial energy sensor (Ingram and Roth 2015).

This chapter summarizes different types of CRMs and highlights their mechanisms of action, such as the activation of AMPK and sirtuin 1 (SIRT1); the inhibition of mammalian target of rapamycin (mTOR) and NF-kB; and the enhancement of mitochondrial function. It also includes several preclinical and clinical studies demonstrating healthy longevity and metabolic benefits and evaluating their safety and efficacy in humans.

19.2 Caloric Restriction Mimetics and Their Mechanism of Action

CRMs elicit anti-aging effects and extend longevity by targeting major metabolic, stress, and epigenetic pathways. CRMs predominantly suppress nutrient-sensing pathways, such as the mTORC1 complex, the AMPK pathway, and the IGF-1 signaling pathway, as these pathways are considered regulators of the aging process (Vitale et al. 2019; Trisal and Singh 2025). AMPK, a sensor protein, regulates cellular metabolism, and a low AMP: ATP ratio triggers its activation through SIRT1-mediated deacetylation (Nassar et al. 2024). Activated SIRT1 downregulates NF-kB and inhibits proinflammatory genes. AMPK and SIRT1 simultaneously attenuate the mTOR and insulin signaling pathways (Ge et al. 2022). CRMs inhibit the IGF-1R-PI3K-AKT pathway, FoxO, and mTORC1, thus regulating autophagy (Yang and Zhang 2020). The mTOR pathway primarily regulates protein synthesis, metabolism, cell proliferation, and autophagy, in addition to regulating lifespan (Kapahi et al. 2004).

CRMs also exhibit antioxidative effects by activating cellular stress response pathways, including AMPK, Nrf2, and SIRT1. Activation of these pathways leads to upregulation of endogenous antioxidant defenses, such as superoxidase dimutase (SOD), glutathione peroxidase, and catalase expression thereby, mitigating ROS induced oxidative stress. These pathways are intrinsically part of the cellular adaptive stress response system. Consequently, the mechanism of action of CRMs aligns with the concept of hormesis, whereby low-level exposure to a stressor elicits a beneficial adaptive response. CRMs play an anti-inflammatory role by regulating the immune response. NAD + -dependent deacetylase SIRT1 activation via several CRMs inhibits the transcription factor NF-kB by deacetylating and diminishing the proinflammatory cytokines TNF-α, IL-6, and IL-1β. expression (Cho et al. 2015).

19.3 Caloric Restriction Mimetic Molecules

CRMs are compounds that replicate the beneficial effects of CR without the need to reduce caloric intake. Molecules such as resveratrol, metformin, and spermidine activate key cellular pathways such as the sirtuin, AMPK, and AKT pathways, which promote longevity, increase stress resistance, and regulate metabolism. CRMs show great potential in supporting healthy aging and preventing age-related diseases. Some of the well-known CRMs are detailed below.

19.3.1 Resveratrol

Resveratrol and polyphenols show potential antiaging and anti-inflammatory properties. Resveratrol stimulates SIRT1, a protein linked to extended lifespan. A study revealed that resveratrol induces autophagy *in vitro* and *in vivo*, resulting in a prolonged lifespan in *C. elegans*. However, this effect was abolished when Sirt1 was knocked down or knocked out, suggesting that resveratrol promotes longevity through Sirt1-dependent activation of autophagy (Morselli et al. 2010). *In vitro* studies on hippocampal astrocytes and age-related glial cells have shown that resveratrol alleviates IL-1β and TNF-α, elevates antioxidative levels, and acts as an antiaging agent (Bellaver et al. 2014). In a recent clinical study of 30 Alzheimer's patients (CTR20151780X), those taking a 500 mg dose of transresveratrol daily for 52 weeks presented stable Aβ40 levels, reduced brain volume ($P = 0.011$), and a 46% decrease in CSF matrix metallopeptidase 9 ($P = 0.033$). Transresveratrol may offer neuroprotection by reducing Aβ accumulation and neuroinflammation (Gu et al. 2021). In a 6-month trial with 192 T2DM patients, 500 mg/day of resveratrol prevented bone loss, unlike lower doses or placebo. Only the high-dose group presented increased vitamin D, and the BMD/BMC ratio remained stable. The benefits were greater in those with low calcium, low vitamin D, or alcohol use (Bo et al. 2018).

19.3.2 Rapamycin

Rapamycin acts as an anti-aging intervention, delays aging and attenuates senescence. It primarily acts by inhibiting mTOR by binding to FKBP12 of mTORC1 and mTORC2 (Fig. 19.2) (Schreiber et al. 2015; Singh et al. 2016). Studies on the mTOR (Δ/Δ) hypomorphic mouse model revealed an increased life span and improved outcomes in aging tissues and organs (Wu et al. 2019). Male CB6F1 mice treated with rapamycin (14 mg/kg) for several weeks presented significantly reduced aging-related ROS generation, mtDNA fragment buildup, mitochondrial protein oxidative damage, and lipofuscin accumulation. A reduction in the RAPTOR level and

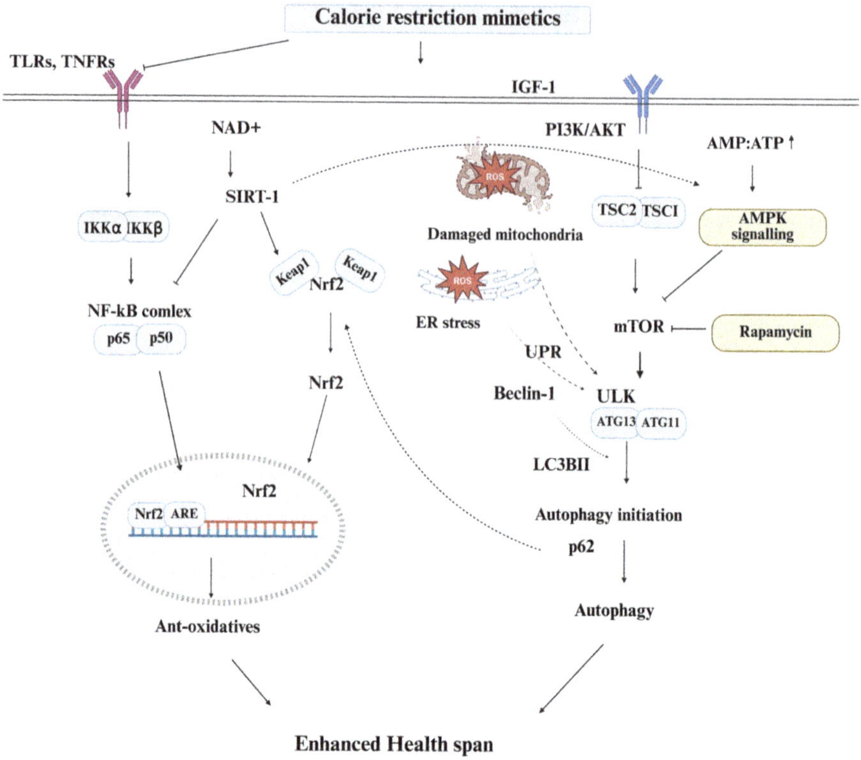

Fig. 19.2 Several calorie restriction mimetics (CRMs) and their associated pathways, including AMPK activation, mTOR inhibition, and sirtuin modulation, contribute to enhanced cellular function, stress resistance, and healthy aging

simultaneous increase in the PGC1α and ATG13 protein levels were observed (Martínez-Cisuelo et al. 2016). A pilot study was conducted to assess the safety of rapamycin in 25 healthy adults aged 70–95 years. The participants took 1 mg of rapamycin or a placebo daily for 8 weeks. Minor side effects, including rash and gastrointestinal issues, were reported. Rapamycin led to slight decreases in RBC parameters without clinical impact. Cognitive function, physical performance, immune markers, and glucose metabolism remained unchanged. Overall, short-term rapamycin use appeared safe, warranting further research (Kraig et al. 2018). A phase 2b and phase 3 trial assessed RTB101 in adults aged 65+ across New Zealand, Australia, and the USA. In phase 2b, 10 mg RTB101 daily reduced the incidence of respiratory infections (19% vs. 28%, $p = 0.02$). Phase 3 had no effect on respiratory illness (26% vs. 25%, $p = 0.65$). RTB101 was well tolerated and boosted antiviral responses. Further research is needed (Mannick et al. 2021).

19.3.3 Metformin

The diabetic medication metformin has attracted considerable interest for its potential to address multiple aging hallmarks. Metformin activates the AMPK pathway by suppressing the mTOR pathway, a crucial regulator of cellular growth, proliferation, autophagy, and aging (Garg et al. 2017a; Howell et al. 2017; Tripathi et al. 2021). It also mitigates mitochondrial oxidative damage while reducing the excessive removal of dysfunctional mitochondria (Garg et al. 2017b; Chen et al. 2022). By activating AMPK and modulating the IIS pathway, mTOR is suppressed, triggering ULK1 through dephosphorylation (Fig. 19.2) (Zhu et al. 2020). Deficiencies in autophagy and disturbances in the mTOR-AMPK-ULK1 signaling pathway hinder autophagy, leading to cellular deterioration, increased apoptosis, and accelerated tissue aging (J. Yuan et al. 2018). Additionally, it reduces inflammation through various mechanisms, primarily by modulating the AMPK/NF-κB, AMPK/mTOR, and ROS pathways (Table 19.1) (Zhang et al. 2025). These findings indicate that metformin supports memory restoration and protects neurons by enhancing neurogenesis and reducing inflammation through the mTOR/AMPK/S6K/Bace1 pathway and the hippocampus NF-κB AMPK/P65 pathway. These processes may help alleviate neurological impairments (Table 19.1) (Ou et al. 2018). Metformin protected substantia nigra compacta dopaminergic neurons from degeneration, increased striatal dopamine levels, improved motor function and reduced microglia-driven neuroinflammation in the substantia nigra compacta (Odubela et al. 2025). Metformin increases neurogenesis, restores mitochondria, and enhances memory and motor skills. It reduces LPS-induced brain damage in the frontal cortex and hippocampus, with 200 mg/kg being more effective than 100 mg/kg. The treated groups had fewer falls in the rotarod test, indicating improved coordination (Blossom et al. 2025).

In humans, prolonged metformin consumption has been linked to a lower risk of dementia and cognitive decline in older individuals with T2DM. In a study of 20 nondiabetic subjects with mild cognitive impairment or mild dementia due to AD, participants received metformin and a placebo for 8 weeks each. Metformin was safe, well tolerated, and detected in cerebrospinal fluid (CSF) at an average concentration of 95.6 ng/ml. It improved executive function, with trends suggesting benefits in memory and attention. No significant changes in cerebral blood flow (CBF) were noted, although a post hoc analysis indicated increased orbitofrontal CBF with metformin. Further research is recommended to study the impact of metformin (Koenig et al. 2017). This study assessed the safety and efficacy of adding metformin to IFNβ-1a therapy in relapsing-remitting multiple sclerosis patients. Eighty participants were split into two groups: one treated with IFNβ-1a and 2 g of metformin daily and the other treated with IFNβ-1a alone. After six months, both groups presented similar levels of IL-17 and IL-22 and similar EDSS scores and MRI findings. However, metformin significantly decreased the level of MDA, an indicator of oxidative stress ($p = 0.038$). Further large-scale research is advised to confirm these results (Abdelgaied et al. 2024).

Table 19.1 CRMs reported are to increase longevity in various model organisms and human clinical trials

Active ingredients	Model organism	Mechanism	Effects	References
Apigenin	Streptozotocin-induced diabetic rats	Regulates blood glucose; Modulates hepatic phase I and II enzymes; Enhances GLUT4 movement; Reduces CD38 expression	Improved glucose metabolism; Restored antioxidant defense; Reduced lipid peroxidation	Hossain (2014)
	Sepsis-induced lung injury model	Blocks TLR7–miR-146a interaction; promotes anti-inflammatory M2 macrophage polarization	Reduces inflammation and lung injury in ALI	Geng et al. (2025)
Mito-Esculetin + Metformin	Human Aortic Endothelial Cells, Mice	Activated AMPK-SIRT1/SIRT6 pathways, reduced NF-κB signaling, enhanced antioxidant defenses	Reduced atherosclerotic lesions, vascular inflammation, and senescence	Pulipaka et al. (2024)
Metformin	Mice	Inhibited p62/RIP1/RIP3 necrosome formation	Reduced myocardial necroptosis and I/R damage	Li et al. (2020)
	LPS-exposed rats	Enhanced neurogenesis, restored mitochondrial function	Improved memory, motor coordination, and reduced neuronal damage in cortex and hippocampus	Blossom et al. (2025)
	Aged Rats	Activated SIRT1/PPAR-γ/irisin pathway; Maintained redox balance	Reduced liver damage and sustained liver function	Muhammad et al. (2024)
	HTR-8 cells in preeclampsia model	Reduces NLRP3 inflammasome-driven pyroptosis; Inhibits TLR4/NF-κB signaling; Prevents NF-κB1 binding to PFKFB3 promoter	Improved glycometabolic alterations; Reduced oxidative stress and inflammation	Zhang et al. (2021)
Resveratrol	Humans, Mice	Reduced NF-κB and p38 MAPK signaling, increased SIRT1 levels	Improved lipid profile, reduced inflammation and oxidative stress	Deng et al. (2015)
	Humans (Clinical Trial) (NCT01244360)	Reduced IL-8, VCAM, and ICAM mRNA expression	Reduced inflammatory markers in coronary artery endothelial cells	Agarwal et al. (2013)

(continued)

Table 19.1 (continued)

Active ingredients	Model organism	Mechanism	Effects	References
	Rats (gas explosion model)	Decreases TNF-α, IL-1β; increases ENaC and PI3K/AKT	Reduced pulmonary edema, oxidative stress, inflammatory cell infiltration	Pai et al. (2024)
	Humans (Dyslipidemia Patients)	Improved lipid metabolism	Reduced total cholesterol and triglyceride levels	Simental-Mendía and Guerrero-Romero (2019)
	Human myotubes (Lean and Obese)	Enhances insulin-driven glucose metabolism; Increases TBC1D4 phosphorylation in lean individuals; Activates AMPK	Improves insulin resistance; No significant effect on insulin-related signaling in obese females	Richter and Ruderman (2009), Park et al. (2024)
	High-fat diet (HFD) induced obese mice	Alleviates oxidative stress and insulin resistance via Nrf2-Keap1 mechanism	Reduced oxidative stress; Improved insulin sensitivity	Shabani et al. (2024)
Rapamycin	Young and Aged Mice	Targeted mTOR pathway; Reduced ribosomal protein turnover	Lowered liver degeneration, improved protein stability, reduced liver pathology	Wilkinson et al. (2012), Y. Zhang et al. (2014)
Spermidine	Wild-type and MAP1S Knockout Mice	Inhibited hepatic stellate cell (HSC) activation; Reduced ECM protein levels	Prevented and reduced liver fibrosis; Increased lipid droplets in HSCs	Shi et al. (2023)
Curcumin & Quercetin	Alzheimer's disease *Drosophila melanogaster* model	Reduced oxidative stress and acetylcholinesterase levels	Enhanced survival, locomotor function, antioxidant activity	Odubela et al. (2025)
Metformin + Glucose Oxidase (GOx)	Cancer cell	Suppresses glycolysis, activates AMPK and p53, impairs Warburg effect	Inhibited cell proliferation, triggered apoptosis	Meng et al. (2022)
MPX Capsules (Muscadine Grape)	Phase I/II clinical trial in prostate cancer patients NCT01317199	Contains quercetin, ellagic acid, and resveratrol	Safe, but no significant effect on PSADT	Paller et al. (2018)

19.3.4 Spermidine

Spermidine is an endogenous polyamine synthesized by cellular metabolism and intestinal microbiota. Spermidine enhances autophagy by hindering mTORC1 activity, simultaneously stimulates AMPK-FOXO3a, and activates the eIF5A factor, promoting autophagy via TEFB transcription factor formation (Fan et al. 2017; Yan et al. 2019). Furthermore, spermidine treatment reduced the high-fat diet-induced increases in the Srebf1c, Fasn, and Pparγ mRNAs, contributing to decreased liver fat accumulation. The expression of the β-oxidation genes PPARα and CPT1a increases with spermidine (Ma et al. 2021). Moreover, it reduces inflammatory responses by inhibiting NF-kB p65 nuclear localization and increasing MAPK and Akt phosphorylation (Table 19.1) (Choi and Park 2012; Singh et al. 2021; S. Singh et al. 2024b). Furthermore, it prevents cardiac degradation during aging by stimulating mitochondrial biogenesis through the SIRT1/PGC-1α pathway in rats (Wang et al. 2020). Dietary spermidine supplementation reduces IGF and insulin signaling pathways, extending lifespan (Tain et al. 2020). This study (NCT05459961) evaluated the safety of a novel spermidine supplement in humans. In a placebo-controlled study, 37 healthy men aged 50–70 years received 40 mg/day hpSPD or placebo for up to 28 days. No adverse effects or significant changes in clinical, metabolic, or blood markers were observed, and the polyamine levels remained stable, suggesting that homeostasis was maintained. These findings suggest that hpSPD at this dose is safe and well tolerated, supporting further research into its potential health benefits (Keohane et al. 2024).

19.3.5 Berberine

Berberine, an alkaloid polyphenol, has anti-inflammatory, antioxidant, and neuroprotective effects. It reduces spinal cord inflammation, minimizes neuronal apoptosis, and enhances autophagy, promoting neuron survival. Berberine also activates the PI3K/Akt pathway, stimulating Nrf2 and HO-1, reducing ROS production, and protecting cells (Table 19.1) (Zhang et al. 2016). In cardiac studies, it reduced arrhythmias, infarct size, and apoptosis, with borneol further enhancing its protective effects (Pan et al. 2024). The randomized clinical trial (NCT06411275) included 113 patients with T2D. HTD1801 significantly reduced HbA1c levels in a dose-dependent manner: −0.4% (500 mg, $p = .04$) and − 0.7% (1000 mg, $p < .001$) compared with placebo at week 12. FPG levels also improved: −13.0 mg/dL (500 mg) and −18.4 mg/dL (1000 mg). The 1000 mg group showed additional benefits in terms of lipids and liver markers. HTD1801 was well tolerated, with mild adverse events in 52.2% of the participants. These findings support HTD1801 as a potential oral treatment for T2D.

19.3.6 Trehalose

A disaccharide consisting of two glucose units connected by an α,α-1,1-glycosidic linkage. Trehalose stimulates autophagy by regulating target genes linked to TFEB. Additionally, it reduces SQSTM1/p62 protein levels and facilitates the conversion of LC3B-I to LC3-II, which is involved in autophagosome formation (Rusmini et al. 2019). The mechanism aids in preventing the buildup of misfolded proteins, suggesting its potential as a treatment for several degenerative conditions. However, the upregulation of autophagy markers in ocular tissues has also been shown to contribute to enhanced retinal function and attenuation of Bruch's membrane thickening in a murine model of age-related macular degeneration (Vessey et al. 2022). Furthermore, trehalose promotes autophagy and decreases the accumulation of α-synuclein in mice exhibiting excessive α-synuclein expression. This process aids in clearing protein aggregates, potentially mitigating the neurodegenerative damage associated with conditions such as PD (Tanji et al. 2015).

19.3.7 Pterostilbene (Pts)

Pts is a natural antioxidant and polyphenol compound found in blueberries and other plants. Chemically similar to resveratrol. Pts therapy decreased the IL-4, IL-13, and IL-5 levels while increasing the serum SOD and CAT levels and lowering the MDA level. It stimulated AMPK phosphorylation in 16HBE cells and suppressed the LPS-induced increase in ROS. Moreover, Pts suppressed p-AMPK, SIRT1, Nrf2, and HO-1, resulting in the mitigation of the AMPK/SIRT1 and Nrf2/HO-1 signaling pathways (Xu et al. 2021; X. Wang et al. 2024a). These findings suggest that Pts helps prevent chondrocyte aging in osteoarthritis by suppressing the PI3K, AKT, and NF-κB pathways while lowering SASP production (Wang et al. 2022).

19.3.8 NAD^+

Aging causes a reduction in SIRT1 and NAD+; the administration of NMN, which serves as a precursor for NAD+; restored NAD+ concentrations; and activated SIRT1 function (Guan et al. 2017). The NAD-dependent protein deacetylase SIRT1 was notably upregulated after Nam treatment, indicating that Nam enhances liver regeneration by stimulating SIRT1 activity in mice (Wan et al. 2019). NMN treatment reversed these effects. Apoptosis, inflammation, and oxidative stress in the hippocampus were notably reduced with NMN. However, the benefits of NMN against memory issues, inflammation, and oxidative harm are blocked by the SIRT1 inhibitor EX-527 (Li et al. 2023). Research has demonstrated that delivering NAD+ protects against neuroinflammation, mitochondrial injury, and ROS generation triggered by hypoxia in BV2 microglia through the Sirt1 and PGC-1α pathways (Zhao et al. 2021).

19.3.9 Apigenin

Apigenin is a flavonoid known for its antiaging effects, which promote osteoblastic differentiation and bone formation while reducing age-related effects in hBMSCs (Ali et al. 2024). In *C. elegans*, apigenin glycosides extend lifespan by 23.9% by lowering ROS levels and activating antiaging genes such as *daf-2*, *daf-16*, and *sir-2.1* (Elkhedir et al. 2022). Additionally, apigenin alleviates oxidative stress, reduces apoptosis, and enhances proteostasis by increasing antioxidant enzyme activity (Oyebode et al. 2020). Additionally, it increased Pten expression and AKT/S6K phosphorylation. Additionally, in mice, apigenin alleviated age-related damage, increased Nrf2 nuclear translocation, and upregulated the expression of antioxidant genes such as HO-1 and NQO1 (Sang et al. 2017; Sun et al. 2025).

19.3.10 2-Deoxy-D-Glucose (2-DG)

2-DG, a synthetic glucose analog glycolytic inhibitor, is transported into cells via glucose transporters and inhibits glucose-6-phosphate isomerase, thereby reducing cellular energy and ATP (Singh et al. 2023). The mitohormesis factor 2-DG enhances ROS production, triggering catalase activity and improving survival rates in *C. elegans*. Suppressing glycolysis and activating AMPK prompts the body to rely on stored fat and mitochondrial respiration for energy (Schulz et al. 2007). Dietary supplementation with a low dose of 2-DG in Wistar rats (young and old) resulted in elevated levels of PMRS, ROS, CAT, and SOD. Treatment with 2-DG alleviates sialic acid in plasma and AGE products (Saraswat et al. 2019). Prolonged consumption of 2-DG leads to cardiac vacuolization and increased mortality rates in rats (Yang et al. 2011).

19.3.11 Acarbose

Acarbose, an alpha-glucosidase inhibitor, has been shown to extend lifespan in preclinical models, with greater longevity benefits observed in males. Additionally, it has been associated with reduced body weight gain, independent of any decrease in caloric intake (Smith et al. 2021). Patients using acarbose for type 2 diabetes (T2D) have experienced lower rates of heart disease, heart attack, and high blood pressure (Hsu et al. 2018). Recent research suggests that acarbose treatment in individuals with T2D may be linked to a lower likelihood of developing or worsening dementia (Tseng 2019). Emerging evidence from both preclinical and clinical models indicates that acarbose may help relieve symptoms and slow the progression of inflammatory conditions such as arthritis (Zhang et al. 2020).

19.4 Conclusion and Future Perspective

Dietary intake significantly influences lifestyle by impacting physical health, mental well-being, energy levels, and overall quality of life. Aging and age-related diseases are major concerns for growing populations. Research has focused on extending lifespan and well-being by slowing down aging-associated pathways. CRMs have been found to influence several signaling pathways related to cellular metabolism, inflammation, oxidative stress, and lifespan. Dietary supplements, phytochemical-based supplements, drug supplements, and CRMs offer promising benefits in promoting healthy aging and longevity. Dietary supplements can be beneficial in addressing nutrient deficiencies, supporting metabolic functions, and promoting overall well-being, particularly in individuals with increased nutritional needs. CRMs, among the most well-researched interventions, have been shown to extend lifespan and improve metabolic resilience by activating cellular repair pathways, reducing inflammation, and enhancing mitochondrial function.

Combining these approaches forms a holistic strategy for maintaining health, preserving cognitive and physical function, and promoting longevity. However, it is essential to personalize these interventions, considering individual health conditions, genetic predispositions, and lifestyle factors. A balanced diet rich in phytochemical-dense foods and mindful supplementation can increase vitality and delay the onset of age-related diseases. Adopting these evidence-based strategies offers a promising path toward healthier and longer lives.

Some CRMs, such as resveratrol, metformin, and berberine, are already used by humans mainly for conditions like diabetes or metabolic syndrome. But they are also being explored for their potential anti-aging effects. Most CRMs are commercially available as dietary supplements, not as drugs. For example, resveratrol, metformin, and spermidine are often sold as dietary supplements, while metformin and rapamycin are primarily used for diabetes and immunosuppression. Rapamycin and NAD+ precursors are still under research, with limited approval. Meanwhile, synthetic CRMs such as sirtuin activators and AMPK modulators remain in preclinical or early clinical trials and are tested primarily in animal models. However, the levels of CRMs in natural foods are lower than those used in studies. Thus, getting the therapeutic doses of these molecules is not easy. In preclinical and clinical studies, spermidine and berberine show tolerability with fewer side effects. A deeper insight into the mechanisms of CRMs is still needed to drive forward many therapeutic breakthroughs aimed at improving human health. Given the rapid pace of research in aging research, it is likely that within the next decade, more CRMs will emerge as validated in clinical trials, and regulated interventions for promoting healthy aging and lifespan in humans, especially as clinical trials expand and safety profiles become clearer. However, no definitive and fully reliable human trials prove that CRMs can extend healthy longevity. While some clinical trials and observational studies have shown promising outcomes, such as improved metabolic health, reduced inflammation, and potential markers of aging. Thus, this evidence remains preliminary. Most of the research is based on animal studies or short-term human trials, and large, long-term, controlled studies in diverse human populations are still needed to confirm their effectiveness and safety for promoting healthy aging.

Future studies should focus on key areas, such as the long-term effects of CRMS on health outcomes, which are crucial in diverse populations. Investigating the unintended effects and chronic risks of hormesis and medication is highly important. The underlying mechanisms by which CRMs exert antiaging and longevity effects must be further explored, including the enhancement of autophagy, oxidative stress, and anti-inflammatory mechanisms. A combination of dietary supplements, phytochemicals, and CRMs must be tested for synergistic effects on healthspan. Identifying optimal combinations that maximize health benefits while minimizing adverse effects is essential. Furthermore, developing CRM compounds that replicate the beneficial effects of calorie restriction without drastic dietary changes will provide accessible and practical interventions for promoting longevity. Emerging evidence suggests that the gut microbiome is critical for mediating the effects of dietary supplements, phytochemicals, and caloric restrictions on health and longevity.

Acknowledgments The authors acknowledge the Manipal Academy of Higher Education (MAHE), Manipal, Karnataka, India, for supporting the study.

Authors' Contributions All the authors contributed to conceptualizing the study, writing, and editing the manuscript. All authors approved the final version of the manuscript.

Ethical Approval Not applicable.

Declaration of Competing Interests The authors have no conflicts of interest to disclose.

Data Availability No data were used for the research described in the article.

References

Abdelgaied MY et al (2024) The impact of metformin use on the outcomes of relapse-remitting multiple sclerosis patients receiving interferon beta 1a: an exploratory prospective phase II open-label randomized controlled trial. J Neurol 271(3):1124–1132. https://doi.org/10.1007/s00415-023-12113-2

Agarwal B et al (2013) Resveratrol for primary prevention of atherosclerosis: clinical trial evidence for improved gene expression in vascular endothelium. Int J Cardiol 166(1):246–248. https://doi.org/10.1016/j.ijcard.2012.09.027

Ali D et al (2024) Apigenin and Rutaecarpine reduce the burden of cellular senescence in bone marrow stromal stem cells. Front Endocrinol 15. https://doi.org/10.3389/fendo.2024.1360054

Bellaver B et al (2014) Resveratrol increases antioxidant defenses and decreases proinflammatory cytokines in hippocampal astrocyte cultures from newborn, adult and aged Wistar rats. Toxicol In Vitro 28(4):479–484. https://doi.org/10.1016/j.tiv.2014.01.006

Bhaduri I, Trisal A, Singh AK (2023) Autophagy as a promising therapeutic target in age-associated neurodegenerative disorders. In: Rizvi SI (ed) Emerging Anti-Aging Strategies. Springer Nature, Singapore, pp 41–56. https://doi.org/10.1007/978-981-19-7443-4_3

Blossom V et al (2025) Effect of metformin on the neuronal morphology of frontal cortex and hippocampal regions in lipopolysaccharide induced neuroinflammation in male *Wistar* rats. Transl Res Anat 38:100365. https://doi.org/10.1016/j.tria.2024.100365

Bo S et al (2018) Effects of resveratrol on bone health in type 2 diabetic patients. A double-blind randomized-controlled trial. Nutr Diabetes 8(1):1–10. https://doi.org/10.1038/s41387-018-0059-4

Chen Q et al (2022) Metformin attenuates UVA-induced skin photoaging by suppressing mitophagy and the PI3K/AKT/mTOR pathway. Int J Mol Sci 23(13):6960. https://doi.org/10.3390/ijms23136960

Cho S-H et al (2015) SIRT1 deficiency in microglia contributes to cognitive decline in aging and neurodegeneration via epigenetic regulation of IL-1β. J Neurosci 35(2):807–818. https://doi.org/10.1523/JNEUROSCI.2939-14.2015

Choi YH, Park HY (2012) Anti-inflammatory effects of spermidine in lipopolysaccharide-stimulated BV2 microglial cells. J Biomed Sci 19(1):31. https://doi.org/10.1186/1423-0127-19-31

Deng ZY et al (2015) Resveratrol alleviates vascular inflammatory injury by inhibiting inflammasome activation in rats with hypercholesterolemia and vitamin D2 treatment. Inflamm Res 64(5):321–332. https://doi.org/10.1007/s00011-015-0810-4

Elkhedir AE et al (2022) Apigenin glycosides from green pepper enhance longevity and stress resistance in *Caenorhabditis elegans*. Nutr Res 102:23–34. https://doi.org/10.1016/j.nutres.2022.02.003

Fan J et al (2017) Spermidine coupled with exercise rescues skeletal muscle atrophy from D-gal-induced aging rats through enhanced autophagy and reduced apoptosis via AMPK-FOXO3a signal pathway. Oncotarget 8(11):17475–17490. https://doi.org/10.18632/oncotarget.15728

Garg G et al (2017a) Antiaging effect of metformin on brain in naturally aged and accelerated senescence model of rat. Rejuvenation Res 20(3):173–182. https://doi.org/10.1089/rej.2016.1883

Garg G et al (2017b) Metformin alleviates altered erythrocyte redox status during aging in rats. Rejuvenation Res 20(1):15–24. https://doi.org/10.1089/rej.2016.1826

Geng Y et al (2025) Apigenin attenuated sepsis induced acute lung injury via polarizing macrophage towards M2 by blocking miR-146a →TLR7 interaction. Int Immunopharmacol 152:114446. https://doi.org/10.1016/j.intimp.2025.114446

Ge Y et al (2022) Role of AMPK mediated pathways in autophagy and aging. Biochimie 195:100–113. https://doi.org/10.1016/j.biochi.2021.11.008

Grajower MM, Horne BD (2019) Clinical management of intermittent fasting in patients with diabetes mellitus. Nutrients 11(4):873. https://doi.org/10.3390/nu11040873

Gu J et al (2021) Neuroprotective effect of trans-resveratrol in mild to moderate Alzheimer disease: a randomized, double-blind trial. Neurol Ther 10(2):905–917. https://doi.org/10.1007/s40120-021-00271-2

Guan Y et al (2017) Nicotinamide mononucleotide, an NAD+ precursor, rescues age-associated susceptibility to AKI in a Sirtuin 1–dependent manner. J Am Soc Nephrol 28(8):2337–2352. https://doi.org/10.1681/ASN.2016040385

Hepsomali P, Groeger JA (2021) Diet, sleep, and mental health: insights from the UK biobank study. Nutrients 13(8):2573. https://doi.org/10.3390/nu13082573

Hossain (2014) APIGENIN causes biochemical modulation, GLUT4 and CD38 alterations to improve diabetes and to protect damages of some vital organs in experimental diabetes. Am J Pharmacol Toxicol 9(1):39–52. https://doi.org/10.3844/ajptsp.2014.39.52

Howell JJ et al (2017) Metformin inhibits hepatic mTORC1 signaling via dose-dependent mechanisms involving AMPK and the TSC complex. Cell Metab 25(2):463–471. https://doi.org/10.1016/j.cmet.2016.12.009

Hsu P-F et al (2018) Cardiovascular benefits of Acarbose vs Sulfonylureas in patients with type 2 diabetes treated with metformin. J Clin Endocrinol Metab 103(10):3611–3619. https://doi.org/10.1210/jc.2018-00040

Ingram DK, Roth GS (2015) Calorie restriction mimetics: can you have your cake and eat it, too? Ageing Res Rev 20:46–62. https://doi.org/10.1016/j.arr.2014.11.005

Kapahi P et al (2004) Regulation of lifespan in drosophila by modulation of genes in the TOR signaling pathway. Curr Biol 14(10):885–890. https://doi.org/10.1016/j.cub.2004.03.059

Keohane P et al (2024) Supplementation of spermidine at 40 mg/day has minimal effects on circulating polyamines: an exploratory double-blind randomized controlled trial in older men. Nutr Res 132:1–14. https://doi.org/10.1016/j.nutres.2024.09.012

Koenig AM et al (2017) Effects of the insulin sensitizer metformin in Alzheimer's disease: pilot data from a randomized placebo-controlled crossover study. Alzheimer Dis Assoc Disord 31(2): 107–113. https://doi.org/10.1097/WAD.0000000000000202

Kraig E et al (2018) A randomized control trial to establish the feasibility and safety of rapamycin treatment in an older human cohort: immunological, physical performance, and cognitive effects. Exp Gerontol 105:53–69. https://doi.org/10.1016/j.exger.2017.12.026

Li C et al (2020) Metformin mediates cardioprotection against aging-induced ischemic necroptosis. Aging Cell 19(2):e13096. https://doi.org/10.1111/acel.13096

Li H et al (2023) β-Nicotinamide mononucleotide activates NAD+/SIRT1 pathway and attenuates inflammatory and oxidative responses in the hippocampus regions of septic mice. Redox Biol 63:102745. https://doi.org/10.1016/j.redox.2023.102745

Ma L et al (2021) Spermidine ameliorates high-fat diet-induced hepatic steatosis and adipose tissue inflammation in preexisting obese mice. Life Sci 265:118739. https://doi.org/10.1016/j.lfs.2020.118739

Mannick JB et al (2021) Targeting the biology of ageing with mTOR inhibitors to improve immune function in older adults: phase 2b and phase 3 randomised trials. Lancet Healthy Longev 2(5): e250–e262. https://doi.org/10.1016/S2666-7568(21)00062-3

Martínez-Cisuelo V et al (2016) Rapamycin reverses age-related increases in mitochondrial ROS production at complex I, oxidative stress, accumulation of mtDNA fragments inside nuclear DNA, and lipofuscin level, and increases autophagy, in the liver of middle-aged mice. Exp Gerontol 83:130–138. https://doi.org/10.1016/j.exger.2016.08.002

Meng X et al (2022) Tumor metabolism destruction via metformin-based glycolysis inhibition and glucose oxidase-mediated glucose deprivation for enhanced cancer therapy. Acta Biomater 145: 222–234. https://doi.org/10.1016/j.actbio.2022.04.022

Morselli E et al (2010) Caloric restriction and resveratrol promote longevity through the Sirtuin-1-dependent induction of autophagy. Cell Death Dis 1(1):e10. https://doi.org/10.1038/cddis.2009.8

Muhammad MH et al (2024) An AMPK activator, Metformin, ameliorates aging-related oxidative hepatic injury induced by D-galactose via Irisin/ SIRT1 pathway, a possible role of PPAR-γ besides its antioxidant potency in rats. Bull Egypt Soc Physiol Sci 44(4):212–227. https://doi.org/10.21608/besps.2024.303088.1170

Nassar K et al (2024) The significance of caloric restriction mimetics as anti-aging drugs. Biochem Biophys Res Commun 692:149354. https://doi.org/10.1016/j.bbrc.2023.149354

Odubela OO, Olanrewaju AJ, Daniel ED (2025) Synergistic therapeutic potential of curcumin and quercetin co-administration in Alzheimer's disease: insights from transgenic Drosophila melanogaster models. IPS J Drug Discov Res Rev 3(1):5–14. https://doi.org/10.54117/ijddrr.v3i1.25

Ou Z et al (2018) Metformin treatment prevents amyloid plaque deposition and memory impairment in APP/PS1 mice. Brain Behav Immun 69:351–363. https://doi.org/10.1016/j.bbi.2017.12.009

Oyebode OT et al (2020) Apigenin ameliorates D-galactose-induced lifespan shortening effects via antioxidative activity and inhibition of mitochondrial-dependent apoptosis in *Drosophila melanogaster*. J Funct Foods 69:103957. https://doi.org/10.1016/j.jff.2020.103957

Pai V, Singh BN, Singh AK (2024) Insights into advances and applications of biomaterials for nerve tissue injuries and neurodegenerative disorders. Macromol Biosci 24(12):2400150. https://doi.org/10.1002/mabi.202400150

Paller CJ et al (2018) Muscadine grape skin extract (MPX) in men with biochemically recurrent prostate cancer: a randomized, multicenter, placebo-controlled clinical trial. Clin Cancer Res 24(2):306–315. https://doi.org/10.1158/1078-0432.CCR-17-1100

Pan X et al (2024) Borneol promotes berberine-induced cardioprotection in a rat model of myocardial ischemia/reperfusion injury via inhibiting P-glycoprotein expression. Eur J Pharmacol 983:177009. https://doi.org/10.1016/j.ejphar.2024.177009

Park S et al (2024) Effect of resveratrol on insulin action in primary myotubes from lean individuals and individuals with severe obesity. Am J Physiol 326(3):E398–E406. https://doi.org/10.1152/ajpendo.00299.2023

Pulipaka S et al (2024) Therapeutic efficacies of mitochondria-targeted esculetin and metformin in the improvement of age-associated atherosclerosis via regulating AMPK activation. GeroScience 46(2):2391–2408. https://doi.org/10.1007/s11357-023-01015-w

Richter EA, Ruderman NB (2009) AMPK and the biochemistry of exercise: implications for human health and disease. Biochem J 418(2):261–275. https://doi.org/10.1042/BJ20082055

Rusmini P et al (2019) Trehalose induces autophagy via lysosomal-mediated TFEB activation in models of motoneuron degeneration. Autophagy 15(4):631–651. https://doi.org/10.1080/15548627.2018.1535292

Sang Y et al (2017) Apigenin exhibits protective effects in a mouse model of D-galactose-induced aging via activating the Nrf2 pathway. Food Funct 8(6):2331–2340. https://doi.org/10.1039/C7FO00037E

Saraswat K, Kumar R, Rizvi SI (2019) Glycolytic inhibitor 2-Deoxy- D -glucose at chronic low dose mimics calorie restriction in rats through Mitohormetic induction of reactive oxygen species. Rejuvenation Res 22(5):377–384. https://doi.org/10.1089/rej.2018.2125

Schreiber KH et al (2015) Rapamycin-mediated mTORC2 inhibition is determined by the relative expression of FK506-binding proteins. Aging Cell 14(2):265–273. https://doi.org/10.1111/acel.12313

Schulz TJ et al (2007) Glucose restriction extends Caenorhabditis elegans life span by inducing mitochondrial respiration and increasing oxidative stress. Cell Metab 6(4):280–293. https://doi.org/10.1016/j.cmet.2007.08.011

Shabani M et al (2024) Resveratrol relieves HFD-induced insulin resistance in skeletal muscle tissue through antioxidant capacity enhancement and the Nrf2-Keap1 signaling pathway. Mol Biol Rep 51(1):516. https://doi.org/10.1007/s11033-024-09434-4

Sharma A, Singh AK (2023) Molecular mechanism of caloric restriction mimetics-mediated neuroprotection of age-related neurodegenerative diseases: an emerging therapeutic approach. Biogerontology 24(5):679–708. https://doi.org/10.1007/s10522-023-10045-y

Shi B et al (2023) Spermidine suppresses the activation of hepatic stellate cells to cure liver fibrosis through autophagy activator MAP1S. Liver Int 43(6):1307–1319. https://doi.org/10.1111/liv.15558

Simental-Mendía LE, Guerrero-Romero F (2019) Effect of resveratrol supplementation on lipid profile in subjects with dyslipidemia: a randomized double-blind, placebo-controlled trial. Nutrition 58:7–10. https://doi.org/10.1016/j.nut.2018.06.015

Singh AK et al (2016) Rapamycin alleviates oxidative stress-induced damage in rat erythrocytes. Biochem Cell Biol 94(5):471–479. https://doi.org/10.1139/bcb-2016-0048

Singh S et al (2018) Fisetin as a caloric restriction mimetic protects rat brain against aging induced oxidative stress, apoptosis and neurodegeneration. Life Sci 193:171–179. https://doi.org/10.1016/j.lfs.2017.11.004

Singh S et al (2021) Spermidine, a caloric restriction mimetic, provides neuroprotection against normal and d-galactose-induced oxidative stress and apoptosis through activation of autophagy in male rats during aging. Biogerontology 22(1):35–47. https://doi.org/10.1007/s10522-020-09900-z

Singh A et al (2023) Glycolytic inhibitors as Caloric Restriction Mimetics (CRM). In: Rizvi SI (ed) Emerging anti-aging strategies. Springer Nature, Singapore, pp 57–78. https://doi.org/10.1007/978-981-19-7443-4_4

Singh I et al (2024a) Caloric restriction mimetics improve gut microbiota: a promising neurotherapeutics approach for managing age-related neurodegenerative disorders. Biogerontology 25(6):899–922. https://doi.org/10.1007/s10522-024-10128-4

Singh S et al (2024b) Spermidine protects cellular redox status and ionic homeostasis in D-galactose induced senescence and natural aging rat models. Z Naturforsch C [Preprint]. https://doi.org/10.1515/znc-2024-0181

Smith DL et al (2021) Diabetes medications as potential calorie restriction mimetics—a focus on the alpha-glucosidase inhibitor acarbose. Geroscience 43(3):1123–1133. https://doi.org/10.1007/s11357-020-00278-x

Strilbytska O et al (2024) Intermittent fasting and longevity: from animal models to implication for humans. Ageing Res Rev 96:102274. https://doi.org/10.1016/j.arr.2024.102274

Sun Z, Li L, Zhang L (2025) Apigenin enhancing oxidative resistance and proteostasis to extend lifespan via PTEN-mediated AKT signalling pathway. Biochim Biophys Acta 1871(3):167670. https://doi.org/10.1016/j.bbadis.2025.167670

Tain LS et al (2020) Longevity in response to lowered insulin signaling requires glycine N-methyltransferase-dependent spermidine production. Aging Cell 19(1):e13043. https://doi.org/10.1111/acel.13043

Tanji K et al (2015) Trehalose intake induces chaperone molecules along with autophagy in a mouse model of Lewy body disease. Biochem Biophys Res Commun 465(4):746–752. https://doi.org/10.1016/j.bbrc.2015.08.076

Tripathi SS et al (2021) Metformin protects red blood cells against rotenone induced oxidative stress and cytotoxicity. *Arch Physiol Biochem* 127(2):102–111. https://doi.org/10.1080/13813455.2019.1620288

Trisal A, Singh AK (2024) Clinical insights on caloric restriction Mimetics for mitigating brain aging and related neurodegeneration. Cell Mol Neurobiol 44(1):67. https://doi.org/10.1007/s10571-024-01493-2

Trisal A, Singh AK (2025) Mechanisms and early efficacy data of caloric restriction and caloric restriction mimetics in neurodegenerative disease. Neuroscience 567:235–248. https://doi.org/10.1016/j.neuroscience.2025.01.004

Tseng C-H (2019) Dementia risk in type 2 diabetes patients: Acarbose use and its joint effects with metformin and pioglitazone. Aging Dis 11(3):658–667. https://doi.org/10.14336/AD.2019.0621

Vessey KA et al (2022) Treatments targeting autophagy ameliorate the age-related macular degeneration phenotype in mice lacking APOE (apolipoprotein E). Autophagy 18(10):2368–2384. https://doi.org/10.1080/15548627.2022.2034131

Vitale G et al (2019) ROLE of IGF-1 system in the modulation of longevity: controversies and new insights from a centenarians' perspective. Front Endocrinol 10. https://doi.org/10.3389/fendo.2019.00027

Wan H-F et al (2019) Nicotinamide induces liver regeneration and improves liver function by activating SIRT1. Mol Med Rep 19(1):555–562. https://doi.org/10.3892/mmr.2018.9688

Wang J et al (2020) Spermidine alleviates cardiac aging by improving mitochondrial biogenesis and function. Aging (Albany NY) 12(1):650–671. https://doi.org/10.18632/aging.102647

Wang Y et al (2022) Senomorphic agent pterostilbene ameliorates osteoarthritis through the PI3K/AKT/NF-κB axis: an in vitro and in vivo study. Am J Transl Res 14(8):5243–5262. https://www.ncbi.nlm.nih.gov/pmc/articles/PMC9452324/. Accessed 26 Mar 2025

Wang X et al (2024a) Pterostilbene, a resveratrol derivative, improves ovary function by upregulating antioxidant defenses in the aging chickens via increased SIRT1/Nrf2 expression. Antioxidants 13(8):935. https://doi.org/10.3390/antiox13080935

Wang Y et al (2024b) Unveiling the longevity potential of natural phytochemicals: a comprehensive review of active ingredients in dietary plants and herbs. J Agric Food Chem 72(45):24908–24927. https://doi.org/10.1021/acs.jafc.4c07756

Wilkinson JE et al (2012) Rapamycin slows aging in mice. Aging Cell 11(4):675–682. https://doi.org/10.1111/j.1474-9726.2012.00832.x

Wu J et al (2019) Rapamycin improves bone mass in high-turnover osteoporosis with iron accumulation through positive effects on osteogenesis and angiogenesis. Bone 121:16–28. https://doi.org/10.1016/j.bone.2018.12.019

Xu C et al (2021) Pterostilbene suppresses oxidative stress and allergic airway inflammation through AMPK/Sirt1 and Nrf2/HO-1 pathways. Immun Inflamm Dis 9(4):1406–1417. https://doi.org/10.1002/iid3.490

Yan J et al (2019) Spermidine-enhanced autophagic flux improves cardiac dysfunction following myocardial infarction by targeting the AMPK/mTOR signalling pathway. Br J Pharmacol 176(17):3126–3142. https://doi.org/10.1111/bph.14706

Yang Y, Zhang L (2020) The effects of caloric restriction and its mimetics in Alzheimer's disease through autophagy pathways. Food Funct 11(2):1211–1224. https://doi.org/10.1039/C9FO02611H

Yang N-C et al (2011) Effects of 2-deoxyglucose and dehydroepiandrosterone on intracellular NAD+ level, SIRT1 activity and replicative lifespan of human Hs68 cells. Biogerontology 12(6):527–536. https://doi.org/10.1007/s10522-011-9342-7

Yuan J et al (2018) Autophagy regulates the degeneration of the auditory cortex through the AMPK-mTOR-ULK1 signaling pathway. Int J Mol Med [Preprint]. https://doi.org/10.3892/ijmm.2018.3393

Zhang Y et al (2014) Rapamycin extends life and health in C57BL/6 mice. J Gerontol: Series A 69A (2):119–130. https://doi.org/10.1093/gerona/glt056

Zhang X et al (2016) Protective effects of Berberine on renal injury in Streptozotocin (STZ)-induced diabetic mice. Int J Mol Sci 17(8):1327. https://doi.org/10.3390/ijms17081327

Zhang L et al (2020) Alpha-glucosidase inhibitors alter gut microbiota and ameliorate collagen-induced arthritis. Front Pharmacol 10. https://doi.org/10.3389/fphar.2019.01684

Zhang Y et al (2021) Metformin corrects glucose metabolism reprogramming and NLRP3 Inflammasome-induced Pyroptosis via inhibiting the TLR4/NF-κB/PFKFB3 signaling in trophoblasts: implication for a potential therapy of preeclampsia. Oxid Med Cell Longev 2021(1):1806344. https://doi.org/10.1155/2021/1806344

Zhang T et al (2025) The anti-aging mechanism of metformin: from molecular insights to clinical applications. Molecules 30(4):816. https://doi.org/10.3390/molecules30040816

Zhao Y et al (2021) NAD+ improves cognitive function and reduces neuroinflammation by ameliorating mitochondrial damage and decreasing ROS production in chronic cerebral hypoperfusion models through Sirt1/PGC-1α pathway. J Neuroinflammation 18(1):207. https://doi.org/10.1186/s12974-021-02250-8

Zhu J et al (2020) Exogenous NADPH ameliorates myocardial ischemia–reperfusion injury in rats through activating AMPK/mTOR pathway. Acta Pharmacol Sin 41(4):535–545. https://doi.org/10.1038/s41401-019-0301-1

Chapter 20
Dasatinib and Quercetin Senolytics for Health and Longevity

Verónica Salas-Venegas, Rosa Pamela Flores-Torres, Adriana Alarcón-Aguilar, Norma Edith López-Díazguerrero, and Mina Königsberg

Abstract Dasatinib and Quercetin (D+Q) are the most studied senolytic agents. Although they can effectively act as single drugs, their combined use has been shown to selectively target a wider range of senescent cells. In this way, together and administered intermittently, they have shown positive results in inducing apoptosis in senescent cells. This chapter comprehensively examines their efficacy, selectivity, as well as their positive and adverse effects in different models, revealing both their therapeutic potential and limitations. While D+Q treatment has demonstrated benefits in certain age-related conditions and specific cell types, its effects vary significantly depending on the senescence inducer, tissue type, and age. Moreover, their efficacy is often cell-type dependent, producing different outcomes between males and females. Before drugs like D+Q can be used to eliminate senescent cells, it is crucial to consider issues of heterogeneity, including individual drug response and sex-dependent differences. Additional clinical studies must be carried out before these drugs can be made available to the public.

Keywords Aging · Senescence · Senolysis · Senolytics

V. Salas-Venegas
Unidad de Investigación en Medicina Experimental, Facultad de Medicina, Universidad Nacional Autónoma de México, Mexico, Ciudad de México

R. P. Flores-Torres
Departamento de Ciencias de la Salud, Laboratorio de Bioenergética y Envejecimiento Celular, Universidad Autónoma Metropolitana-Iztapalapa, Ciudad de México, Mexico

Posgrado en Biología Experimental, División de Ciencias Biológicas y de la Salud, Universidad Autónoma Metropolitana-Iztapalapa, Ciudad de México, Mexico

A. Alarcón-Aguilar · N. E. López-Díazguerrero · M. Königsberg (✉)
Departamento de Ciencias de la Salud, Laboratorio de Bioenergética y Envejecimiento Celular, Universidad Autónoma Metropolitana-Iztapalapa, Ciudad de México, Mexico
e-mail: norm@xanum.uam.mx; mkf@xanum.uam.mx

S. I. S. Rattan, C. Gezer (eds.), *Supplements and Drugs for Healthy Longevity*, Healthy Ageing and Longevity 22, https://doi.org/10.1007/978-3-032-11608-6_20

20.1 Introduction

One of the hallmarks of aging is cellular senescence (López-Otín et al. 2013, 2023), a state in which cells stop proliferating but remain metabolically active. Senescent cells are characterized by overexpressing cyclin-dependent kinase inhibitors, such as p16 and p21, and cellular and nuclear morphology changes. Cells can enter senescence in response to different stressors (Rodier et al. 2009; Gorgoulis et al. 2019), and its most important feature is the secretion of a wide variety of cytokines and chemokines, which over time become increasingly proinflammatory and are collectively referred to as senescent-associated secretory phenotype (SASP) (Coppé et al. 2010; Hernandez-Segura et al. 2018). The SASP molecules can have autocrine or paracrine effects, affecting neighboring cells by activating various cell surface receptors and signal transduction pathways (Coppé et al. 2010). The components of SASP can vary depending on the senescence inducer (Maciel-Barón et al. 2016), time (Acosta et al. 2013), and tissue involved, and can have both beneficial and detrimental effects (Ritschka et al. 2017; Antelo-Iglesias et al. 2021; Wang et al. 2024). Among the SASP's beneficial effects, tissue remodeling during embryogenesis (Storer et al. 2013) and the wound-healing process (Demaria et al. 2014) have been reported. Moreover, senescent cells can act as a tumor suppressor mechanism that prevents the damaged cells from further proliferation and malignization (Collado and Serrano 2010). SASP also contains chemokines that attract the immune system to remove senescent cells when they have fulfilled their function. However, the immune system fails to eliminate them with age, and senescent cells accumulate within the tissue. The SASP becomes more inflammatory and has been associated with aging decline and related diseases (Rodier and Campisi 2011) (Fig. 20.1).

20.1.1 Senolytics Eliminate Senescent Cells

In 2011, the first report was published in which the selective elimination of p16Ink4-positive senescent cells was performed by administering a drug in the transgenic mouse (INK-ATTAC) (Baker et al. 2011). This report demonstrated that eliminating senescent cells delayed the deterioration associated with aging in some tissues. From then on, it was accepted that senescent cells are involved in the decline and pathologies related to aging. Since it is not feasible to eliminate all cells overexpressing p16Ink4 *in vivo*, other molecules were sought as therapeutic targets. Through bioinformatic strategies, it was discovered that several anti-apoptotic pathways within senescent cells are overexpressed and that they are the ones that allow their survival by counteracting the pro-apoptotic pathways, as well as neutralizing the effect of SASP that also promotes death in an autocrine and paracrine manner (Zhu et al. 2015). These molecules and metabolic pathways are called senescent cells antiapoptotic pathways (SCAP). Since then, scientists in the aging field, encouraged by the pharmaceutical industry, have embarked on a quest to find drugs that target

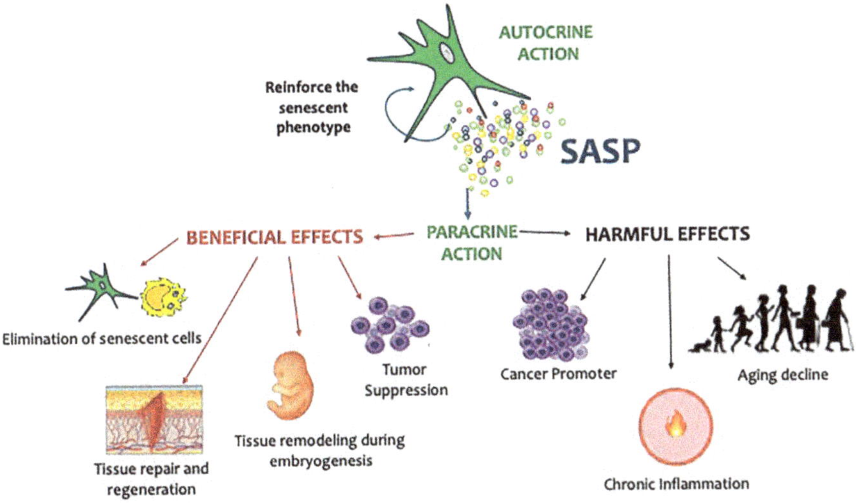

Fig. 20.1 The SASP has beneficial and harmful effects
The SASP has beneficial effects such as tissue remodeling in the embryonic stage and wound repair, and it is a signaling mediator that attracts the immune system to eliminate senescent cells when they have completed their function. However, if senescent cells accumulate in the tissues, their SASP can become more inflammatory and harmful by increasing inflammation and favoring the appearance of tumors and other diseases

the SACPs to selectively eliminate senescent cells. These molecules have been called senolytics.

One of the most used strategies, but not the only one, to find senolytics is repositioning drugs already approved and used to treat other conditions (Olascoaga-Del Ángel et al. 2022). The main senolytics include Fisetin, Navitoclax, and FOXO4-DRI, but the most widely studied in animal models and human pilot studies are Dasatinib and Quercetin, which will be described below.

20.2 Dasatinib and Quercetin: Structure and Function

Dasatinib (D) is a small tyrosine kinase inhibitor approved by the Food and Drug Administration in United States in 2006 (Schade et al. 2008; Suresh et al. 2017) (Fig. 20.2a). This molecule potently inhibits BCR/ABL competing for the ATP binding site and the SRC family of kinases (SRC, LCK, HCK, YES, FYN, FGR, BLK, LYN, FRK), as well as c-KIT, PDGFR-α, PDGFR-β, and ephrin receptor kinase. Dasatinib has also been reported to block the G1/S transition and inhibit cell growth (Chen and Chen 2015). It was first used for the treatment of chronic myeloid leukemia and Philadelphia chromosome-positive ALL, particularly in patients resistant to imatinib (Tokarski et al. 2006); and was recently reported to inhibit the production of proinflammatory cytokines, the recruitment of immune system cells

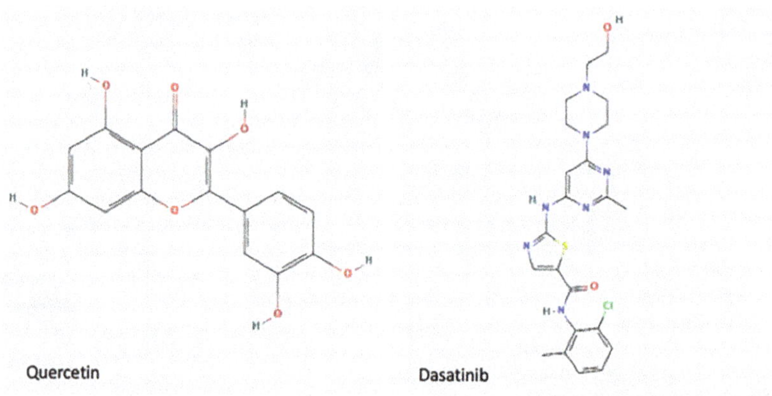

The 2D structures of the quercetin molecule (2-(3,4-dihydroxyphenyl)-3,5,7-trihydroxychromen-4-one), and dasatinib molecule (N-(2-chloro-6-methylphenyl)-2-[[6-[4-(2-hydroxyethyl)piperazin-1-yl]-2-methylpyrimidin-4-yl]amino]-1,3-thiazole-5-carboxamide), are taken from the PubChem database (CID: 5280343) and (CID: 3062316), respectively. https://pubchem.ncbi.nlm.nih.gov/

Fig. 20.2 Molecular structure of the senolytics dasatinib and quercetin
Dasatinib: Yellow (thiazole), green (pyrimidine), red (piperazine)
Quercetin: blue (OH electron donor groups that give it its antioxidant capacity). (Figure created in ChemDraw, adapted from Tokarski et al. (2006), Magar and Sohong (2020))

and tissue destruction, thus avoiding the chronic inflammation seen during aging (Chen and Chen 2015). Its senolytic effect is based on multiple tyrosine kinase inhibitors, in addition to acting on the FOXO-p53 pathway (Park and Shin 2022).

Quercetin (Q) is a naturally occurring flavonoid present in fruits and vegetables, such as onions, mangoes (Magar and Sohng 2020), tea, nuts, wine, and seeds (Kim and Park 2018). Quercetin is also found in medicinal plants such as *ginkgo biloba*, *Hypericum perforatum*, and elderberry (Yang et al. 2020) (Fig. 20.2b). Its name is derived from quercetum (oak woodland), in honor of Quercus (Aghababaei and Hadidi 2023). Due to its chemical structure, it has several biological properties, such as antioxidant, anti-inflammatory, antibacterial, antiviral, gastroprotective, and immunomodulatory, and has been reported to have low toxicity (Kim and Park 2018; Aghababaei and Hadidi 2023). It has also been shown to have anticancer properties by inhibiting cathepsin B in humans (Magar and Sohng 2020). Its effects as a senolytic are based on quercetin acting on PI3K, AKT, p53, and HIF1-alpha pathways, promoting senescent cell apoptosis (Malavolta et al. 2018).

20.3 Effect of D+Q on Cell Cultures: First Studies

Zhu et al. (2015) performed a pioneering study to demonstrate the senolytic effect of dasatinib and quercetin in cell culture. They showed that D has a greater senolytic effect on ionizing radiation-induced senescent human abdominal subcutaneous pre-adipocyte cells (PASAH) than quercetin. While quercetin had higher senolytic activity in senescent HUVEC cells, induced to senescence with the stimulus. Furthermore, the combination of Dasatinib + Quercetin (D+Q) increased the senolytic activity in PASAH and HUVEC cells, decreasing the levels of p21, PAI-2, and Bcl-xL (Zhu et al. 2015), showing for the first time that the combination of D+Q potentiates the senolytic effects and its cell-type dependence.

Schafer et al. 2017, using a different cell type, IMR90 cells (human lung fibroblasts primary culture), confirmed that the combination of D+Q eliminated cells induced to senescence by radiation, but also induced with etoposide (Schafer et al. 2017). However, not all studies were successful. The senolytic effectiveness of quercetin was tested in Murine Embryonic Fibroblasts (MEF) induced to senescence by hypoxia, which generates a state of oxidative stress. The results showed that quercetin lacked senolytic activity in MEF, suggesting that its senolytic activity might depend on the senescence induction mechanism (Yousefzadeh et al. 2018). The combination of D+Q has also been tested in cancer cell lines induced by senescence. Hepatocarcinoma cell lines HepG2 and Huh-7 were induced to senescence with doxorubicin and subsequently exposed to D+Q. Interestingly, the combination of senolytics did not affect these cells, since no decrease in β-Gal activity, nor in the classical senescence biomarkers, such as p16 and histone γ-H2AX, was observed. Corroborating that the senolytic effect of D+Q might depend on the senescence induction mechanism and cell type (Kovacovicova et al. 2018).

Conversely, D+Q treatments might have different outcomes besides their senolytic potential. A recent study showed that in bone marrow-derived mesenchymal stem cells (BMSC) from aged organisms, D+Q treatment decreased the concentrations of several SASP components, including IL-6, CXL-1, and MCP1, thus also describing a senomorphic function for D+Q treatment (i.e., modulation of SASP). Interestingly, BMSC cells from aged organisms increased their proliferation and differentiation in a mechanism involving the over-expression of IBSP (bone sialoprotein) and BGLAP (osteocalcin), two key factors for osteogenesis (Zhou et al. 2021). In another study, D+Q treatment in vascular smooth muscle cells (VSMC) showed an unusual effect on the chromatin structure of control and senescent cells. After D+Q treatment, the chromatin structure of non-senescent cells acquired some characteristics typical of senescent cells. Suggesting that the beneficial effect of D+Q may result, not only from the elimination of senescent cells but also from the "rejuvenation" of their chromatin structure and gene expression, including the decreased expression of genes encoding proteins responsible for SASP, while the effect of D+Q treatment in non-senescent cells induced a transient acquisition of a chromatin-phenotype similar to that of senescent cells, which was lost when removing the senolytics (Gadecka et al. 2025).

A recent study utilizing single-cell analysis and bioinformatics revealed that primary fibroblast cultures derived from young and aged individuals exhibit distinct senescent cell subtypes when senescence is induced with doxorubicin. Moreover, these senescent subtypes responded differently to treatment with D+Q. Single-cell morphological analysis showed variability in biomarker expression among senescent subtypes. Interestingly, this study found that not all cells exposed to senescence-inducing agents become senescent. Some populations can give rise to resistant subpopulations capable of proliferating and restoring cell numbers, as observed in hydrogen peroxide-induced senescence (Kamat et al. 2025).

Additionally, not all senescent subtypes exhibit the same degree of age dependency, and chronological age alone does not fully determine a cell's susceptibility to senescence induction (Kamat et al. 2025). The researchers identified three senescence subtypes, each showing distinct responses to D+Q treatment. One subtype was highly sensitive, with cells undergoing death or morphological instability during treatment, while another was notably resistant. This differential response was attributed to various characteristics of senescent cells: the donor's age, the senescence-inducing stimulus, and the time elapsed since induction. These findings underscore the potential of ex vivo morphological profiling to identify senescence subtypes and predict their therapeutic responses, making it a valuable tool for screening and mapping senolytic treatment outcomes (Kamat et al. 2025).

20.4 Preclinical Studies Using D+Q

Since the pioneering studies of Baker et al. (2011), where p16-positive cells were eliminated in the INK-ATTAK transgenic mouse model, and the studies of Kirkland and his group evaluating molecules with possible senolytic effects (Zhu et al. 2015), multiple studies in animal models have demonstrated that D+Q treatment reduces the burden of senescent cells and inflammation associated with aging. A large number of studies have been performed in transgenic mouse models such as Tau+ mice, where D+Q administered every 2 weeks for 3 months, reduced neurofibrillary tangles, neuroinflammation, gliosis, small vessel hypoperfusion, and ventricular enlargement and increased cortical volume and neurogenesis, thus alleviating Alzheimer's-Like pathology (Musi et al. 2018). In the obesity-induced model and db/db transgenic mice, targeting senescent cells with D+Q alleviates obesity-induced metabolic dysfunction (Palmer et al. 2019). Hence, D+Q decreases the deterioration associated with aging, increasing life expectancy, and delaying or correcting specific outcomes of diseases such as IPF, arthritis, Tau+ Alzheimer's, and diabetes in transgenic mice models.

In addition, there are also studies in other animal models during normal aging, such as in 21-month-old male C57BL/6 mice, where D+Q decreased senescent cells, proinflammatory cytokines, and immune cell infiltration in perigonadal adipose tissue, leading to improved metabolic function (Islam et al. 2023). Similarly, in middle-aged male and female cynomolgus macaques, D+Q reduced senescence markers (p16, p21, PAI-1, and MMP-9) in subcutaneous adipose tissue and promoted an anti-inflammatory immune phenotype (Ruggiero et al. 2023). In aged male mice, D+Q selectively cleared senescent microglia in the hippocampus, improving memory and learning (Ogrodnik et al. 2021). Aged male Wistar rats treated with D +Q showed improved short-term memory, reduced TNF-α, and increased IL-10 expression, suggesting an anti-inflammatory shift (Krzystyniak et al. 2022).

Similarly, cognitive and anti-inflammatory benefits were observed in aged F344 male rats, along with preservation of blood-brain barrier integrity and synaptic function (Budamagunta et al. 2023). In SAMP10 mice, D+Q reduced β-galactosidase-positive cells and p16 expression in the hippocampus, improving physical and cognitive function (Ota and Kodama 2022). Beyond the brain, D+Q also showed benefits in age-related musculoskeletal conditions. In a male murine model of facet joint osteoarthritis, D+Q reversed chondrocyte senescence, SASP expression, and joint degeneration (Zhao et al. 2025). In another study, aged mice exhibited senescence markers and SASP factors in intervertebral discs; long-term D+Q treatment reduced these markers and prevented disc degeneration (Novais et al. 2021).

Collectively, these findings support the therapeutic potential of D+Q in alleviating various age-associated dysfunctions. However, there are also reports where the removal of senescent cells generated deleterious effects in animals (Chaib et al. 2022; Lushchak et al. 2023). The group of Bulavin reported that removal of senescent liver sinusoid cells (LSEC) did not induce cell replacement, generating fibrosis and premature death in mice (Grosse et al. 2020). Moreover, senotherapy's effects have been reported in various models with different effects regarding the animal's sex. As exemplified by the reports described above, D+Q treatment mostly has beneficial effects in males, decreasing senescent markers and proinflammatory factors and improving behavioral test responses. However, in females, the use of senolytics has sometimes shown adverse effects. Fang and collaborators (2023), reported in female C57BL/6 mice that D+Q oral administration once a month (5 mg/kg and 50 mg/kg respectively) from 4 to 13 months of age was associated with memory decline and increased SASP factors (IL-6 and MCP-1) due to an increase in WAT (Fang et al. 2023, 2024). In another study, using a model of middle-aged female Wistar rats with chronic obesity, D+Q treatment (5 mg/kg and 50 mg/kg respectively, monthly from 12 to 14 months of age), failed to prevent cognitive deficits or modulate gut microbial composition; moreover, treating standard diet-fed middle-aged female rats with D+Q was associated with increased proinflammatory cytokines and worse performance on memory and learning tests (Flores-Torres et al. 2025). Hence, it has been proposed that senolytic treatment in middle-aged females could have counterproductive effects by accelerating the drop in estrogenic levels (Rani et al. 2024).

20.5 Clinical Trials Using D+Q

Following the initial investigations that reported promising results in preclinical studies, progress has been made toward clinical trials evaluating the safety and efficacy of senolytics. More than 30 clinical trials of senolytic and senomorphic agents have already been completed, underway, or planned (Suda et al. 2024).

The combination of D+Q is still the most studied senolytic drug. D+Q treatment has been demonstrated to facilitate the selective elimination of senescent cells (Fig. 20.1), which has been validated in many diseases (Guo et al. 2024). Table 20.1 summarizes senolytic clinical trials in different phases: completed and

Table 20.1 Clinical trials using D+Q: completed and current

Name of study	Clinic trial ID	Status	Description	Publications derived from the study
Targeting Pro-inflammatory Cells in Idiopathic Pulmonary Fibrosis (IPF): a Human Trial	NCT02874989	Completed	IPF is a progressive, fatal cellular senescence-associated disease. In this study the researchers applied an intermittent senolitycs administration schedule (D:100 mg/ day, Q:1250 mg/day, three-days/week over three-weeks) and was conducted in participants with IPF (n = 14) to evaluate feasibility of implementing a senolytic intervention. This study reported D +Q effects on circulating SASP factors were inconclusive, but correlations were observed between change in function and change in SASP-related matrix-remodeling proteins, microRNAs, and proinflammatory cytokines. The results observed in this first-in-humans open-label pilot supports study feasibility and provides initial evidence that senolytics may alleviate physical dysfunction in IPF, warranting evaluation of D+Q in larger randomized controlled trials for senescence-related diseases.	Justice et al. (2019b)
Senolytic Therapy to Modulate Progression of Alzheimer's Disease (SToMP-AD)	NCT04063124	Current	This research analyzed an intermittent administration senolytic schedule combination therapy of D+Q in older adults with early-stage Alzheimer's disease. The purpose of this pilot study is to evaluate	Gonzales (2021, 2022)

(continued)

Table 20.1 (continued)

Name of study	Clinic trial ID	Status	Description	Publications derived from the study
			whether a combination of D+Q penetrate the brain using cerebrospinal fluid (CSF) in older adults with early-stage Alzheimer's disease. This research analyzes an intermittent administration senolytic schedule combination therapy 100 mg of D (one 100 mg capsule, Sprycel, Bristol Myers Squibb) and 1000 mg of Q (four 250 mg capsules, Thorne Research) daily for two consecutive days followed by a 14-day (+/– 2 day) no-drug period to complete a single cycle. Six cycles across 12 consecutive weeks. The researchers team want to know if this combination of medications will reach the brain in order to evaluate if this intervention may be effective for treating AD symptoms in future studies.	
Hematopoietic Stem Cell Transplant Survivors Study	NCT02652052	Current	The survivor's patients of hematopoietic stem cell transplant (HSCT) show a higher risk for premature aging. At the moment, there are no studies that evaluate the biologic markers of premature aging and senescence in HSCT survivors and their correlation with clinical outcomes, lifestyle, and nutrition. The objective of this study is evaluate age-related changes in HSCT survivors, with specified measures of premature aging, and	No results posted

(continued)

Table 20.1 (continued)

Name of study	Clinic trial ID	Status	Description	Publications derived from the study
			employ therapeutic opportunities based on targeting senescent cells.	
ALSENLITE: Senolytics for Alzheimer's Disease	NCT04785300	Current	The aim of this study is to evaluate the safety and feasibility of using an intermittent administration of D+Q in patients with Mild Cognitive Impairment (MCI) or Alzheimer's disease. The subjects contemplated are adults over 55 with clinical diagnosis of probable Alzheimer's Disease and Alzheimer's biomarker positivity by tau-PET.	Guerrero (2021)
Senescence in Chronic Kidney Disease	NCT02848131	Current	Idiopathic pulmonary fibrosis (IPF) is a fatal senescence-associated disease. In an open label Phase 1 pilot study, the protocol of administration consisted of 3 days of oral doses D 100 mg and Q 1000 mg to subjects with diabetic kidney disease. The first results reported the D+Q reduced adipose tissue senescent cell burden within 11 days, with decreases in p16INK4A-and p21CIP1-expressing cells, cells with senescence-associated β-galactosidase activity. Skin epidermal p16INK4A+ and p21CIP1+ cells were reduced, as were circulating SASP factors, including IL-1α, IL-6, and MMPs-9 and −12.	Hickson et al. (2019b)

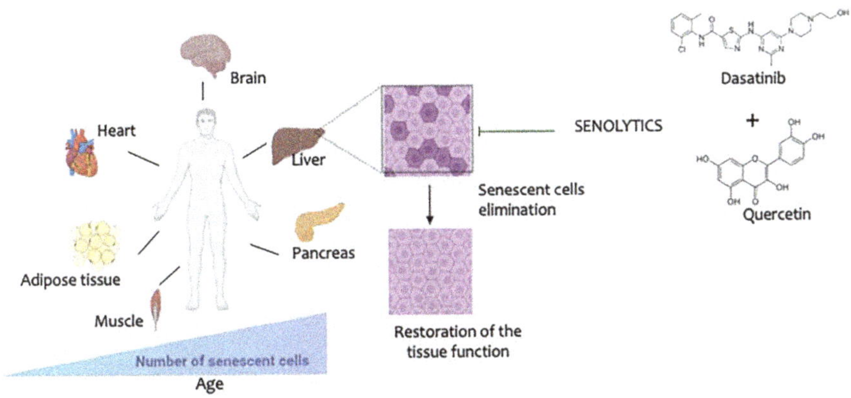

Fig. 20.3 Positive effect of senolytic therapy
With age, the number of senescent cells increases in tissues, and treatment with senolytics, such as D+Q, selectively eliminates senescent cells, allowing tissue regeneration so that other cells can take their place

current using the D+Q combination. It is important to mention that other clinical trials are using D+Q in combination with other drugs such as Fisetin (F). However, adding another drug increases the complexity of determining the signaling pathways involved and the results observed.

The active role of cellular senescence in the development and progression of various pathologies is evident. However, since the discovery of senolytics, it has become clear that, due to the differences between senescent cells, the senolytic effect is not homogeneous. As mentioned above, it is cell type-specific and depends on the senescence inducer. In addition, it has recently been shown that the metabolic state of the cell, which can vary in different tissues (such as oxygen supply and nutrient availability), may induce different senescent subtypes and modify the molecules contained in the SASP (Rad and Grillari 2024). Hence, the use of senolytics may result in different outcomes (Fig. 20.3).

20.6 Side Effects or Adverse Effects of D+Q

It is known that D and Q in individual treatments can have adverse effects when used at high doses and in long-term treatments. Dasatinib's chronic use at high doses for oncological treatments has shown adverse effects, such as neutropenia, thrombocytopenia, anemia, diarrhea, nausea, edema, pleural effusion, pulmonary arterial hypertension, and cardiovascular alterations (Cortes et al. 2016; Al-Ali et al. 2022). Quercetin is a flavonoid commonly used in nutritional therapies, due to its anti-antioxidant and anti-inflammatory effect; however, quercetin can be toxic at high doses or prolonged exposure (Xu et al. 2018; Andres et al. 2018).

The rationale for senotherapy is to eliminate senescent cells without causing systemic toxicity, so the D+Q cocktail has been well tolerated in clinical and preclinical trials because it is administered at intermittent, short-term doses, in a regimen known as hit-and-run (Hickson et al. 2019a; Justice et al. 2019a). However, despite these low doses, adverse effects have been reported as "mild" or "expected", with symptoms, such as gastrointestinal disturbances, respiratory symptoms, irritation or bruising, and fatigue. For example, in an open-label pilot study of 65 patients with idiopathic pulmonary fibrosis (IPF), some participants experienced mild insomnia and anxiety after using D+Q (Justice et al. 2019a). In another study of 14 patients with IPF, one case of multifocal bacterial pneumonia and pulmonary edema was reported, which was resolved after hospitalization. Contrariwise, no liver or kidney toxicity was found, nor were there any changes in lung function or quality of life reported by most participants (Hickson et al. 2019a; Walaszczyk et al. 2023).

An important aspect to consider is that the effects of D+Q are tissue-specific, and some may be positive, negative, or not observed at all. Randomized preclinical and clinical studies should be continued with larger groups of individuals for a longer period to eliminate suspicions about the deleterious effects in a careful way, considering safety and symptoms.

It is important not to overlook that a patient's comorbidities could increase the risk of adverse effects associated with the use of D+Q. Therefore, it is also important to conduct personalized follow-ups, make dose adjustments, check for interactions with other medications, maintain constant monitoring and, if necessary, contraindicate treatment.

20.7 Considerations Before Using D+Q in Humans

Although the use of senolytics in treating various pathologies is gaining relevance, and numerous clinical trials are already underway, there are important factors to consider before safely administering these compounds to patients. The similarities and differences in senolytic responses depend on the age at which treatment is given, sex-based differences, the patient's overall health status, and the treatment regimen.

1. **Age.** The number of senescent cells in an organism largely depends on age. Older organisms naturally carry a higher burden of senescent cells compared to younger ones, as cellular senescence is a hallmark of aging (López-Otín et al. 2023).
2. **Sexual dimorphism.** Males may exhibit a greater accumulation of senescent cells in several tissues. The immune system's aging and its response show sex-specific differences across species. Immune senescence tends to occur earlier and more extensively in males, with senescent immune cells contributing to elevated SASP release. So, their elimination may benefit the tissue, which might not be the case for females. Additionally, drug responsiveness may vary due to sex differences in liver function and drug metabolism. Variations in liver enzyme activity, hormone levels, and body composition can result in different drug exposures between

males and females, as demonstrated in the preclinical models (Foster and Kumar 2025).

3. **Health status of the patient.** While senescent cells contribute to disease pathogenesis, they also play beneficial roles in tissue regeneration and wound healing. For example, elderly patients with advanced diabetes (who are already at high risk for infections, skin ulcers, and amputations) could suffer impaired tissue repair if senescent cells are depleted, underscoring the importance of senescence in wound healing and tissue homeostasis (Al-Naggar et al. 2020).
4. **Treatment regimen.** Since cellular senescence is involved in processes, such as development and tissue repair, prolonged senolytic treatment may be harmful. To reduce adverse effects, senolytics are typically administered in cycles (e.g., 3–5 days of treatment followed by a 1 to 2-week break). Nevertheless, even intermittent dosing can provoke a transient inflammatory response due to the sudden clearance of senescent cells (Garrido et al. 2022).

The clinical use of senolytics, such as D+Q, holds a significant promise for addressing age-related and chronic diseases. However, several critical factors must be considered to ensure their safe and effective use in humans. Age-related accumulation of senescent cells, sex-specific differences in immune and hormonal aging, drug metabolism, and the overall health status of patients can all influence therapeutic outcomes. Moreover, the treatment regimen must be carefully adjusted to avoid disrupting the beneficial roles of senescent cells, particularly in tissue repair and immune responses. These considerations highlight the need for personalized senolytic strategies that account for patient-specific variables to maximize therapeutic benefit while minimizing risk.

20.8 Conclusion and Perspectives

1. The combination of D+Q for senolytic treatments is not yet available to the public; nevertheless, each is sold separately. Dasatinib requires a prescription, but quercetin can be purchased in drug stores and on the Internet. The tablets are called senolytic complex or activator, almost always in conjunction with fisetin, curcumin, or resveratrol. However, the doses offered for this flavonoid are very low compared to those used in clinical protocols. Furthermore, its purity and formulation are not regulated.
2. There is still controversy regarding the existence of definitive and reliable human studies that support the use of these compounds as supplements and/or drugs for healthy longevity. Despite the beneficial effects of D+Q in improving health in various specific conditions, such as idiopathic pulmonary fibrosis and diabetic nephropathy, no reliable long-term studies have been conducted evaluating aging and longevity, much less have they been conducted in healthy people for this purpose. The results seen in animal models cannot be extrapolated to humans because most are performed in transgenic or manipulated animal models, and

there are very few studies in females. So, controlled, randomized trials with long-term follow-up are required.

3. Their use as supplements in humans should not be recommended yet, due to the lack of studies. Dasatinib has adverse effects, such as pleurisy, pulmonary hypertension, liver and kidney toxicity, and immunosuppression, among others, which are dangerous. In the case of quercetin, although it is a flavonoid of "natural origin", it is not harmless, and the lack of regulation of commercial formulations, and the absence of knowledge about its interaction with other medications pose a considerable risk. Their use is only justified in clinical trials and under supervision.
4. Quercetin is naturally found in onions, apples, berries, grapes, broccoli, and tea. Normal human dietary intake is 5–40 mg/day, so the doses used in the studies (500–1000 mg/day) are not achieved. Dasatinib is a synthetic product and is not obtained from food.
5. The most important knowledge gap in D+Q use as senolytics is the lack of robust, randomized, controlled, and long-term clinical studies demonstrating long-term safety and side effects that lead to this gap in the context of aging, human longevity, and related diseases. To date, there is even less information about optimal doses, frequency, duration of treatment, and interactions with other medications.

In summary, it is crucial to consider the heterogeneity in the progression of aging due to many factors, including individual drug response and sex-dependent differences (Rattan 2024), when using senolytics to eliminate senescent cells, such as D+Q treatment.

Acknowledgments This work was supported by Consejo Nacional de Humanidades, Ciencia y Tecnología (CONAHCyT) grant FORDECYT-PRONACES/263957/2020. Salas-Venegas is a fellow in UNAM Posdoctoral-Program (POSDOC) and Flores-Torres is CONAHCyT scholarship holder.

References

Acosta JC, Banito A, Wuestefeld T, Georgilis A, Janich P, Morton JP, Athineos D, Kang TW, Lasitschka F, Andrulis M, Pascual G, Morris KJ, Khan S, Jin H, Dharmalingam G, Snijders AP, Carroll T, Capper D, Pritchard C, Inman GJ, Longerich T, Sansom OJ, Benitah SA, Zender L, Gil J (2013) A complex secretory program orchestrated by the inflammasome controls paracrine senescence. Nat Cell Biol 15(8):978–990. https://doi.org/10.1038/ncb2784

Aghababaei F, Hadidi M (2023) Recent advances in potential health benefits of quercetin. Pharmaceuticals 16(7):1020. https://doi.org/10.3390/ph16071020

Al-Ali H, Shafran I, Patel K, Lee K (2022) Adverse reactions after treatment with Dasatinib in chronic myeloid leukemia: characteristics, potential mechanisms, and clinical management strategies. J Oncol Pharm Pract 28(6):1342–1348

Al-Naggar IMA, Kuchel GA, Xu M (2020) Senolytics: targeting senescent cells for age-associated diseases. Curr Mol Biol Rep 6(4):161–172. https://doi.org/10.1007/s40610-020-00140-1

Andres S, Pevny S, Ziegenhagen R, Bakhiya N, Schäfer B, Hirsch-Ernst KI, Lampen A (2018) Safety aspects of the use of quercetin as a dietary supplement. Mol Nutr Food Res 62(1): 1700447

Antelo-Iglesias L, Picallos-Rabina P, Estévez-Souto V, Da Silva-Álvarez S, Collado M (2021) The role of cellular senescence in tissue repair and regeneration. Mech Ageing Dev 198:111528

Baker DJ, Wijshake T, Tchkonia T, LeBrasseur NK, Childs BG, van de Sluis B, Kirkland JL, van Deursen JM (2011) Clearance of p16Ink4a-positive senescent cells delays ageing-associated disorders. Nature 479(7372):232–236

Budamagunta V, Kumar A, Rani A, Bean L, Manohar-Sindhu S, Yang Y, Zhou D, Foster TC (2023) Effect of peripheral cellular senescence on brain aging and cognitive decline. Aging Cell 22(5):e13817. https://doi.org/10.1111/acel.13817

Chaib S, Tchkonia T, Kirkland JL (2022) Cellular senescence and senolytics: the path to the clinic. Nat Med 28(8):1556–1568. https://doi.org/10.1038/s41591-022-01923-y

Chen B-A, Chen R (2015) The role of dasatinib in the management of chronic myeloid leukemia. Drug Des Devel Ther 9:773–779. https://doi.org/10.2147/DDDT.S80207

Collado M, Serrano M (2010) Senescence in tumours: evidence from mice and humans. Nat Rev Cancer 10(1):51–57. https://doi.org/10.1038/nrc2772

Coppé JP, Desprez PY, Krtolica A, Campisi J (2010) The senescence-associated secretory phenotype: the dark side of tumor suppression. Annu Rev Pathol 5:99–118

Cortes JE, Saglio G, Kantarjian HM et al (2016) Long-term safety and efficacy of Dasatinib in treating chronic-phase chronic myeloid leukemia patients resistant or intolerant to imatinib. Am J Hematol 91(9):869–874

Demaria M, Ohtani N, Youssef SA, Rodier F, Toussaint W, Mitchell JR, Laberge RM, Vijg J, Van Steeg H, Dollé ME, Hoeijmakers JH, de Bruin A, Hara E, Campisi J (2014) An essential role for senescent cells in optimal wound healing through secretion of PDGF-AA. Dev Cell 31(6): 722–733. https://doi.org/10.1016/j.devcel.2014.11.012

Fang Y, Medina D, Stockwell R, McFadden S, Quinn K, Peck MR, Bartke A, Hascup KN, Hascup ER (2023) Sexual dimorphic metabolic and cognitive responses of C57BL/6 mice to Fisetin or Dasatinib and quercetin cocktail oral treatment. GeroScience 45(5):2835–2850

Fang Y, Peck MR, Quinn K, Chapman JE, Medina D, McFadden SA, Bartke A, Hascup ER, Hascup KN (2024) Senolytic intervention improves cognition, metabolism, and adiposity in female APPNL-F/NL-F mice. Geroscience [e-pub ahead of print]

Foster TC, Kumar A (2025) Sex, senescence, senolytics, and cognition. Front Aging Neurosci 17: 1555872. https://doi.org/10.3389/fnagi.2025.1555872

Flores-Torres RP, Salas-Venegas V, Santín-Márquez R, De la Vega-Tinajar M, Ramírez-Carreto RJ, Librado-Osorio R, Luna-López A, Alarcón-Aguilar A, López-Díazguerrero NE, Morán-Ramos S, Chavarría A, Gómez-González B Königsberg M (2025) Senotherapy as a multitarget intervention in chronic obesity: modulation of senescence, neuroinflammation, dysbiosis, and synaptic integrity in middle-aged female Wistar rats. Exp Neurol 392:115331. https://doi.org/10.1016/j.expneurol.2025.115331

Gadecka A, Nowak N, Bulanda E, Janiszewska D, Dudkowska M, Sikora E, Bielak-Zmijewska A (2025) The senolytic cocktail, dasatinib and quercetin, impacts the chromatin structure of both young and senescent vascular smooth muscle cells. Geroscience [online]. Available at: https://doi.org/10.1007/s11357-024-01504-6 [Accessed 16 May 2025]

Garrido AM, Kaistha A, Uryga AK, Oc S, Foote K, Shah A, Finigan A, Figg N, Dobnikar L, Jørgensen H, Bennett M (2022) Efficacy and limitations of senolysis in atherosclerosis. Cardiovasc Res 118(7):1713–1727. https://doi.org/10.1093/cvr/cvab208

Gonzales MM, Krishnamurthy S, Garbarino V, Daeihagh AS, Gillispie GJ, Deep G, Craft S, Orr ME (2021) A geroscience motivated approach to treat Alzheimer's disease: senolytics move to clinical trials. Mech Ageing Dev. 200:111589. https://doi.org/10.1016/j.mad.2021.111589. Epub 2021 Oct 21

Gonzales MM, Garbarino VR, Marques Zilli E, Petersen RC, Kirkland JL, Tchkonia T, Musi N, Seshadri S, Craft S, Orr ME (2022) Senolytic Therapy to Modulate the Progression of

Alzheimer's Disease (SToMP-AD): a pilot clinical trial. J Prev Alzheimers Dis. 9(1):22–29. https://doi.org/10.14283/jpad.2021.62

Gorgoulis V, Adams PD, Alimonti A, Bennett DC, Bischof O, Bishop C, Campisi J, Collado M, Evangelou K, Ferbeyre G, Gil J, Hara E, Krizhanovsky V, Jurk D, Maier AB, Narita M, Niedernhofer L, Passos JF, Robbins PD, Schmitt CA, Sedivy J, Vougas K, von Zglinicki T, Zhou D, Serrano M, Demaria M (2019) Cellular senescence: defining a path forward. Cell 179(4):813–827

Grosse L, Wagner N, Emelyanov A, Molina C, Lacas-Gervais S, Wagner KD, Bulavin DV (2020) Defined p16High senescent cell types are indispensable for mouse healthspan. Cell Metab 32(1):87–99.e6

Guerrero A, De Strooper B, Arancibia-Carcamo IL (2021) Cellular senescence at the crossroads of inflammation and Alzheimer's disease. Trends Neurosci. 44(9):714–727. https://doi.org/10.1016/j.tins.2021.06.007. Epub 2021 Aug 5

Guo X, Wen S, Wang J, Zeng X, Yu H, Chen Y, Zhu X, Xu L (2024) Senolytic combination of dasatinib and quercetin attenuates renal damage in diabetic kidney disease. Phytomedicine 130: 155705. https://doi.org/10.1016/j.phymed.2024.155705

Hernandez-Segura A, Nehme J, Demaria M (2018) Hallmarks of cellular senescence. Trends Cell Biol 28:436–453

Hickson LJ, Prata LGP, Bobart SA et al (2019a) Senolytics decrease senescent cells in humans: preliminary report from a clinical trial of Dasatinib plus Quercetin in individuals with diabetic kidney disease. EBioMedicine 47:446–456. https://doi.org/10.1016/j.ebiom.2019.08.069

Hickson LJ, Langhi Prata LGP, Bobart SA, Evans TK, Giorgadze N, Hashmi SK, Herrmann SM, Jensen MD, Jia Q, Jordan KL, Kellogg TA, Khosla S, Koerber DM, Lagnado AB, Lawson DK, LeBrasseur NK, Lerman LO, McDonald KM, McKenzie TJ, Passos JF, Pignolo RJ, Pirtskhalava T, Saadiq IM, Schaefer KK, Textor SC, Victorelli SG, Volkman TL, Xue A, Wentworth MA, Wissler Gerdes EO, Zhu Y, Tchkonia T, Kirkland JL (2019b) Senolytics decrease senescent cells in humans: preliminary report from a clinical trial of Dasatinib plus Quercetin in individuals with diabetic kidney disease. EBioMedicine 47:446–456. https://doi.org/10.1016/j.ebiom.2019.08.069. Epub 2019 Sep 18. Erratum In: EBioMedicine. 2020 Feb;52: 102595. doi: 10.1016/j.ebiom.2019.12.004

Islam MT, Tuday E, Allen S, Kim J, Trott DW, Holland WL, Donato AJ, Lesniewski LA (2023) Senolytic drugs, dasatinib and quercetin, attenuate adipose tissue inflammation, and ameliorate metabolic function in old age. Aging Cell 22(2):e13767. https://doi.org/10.1111/acel.13767

Justice JN, Nambiar AM, Tchkonia T et al (2019a) Senolytics in idiopathic pulmonary fibrosis: results from a first-in-human, open-label, pilot study. EBioMedicine 40:554–563. https://doi.org/10.1016/j.ebiom.2018.12.052

Justice JN, Nambiar AM, Tchkonia T, LeBrasseur NK, Pascual R, Hashmi SK, Prata L, Masternak MM, Kritchevsky SB, Musi N, Kirkland JL (2019b) Senolytics in idiopathic pulmonary fibrosis: results from a first-in-human, open-label, pilot study. EBioMedicine. 40:554–563. https://doi.org/10.1016/j.ebiom.2018.12.052. Epub 2019 Jan 5

Kamat P, Macaluso N, Li Y, Agrawal A, Winston A, Pan L, Stewart T, Starich B, Milcik N, Min C, Wu PH, Walston J, Fan J, Phillip JM (2025) Single-cell morphology encodes functional subtypes of senescence in aging human dermal fibroblasts. Sci Adv 11(17):eads1875. https://doi.org/10.1126/sciadv.ads1875

Kim JK, Park SU (2018) Quercetin and its role in biological functions: an updated review. EXCLI J 17:856–872. https://doi.org/10.17179/EXCLI2018-1538

Kovacovicova K, Skolnaja M, Heinmaa M, Mistrik M, Pata P, Pata I, Bartek J, Vinciguerra M (2018) Senolytic cocktail Dasatinib+Quercetin (D+Q) does not enhance the efficacy of senescence-inducing chemotherapy in liver cancer. Front Oncol 8:459. https://doi.org/10.3389/fonc.2018.00459

Krzystyniak A, Wesierska M, Petrazzo G, Gadecka A, Dudkowska M, Bielak-Zmijewska A, Mosieniak G, Figiel I, Wlodarczyk J, Sikora E (2022) Combination of dasatinib and quercetin improves cognitive abilities in aged male Wistar rats, alleviates inflammation and changes hippocampal synaptic plasticity and histone H3 methylation profile. Aging 14(2):572–595. https://doi.org/10.18632/aging.203835

López-Otín C, Blasco MA, Partridge L, Serrano M, Kroemer G (2013) The hallmarks of aging. Cell 153(1):1194–1217

López-Otín C, Blasco MA, Partridge L, Serrano M, Kroemer G (2023) Hallmarks of aging: an expanding universe. Cell 186(1):243–278

Lushchak O, Schosserer M, Grillari J (2023) Senopathies—diseases associated with cellular senescence. Biomolecules 13(6):966. https://doi.org/10.3390/biom13060966

Maciel-Barón LA, Morales-Rosales SL, Aquino-Cruz AA, Triana-Martínez F, Galván-Arzate S, Luna-López A, González-Puertos VY, López-Díazguerrero NE, Torres C, Königsberg M (2016) Senescence-associated secretory phenotype profile from primary lung mice fibroblasts depends on the senescence induction stimuli. Age 38(1):26

Magar RT, Sohng JK (2020) A review on structure, modifications and structure-activity relation of quercetin and its derivatives. J Microbiol Biotechnol 30(1):11–20. https://doi.org/10.4014/jmb.1907.07003

Malavolta M, Bracci M, Santarelli L, Sayeed MA, Pierpaoli E, Giacconi R, Costarelli L, Piacenza F, Basso A, Cardelli M, Provinciali M (2018) Inducers of senescence, toxic compounds, and senolytics: the multiple faces of Nrf2-activating phytochemicals in cancer adjuvant therapy. Mediators Inflamm 2018:4159013. https://doi.org/10.1155/2018/4159013

Musi N, Valentine JM, Sickora KR, Baeuerle E, Thompson CS, Shen Q, Orr ME (2018) Tau protein aggregation is associated with cellular senescence in the brain. Aging Cell 17(6):e12840. https://doi.org/10.1111/acel.12840

Novais EJ, Tran VA, Johnston SN, Darris KR, Roupas AJ, Sessions GA, Shapiro IM, Diekman BO, Risbud MV (2021) Long-term treatment with senolytic drugs Dasatinib and Quercetin ameliorates agedependent intervertebral disc degeneration in mice. Nat Commun 12(1):5213. https://doi.org/10.1038/s41467-021-25453-2

Ogrodnik M, Evans SA, Fielder E, Victorelli S, Kruger P, Salmonowicz H, Weigand BM, Patel AD, Pirtskhalava T, Inman CL, Johnson KO, Dickinson SL, Rocha A, Schafer MJ, Zhu Y, Allison DB, von Zglinicki T, LeBrasseur NK, Tchkonia T, Jurk D (2021) Whole-body senescent cell clearance alleviates age-related brain inflammation and cognitive impairment in mice. Aging Cell 20(2):e13296. https://doi.org/10.1111/acel.13296

Olascoaga-Del Angel KS, Gutierrez H, Königsberg M, Pérez-Villanueva J, López-Diazguerrero NE (2022) Exploring the fuzzy border between senolytics and senomorphics with chemoinformatics and systems pharmacology. Biogerontology 23(4):453–471. https://doi.org/10.1007/s10522-022-09974-x

Ota H, Kodama A (2022) Dasatinib plus quercetin attenuates some frailty characteristics in SAMP10 mice. Sci Rep 12(1):2425. https://doi.org/10.1038/s41598-022-06448-5

Palmer AK, Xu M, Zhu Y, Pirtskhalava T, Weivoda MM, Hachfeld CM, Prata LG, van Dijk TH, Verkade E, Casaclang-Verzosa G, Johnson KO, Cubro H, Doornebal EJ, Ogrodnik M, Jurk D, Jensen MD, Chini EN, Miller JD, Matveyenko A, Stout MB, Schafer MJ, White TA, Hickson LJ, Demaria M, Garovic V, Grande J, Arriaga EA, Kuipers F, von Zglinicki T, LeBrasseur NK, Campisi J, Tchkonia T, Kirkland JL (2019) Targeting senescent cells alleviates obesity-induced metabolic dysfunction. Aging Cell 18(3):e12950. https://doi.org/10.1111/acel.12950

Park J, Shin DW (2022) Senotherapeutics and their molecular mechanism for improving aging. Biomol Ther 30(6):490–500. https://doi.org/10.4062/biomolther.2022.114

Rad AN, Grillari J (2024) Current senolytics: mode of action, efficacy and limitations, and their future. Mech Ageing Dev 217:111888. https://doi.org/10.1016/j.mad.2023.111888

Rani A, Bean L, Budamagunta V, Kumar A, Foster TC (2024) Failure of senolytic treatment to prevent cognitive decline in a female rodent model of aging. Front Aging Neurosci 16:1384554

Rattan SIS (2024) Seven knowledge gaps in modern biogerontology. Biogerontology 25:1–8. https://doi.org/10.1007/s10522-023-10089-0

Ritschka B, Storer M, Mas A, Heinzmann F, Ortells MC, Morton J, Sansom O, Zender L, Keyes W (2017) The senescence-associated secretory phenotype induces cellular plasticity and tissue regeneration. Genes Dev 31:172–183

Rodier F, Campisi J (2011) Four faces of cellular senescence. J Cell Biol 192(4):547–556

Rodier F, Coppé JP, Patil CK, Hoeijmakers WA, Muñoz DP, Raza SR, Campisi J (2009) Persistent DNA damage signalling triggers senescence-associated inflammatory cytokine secretion. Nat Cell Biol 11(8):973–979

Ruggiero AD, Vemuri R, Blawas M, Long M, DeStephanis D, Williams AG, Chen H, Justice JN, Macauley SL, Day SM, Kavanagh K (2023) Long-term dasatinib plus quercetin effects on aging outcomes and inflammation in nonhuman primates: implications for senolytic clinical trial design. Geroscience 45(5):2785–2803. https://doi.org/10.1007/s11357-023-00830-5

Schade AE, Schieven GL, Townsend R, Jankowska AM, Susulic V, Zhang R, Szpurka H, Maciejewski JP (2008) Dasatinib, a small-molecule protein tyrosine kinase inhibitor, inhibits T-cell activation and proliferation. Blood 111(3):1366–1377. https://doi.org/10.1182/blood-2007-04-084814

Schafer MJ, White TA, Iijima K, Haak AJ, Ligresti G, Atkinson EJ, Oberg AL, Birch J, Salmonowicz H, Zhu Y, Mazula DL, Brooks RW, Fuhrmann-Stroissnigg H, Pirtskhalava T, Prakash YS, Tchkonia T, Robbins PD, Aubry MC, Passos JF, Kirkland JL, Tschumperlin DJ, Kita H, LeBrasseur NK (2017) Cellular senescence mediates fibrotic pulmonary disease. Nat Commun 8:14532. https://doi.org/10.1038/ncomms14532

Storer M, Mas A, Robert-Moreno A, Pecoraro M, Ortells MC, Di Giacomo V, Yosef R, Pilpel N, Krizhanovsky V, Sharpe J, Keyes WM (2013) Senescence is a developmental mechanism that contributes to embryonic growth and patterning. Cell 155(5):1119–1130. https://doi.org/10.1016/j.cell.2013.10.041

Suda M, Paul KH, Tripathi U, Minamino T, Tchkonia T, Kirkland JL (2024) Targeting cell senescence and senolytics: novel interventions for age-related endocrine dysfunction. Endocr Rev 45(5):655–675. https://doi.org/10.1210/endrev/bnae010

Suresh G, Nadh RV, Srinivasu N, Yennity D (2017) A convenient new and efficient commercial synthetic route for dasatinib (Sprycel®). Synth Commun 47(17):1610–1621. https://doi.org/10.1080/00397911.2017.1337150

Tokarski JS, Newitt JA, Chang CYJ, Cheng JD, Wittekind M, Kiefer SE, Kish K, Lee FYF, Borzilleri R, Lombardo LJ, Xie D, Zhang Y, Klei HE (2006) The structure of Dasatinib (BMS-354825) bound to activated ABL kinase domain elucidates its inhibitory activity against Imatinib-resistant ABL mutants. Cancer Res 66(11):5790–5797. https://doi.org/10.1158/0008-5472.CAN-05-4187

Walaszczyk A, Dookun E, Redgrave R et al (2023) Senolytics dasatinib and quercetin in idiopathic pulmonary fibrosis: results of a phase I, single-blind, single-center, randomized, placebo-controlled pilot trial on feasibility and tolerability. Lancet Healthy Longev 4(1):E13–E23

Wang B, Han J, Elisseeff JH, Demaria M (2024) The senescence-associated secretory phenotype and its physiological and pathological implications. Nat Rev Mol Cell Biol 25(12):958–978. https://doi.org/10.1038/s41580-024-00727-x

Xu M, Palmer AK, Ding H et al (2018) Senolytic drugs, dasatinib, and quercetin, attenuate adipose tissue inflammation and ameliorate metabolic function in old age. Aging Cell 17(4):e12765. https://doi.org/10.1111/acel.12765

Yang D, Wang T, Long M, Li P (2020) Quercetin: its main pharmacological activity and potential application in clinical medicine. Oxid Med Cell Longev 2020:1–13. https://doi.org/10.1155/2020/8825387

Yousefzadeh MJ, Zhu Y, McGowan SJ, Angelini L, Fuhrmann-Stroissnigg H, Xu M, Ling YY, Melos KI, Pirtskhalava T, Inman CL, McGuckian C, Wade EA, Kato JI, Grassi D, Wentworth M, Burd CE, Arriaga EA, Ladiges WL, Tchkonia T, Kirkland JL, Robbins PD, Niedernhofer LJ (2018) Fisetin is a senotherapeutic that extends health and lifespan. EBioMedicine 36:18–28. https://doi.org/10.1016/j.ebiom.2018.09.015

Zhao J, Zheng L, Dai G, Sun Y, He R, Liu Z, Jin Y, Wu T, Hu J, Cao Y, Duan C (2025) Senolytics cocktail dasatinib and quercetin alleviate chondrocyte senescence and facet joint osteoarthritis in mice. Spine J 25(1):184–198. https://doi.org/10.1016/j.spinee.2024.09.017

Zhou Y, Xin X, Wang L, Wang B, Chen L, Liu O, Rowe DW, Xu M (2021) Senolytics improve bone forming potential of bone marrow mesenchymal stem cells from aged mice. NPJ Regen Med 6(1):34. https://doi.org/10.1038/s41536-021-00145-z

Zhu Y, Tchkonia T, Pirtskhalava T, Gower AC, Ding H, Giorgadze N, Palmer AK, Ikeno Y, Hubbard GB, Lenburg M, O'Hara SP, LaRusso NF, Miller JD, Roos CM, Verzosa GC, LeBrasseur NK, Wren JD, Farr JN, Khosla S, Stout MB, McGowan SJ, Fuhrmann-Stroissnigg-H, Gurkar AU, Zhao J, Colangelo D, Dorronsoro A, Ling YY, Barghouthy AS, Navarro DC, Sano T, Robbins PD, Niedernhofer LJ, Kirkland JL (2015) The Achilles' heel of senescent cells: from transcriptome to senolytic drugs. Aging Cell 14(4):644–658. https://doi.org/10.1111/acel.12344

Chapter 21
Carotenoids for Healthy Longevity

Ezgi Bellikci Koyu

Abstract Carotenoids are widely distributed pigments responsible for yellow, orange, and red colors in nature. As potent antioxidants, carotenoids can counteract oxidative stress by directly scavenging or quenching reactive oxygen species. They can also enhance endogenous antioxidant defense mechanisms, particularly through activation of the nuclear factor erythroid 2–related factor 2 signaling pathway. Additionally, by the inhibition of nuclear factor κB pathway they prevent the production and release of inflammatory cytokines. Epidemiological studies indicate that increased consumption of carotenoid-rich foods or elevated blood levels of carotenoids may be associated with a reduced risk of various aging-related conditions, such as cardiovascular disease, type 2 diabetes mellitus, cancer, neurodegenerative diseases, sarcopenia, and skin aging. However, the results of clinical trials investigating the effects of carotenoid supplementation have yielded conflicting results. This chapter aims to summarize the current evidence on the role of carotenoids in preventing or delaying the development of aging-related diseases and supporting healthy longevity.

Keywords Carotenoids · Xanthophylls · Aging related diseases · Longevity

21.1 Introduction

Carotenoids are widely distributed, a large group of lipophilic pigments responsible for the yellow, orange, and red colors in nature. These are synthesized by both photosynthetic organisms, including plants, cyanobacteria and algae, as well as by some non-photosynthetic bacteria, fungi and insects (Martínez-Cámara et al. 2021; Maoka 2020; Terao 2023). While animal tissues generally cannot synthesize carotenoids, they can accumulate them through dietary intake (Maoka 2020). Carotenoids are typically found as tetraterpenoids with a 40-carbon skeleton, although higher

E. B. Koyu (✉)
Department of Nutrition and Dietetics, Faculty of Health Sciences, İzmir Katip Celebi University, İzmir, Türkiye
e-mail: ezgi.bellikci@ikcu.edu.tr

S. I. S. Rattan, C. Gezer (eds.), *Supplements and Drugs for Healthy Longevity*, Healthy Ageing and Longevity 22, https://doi.org/10.1007/978-3-032-11608-6_21

carotenoids with 45- or 50-carbon skeletons have also been identified in certain bacterial species (Yabuzaki 2017). Based on their chemical structures, carotenoids are classified as carotenes (hydrocarbon carotenoids such as α-carotene, β-carotene, γ-carotene, and lycopene) and xanthophylls (oxygenated derivatives of carotenes such as lutein, astaxanthin, cryptoxanthin, and zeaxanthin) (Maoka 2020).

Although more than 1200 carotenoids have been identified and characterized (Yabuzaki 2017), only about 50 are present in human diet and are capable of being absorbed and metabolized by humans (Maiani et al. 2009). The primary carotenoids in the human blood and diet are α-carotene, β-carotene, β-cryptoxanthin, zeaxanthin, lutein, and lycopene and they are accounted for more than 95% of total blood carotenoids (Maiani et al. 2009).

The most important dietary sources of carotenoids are fruits and vegetables. β-carotene is the most widely distributed carotenoid with the highest provitamin A activity. Green vegetables like spinach, broccoli, kale, and lettuce, as well as other vegetables including carrots, red peppers, potatoes, and fruits such as apricots, mangoes, pumpkin, and oranges are rich in β-carotene. α-carotene is usually found together with β-carotene, and its major sources are carrots, pumpkin, and winter squash. Tomatoes and tomato-based products, including ketchup, puree, and sauces are rich sources of lycopene. Other lycopene-rich foods include pink grapefruit, watermelon, papaya, and guava. Green leafy vegetables such as spinach, kale, broccoli, Brussels sprouts, parsley, and lettuce are rich in lutein. Egg yolk and other green vegetables such as peas are also important sources of lutein. β-cryptoxanthin is found in red peppers and fruits such as papaya, orange, mandarin, mango, and persimmon. Egg yolk, maize, red peppers, orange, and goji berries are good sources of zeaxanthin (Maiani et al. 2009; Meléndez-Martínez et al. 2022). Apart from these six main carotenoids, another member of the xanthophyll family is astaxanthin, which is produced primarily by algae and is also responsible for the pink colors of salmon and crustaceans (Sztretye et al. 2019). Fucoxanthin, another marine carotenoid, is found in brown microalgae and seaweeds. Capsanthin and capsorubin are found in paprika, and the stigmas of saffron contain crocetin (Saini et al. 2022).

21.2 The Effects of Carotenoids on Lifespan and Hallmarks of Aging

One widely accepted framework for understanding the biological basis of aging is the hallmarks of aging, as formulated by López-Otín et al. (2013, 2023). These hallmarks refer to a series of progressive molecular and cellular alterations that accumulate over time and offer potential targets for therapeutic interventions to slow-down age-related decline. The hallmarks of aging include DNA instability, telomere attrition, epigenetic alterations, loss of proteostasis, deregulated nutrient-sensing, mitochondrial dysfunction, cellular senescence, stem cell exhaustion, altered intercellular communication, disabled macroautophagy, chronic inflammation, and age-associated dysbiosis (López-OtíN et al. 2023). In recent years, growing interest has emerged in identifying natural compounds that may influence hallmarks of aging. Among these, carotenoids are potential agents for promoting longevity and mitigating aging-associated changes. Studies conducted on various model organisms, including *Caenorhabditis elegans*, *Drosophila melanogaster*, and mice, have shown that carotenoids such as lutein, astaxanthin, fucoxanthin, and crocin can extend lifespan in these organisms (Lee et al. 2022; Zhang et al. 2014; Yazaki et al. 2011; Lashmanova et al. 2015; Li et al. 2025; Harrison et al. 2024). Observational studies in humans also indicated higher serum carotenoids were associated with lower risk of all-cause mortality (Zhu et al. 2023; Huang et al. 2018).

Several studies have investigated the relationship between carotenoids and hallmarks of aging. In American National Health and Nutrition Examination Survey (NHANES) cohorts, higher serum carotenoid levels were associated with longer leukocyte telomere length (Wang et al. 2024; Nomura et al. 2017; Min and Min 2017). Intervention studies have indicated that carotenoid supplementation significantly reduces inflammatory markers (Hajizadeh-Sharafabad et al. 2022). Recently, it has been suggested that carotenoids may play a role in modulating gut dysbiosis. While carotenoid absorption in the small intestine is limited, a considerable amount of carotenoids reaches the colon. It is suggested that carotenoids may exert bactericidal activity, regulate secretory immunoglobulins, and modulate oxidative stress in the colon (Eroglu et al. 2023). Notably, certain carotenoids such as lycopene have been shown to impact the abundance of beneficial bacterial strains in colon, including *Bifidobacterium adolescentis* and *Bifidobacterium longum* (Wiese et al. 2019).

Carotenoids have also been studied in relation to various other biomarkers of biological aging. Cross-sectional data from the NHANES 1999–2018 cohort, involving 27,338 participants, investigated the associations between dietary carotene intake and several biological aging indicators, including allostatic load, the Klemera-Doubal method, PhenoAge, and homeostatic dysregulation. The findings demonstrated that total carotene, along with specific carotenoids such as α-carotene, β-carotene, β-cryptoxanthin, lycopene, and lutein/zeaxanthin, showed significant negative correlations with most biological aging metrics (Qi et al. 2025). Another study showed a link between carotenoids and the Klotho gene. Higher consumption of total carotenoids, as well as the intake of specific carotenoids, were associated with increased S-Klotho levels, which is accepted as a biomarker of healthy aging (He et al. 2023).

21.3 Carotenoids and Aging-Related Diseases

Aging is a major risk factor for the onset of several chronic conditions, including cardiovascular diseases (CVD), type-2 diabetes mellitus (T2DM), cancer, and neurodegenerative disease. Although multiple factors may contribute to the development of chronic diseases, oxidative stress and chronic inflammation are the key mechanisms underlying aging-related diseases significantly causing cellular and tissue damage (Bakac et al. 2023). As potent antioxidants, carotenoids can counteract oxidative stress by directly scavenging or quenching reactive oxygen species (ROS). Additionally, they can enhance endogenous antioxidant defense mechanisms, particularly through activation of the nuclear factor erythroid 2–related factor 2 (Nrf2) signaling pathway. Upon activation, Nrf2 translocates into the nucleus and induces transcription of antioxidant and detoxifying enzymes in body (Terao 2023). During the inflammatory process, carotenoids also play a role by inhibiting the translocation of nuclear factor κB (NFκB) to the nucleus, thereby preventing the downstream production and release of inflammatory cytokines (Kaulmann and Bohn 2014). Consequently, carotenoids may contribute to the prevention of aging-related diseases by mitigating oxidative damage and chronic inflammation.

21.3.1 Carotenoids and CVD

CVD are the leading cause of mortality worldwide and significantly contribute to disability (Vaduganathan et al. 2022). Five modifiable risk factors, body mass index, systolic blood pressure (SBP), low-density lipoprotein cholesterol (LDL-C), tobacco smoking, and diabetes, account for over half of the global incidence of CVD (Global Cardiovascular Risk Consortium 2023). Furthermore, high blood pressure, dietary risks, and elevated LDL-C were associated with approximately 10.8 million, 6.58 million, and 3.81 million cardiovascular deaths in 2021, respectively (Vaduganathan et al. 2022). Atherosclerosis is the primary underlying cause of CVD (FrąK et al. 2022), and carotenoids have several protective effects on vascular health by inhibiting LDL oxidation, increasing nitric oxide bioavailability, reducing carotid intima-media thickness, maintaining of the structural integrity of blood vessels, and improving endothelial dysfunction (Sumalla-Cano et al. 2024).

Higher serum levels of various carotenoids have been associated with a reduced risk of coronary heart disease, overall cardiovascular disease (Aune et al. 2018), advanced cardiovascular-kidney-metabolic syndrome (Chen et al. 2025) and cardiovascular mortality (Zhu et al. 2023). Additionally, higher dietary intake of six main carotenoids has been linked to reduced risks of stroke and other cardiovascular events (Bahonar et al. 2017). Lycopene is one of the extensively studied carotenoids in relation to cardiovascular diseases. Cheng et al. (2017, 2019) conducted two meta-analyses to investigate the cardiovascular effects of lycopene. According to meta-analysis of observational studies, high dietary intakes or high serum concentrations of lycopene were associated with a 26% lower risk of stroke and a 14% lower risk of CVD (Cheng et al. 2019). In the other meta-analysis, they investigated the effects of tomato products (ranging from 70 to 400 g/day) and lycopene supplementation (ranging from 4 to 30 mg/day) on cardiovascular risk factors. While lycopene supplementation significantly reduced SBP by 5.66 mmHg, no significant effects were observed on other outcomes such as blood lipids, pulse wave velocity, inflammatory markers or adhesion molecules after lycopene supplementation. In the same study, tomato supplementation was associated with a decrease in LDL-C, interleukin-6 (IL-6) and flow-mediated dilation (Cheng et al. 2017). In a double-blind, randomized, placebo-controlled trial Wolak et al. (2019) investigated the effects of different doses of Tomato Nutrient Complex (TNC) on blood pressure in untreated hypertensive individuals. Participants received either TNC containing 5 mg, 15 mg, or 30 mg of lycopene, 15 mg of synthetic lycopene, or a placebo for 8 weeks. The TNC formulation also included phytoene, phytofluene, β-carotene, vitamin E, and other fat-soluble contituents naturally present in tomatoes. The results indicated that synthetic lycopene and TNC containing 5 mg of lycopene had no significant effect; only TNC containing 15 mg or 30 mg of lycopene led to a significant reduction in systolic blood pressure. These findings suggest that tomatoes, as a whole food, may be more effective than isolated lycopene.

Lutein also shows promising results against the risk of CVD. Plasma lutein levels were inversely associated with IL-6 concentrations in coronary artery disease patients (Chung et al. 2017). A meta-analysis has reported that higher lutein intake or higher blood concentrations were associated with a lower risk of coronary heart disease and stroke (Leermakers et al. 2016). However, the effects of β-carotene on CVD have been conflicting. While some studies have demonstrated a protective effect of β-carotene on cardiovascular outcomes, other research has failed to find such an association or have reported adverse effects. In the Nurses' Health Study, dietary intake of β-carotene was inversely associated with coronary artery disease (Osganian et al. 2003). However, a prospective analysis of over 3000 adults with T2DM from NHANES cohort has shown that higher serum β-carotene concentrations were significantly associated with increased cardiovascular mortality over a 14-year follow-up period (Qiu et al. 2022). Conversely, another study including 8390 hypertensive adults from the same cohort reported no association between serum β-carotene levels and CVD mortality (Zhu et al. 2023). When supplementation studies were examined, meta-analysis indicated that β-carotene supplementation increased the risk of CVD incidence, CVD mortality and stroke (An et al. 2022; Yang et al. 2022).

Astaxanthin is a powerful antioxidant, demonstrating greater activity than well-known antioxidants such as vitamins C and vitamin E, and β-carotene. Consumption of natural astaxanthin has not been associated with any adverse effects even at high doses (Brendler and Williamson 2019). Regarding cardiovascular risk factors, a meta-analysis, including seven randomized controlled trial, investigated the effects of astaxanthin on metabolic syndrome parameters. Among these, a significant reduction was observed in LDL-C levels. Total cholesterol and SBP also showed reductions, although these changes were of statically borderline significance (Leung et al. 2022).

21.3.2 Carotenoids and Cancer

Cancer is a group of diseases that may originate in almost any organ or tissue of the body, characterized by the uncontrolled growth of abnormal cells. Globally, cancer ranked as the second leading cause of death in 2018, responsible for approximately 9.6 million deaths. Between 30% and 50% of cancer-related deaths are considered preventable through the reduction of major risk factor exposure (WHO 2019). It is suggested that carotenoids exert cytotoxic and antiproliferative effects on cancer cells through diverse molecular pathways, including modulation of oxidative stress, cellular signaling proteins, apoptosis, cell cycle regulation, angiogenesis, metastasis, intercellular communication, and multidrug resistance (Saini et al. 2020).

During the 1980s, epidemiological studies suggested that carotenoid intake might reduce the risk of certain cancers, particularly lung cancer (Shekelle et al. 1981; Colditz et al. 1985). Based on these findings, β-carotene was proposed as a non-toxic and potentially "ideal" agent for use in human chemoprevention trials (Toma et al. 1995). To further investigate these promising observations, large-scale intervention studies were conducted (Blot et al. 1993; ATBC Study Group 1994; Hennekens et al. 1996; Omenn et al. 1996). However, some of these intervention trials yielded unexpected results. The Alpha-Tocopherol Beta-Carotene Cancer Prevention (ATBC) study included 29,133 male smokers aged 50–69, found that daily supplementation with 20 mg of β-carotene was associated with a significantly increased incidence and mortality of lung cancer (ATBC Study Group 1994). Similar results were observed in the Beta-Carotene and Retinol Efficacy Trial (CARET), conducted with 18,314 participants at high risk for lung cancer due to a smoking history or asbestos exposure. CARET study reported an increased incidence and mortality of lung cancer following the daily supplementation with 30 mg β-carotene and 25,000 IU retinyl palmitate. Due to these adverse outcomes, the intervention was terminated 21 months before its planned (Omenn et al. 1996). A recent systematic review and meta-analysis of randomized controlled trials involving 167,141 participants also confirmed that β-carotene supplementation was associated with an increased risk of lung cancer, particularly among smokers and asbestos workers (Kordiak et al. 2022). It has been suggested that excessive β-carotene accumulation in the plasma following supplementation may act as a pro-oxidant rather than an antioxidant in certain conditions, potentially contributing to lung carcinogenesis. Bates et al. (2024) stated that high supplemental doses of β-carotene, when combined with smoke exposure, can increase cellular proliferation signaling, potentially promoting the progression of lung cancer. On the contrary, dietary levels of β-carotene may reduce cellular proliferation signaling even in individuals exposed to smoke.

Another carotenoid, lycopene, has been particularly investigated for its protective effects against prostate cancer. Higher dietary tomato or lycopene intake, and as well as serum lycopene concentrations, have been associated with reduced risk of prostate cancer. Every additional 2 mg of dietary lycopene intake was linked with a 1% decrease in prostate cancer risk (Rowles 3rd et al. 2017). However, in a meta-analysis of six studies supplemental lycopene or tomato extracts containing lycopene showed no significant effect on prostate-specific antigen levels (Sharifi-Zahabi et al. 2022). The authors indicated that the limited number of studies, variations regarding the source and dose of lycopene, baseline health status, and prostate specific antigen levels of participants complicate the interpretation of findings and warrant a cautious approach to draw conclusions.

An umbrella review conducted by Sui et al. (2024) examined the relationship between dietary intake, blood concentrations, and supplementation of carotenoids and cancer risk. This review, which included 51 articles encompassing 198 meta-analyses, found that higher dietary intake and higher serum levels of carotenoids were associated with a reduced risk of total cancer, gastrointestinal cancers, prostate cancer, breast cancer, head and neck cancer, bladder cancer, and other cancer types. Subgroup analyses further revealed that specific carotenoids, such as α-carotene, β-carotene, β-cryptoxanthin, lutein, zeaxanthin, and lycopene, may confer protective effects against certain cancers. However, carotenoid supplementation was not associated with a reduced risk for any type of cancer; on the contrary, it was linked to an increased risk of total cancer, lung cancer, and bladder cancer. These findings suggest that a high intake of carotenoids through whole foods may be more beneficial in lowering cancer risk.

21.3.3 Carotenoids and Diabetes Mellitus

T2DM is a chronic metabolic disorder characterized by insulin resistance and β-cell dysfunction. Hyperglycemia-induced ROS production and oxidative stress are associated with the pathogenesis and progression of T2DM as well as its micro- and macro-vascular complications (Bhatti et al. 2022). Carotenoids may modulate T2DM and its complications through several interrelated mechanisms. As potent natural antioxidants, they can reduce oxidative stress by scavenging ROS, which are known to impair insulin secretion and promote insulin resistance. They also exert anti-inflammatory effects by downregulating pro-inflammatory cytokines such as tumor necrosis factor alpha (TNF-α), IL-1β, and IL-6, through inhibition of the NF-κB pathway. This provides a protection for pancreatic β-cells and peripheral tissues from inflammation-induced damage. Besides, certain carotenoids such as crocin, bixin, astaxanthin, and fucoxanthin can enhance insulin sensitivity by activating proliferator-activated receptor gamma (PPAR-γ) and upregulating glucose transporter 4 (GLUT4) expression in skeletal muscle. Additionally, carotenoids have demonstrated protective effects against diabetes-related complications by mitigating oxidative stress and inflammation in the kidneys, nervous system and ocular tissues (Roohbakhsh et al. 2017).

A systematic review and meta-analysis of prospective observational studies indicated that higher dietary intake and circulating levels of total carotenoids, particularly β-carotene and lutein, were inversely associated with T2DM risk. Although findings for other carotenoids were less consistent, there was a generally protective trend (Jiang et al. 2021). While dietary intake of carotenoids has been associated with a lower risk of T2DM, the evidence for the carotenoid supplementation is limited. Lampousi et al. (2024) reported no significant effect of β-carotene supplementation on the incidence of T2DM. Shokri-Mashhadi et al. (2025) investigated the effects of carotenoid supplementation on glycemic control and based on 36 randomized controlled trials, they found that carotenoid supplementation led to a statistically significant reduction in fasting blood glucose by 4.54 mg/dL and in hemoglobin A1c (HbA1c) by 0.25%. Subgroup analyses revealed that these effects were more pronounced among individuals with T2DM, particularly with specific carotenoids such as crocin, astaxanthin, and fucoxanthin. By contrast, the effects of lutein, zeaxanthin, lycopene, and β-carotene were minimal or non-significant. The authors noted that although certain carotenoids may support glycemic regulation in individuals with T2DM, the overall certainty of the evidence was low.

21.3.4 Carotenoids and Age-Related Macular Degeneration

Visual impairments are common among elderly, and one of the leading causes of blindness in people aged 50 years and older is age-related macular degeneration (AMD) (GBD 2019 Blindness and vision impairment collaborators 2021). It primarily affects people over 55 years of age and become an important public health issue with ageing populations. It is estimated that 8.69% of the population is affected by AMD. It is a progressive eye disease that primarily affects the macula, the central part of the retina responsible for sharp, straight-ahead vision, although the disease is not limited with this area. While advanced age is the single biggest risk factor for AMD, smoking, diet and physical activity are involve among the modifiable risk factors (Guymer and Campbell 2023). Chronic inflammation, oxidative stress, lipid deposition and impaired extracellular matrix maintenance are involved in the pathogenesis of AMD (Fleckenstein et al. 2021).

Carotenoids, especially lutein and zeaxanthin, have beneficial effects on macular health and function. These two carotenoids are specifically distributed in the human retina. Zeaxanthin is primarily present in the center of the fovea, an area with a high density of cone photoreceptors that is exposed to bright light, whereas lutein is dominant in the peripheral part of the retina. Another macular xanthophyll, mesozeaxanthin, is a non-dietary xanthophyll formed from lutein in the retina. These macular pigments may help protect the retina from light-induced oxidative stress, by absorbing blue light (400–500 nm), which is highly reactive and can lead to oxidative damage in retinal tissues. They can also neutralize ROS that generated in response to short-wavelength visible light and ultraviolet (UV) radiation (Mrowicka et al. 2022; Terao 2023).

Supplementation with lutein, zeaxanthin and meso-zeaxanthin increases macular pigment optical density in a dose-dependent manner (Ma et al. 2016) and elevated macular optic pigment density provides a protection against the onset and progression of AMD (Masri et al. 2024). Epidemiological studies have reported higher intake or higher blood levels of lutein and zeaxanthin are associated with slower progression of AMD (Lem et al. 2021; Ma et al. 2012).

The effects of carotenoids and antioxidant supplementation on AMD have been investigated in large-scale clinical trials (AREDS Group 2001; AREDS2 Group 2013). The Age-Related Eye Disease Study (AREDS) followed 3640 elderly participants with early or worse AMD for an average 6.3 year (AREDS Group 2001). Participants received daily supplementation with 15 mg of β-carotene along with a combination of vitamins (500 mg vitamin C and 400 IU vitamin E and 80 mg zinc, and 2 mg copper). The subsequent AREDS2 study evaluated the impact of adding lutein (10 mg) and zeaxanthin (2 mg) to the formulation used in the initial AREDS. Although neither formulation prevented the initial onset of AMD, the AREDS studies demonstrated that antioxidant supplementation reduced the risk of progression from intermediate to advanced AMD. While adding lutein and zeaxanthin to the original formula did not further enhance the protective effect, these carotenoids were effective in reducing the risk of disease progression when used instead of β-carotene (AREDS2 Group 2013). Furthermore, a long-term epidemiological follow-up study of AREDS2 cohort reported an increased risk of lung cancer associated with β-carotene use, supporting its replacement with lutein and zeaxanthin (Chew et al. 2022). Although antioxidant supplements, particularly those containing lutein and zeaxanthin, may slow the progression of AMD, a Cochrane review concluded that there is no evidence supporting their use for preventing the onset of AMD (Evans and Lawrenson 2017).

21.3.5 Carotenoids and Neurodegenerative Disease

Neurodegenerative diseases are characterized by the progressive loss of neurons and the accumulation of protein aggregates. Multiple factors, including protein misfolding, oxidative stress, free radical damage, and mitochondrial dysfunction, contribute to their development. Neurodegenerative diseases typically manifest with symptoms such as memory loss, cognitive decline, motor impairments, and behavioral changes (Samanta et al. 2024). In cellular and animal models, mostly targeting Alzheimer's and Parkinson's diseases, carotenoids have demonstrated potential neuroprotective effects through antioxidant and anti-neuroinflammatory activities, as well as modulation of autophagy (Cho et al. 2018). Case-control studies have indicated that serum concentrations of α-carotene, β-carotene, α-cryptoxanthin, lutein, lycopene, and zeaxanthin were significantly lower in patients with dementia (Wang et al. 2023). A 12-month intervention study indicated that supplementation with carotenoids (10 mg lutein, 10 mg meso-zeaxanthin, 2 mg zeaxanthin) along with omega-3 and vitamin E has a positive effect on the symptoms and progression of Alzheimer's disease (Nolan et al. 2022). The effects of carotenoids on cognitive function in healthy populations have also been studied. In epidemiological studies, higher carotenoid intake has been associated with better cognitive performance in adults (Christensen et al. 2020; Yuan et al. 2020). In a meta-analysis of nine studies, mostly with small sample sizes, carotenoid interventions (mostly lutein and zeaxanthin) improved cognitive outcomes in non-demented individuals aged 45–78 (Davinelli et al. 2021).

21.3.6 Carotenoids and Sarcopenia

Sarcopenia is a progressive muscle disease rooted in adverse muscle changes that progressively accumulate throughout life. It is common among the elderly but can also occur earlier in life. Sarcopenia is defined by low levels of muscle strength, muscle quantity/quality, and physical performance, which serves as an indicator of severity. The prevention of sarcopenia is a goal of healthy aging because it is associated with an increased risk of adverse outcomes such as falls, fractures, physical disability and mortality (Cruz-Jentoft et al. 2019). Oxidative stress is central to the age-related loss of muscle and physical function. Biological mechanisms proposed to contribute to sarcopenia, such as disrupted fatty acid metabolism, impaired mitochondrial ATP production, elevated ROS generation, and telomere shortening, have all been associated with carotenoids (Bartali and Semba 2021).

Findings from the six-year follow-up of the Aging in the Chianti Area (InCHIANTI) study, a population-based cohort in Italy, have indicated that older adults with lower plasma carotenoid levels were at greater risk of developing poor hip, knee, and grip strength (Lauretani et al. 2008a), and high plasma carotenoid concentrations may offer protection against the decline in walking speed and the onset of severe walking disability (Lauretani et al. 2008b). Similar results were also observed in the Framingham Offspring Study, where the higher intakes of total carotenoids, lycopene, and lutein + zeaxanthin were significantly associated with improved grip strength and gait speed (Sahni et al. 2021). In contrast, a cross-sectional analysis of 1172 adults aged 50–85 years from NHANES 2001–2002 data reported that only serum α-carotene, not other carotenoids, was positively associated with muscle strength (Bruno et al. 2022). While the individual effects of carotenoids remain unclear, these results still suggest a potentially beneficial role of dietary carotenoids in maintaining muscle strength and mobility in aging populations, highlighting the need for further research through randomized controlled trials.

21.3.7 Carotenoids and Skin Health

Human skin is exposed to sunlight throughout life, and although sunlight offers certain benefits, exposure to UV radiation can lead to phototoxic effects (Flieger et al. 2024). Carotenoids accumulate in the epidermis and dermis and may help to defend the skin against radiation-induced damage (Flieger et al. 2024). In this regard, carotenoids may act as natural photoprotective agents, similar to UV filters, and they also exhibit antioxidant and anti-inflammatory properties to help mitigate skin damage caused by UV exposure.

Lycopene is considered one of the most potent carotenoids for photoprotection. Several studies have assessed the effects of lycopene and tomato-based products on skin deterioration. A systematic review of 21 intervention studies found that consuming tomato products or lycopene significantly reduced markers of skin inflammation and pigmentation associated with UV exposure. Supplementation also improved minimal erythema dose, as well as skin thickness and density (Flieger et al. 2024). The protective effects of β-carotene on skin health have also been investigated for many years. A meta-analysis indicated that β-carotene supplementation may help protect the skin from sunburn, provided that the intervention lasts for at least 10 weeks (Köpcke and Krutmann 2008). It has also been reported that achieving effective photoprotection with β-carotene requires relatively high doses (12 mg/day or more) (Flieger et al. 2024). In a 12-week intervention trial, a combination of lutein and zeaxanthin was found to be effective in improving skin hydration, elasticity, and protection against UV-induced damage (Palombo et al. 2007). Astaxanthin is another carotenoid with notable anti-photoaging properties. A systematic review of intervention studies indicated that astaxanthin supplementation (3–6 mg/day) improved skin texture, reduced the appearance of wrinkles, and enhanced moisture content (Ng et al. 2021). However, most of the included studies had relatively small sample size, highlighting the need for further investigation.

21.4 Conclusion

Current research suggests that a well-balanced diet with high carotenoid intake may have beneficial effects on aging-related diseases and healthy longevity. In this context, the inclusion of carotenoid-rich foods in the diet, such as green leafy vegetables, citrus fruits, apricots, and tomato-based products is considered valuable.

Various forms of dietary supplements, either in isolated or complex formulations of different carotenoids, are commercially available. However, apart from specific conditions, such as the treatment of age-related macular degeneration where lutein and zeaxanthin supplementation may be recommended, there is insufficient evidence to support the routine use of these supplements for the prevention of aging-related diseases. Notably, while high doses of β-carotene have been shown to exert pro-oxidant effects and increase the risk of cancer and cardiovascular mortality, particularly in smokers, its use is not recommended in certain populations (Kordiak et al. 2022; Yang et al. 2022; Bates et al. 2024). Indeed, the U.S. Preventive Services Task Force (2022) recommends against the use of β-carotene supplements for the prevention of cardiovascular disease or cancer. Therefore, based on current evidence, obtaining carotenoids from an adequate and diverse diet can be recommended rather than from supplementation. Nonetheless, carotenoids remain promising natural compounds, and future studies with different carotenoids and large sample sizes are needed to further investigate their potential benefits.

References

Age-Related Eye Diseases Study 2 Research Group (AREDS2 Group) (2013) Lutein + zeaxanthin and omega-3 fatty acids for age-related macular degeneration: the Age-Related Eye Disease Study 2 (AREDS2) randomized clinical trial. JAMA 309:2005–2015

Age-Related Eye Diseases Study Research Group (AREDS Group) (2001) A randomized, placebo-controlled, clinical trial of high-dose supplementation with vitamins C and E, beta carotene, and zinc for age-related macular degeneration and vision loss: AREDS report no. 8. Arch Ophthalmol 119:1417–1436

Alpha-Tocopherol, Beta Carotene Cancer Prevention Study Group (1994) The effect of vitamin E and beta carotene on the incidence of lung cancer and other cancers in male smokers. N Engl J Med 330:1029–1035

An P, Wan S, Luo Y, Luo J, Zhang X, Zhou S, Xu T, He J, Mechanick JI, Wu WC, Ren F, Liu S (2022) Micronutrient supplementation to reduce cardiovascular risk. J Am Coll Cardiol 80: 2269–2285

Aune D, Keum N, Giovannucci E, Fadnes LT, Boffetta P, Greenwood DC, Tonstad S, Vatten LJ, Riboli E, Norat T (2018) Dietary intake and blood concentrations of antioxidants and the risk of cardiovascular disease, total cancer, and all-cause mortality: a systematic review and dose-response meta-analysis of prospective studies. Am J Clin Nutr 108:1069–1091

Bahonar A, Saadatnia M, Khorvash F, Maracy M, Khosravi A (2017) Carotenoids as potential antioxidant agents in stroke prevention: a systematic review. Int J Prev Med 8:70

Bakac ER, Percin E, Gunes-Bayir A, Dadak A (2023) A narrative review: the effect and importance of carotenoids on aging and aging-related diseases. Int J Mol Sci 24:15199

Bartali B, Semba RD (2021) Carotenoids and healthy aging: the fascination continues. Am J Clin Nutr 113:259–260

Bates CA, Vincent MJ, Buerger AN, Santamaria AB, Maier A, Jack M (2024) Investigating the relationship between β-carotene intake from diet and supplements, smoking, and lung cancer risk. Food Chem Toxicol 194:115104

Bhatti JS, Sehrawat A, Mishra J, Sidhu IS, Navik U, Khullar N, Kumar S, Bhatti GK, Reddy PH (2022) Oxidative stress in the pathophysiology of type 2 diabetes and related complications: current therapeutics strategies and future perspectives. Free Radic Biol Med 184:114–134

Blot WJ, Li JY, Taylor PR, Guo W, Dawsey S, Wang GQ, Yang CS, Zheng SF, Gail M, Li GY et al (1993) Nutrition intervention trials in Linxian, China: supplementation with specific vitamin/mineral combinations, cancer incidence, and disease-specific mortality in the general population. J Natl Cancer Inst 85:1483–1492

Brendler T, Williamson EM (2019) Astaxanthin: how much is too much? A safety review. Phytother Res 33:3090–3111

Bruno RR, Rosa FC, Nahas PC, De Branco FMS, De Oliveira EP (2022) Serum α-Carotene, but not other antioxidants, is positively associated with muscle strength in older adults: NHANES 2001–2002. Antioxidants 11:2386

Chen M, Cai S, Jia Q, Suo Y, Tang Y, Shi Y, Zhu X, Zhang H (2025) Inverse relationship between serum carotenoid levels and cardiovascular-kidney-metabolic syndrome among the general adult population. J Diabetes 17:e70046

Cheng HM, Koutsidis G, Lodge JK, Ashor A, Siervo M, Lara J (2017) Tomato and lycopene supplementation and cardiovascular risk factors: a systematic review and meta-analysis. Atherosclerosis 257:100–108

Cheng HM, Koutsidis G, Lodge JK, Ashor AW, Siervo M, Lara J (2019) Lycopene and tomato and risk of cardiovascular diseases: a systematic review and meta-analysis of epidemiological evidence. Crit Rev Food Sci Nutr 59:141–158

Chew EY, Clemons TE, AgróN E, Domalpally A, Keenan TDL, Vitale S, Weber C, Smith DC, Christen W (2022) Long-term outcomes of adding lutein/zeaxanthin and ω-3 fatty acids to the AREDS supplements on age-related macular degeneration progression: AREDS2 report 28. JAMA Ophthalmol 140:692–698

Cho KS, Shin M, Kim S, Lee SB (2018) Recent advances in studies on the therapeutic potential of dietary carotenoids in neurodegenerative diseases. Oxid Med Cell Longev 2018:4120458

Christensen K, Gleason CE, Mares JA (2020) Dietary carotenoids and cognitive function among US adults, NHANES 2011-2014. Nutr Neurosci 23:554–562

Chung RWS, Leanderson P, Lundberg AK, Jonasson L (2017) Lutein exerts anti-inflammatory effects in patients with coronary artery disease. Atherosclerosis 262:87–93

Colditz GA, Branch LG, Lipnick RJ, Willett WC, Rosner B, Posner BM, Hennekens CH (1985) Increased green and yellow vegetable intake and lowered cancer deaths in an elderly population. Am J Clin Nutr 41:32–36

Cruz-Jentoft AJ, Bahat G, Bauer J, Boirie Y, Bruyère O, Cederholm T, Cooper C, Landi F, Rolland Y, Sayer AA, Schneider SM, Sieber CC, Topinkova E, Vandewoude M, Visser M, Zamboni M (2019) Sarcopenia: revised European consensus on definition and diagnosis. Age Ageing 48:16–31

Davinelli S, Ali S, Solfrizzi V, Scapagnini G, Corbi G (2021) Carotenoids and cognitive outcomes: a meta-analysis of randomized intervention trials. Antioxidants (Basel) 10:223

Eroglu A, Al'abri IS, Kopec RE, Crook N, Bohn T (2023) Carotenoids and their health benefits as derived via their interactions with gut microbiota. Adv Nutr 14:238–255

Evans JR, Lawrenson JG (2017) Antioxidant vitamin and mineral supplements for preventing age-related macular degeneration. Cochrane Database Syst Rev 7:Cd000253

Fleckenstein M, Keenan TDL, Guymer RH, Chakravarthy U, Schmitz-Valckenberg S, Klaver CC, Wong WT, Chew EY (2021) Age-related macular degeneration. Nat Rev Dis Primers 7:31

Flieger J, Raszewska-Famielec M, Radzikowska-Büchner E, Flieger W (2024) Skin protection by carotenoid pigments. Int J Mol Sci 25:1431

FrąK W, Wojtasińska A, Lisińska W, Młynarska E, Franczyk B, Rysz J (2022) Pathophysiology of cardiovascular diseases: new insights into molecular mechanisms of atherosclerosis, arterial hypertension, and coronary artery disease. Biomedicine 10:1938

GBD 2019 Blindness and Vision Impairment Collaborators; Vision Loss Expert Group of the Global Burden of Disease Study (2021) Causes of blindness and vision impairment in 2020 and trends over 30 years, and prevalence of avoidable blindness in relation to VISION 2020: the right to sight: an analysis for the Global Burden of Disease Study. Lancet Glob Health 9:e144–e160

Global Cardiovascular Risk Consortium (2023) Global effect of modifiable risk factors on cardiovascular disease and mortality. N Engl J Med 389:1273–1285

Guymer RH, Campbell TG (2023) Age-related macular degeneration. Lancet 401:1459–1472

Hajizadeh-Sharafabad F, Zahabi ES, Malekahmadi M, Zarrin R, Alizadeh M (2022) Carotenoids supplementation and inflammation: a systematic review and meta-analysis of randomized clinical trials. Crit Rev Food Sci Nutr 62:8161–8177

Harrison DE, Strong R, Reifsnyder P, Rosenthal N, Korstanje R, Fernandez E, Flurkey K, Ginsburg BC, Murrell MD, Javors MA, Lopez-Cruzan M, Nelson JF, Willcox BJ, Allsopp R, Watumull DM, Watumull DG, Cortopassi G, Kirkland JL, Tchkonia T, Choi YG, Yousefzadeh MJ, Robbins PD, Mitchell JR, Acar M, Sarnoski EA, Bene MR, Salmon A, Kumar N, Miller RA (2024) Astaxanthin and meclizine extend lifespan in UM-HET3 male mice; fisetin, SG1002 (hydrogen sulfide donor), dimethyl fumarate, mycophenolic acid, and 4-phenylbutyrate do not significantly affect lifespan in either sex at the doses and schedules used. Geroscience 46:795–816

He X, Yin X, Chen X, Chen X (2023) Aging and antioxidants: the impact of dietary carotenoid intakes on soluble klotho levels in aged adults. Front Endocrinol (Lausanne) 14:1283722

Hennekens CH, Buring JE, Manson JE, Stampfer M, Rosner B, Cook NR, Belanger C, Lamotte F, Gaziano JM, Ridker PM, Willett W, Peto R (1996) Lack of effect of long-term supplementation with beta carotene on the incidence of malignant neoplasms and cardiovascular disease. N Engl J Med 334:1145–1149

Huang J, Weinstein SJ, Yu K, Männistö S, Albanes D (2018) Serum beta carotene and overall and cause-specific mortality. Circ Res 123:1339–1349

Jiang YW, Sun ZH, Tong WW, Yang K, Guo KQ, Liu G, Pan A (2021) Dietary intake and circulating concentrations of carotenoids and risk of type 2 diabetes: a dose-response meta-analysis of prospective observational studies. Adv Nutr 12:1723–1733

Kaulmann A, Bohn T (2014) Carotenoids, inflammation, and oxidative stress--implications of cellular signaling pathways and relation to chronic disease prevention. Nutr Res 34:907–929

Köpcke W, Krutmann J (2008) Protection from sunburn with beta-carotene--a meta-analysis. Photochem Photobiol 84:284–288

Kordiak J, Bielec F, JabłOński S, Pastuszak-Lewandoska D (2022) Role of beta-carotene in lung cancer primary chemoprevention: a systematic review with meta-analysis and meta-regression. Nutrients 14:1361

Lampousi AM, Lundberg T, Löfvenborg JE, Carlsson S (2024) Vitamins C, E, and β-Carotene and risk of type 2 diabetes: a systematic review and meta-analysis. Adv Nutr 15:100211

Lashmanova E, Proshkina E, Zhikrivetskaya S, Shevchenko O, Marusich E, Leonov S, Melerzanov A, Zhavoronkov A, Moskalev A (2015) Fucoxanthin increases lifespan of Drosophila melanogaster and Caenorhabditis elegans. Pharmacol Res 100:228–241

Lauretani F, Semba RD, Bandinelli S, Dayhoff-Brannigan M, Giacomini V, Corsi AM, Guralnik JM, Ferrucci L (2008a) Low plasma carotenoids and skeletal muscle strength decline over 6 years. J Gerontol A Biol Sci Med Sci 63:376–383

Lauretani F, Semba RD, Bandinelli S, Dayhoff-Brannigan M, Lauretani F, Corsi AM, Guralnik JM, Ferrucci L (2008b) Carotenoids as protection against disability in older persons. Rejuvenation Res 11:557–563

Lee S-A, Lim W-H, Van Le V, Ko S-R, Kim B, Oh H-M, Ahn C-Y (2022) Lifespan extension and anti-oxidant effects of carotenoid pigments in Caenorhabditis elegans. Bioresour Technol Rep 17:100962

Leermakers ETM, Darweesh SKL, Baena CP, Moreira EM, Melo Van Lent D, Tielemans MJ, Muka T, Vitezova A, Chowdhury R, Bramer WM, Kiefte-De Jong JC, Felix JF, Franco OH (2016) The effects of lutein on cardiometabolic health across the life course: a systematic review and meta-analysis. Am J Clin Nutr 103:481–494

Lem DW, Davey PG, Gierhart DL, Rosen RB (2021) A systematic review of carotenoids in the management of age-related macular degeneration. Antioxidants 10:1255

Leung LY-L, Chan SM-N, Tam H-L, Wong ES-W (2022) Astaxanthin influence on health outcomes of adults at risk of metabolic syndrome: a systematic review and meta-analysis. Nutrients 14:2050

Li AP, Li D, Tan X, Xu R, Mao LX, Kang JJ, Li SH, Liu Y (2025) Crocin extends lifespan by mitigating oxidative stress and regulating lipid metabolism through the DAF-16/FOXO pathway. Food Funct 16:3369–3383

López-Otín C, Blasco M, Partridge L, Serrano M, Kroemer G (2013) The hallmarks of aging. Cell 153:1194–1217

López-OtíN C, Blasco MA, Partridge L, Serrano M, Kroemer G (2023) Hallmarks of aging: an expanding universe. Cell 186:243–278

Ma L, Dou HL, Wu YQ, Huang YM, Huang YB, Xu XR, Zou ZY, Lin XM (2012) Lutein and zeaxanthin intake and the risk of age-related macular degeneration: a systematic review and meta-analysis. Br J Nutr 107:350–359

Ma L, Liu R, Du JH, Liu T, Wu SS, Liu XH (2016) Lutein, Zeaxanthin and Meso-zeaxanthin supplementation associated with macular pigment optical density. Nutrients 8:426

Maiani G, Periago CastóN MJ, Catasta G, Toti E, CambrodóN IG, Bysted A, Granado-Lorencio F, Olmedilla-Alonso B, Knuthsen P, Valoti M, Böhm V, Mayer-Miebach E, Behsnilian D, Schlemmer U (2009) Carotenoids: actual knowledge on food sources, intakes, stability and bioavailability and their protective role in humans. Mol Nutr Food Res 53:S194–S218

Maoka T (2020) Carotenoids as natural functional pigments. J Nat Med 74:1–16

Martínez-Cámara S, Ibañez A, Rubio S, Barreiro C, Barredo J-L (2021) Main carotenoids produced by microorganisms. Encyclopedia 1:1223–1245

Masri A, Armanazi M, Inouye K, Geierhart DL, Davey PG, Vasudevan B (2024) Macular pigment optical density as a measurable modifiable clinical biomarker. Nutrients 16:3273

Meléndez-Martínez AJ, Mandić AI, Bantis F, Böhm V, Borge GIA, Brnčić M, Bysted A, Cano MP, Dias MG, Elgersma A, Fikselová M, García-Alonso J, Giuffrida D, Gonçalves VSS, Hornero-Méndez D, Kljak K, Lavelli V, Manganaris GA, Mapelli-Brahm P, Marounek M, Olmedilla-Alonso B, Periago-Castón MJ, Pintea A, Sheehan JJ, Tumbas Šaponjac V, Valšíková-Frey M, Meulebroek LV, O'Brien N (2022) A comprehensive review on carotenoids in foods and feeds: status quo, applications, patents, and research needs. Crit Rev Food Sci Nutr 62:1999–2049

Min KB, Min JY (2017) Association between leukocyte telomere length and serum carotenoid in US adults. Eur J Nutr 56:1045–1052

Mrowicka M, Mrowicki J, Kucharska E, Majsterek I (2022) Lutein and zeaxanthin and their roles in age-related macular degeneration-neurodegenerative disease. Nutrients 14:827

Ng QX, De Deyn M, Loke W, Foo NX, Chan HW, Yeo WS (2021) Effects of astaxanthin supplementation on skin health: a systematic review of clinical studies. J Diet Suppl 18:169–182

Nolan JM, Power R, Howard AN, Bergin P, Roche W, Prado-Cabrero A, Pope G, Cooke J, Power T, Mulcahy R (2022) Supplementation with carotenoids, omega-3 fatty acids, and vitamin e has a positive effect on the symptoms and progression of lzheimer's sisease. J Alzheimers Dis 90:233–249

Nomura SJ, Robien K, Zota AR (2017) Serum folate, vitamin B-12, vitamin A, γ-tocopherol, α-tocopherol, and carotenoids do not modify associations between cadmium exposure and leukocyte telomere length in the general US adult population. J Nutr 147:538–548

Omenn GS, Goodman GE, Thornquist MD, Balmes J, Cullen MR, Glass A, Keogh JP, Meyskens FL, Valanis B, Williams JH, Barnhart S, Hammar S (1996) Effects of a combination of beta carotene and vitamin A on lung cancer and cardiovascular disease. N Engl J Med 334:1150–1155

Osganian SK, Stampfer MJ, Rimm E, Spiegelman D, Manson JE, Willett WC (2003) Dietary carotenoids and risk of coronary artery disease in women123. Am J Clin Nutr 77:1390–1399

Palombo P, Fabrizi G, Ruocco V, Ruocco E, Fluhr J, Roberts R, Morganti P (2007) Beneficial long-term effects of combined oral/topical antioxidant treatment with the carotenoids lutein and zeaxanthin on human skin: a double-blind, placebo-controlled study. Skin Pharmacol Physiol 20:199–210

Qi X, Wang X, Cheng L, Li Y, Dang K, Yang S, Wang Y, Zhou R, Zhang C, Li Y (2025) Dietary carotenoid intakes and biological aging among US adults, NHANES 1999-2018. Nutr J 24:9

Qiu Z, Chen X, Geng T, Wan Z, Lu Q, Li L, Zhu K, Zhang X, Liu Y, Lin X, Chen L, Shan Z, Liu L, Pan A, Liu G (2022) Associations of serum carotenoids with risk of cardiovascular mortality among individuals with type 2 diabetes: results from NHANES. Diabetes Care 45:1453–1461

Roohbakhsh A, Karimi G, Iranshahi M (2017) Carotenoids in the treatment of diabetes mellitus and its complications: a mechanistic review. Biomed Pharmacother 91:31–42

Rowles JL 3rd, Ranard KM, Smith JW, An R, Erdman JW Jr (2017) Increased dietary and circulating lycopene are associated with reduced prostate cancer risk: a systematic review and meta-analysis. Prostate Cancer Prostatic Dis 20:361–377

Sahni S, Dufour AB, Fielding RA, Newman AB, Kiel DP, Hannan MT, Jacques PF (2021) Total carotenoid intake is associated with reduced loss of grip strength and gait speed over time in adults: The Framingham Offspring Study. Am J Clin Nutr 113:437–445

Saini RK, Keum YS, Daglia M, Rengasamy KR (2020) Dietary carotenoids in cancer chemoprevention and chemotherapy: a review of emerging evidence. Pharmacol Res 157:104830

Saini RK, Prasad P, Lokesh V, Shang X, Shin J, Keum YS, Lee JH (2022) Carotenoids: dietary sources, extraction, encapsulation, bioavailability, and health benefits-a review of recent advancements. Antioxidants (Basel) 11:795

Samanta S, Chakraborty S, Bagchi D (2024) Pathogenesis of neurodegenerative diseases and the protective role of natural bioactive components. J Am Nutr Assoc 43:20–32

Sharifi-Zahabi E, Soltani S, Malekahmadi M, Rezavand L, Clark CCT, Shidfar F (2022) The effect of lycopene supplement from different sources on prostate specific antigen (PSA): a systematic review and meta-analysis of randomized controlled trials. Complement Ther Med 64:102801

Shekelle R, Liu S, Raynor W, Lepper M, Maliza C, Rossof A, Paul O, Shryock A, Stamler J (1981) Dietary vitamin A and risk of cancer in the western electric study. Lancet 318:1185–1190

Shokri-Mashhadi N, Baechle C, Schiemann T, Schaefer E, Barbaresko J, Schlesinger S (2025) Effects of carotenoid supplementation on glycemic control: a systematic review and meta-analysis of randomized clinical trials. Eur J Clin Nutr 79:113–125

Sui J, Guo J, Pan D, Wang Y, Xu Y, Sun G, Xia H (2024) The efficacy of dietary intake, supplementation, and blood concentrations of carotenoids in cancer prevention: insights from an umbrella meta-analysis. Foods 13:1321

Sumalla-Cano S, Eguren-GarcíA I, Lasarte-GarcíA Á, Prola TA, Martínez-Díaz R, ElíO I (2024) Carotenoids intake and cardiovascular prevention: a systematic review. Nutrients 16:3859

Sztretye M, Dienes B, Gönczi M, CzirjáK T, Csernoch L, Dux L, Szentesi P, Keller-PintéR A (2019) Astaxanthin: a potential mitochondrial-targeted antioxidant treatment in diseases and with aging. Oxid Med Cell Longev 2019:3849692

Terao J (2023) Revisiting carotenoids as dietary antioxidants for human health and disease prevention. Food Funct 14:7799–7824

Toma S, Losardo PL, Vincent M, Palumbo R (1995) Effectiveness of beta-carotene in cancer chemoprevention. Eur J Cancer Prev 4:213–224

US Preventive Services Task Force, Mangione CM, Barry MJ, Nicholson WK, Cabana M, Chelmow D, Coker TR, Davis EM, Donahue KE, Doubeni CA, Jaén CR, Kubik M, Li L, Ogedegbe G, Pbert L, Ruiz JM, Stevermer J, Wong JB (2022) Vitamin, mineral, and multivitamin supplementation to prevent cardiovascular disease and cancer: US Preventive Services Task Force Recommendation Statement. JAMA 327:2326–2333

Vaduganathan M, Mensah GA, Turco JV, Fuster V, Roth GA (2022) The global burden of cardiovascular diseases and risk: a compass for future health. J Am Coll Cardiol 80:2361–2371

Wang L, Zhao T, Zhu X, Jiang Q (2023) Low blood carotenoid status in dementia and mild cognitive impairment: a systematic review and meta-analysis. BMC Geriatr 23:195

Wang J, Xie F, Zhu W, Ye D, Xiao Y, Shi M, Zeng R, Bian J, Xu X, Chen L, Zhu A, Zhu K, Fan T, Liu B, Xiao L, Zhang X (2024) Relationship between serum carotenoids and telomere length in overweight or obese individuals. Front Nutr 11:1479994

Wiese M, Bashmakov Y, Chalyk N, Nielsen DS, Krych Ł, Kot W, Klochkov V, Pristensky D, Bandaletova T, Chernyshova M (2019) Prebiotic effect of lycopene and dark chocolate on gut microbiome with systemic changes in liver metabolism, skeletal muscles and skin in moderately obese persons. Biomed Res Int 2019:4625279

Wolak T, Sharoni Y, Levy J, Linnewiel-Hermoni K, Stepensky D, Paran E (2019) Effect of tomato nutrient complex on blood pressure: a double blind, randomized dose¯response study. Nutrients 11:950

World Health Organization, Cancer (2019) Available online: https://www.who.int/cancer/en/. Accessed on 18 May 2025

Yabuzaki J (2017) Carotenoids database: structures, chemical fingerprints and distribution among organisms. Database 2017

Yang J, Zhang Y, Na X, Zhao A (2022) β-Carotene supplementation and risk of cardiovascular disease: a systematic review and meta-analysis of randomized controlled trials. Nutrients 14: 1284

Yazaki K, Yoshikoshi C, Oshiro S, Yanase S (2011) Supplemental cellular protection by a carotenoid extends lifespan via Ins/IGF-1 signaling in Caenorhabditis elegans. Oxid Med Cell Longev 2011:596240

Yuan C, Fondell E, Ascherio A, Okereke OI, Grodstein F, Hofman A, Willett WC (2020) Long-term intake of dietary carotenoids is positively associated with late-life subjective cognitive function in a prospective study in US women. J Nutr 150:1871–1879

Zhang Z, Han S, Wang H, Wang T (2014) Lutein extends the lifespan of Drosophila melanogaster. Arch Gerontol Geriatr 58:153–159

Zhu X, Cheang I, Tang Y, Shi M, Zhu Q, Gao R, Liao S, Yao W, Zhou Y, Zhang H, Li X (2023) Associations of serum carotenoids with risk of all-cause and cardiovascular mortality in hypertensive adults. J Am Heart Assoc 12:e027568

Chapter 22
Curcumin and Resveratrol for Healthy Longevity

Cemre Elmas and Ceren Gezer

Abstract Curcumin, primarily derived from turmeric, and resveratrol, found mainly in grapes, peanuts, and blueberries, are prominent candidates as dietary supplements for promoting healthy longevity. Curcumin may confer health benefits by modulating key biological pathways that reduce oxidative stress, inhibit inflammation and cellular senescence, enhance telomere length, adjust dysregulated nutrient sensing and intercellular communication, and influence epigenetic regulation. In addition to the effects of curcumin, resveratrol is notable for its role in enhancing autophagy and mitochondrial function. Commercial curcumin and resveratrol supplements often display significant variability in their composition, while the concentrations of these compounds in natural food sources are generally much lower than those present in supplemental forms. Although certain regulatory agencies have approved higher safe intake limits due to the low bioavailability of these compounds, definitive guidelines on supplementation dosages for healthy longevity remain lacking. Comprehensive clinical studies that incorporate pharmacokinetic and nutrigenetic analyses are crucial to clarify their potential effects on lifespan extension. In conclusion, until optimal supplementation doses of curcumin and resveratrol for promoting healthy longevity are established, their consumption as part of a balanced dietary model—taking into account the synergistic effects of the food matrix—may offer beneficial outcomes.

Keywords Curcumin · Resveratrol · Nutrition · Longevity · Ageing · Health · Lifespan

---

C. Elmas (✉)
Department of Nutrition and Dietetics, Faculty of Health Sciences, European University of Lefke, Lefke, Cyprus
e-mail: celmas@eul.edu.tr

C. Gezer
Department of Nutrition and Dietetics, Faculty of Health Sciences, Eastern Mediterranean University, Famagusta, Cyprus

S. I. S. Rattan, C. Gezer (eds.), *Supplements and Drugs for Healthy Longevity*, Healthy Ageing and Longevity 22, https://doi.org/10.1007/978-3-032-11608-6_22

Abbreviations

ADI	Acceptable Daily Intake
AMPK	Adenosine monophosphate-activated protein kinase
CAT	Catalase
CCL-2	Chemokine (C-C motif) ligand 2
CRP	C-reactive protein
EFSA	The European Food Safety Authority
FDA	Food and Drug Administration
FoxO	Forkhead box protein O
GPx	Glutathione peroxidase
GRAS	Generally recognized as safe
GSH	Glutathione
GST	Glutathione-S-transferase
hADSCs	Human adipose tissue-derived stem cells
HO-1	Heme oxygenase-1
IFN-γ	Interferon-gamma
IL-10	Interleukin-10
IL-4	Interleukin-4
IL-6	Interleukin-6
IL-8	Interleukin-8
JECFA	The Joint FAO/WHO Expert Committee on Food Additives
MCP-1	Monocyte chemotactic protein-1
mTOR	Mammalian target of rapamycin
NF-κB	Nuclear factor kappa B
Nrf2	Nuclear factor erythroid 2-related factor 2
PBMCs	Human peripheral blood mononuclear cells
PGC-1α	Peroxisome proliferator-activated receptor gamma coactivator 1 alpha
PPAR-γ	Peroxisome proliferator-activated receptor gamma
ROS	Reactive oxygen species
SASP	Senescence-associated secretory phenotype
SIRT1	Sirtuin 1
SOD	Superoxide dismutase
TERF2	Telomeric repeat-binding factor 2
TNF-α	Tumor necrosis factor-alpha

22.1 Introduction

The concept of longevity encompasses not only lifespan but also the quality of life maintained throughout that period. Healthy aging emphasizes the importance of preserving physical function and sustaining cognitive and emotional well-being across the aging trajectory. Achieving a long and healthy life requires a comprehensive understanding of age-related metabolic changes and the various factors influencing these alterations (Addo and Khan 2024). Aging is generally defined as a complex biological process marked by a progressive decline in physiological functions across multiple organ systems and a diminished cellular capacity to respond to injury. The recognized hallmarks of aging include genomic instability, telomere attrition, epigenetic alterations, loss of proteostasis, dysregulated nutrient sensing, mitochondrial dysfunction, cellular senescence, stem cell exhaustion, and disrupted intercellular communication (López-Otín et al. 2023). For these reasons, aging is regarded as a fundamental underlying factor in the development of numerous age-related diseases, including, but not limited to, musculoskeletal disorders, neurodegenerative conditions, cardiovascular disease, cancer, and arthritis (Bautmans et al. 2022; Izadi et al. 2024; Pyo et al. 2020).

Curcumin and resveratrol have been recognized for their capacity to modulate key biological processes associated with aging, thereby supporting healthy longevity. Curcumin exerts its effects through antioxidant and anti-inflammatory mechanisms, stabilization of telomeres, regulation of nutrient-sensing and epigenetic pathways, and inhibition of cellular senescence (Zia et al. 2021). Resveratrol complements these effects by promoting autophagy and improving mitochondrial function, contributing to its protective role against age-related diseases (Zhou et al. 2021a, b). Both compounds are also recognized as caloric restriction mimetics, further aligning their effects with well-established longevity-promoting interventions (Hofer et al. 2021). Moreover, the identification of hormetic dose–response effects of resveratrol in the late 1990s represented a significant advancement in the understanding of polyphenol-induced hormesis, subsequently stimulating extensive research into similar properties in other natural compounds, including curcumin (Calabrese 2021). Therefore, their significance lies in their diverse biological activities and the relatively extensive body of research conducted in both animal and human models, distinguishing them from other polyphenols. This chapter aims to critically examine the roles of curcumin and resveratrol in supporting healthy aging and promoting longevity.

22.2 Curcumin

Curcumin, a natural polyphenolic compound, is derived from the rhizomes of *Curcuma longa L.* Its chemical name is diferuloylmethane, and its molecular formula is 1,7-bis(4-hydroxy-3-methoxyphenyl)-1,6-heptadiene-3,5-dione. As the principal bioactive constituent of turmeric, curcumin is widely utilized in the dietary practices of various cultures. Curcumin is present at a concentration of 2–5% in dried turmeric powder derived from the rhizome of *Curcuma longa L.* In commercially available turmeric powder, the primary bioactive constituents are curcumin (77%), demethoxycurcumin (17%), and bisdemethoxycurcumin (3%) (Bertoncini-Silva et al. 2024). In the food industry, turmeric is utilized in various applications, including as a natural colorant—producing hues from bright yellow to orange—

and as a flavor enhancer (Zheng et al. 2024). Curcumin, the principal bioactive component of turmeric, is incorporated into dietary regimens for various purposes due to its well-documented health-promoting effects (Yang et al. 2022). These effects include antioxidant, anti-inflammatory, anti-tumor, anti-microbial, anti-hypertensive, anti-hyperlipidemic, anti-thrombotic, anti-diabetic, and anti-hepatotoxic activities (Garodia et al. 2023). Owing to these properties, curcumin is widely utilized in the formulation of functional foods, pharmaceuticals, cosmetics, and healthcare products (Chang et al. 2023).

A review of studies on the safe intake of curcumin indicates that the U.S. Food and Drug Administration (FDA) has classified it as generally recognized as safe (GRAS) (No 2018). One study reported that the optimal daily dosage for achieving maximal efficacy ranges from 4 to 8 g, with a documented human tolerance of up to 12 g per day (Barati et al. 2019). Similarly, another study demonstrated that daily consumption of curcumin up to 8 g for a duration of 3 months is safe and does not produce adverse effects in humans (Izadi et al. 2024). According to the Joint FAO/WHO Expert Committee on Food Additives (JECFA) and the European Food Safety Authority (EFSA), the Acceptable Daily Intake (ADI) for curcumin has been established at 0–3 mg/kg body weight (EFSA 2010; JECFA 2004; Lee et al. 2023). Epidemiological findings on curcumin intake show considerable variability (Kwon 2014; EFSA 2010; Hashemian et al. 2019). In a large-scale study conducted in Korea (n = 10,000), curcumin intake was found to be relatively low, ranging from 2.7 to 14.8 mg/day, based on an average turmeric consumption of 0.47 g/day (Kwon 2014). In the United Kingdom, the average adult exposure to curcumin has been reported as 0.9 mg/kg per day (EFSA 2010). Similarly, a study conducted in Iran involving 44,398 individuals aged 40–75 years identified the average turmeric intake as 100 mg/day (Hashemian et al. 2019). Preliminary findings suggest that curcumin may have the potential to inhibit the progression of various chronic diseases and promote longevity, particularly through strategies aimed at enhancing its bioavailability. Its strong antioxidant and anti-inflammatory properties have positioned curcumin as a promising candidate in addressing age-related health conditions (Izadi et al. 2024).

22.2.1 Curcumin—Healthy Longevity Mechanisms

The potential beneficial effects of curcumin may be attributed to its regulatory influence on multiple biological processes, including oxidative stress, inflammation, telomere maintenance, nutrient sensing, intracellular communication, cellular senescence, and epigenetic programming (Zia et al. 2021; Izadi et al. 2024; Bielak-Zmijewska et al. 2019; Pereira et al. 2023).

22.2.1.1 Antioxidative Effects of Curcumin

Curcumin demonstrates potent antioxidant activity both directly, through hydrogen donation by its phenolic groups, and indirectly, by inducing a hormetic stress response. Its hormetic effects are mediated by the stabilization of nuclear factor erythroid 2-related factor 2 (Nrf2) and the upregulation of heme oxygenase-1 (HO-1). Activation of the Nrf2 and HO-1 pathways leads to the induction of thioredoxin reductase, Hsp70, sirtuins, and several antioxidant enzymes, including superoxide dismutase (SOD), glutathione-S-transferase (GST), glutathione (GSH), glutathione peroxidase (GPx), and catalase (CAT) (Zia et al. 2021; Calabrese 2021; Izadi et al. 2024; Aranda-Rivera et al. 2022). A study on antioxidant responses demonstrated that the administration of curcumin at a dose of 10 mg/kg/day effectively prevented oxidative stress-induced auditory hair cell degeneration by activating Nrf2 in the cochlear hair cells of C57BL/6 J mice (Li et al. 2023a). Additionally, curcumin was shown to mitigate the harmful effects of oxidative stress in red blood cells by increasing glutathione levels and reducing malondialdehyde concentrations (Kumar and Maurya 2023). A study conducted using *Caenorhabditis elegans* demonstrated that curcumin acetylsalicylate induced a dose-dependent extension of lifespan and an increase in SOD expression (Zhou et al. 2021a, b). In another study, curcumin administered at doses of 0.2–0.5 mg/kg effectively mitigated Cu^{2+}-induced oxidative damage in *Drosophila melanogaster*, leading to improved antioxidant status and increased survival rates (Abolaji et al. 2020). In a randomized, double-blind, placebo-controlled clinical study involving men with infertility ($n = 28$), daily supplementation with 80 mg of curcumin nanomicelles for 10 weeks significantly improved plasma total antioxidant capacity and reduced levels of malondialdehyde, C-reactive protein (CRP), and tumor necrosis factor-alpha (TNF-α) (Alizadeh et al. 2018). In a pilot study, supplementation with curcumin (Meriva®, 200 mg/day) was found to significantly reduce lipid peroxidation and plasma levels of pro-inflammatory mediators, including chemokine (C-C motif) ligand 2 (CCL-2), interferon-gamma (IFN-γ), and interleukin-4 (IL-4), in patients aged ≥18 years with chronic kidney disease (Pivari et al. 2022). Oral administration of 1 g/day curcumin with piperine for 4 weeks in 46 individuals with chronic pruritus was shown to affect serum levels and activities of CAT, glutathione, and SOD, while reducing markers of inflammation. The combination of curcumin with certain compounds, including piperine, has been reported to enhance its bioavailability. It was observed that the bioavailability of curcumin increased by 2000% when co-administered with piperine, compared to curcumin alone (Vaughn et al. 2016; Shoba et al. 1998).

22.2.1.2 Anti-Inflammatory Effects of Curcumin

Inflammation is defined as the response of an organism's defense system to tissue alterations resulting from various factors, including infection and injury. Chronic inflammation represents a localized and systemic pathological condition that contributes to disease pathogenesis. It is characterized by elevated levels of pro-inflammatory cytokines, such as CRP, interleukin-8 (IL-8), interleukin-6 (IL-6), and TNF-α (Gusev and Zhuravleva 2022). Chronic inflammation is also recognized as a major contributor to age-related physiological changes, with the potential to cause molecular and cellular damage (Baechle et al. 2023).

The phenomenon of senescence-associated secretory phenotype (SASP) secretion by senescent cells has been shown to promote chronic inflammation. Simultaneously, chronic inflammation has been demonstrated to accelerate the senescence of immune cells. As a result, impaired immune cells fail to function effectively, leading to a sustained elevation of inflammatory mediators. Prolonged and uncontrolled inflammatory activity can lead to organ damage and contribute to the development of age-related diseases. Accordingly, inflammation is considered an endogenous factor in the aging process (Li et al. 2023b). Curcumin is known to inhibit the pro-inflammatory nuclear factor kappa B (NF-κB) pathway, resulting in reduced levels of inflammatory markers such as CRP and IL-8. Overall, curcumin plays a significant role in promoting longevity by suppressing inflammation and enhancing immune system function via hormesis (Izadi et al. 2024; Zia et al. 2021).

Amyloidosis and chronic inflammation have been identified as the primary pathological mechanisms underlying alzheimer's disease. In one study, the effects of miR-101, curcumin, and their combination were investigated in a cellular model of Alzheimer's disease over a 12-hour observation period. The combination treatment (miR-101 + curcumin) effectively suppressed the increase in TNF-α and interleukin-10 (IL-10) concentrations, while also reducing IL-6 levels. This combined administration enhanced the anti-amyloidogenic and anti-inflammatory effects of each agent (Sokolik and Berchenko 2023). In a four-week study involving male Wistar rats, curcumin supplementation (200 mg/kg/day) significantly reduced inflammatory markers, including TNF-α and IL-6, in both young and aged groups (Singh et al. 2025). In another study conducted with aged (18–20 months old) male C57BL/6 mice, curcumin supplementation at 4% (w/w), administered alongside both a normal diet and a high-fat, sugary diet, was shown to decrease p38 protein expression in the liver and modulate inflammatory pathways such as NF-κB (Lee 2022). A randomized, double-blind, placebo-controlled study demonstrated that daily intake of 160 mg of nano-curcumin in individuals aged 19–69 years significantly modulated the ratio of inflammatory cytokines by reducing the expression and secretion of IL-6 and interleukin-1β (IL-1β) in serum (Valizadeh et al. 2020). Another pilot randomized study involving older individuals aged 66–94 years found that supplementation with Curcumin C3 Complex (1 g/day) for 12 weeks resulted in lower levels of galectin-3 and IL-6 in the curcumin group compared to the placebo group (Mankowski et al. 2023). In a prospective cohort study involving 4551 adults over the age of 55, curcumin intake from curry was associated with reductions in systemic and immune inflammation indices. After adjusting for baseline covariates, curcumin-rich dietary consumption was linked to a 39% reduction in total mortality and an increase of 1.0 year in life expectancy (Ng et al. 2024). Additionally, a recent umbrella meta-analysis of randomized clinical trials was conducted to evaluate the overall effects of curcumin on inflammatory biomarkers. The meta-analysis, which included 10 studies encompassing a total of 5870 participants, demonstrated a significant reduction in CRP, IL-6, and TNF-α levels (Naghsh et al. 2023). In a randomized, open-label, parallel, active-controlled clinical trial involving 70 patients with knee osteoarthritis aged 38–65 years, administration of 500 mg curcumin (BCM-95®) capsules three times daily for 28 days produced anti-inflammatory effects comparable to conventional medications (Shep et al. 2019). These findings suggest that curcumin may serve as an effective adjunct therapeutic agent in conditions associated with elevated inflammatory biomarkers (Naghsh et al. 2023).

22.2.1.3 Effect of Curcumin on Telomere Length

Curcumin has been shown to promote telomere elongation by enhancing telomerase activity (Izadi et al. 2024). In a cellular *in vitro* study using human adipose tissue-derived stem cells (hADSCs), treatment with curcumin resulted in a significant increase in telomere length (Serati-Nouri et al. 2021). Another study demonstrated that the combination of endurance training and nano-curcumin supplementation (80 mg/kg) increased telomeric repeat-binding factor 2 (TERF2) gene expression and may help mitigate telomere shortening and aging in the muscle tissue of male Wistar rats (Pourjafarian and Kandi 2023).

22.2.1.4 Effect of Curcumin on Nutrient Sensing, Intracellular Communication Regulation, and Cellular Senescence

Curcumin has been shown to activate adenosine monophosphate-activated protein kinase (AMPK) and sirtuins while suppressing mammalian target of rapamycin (mTOR) and NF-κB, thereby exerting anti-aging effects (He et al. 2025; Muzammil et al. 2021). In a study involving male Wistar rats, simultaneous supplementation with D-galactose (300 mg/kg, subcutaneously) and curcumin (200 mg/kg, orally) over 6 weeks resulted in a significant increase in antioxidant levels and a marked reduction in oxidative stress markers. Curcumin has been proposed as a potential neuroprotective agent in aging and age-related disorders through its regulation of sirtuin-1 (SIRT1) and autophagy-related genes (Singh et al. 2024). In a separate study, septic male C57BL/6 mice administered equivalent doses of curcumin and nanocurcumin (12.5 mg/kg) exhibited a reduction in sepsis-induced cardiac lesions. The mTOR pathway has been directly implicated in sepsis-related cardiac damage, and its appropriate modulation may function as a protective mechanism (Rattis et al. 2022).

22.2.1.5 Effect of Curcumin on Regulating Epigenetic Programming

A study conducted on the fruit fly *Drosophila melanogaster* involved administering various doses of curcumin during the larval stage. The findings demonstrated a significant increase in both average and maximum lifespan, suggesting a potential role of curcumin in the epigenetic regulation of lifespan (Pisaruk et al. 2020).

22.3 Resveratrol

Resveratrol (3,5,4′-trihydroxy-stilbene) is a naturally occurring stilbene polyphenol present in a wide variety of plant species, including peanuts, grapes, pine, berries, and numerous other fruits. Classified as a stilbenoid compound, it is a secondary metabolite that also functions as a phytoalexin (Bejenaru et al. 2024; Meng et al. 2023). Resveratrol is synthesized from tyrosine through the sequential action of tyrosine ammonia lyase (deamination), 4-hydroxycinnamoyl-CoA ligase, and stilbene synthase, which catalyzes the condensation with three malonyl-CoA molecules. It exists in two isomeric forms: cis-resveratrol and trans-resveratrol. Among these, trans-resveratrol is more prominent in terms of biological activity and pharmaceutical applications (Salla et al. 2024; Li et al. 2024).

Resveratrol has been shown to exhibit a wide range of biological activities, including cardioprotective, neuroprotective, anticancer, antidiabetic, antimicrobial, and anti-aging effects. These effects are largely attributed to its antioxidant, anti-inflammatory, and metabolic regulatory properties, which represent common underlying mechanisms of action (Brown et al. 2024). Due to these effects, various forms of resveratrol are incorporated into pharmaceutical products and food items to preserve quality and extend shelf life (Salla et al. 2024). However, these findings also indicate that resveratrol exhibits suboptimal pharmacokinetics, making it difficult to achieve the necessary dosage to elicit its established biological effects, even with varying levels of intake. The therapeutic application of resveratrol derived from natural sources is limited by these pharmacokinetic challenges, restricting its ability to exert its diverse health benefits. Although ongoing research seeks to overcome these limitations, FDA approval remains pending due to a lack of robust clinical studies specifically addressing its medical use, optimal dosage, and administration methods (Salla et al. 2024; Khattar et al. 2022). According to the EFSA, a daily intake of 150 mg of resveratrol in adults does not pose any safety concerns (EFSA 2016). Health benefits have been associated with daily intake levels ranging from 30 to 150 mg. Trans-resveratrol is commercially available in dietary supplements at dosages ranging from 50 to 500 mg. Findings from human clinical trials indicate that doses of up to 5 g of trans-resveratrol are well tolerated and do not result in adverse effects (Xu et al. 2024; Khattar et al. 2022; Hoca et al. 2023). However, studies examining daily resveratrol intake report considerable variability in consumption levels (Xu et al. 2024; Pastor et al. 2019; Zamora-Ros et al. 2008, 2011;

Tresserra-Rimbau et al. 2013; Perez-Jimenez et al. 2011; Grosso et al. 2014). In the European Prospective Investigation into Cancer and Nutrition (EPIC)-Spain cohort study, which included 40,000 individuals aged 35–64, the estimated average intake of resveratrol was 0.93 mg/day (Zamora-Ros et al. 2008). A smaller study involving 7000 participants in Spain reported a mean stilbene intake of 1.84 mg/day (Tresserra-Rimbau et al. 2013). In a study conducted in France, the average daily intake of resveratrol was reported as 0.45 mg/day (Perez-Jimenez et al. 2011). In Poland, a study involving 10,477 participants found the mean stilbene intake to be 0.20 mg/day (Grosso et al. 2014). Similarly, in a study conducted in Italy with 1000 participants, resveratrol intake was determined to be 0.29 mg/day (Zamora-Ros et al. 2011). In China, resveratrol intake was estimated at approximately 0.78 mg/day (Xu et al. 2024).

It is imperative to determine the optimal dosage of resveratrol that elicits a therapeutic effect and to identify the most effective route of administration to enhance its bioavailability. If validated by further clinical research, resveratrol holds considerable potential as a therapeutic agent for health promotion and the prevention of chronic diseases (Faisal et al. 2024).

22.3.1 Resveratrol—Healthy Longevity Mechanisms

A substantial body of research has demonstrated the potential of resveratrol to extend lifespan in experimental model systems (Zhou et al. 2021a, b; Kohandel et al. 2024). A meta-analysis of 19 published studies involving six species—yeast, nematodes, mice, fruit flies, Mexican fruit flies, and turquoise killifish—found that resveratrol exerts life-extending effects across species through multiple distinct pathways (Hector et al. 2012). The mechanisms underlying the effects of resveratrol on healthy longevity are rooted in its antioxidative, anti-inflammatory, telomere-sparing properties, regulation of cellular senescence, and modulation of nutrient-sensing pathways—similar to other polyphenols such as curcumin. In addition to these, resveratrol's roles in promoting autophagy and enhancing mitochondrial function are also prominently observed (Zhou et al. 2021a, b; Pereira et al. 2023).

22.3.1.1 Antioxidative Effects of Resveratrol

Resveratrol has been shown to reduce reactive oxygen species (ROS) production by stimulating Nrf2 activity in a biphasic, dose-dependent manner, thereby enhancing the activity of endogenous antioxidant enzymes such as GPx, SOD, and CAT with a hormetic manner (Bastianetto et al. 2015; Calabrese et al. 2010). This antioxidant action also contributes to mitochondrial protection and exhibits significant neuroprotective effects, which are particularly relevant in the context of aging (Kung et al. 2021). A study conducted on human peripheral blood mononuclear cells (PBMCs) demonstrated that varying concentrations of resveratrol reduced ROS levels through the activation of SIRT1 and AMPK signaling pathways in cells exposed to oxidative stress (Caldeira et al. 2021). A study conducted on *Bombyx mori* silkworms demonstrated a significant upregulation in the expression of genes associated with signaling pathways, including AMPK and forkhead box protein O (FoxO). Furthermore, the antioxidant capacity of the silkworms was enhanced, and their survival rate increased by 8.6–12.1% (Chen et al. 2024). In a similar study, antioxidant activity was found to increase and lifespan was significantly extended in silkworms administered resveratrol at escalating doses: 3 mL during the third larval stage, 4 mL during the fourth, and 5 mL during the fifth larval stage (Song et al. 2021). A randomized, double-blind, placebo-controlled study involving patients with type 2 diabetes and coronary heart disease ($n = 28$) demonstrated that supplementation with 500 mg/day of resveratrol for 4 weeks increased total antioxidant capacity, reduced malondialdehyde levels, and upregulated Peroxisome proliferator-activated receptor gamma (PPAR-γ) and SIRT1 expression in PBMCs. However, no significant effect was observed on inflammatory markers (Hoseini et al. 2019).

22.3.1.2 Anti-Inflammation Effects of Resveratrol

Resveratrol has been shown to reduce the levels of several inflammatory markers, including IL-1β, TNF-α, and monocyte chemotactic protein-1 (MCP-1). This reduction is mediated through hormetic mode of action of multiple signaling pathways, including the arachidonic acid pathway, NF-κB, mitogen-activated protein kinase, activator protein-1, and SIRT1. As such, resveratrol modulates the inflammatory response and may exert anti-inflammatory effects that contribute to increased longevity (Meng et al. 2021; Zhou et al. 2021a, b). In a study involving PBMCs from donors aged 40–59 years and 60–80 years, resveratrol (5 μM) and 0.64% v/v H_2O_2 were administered, and resveratrol was found to be more effective in the middle-aged group (40–59 years). It significantly reduced key biomarkers of oxidation and inflammation (Santos et al. 2023). A further investigation on Wistar rats examined the effects of resveratrol supplementation on inflammatory, oxidative, and apoptotic markers associated with aging. In this study, aged rats received 10 mg/kg/day of resveratrol for 10 weeks. The findings indicated that resveratrol effectively reversed age-related alterations in these markers within the cardiac tissue (Torregrosa-Muñumer et al. 2021). In a randomized, double-blind, crossover study involving healthy obese men ($n = 11$), supplementation with 150 mg/day of resveratrol (resVida) for 30 days was found to activate AMPK, increase SIRT1 and peroxisome proliferator-activated receptor gamma coactivator 1 alpha (PGC-1α) protein levels, and reduce inflammatory markers (Timmers et al. 2011). In a separate clinical study, daily intake of 1000 mg of resveratrol for 1 month significantly reduced plasma levels of the pro-inflammatory cytokines TNF-α and MCP-1, while also enhancing plasma antioxidant activity (Espinoza et al. 2017).

22.3.1.3 Effect of Resveratrol on Telomere Length and Cellular Senescence

Resveratrol has been shown to protect telomeres from damage and to attenuate telomere shortening. It has also been observed to significantly enhance telomerase activity by modulating the expression of the catalytic subunit of telomerase reverse transcriptase (Yıldırım 2024; Ismaila et al. 2025). Additionally, resveratrol has been demonstrated to delay cellular aging, a process characterized by a progressive decline in physiological integrity and impaired progression through the cell cycle (Kohandel et al. 2024).

22.3.1.4 Regulatory Effect of Resveratrol on Nutrient Sensing, Enhancing Autophagy and Mitochondrial Function

The potential of resveratrol to promote longevity is attributed to its ability to induce autophagy, a process that is SIRT1-dependent (Zhou et al. 2021a, b). As a potent activator of SIRT1, resveratrol plays a critical role in preserving mitochondrial health, regulating inflammation, and mitigating oxidative stress. These combined functions contribute significantly to its longevity-promoting effects (Yıldırım 2024; Wilson et al. 2023; Meccariello and D'Angelo 2021). Resveratrol has been shown to activate FoxO1 and inhibit p53 through SIRT1, thereby modulating the expression of both apoptotic and antiapoptotic proteins and regulating the process of apoptosis. This regulatory effect on apoptosis underscores resveratrol's potential efficacy in preventing aging and age-related diseases (Zhou et al. 2021a, b). In a randomized study involving 48 individuals aged 55 to 65 years, supplementation with resveratrol (500 mg/day) combined with energy restriction (1000 kcal/day) for 30 days led to increased circulating SIRT1 levels and a comparable reduction in sympathetic activity (Gonçalinho et al. 2023). In Spain, a randomized controlled trial was conducted with nine women aged 32 to 57 years who had abstained from alcohol consumption for at least 2 years. The participants were instructed to consume two 200 mL glasses of red wine daily for 14 days, with one glass consumed at lunch and the other at dinner. By the end of the fourteenth day, increased expression levels of CAT, manganese-SOD, SIRT1, and p53 were observed in peripheral blood mononuclear cells (Gambini et al. 2021).

22.4 Conclusions

Curcumin and resveratrol have been shown to play diverse roles in the prevention of aging and age-related diseases in both model systems and humans. Their effects on healthy longevity are mediated through multiple metabolic pathways, with the most prominent involving the activation of pathways associated with strong antioxidant and anti-inflammatory functions through hormesis. Given the wide dosage ranges employed in clinical studies, the optimal dose and duration of curcumin and resveratrol supplementation remain uncertain. The variability in dosage is associated with their involvement in distinct physiological and pathological functions. Clinical studies reporting beneficial health effects have utilized curcumin doses ranging from 80 to 1500 mg/day and resveratrol doses ranging from 150 to 1000 mg/day (Tables 22.1 and 22.2). The amounts of curcumin and resveratrol in commercially available supplements vary in parallel with the dosages used in clinical studies (Panknin et al. 2023; Fioravanti et al. 2024; Brown et al. 2024; Loke et al. 2024; Drago et al. 2024). The curcumin and resveratrol content in these supplements is generally higher than the levels naturally present in food sources. A critical consideration in this context is the notably low bioavailability of both curcumin and resveratrol, each exhibiting absorption rates below 1% (Chang et al. 2023; Kasprzak-Drozd et al. 2022; Dikmetas et al. 2024). Although certain regulatory agencies have endorsed higher safe intake limits due to this low bioavailability, clear guidelines on supplementation dosages for promoting healthy longevity remain lacking. Rigorous clinical investigations incorporating pharmacokinetic and nutrigenetic analyses are essential to clarify their effects on lifespan extension. In conclusion, until optimal supplementation doses of curcumin and resveratrol for healthy longevity are established, their consumption as part of a healthy dietary model—where the food matrix may confer synergistic benefits—is recommended.

Table 22.1 Clinical trials on curcumin

Type of study	Participants	Dose	Administration	Result(s)	Reference
Randomized, Double-Blind, Placebo-Controlled Clinical Trial	Infertile men (n = 28)	80 mg/day	Curcumin nanomicelles for 10 weeks	Increased plasma total antioxidant capacity, Decreased malondialdehyde, CRP, TNF-α	Alizadeh et al. (2018)
Randomized, double-blind, placebo-controlled study	19–69 y (n = 40)	160 mg/day	Nano-curcumin for 14 days	Decrease in IL-6 and IL-1β expression and secretion in serum	Valizadeh et al. (2020)
Pilot study	≥18 y Chronic kidney disease (n = 24)	200 mg/day	Curcumin supplement (Meriva®) for 6 months.	Reduced lipid peroxidation and plasma CCL-2, IFN-γ, IL-4 levels	Pivari et al. (2022)
Pilot study, randomized trial	Older Adults 66–94 y (n = 9)	1000 mg/day	Curcumin C3 Complex for 12 weeks	Decrease in plasma galectin-3 and IL-6 levels	Mankowski et al. (2023)
Randomized, open-label, parallel, active controlled clinical study	38–65 y patients with knee osteoarthritis (n = 70)	1500 mg/day	Curcumin (BCM-95®) capsule for 28 days	Similar effects to anti-inflammatory drugs by reducing inflammation	Shep et al. (2019)

Table 22.2 Clinical trials on resveratrol

Type of study	Participants	Dose	Administration	Result(s)	Reference
Randomized double-blind crossover study	Healthy, obese men (n = 11)	150 mg/day	Resveratrol supplementation (resVida) for 30 days	Increased SIRT1 and PGC-1a levels, Decreased plasma levels of IL-6 and TNF- α	Timmers et al. (2011)
Randomized, double-blind, placebo-controlled trial	patients with type 2 diabetes mellitus and coronary heart disease (n = 28)	500 mg/day	Resveratrol supplement for 4 weeks	Increase total antioxidant capacity Reduce malondialdehyde levels Upregulate PPAR-γ and SIRT1 in the PBMCs	Hoseini et al. (2019)
Randomized trial	Individuals aged between 55–65 y (n = 48)	500 mg/day Energy restriction (1000 kcal/day)	Resveratrol supplementation for 30 days	Increase circulating SIRT1 levels and a reduction in sympathetic activity	Gonçalinho et al. (2023)
Randomized phase 1 trial	Healthy volunteers (n = 9)	1000 mg/day	Resveratrol capsule for 28 days	Decrease in the plasma levels of TNF-α and MCP-1 Increase in the plasma antioxidant activity	Espinoza et al. (2017)
Randomized, controlled trial	Women aged 32 to 57 years (n = 9)	400 mL/day red wine	400 mL of red wine daily for 14 days	Increase in the expression levels of catalase, superoxide dismutase, SIRT1, and p53	Gambini et al. (2021)

References

Abolaji AO, Fasae KD, Iwezor CE, Aschner M, Farombi EO (2020) Curcumin attenuates copper-induced oxidative stress and neurotoxicity in Drosophila melanogaster. Toxicol Rep 7:261–268

Addo KM, Khan H (2024) Factors affecting healthy aging and its interconnected pathways. Turk J Healthy Aging Med 1:9–24

Alizadeh F, Javadi M, Karami AA, Gholaminejad F, Kavianpour M, Haghighian HK (2018) Curcumin nanomicelle improves semen parameters, oxidative stress, inflammatory biomarkers, and reproductive hormones in infertile men: a randomized clinical trial. Phytother Res 32(3): 514–521

Aranda-Rivera AK, Cruz-Gregorio A, Arancibia-Hernández YL, Hernández-Cruz EY, Pedraza-Chaverri J (2022) RONS and oxidative stress: an overview of basic concepts. Oxygen 2(4): 437–478

Baechle JJ, Chen N, Makhijani P, Winer S, Furman D, Winer DA (2023) Chronic inflammation and the hallmarks of aging. Mol Metabol 74:101755

Barati N, Momtazi-Borojeni AA, Majeed M, Sahebkar A (2019) Potential therapeutic effects of curcumin in gastric cancer. J Cell Physiol 234(3):2317–2328

Bastianetto S, Ménard C, Quirion R (2015) Neuroprotective action of resveratrol. Biochimica et Biophysica Acta (BBA) 1852(6):1195–1201

Bautmans I, Knoop V, Thiyagarajan JA, Maier AB, Beard JR, Freiberger E et al (2022) WHO working definition of vitality capacity for healthy longevity monitoring. Lancet Healthy Longev 3(11):e789–e796

Bejenaru LE, Biţă A, Belu I, Segneanu AE, Radu A, Dumitru A et al (2024) Resveratrol: a review on the biological activity and applications. Appl Sci 14(11):4534

Bertoncini-Silva C, Vlad A, Ricciarelli R, Giacomo Fassini P, Suen VMM, Zingg JM (2024) Enhancing the bioavailability and bioactivity of curcumin for disease prevention and treatment. Antioxidants 13(3):331

Bielak-Zmijewska A, Grabowska W, Ciolko A, Bojko A, Mosieniak G, Bijoch Ł, Sikora E (2019) The role of curcumin in the modulation of ageing. Int J Mol Sci 20(5):1239

Brown K, Theofanous D, Britton RG, Aburido G, Pepper C, Sri Undru S, Howells L (2024) Resveratrol for the management of human health: how far have we come? A systematic review of resveratrol clinical trials to highlight gaps and opportunities. Int J Mol Sci 25(2):747

Calabrese EJ (2021) Hormesis mediates acquired resilience: using plant-derived chemicals to enhance health. Annu Rev Food Sci Technol 12(1):355–381

Calabrese EJ, Mattson MP, Calabrese V (2010) Resveratrol commonly displays hormesis: occurrence and biomedical significance. Hum Exp Toxicol 29(12):980–1015

Caldeira CA, Santos MA, Araújo GR, Lara RC, Franco FN, Chaves MM (2021) Resveratrol: change of SIRT 1 and AMPK signaling pattern during the aging process. Exp Gerontol 146: 111226

Chang R, Chen L, Qamar M, Wen Y, Li L, Zhang J et al (2023) The bioavailability, metabolism and microbial modulation of curcumin-loaded nanodelivery systems. Adv Colloid Interface Sci 318: 102933

Chen L, Zhang J, Ding L, Gu T, Andoh V, Ma A, Yao C (2024) Transcriptomics analyses combined with intestinal microorganism survey suggest resveratrol (RSV) anti-aging and anti-oxidant effects in silkworm (Bombyx mori). Comp Biochem Physiol Part D: Genom Proteom 52: 101344

Dikmetas DN, Yenipazar H, Karaca AC (2024) Recent advances in encapsulation of resveratrol for enhanced delivery. Food Chem:140475

Drago L, Ciprandi G, Brindisi G, Brunese FP, Dinardo G, Gori A et al (2024) Certainty and uncertainty in the biological activities of resveratrol. Food Front 5(3):849–854

EFSA (2010) Scientific Opinion on the re-evaluation of curcumin (E 100) as a food additive. EFSA J 8(9):1679–1724

EFSA Panel on Dietetic Products, Nutrition and Allergies (NDA) (2016) Safety of synthetic trans-resveratrol as a novel food pursuant to regulation (EC) no 258/97. EFSA J 14(1):1–30

EFSA Panel on Food Additives and Nutrient Sources added to Food (ANS) (2010) Statement on the potential risk to human health of turmeric and curcumin supplements. EFSA J 8

Espinoza JL, Trung LQ, Inaoka PT, Yamada K, An DT, Mizuno S et al (2017) The repeated administration of resveratrol has measurable effects on circulating T-cell subsets in humans. Oxid Med Cell Longev 2017(1):6781872

Faisal Z, Mazhar A, Batool SA, Akram N, Hassan M, Khan MU et al (2024) Exploring the multimodal health-promoting properties of resveratrol: a comprehensive review. Food Sci Nutr 12(4):2240–2258

Fioravanti MIA, Pizano FP, Rebellato AP, Milani RF, Morgano MA, Bragotto APA (2024) Turmeric products: evaluation of curcumin and trace elements. Food Res Int 196:115028

Gambini J, Gimeno-Mallench L, Olaso-Gonzalez G, Mastaloudis A, Traber MG, Monleón D et al (2021) Moderate red wine consumption increases the expression of longevity-associated genes in controlled human populations and extends lifespan in Drosophila melanogaster. Antioxidants 10(2):301

Garodia P, Hegde M, Kunnumakkara AB, Aggarwal BB (2023) Curcumin, inflammation, and neurological disorders: how are they linked? Integrat Med Res 12(3):100968

Gonçalinho GHF, Kuwabara KL, Faria NFDO, Goes MFDS, Roggerio A, Avakian SD et al (2023) Sirtuin 1 and vascular function in healthy women and men: a randomized clinical trial comparing the effects of energy restriction and resveratrol. Nutrients 15(13):2949

Grosso G, Stepaniak U, Topor-Mądry R, Szafraniec K, Pająk A (2014) Estimated dietary intake and major food sources of polyphenols in the polish arm of the HAPIEE study. Nutrition 30(11–12): 1398–1403

Gusev E, Zhuravleva Y (2022) Inflammation: a new look at an old problem. Int J Mol Sci 23(9): 4596

Hashemian M, Poustchi H, Murphy G, Etemadi A, Kamangar F, Pourshams A et al (2019) Turmeric, pepper, cinnamon, and saffron consumption and mortality. J Am Heart Assoc 8(18):e012240

He Y, Liu Y, Zhang M (2025) The beneficial effects of curcumin on aging and age-related diseases: from oxidative stress to antioxidant mechanisms, brain health and apoptosis. Front Aging Neurosci 17:1533963

Hector KL, Lagisz M, Nakagawa S (2012) The effect of resveratrol on longevity across species: a meta-analysis. Biol Lett 8(5):790–793

Hoca M, Becer E, Vatansever HS (2023) The role of resveratrol in diabetes and obesity associated with insulin resistance. Arch Physiol Biochem 129(2):555–561

Hofer SJ, Davinelli S, Bergmann M, Scapagnini G, Madeo F (2021) Caloric restriction mimetics in nutrition and clinical trials. Front Nutr 8:717343

Hoseini A, Namazi G, Farrokhian A, Reiner Ž, Aghadavod E, Bahmani F, Asemi Z (2019) The effects of resveratrol on metabolic status in patients with type 2 diabetes mellitus and coronary heart disease. Food Funct 10(9):6042–6051

Ismaila M, Adnan M, Sasidharan S (2025) Exploring medicinal plants and phytochemicals for telomere maintenance and anti-aging: insights into telomerase activation and longevity. Rev Bras:1–16

Izadi M, Sadri N, Abdi A, Zadeh MMR, Jalaei D, Ghazimoradi MM et al (2024) Longevity and anti-aging effects of curcumin supplementation. Geroscience 46(3):2933–2950

JECFA (2004) Joint FAO/WHO Expert Committee on Food Additives. Evaluation of certain food additives and contaminants: sixty-first report of the Joint FAO/WHO Expert Committee on Food Additives, 2004

Kasprzak-Drozd K, Oniszczuk T, Gancarz M, Kondracka A, Rusinek R, Oniszczuk A (2022) Curcumin and weight loss: does it work? Int J Mol Sci 23(2):639

Khattar S, Khan SA, Zaidi SAA, Darvishikolour M, Farooq U, Naseef PP et al (2022) Resveratrol from dietary supplement to a drug candidate: an assessment of potential. Pharmaceuticals 15(8): 957

Kohandel Z, Darrudi M, Naseri K, Samini F, Aschner M, Pourbagher-Shahri AM, Samarghandian S (2024) The role of resveratrol in aging and senescence: a focus on molecular mechanisms. Curr Mol Med 24(7):867–875

Kumar A, Maurya PK (2023) Curcumin ameliorates oxidative stress in red blood cells during ageing. Indian J Nat Prod Resour (IJNPR)[Formerly Natural Product Radiance (NPR)] 14(1): 50–54

Kung HC, Lin KJ, Kung CT, Lin TK (2021) Oxidative stress, mitochondrial dysfunction, and neuroprotection of polyphenols with respect to resveratrol in Parkinson's disease. Biomedicine 9(8):918

Kwon Y (2014) Estimation of curcumin intake in Korea based on the Korea National Health and nutrition examination survey (2008-2012). Nutr Res Pract 8(5):589–594

Lee SJ (2022) Effect of curcumin on hepatic senescence in aged mice. Master's thesis, Oklahoma State University

Lee H, Kuwabara Y, Hirose A, Kakinuma T, Baba A, Takara T (2023) Safety evaluation of high bioavailability curcumin in healthy Japanese adults: a randomized, placebo-controlled, double-blind, parallel-group comparison study. Funct Foods Health Dis 13(12):702–716

Li N, Yan X, Huang W, Chu M, Dong Y, Song H et al (2023a) Curcumin protects against the age-related hearing loss by attenuating apoptosis and senescence via activating Nrf2 signaling in cochlear hair cells. Biochem Pharmacol 212:115575

Li X, Li C, Zhang W, Wang Y, Qian P, Huang H (2023b) Inflammation and aging: signaling pathways and intervention therapies. Signal Transduct Target Ther 8(1):239

Li W, Yuan H, Liu Y, Wang B, Xu X, Xu X et al (2024) Current analytical strategies for the determination of resveratrol in foods. Food Chem 431:137182

Loke WM, Lai JX, Tan R (2024) Evaluating product demographics, composition and health claims of commercial polyphenol supplements: a review of dosage, bioactive compounds and efficacy. Int J Nutraceuticals Funct Foods Novel Foods

López-Otín C, Blasco MA, Partridge L, Serrano M, Kroemer G (2023) Hallmarks of aging: an expanding universe. Cell 186(2):243–278

Mankowski RT, Sibille KT, Leeuwenburgh C, Lin Y, Hsu FC, Qiu P et al (2023) Effects of curcumin C3 complex® on physical function in moderately functioning older adults with low-grade inflammation—a pilot trial. J Frailty Aging 12(2):143–149

Meccariello R, D'Angelo S (2021) Impact of polyphenolic-food on longevity: an elixir of life. An overview. Antioxidants 10(4):507

Meng T, Xiao D, Muhammed A, Deng J, Chen L, He J (2021) Anti-inflammatory action and mechanisms of resveratrol. Molecules 26(1):229

Meng Q, Li J, Wang C, Shan A (2023) Biological function of resveratrol and its application in animal production: a review. J Anim Sci Biotechnol 14(1):25

Muzammil A, Waqas M, Umar A, Sufyan M, Rehman A, Haider A et al (2021) Anti-aging natural compounds and their role in the regulation of metabolic pathways leading to longevity. Mini Rev Med Chem 21(18):2630–2656

Naghsh N, Musazadeh V, Nikpayam O, Kavyani Z, Moridpour AH, Golandam F et al (2023) Profiling inflammatory biomarkers following curcumin supplementation: an umbrella meta-analysis of randomized clinical trials. Evid Based Complement Alternat Med 2023(1):4875636

Ng TP, Nyunt SZ, Gao Q, Gwee X, Chua DQL, Yap KB (2024) Curcumin-rich curry consumption and life expectancy: Singapore longitudinal ageing study. GeroScience 46(1):969–980

No GNG (2018) Notice to US Food and Drug Administration of the conclusion that the intended use of curcumin is generally recognized as safe. https://www.fda.gov/media/132575/download

Panknin TM, Howe CL, Hauer M, Bucchireddigari B, Rossi AM, Funk JL (2023) Curcumin supplementation and human disease: a scoping review of clinical trials. Int J Mol Sci 24(5):4476

Pastor RF, Restani P, Di Lorenzo C, Orgiu F, Teissedre PL, Stockley C et al (2019) Resveratrol, human health and winemaking perspectives. Crit Rev Food Sci Nutr 59(8):1237–1255
Pereira QC, Dos Santos TW, Fortunato IM, Ribeiro ML (2023) The molecular mechanism of polyphenols in the regulation of ageing hallmarks. Int J Mol Sci 24(6):5508
Perez-Jimenez J, Fezeu L, Touvier M, Arnault N, Manach C, Hercberg S et al (2011) Dietary intake of 337 polyphenols in French adults. Am J Clin Nutr 93(6):1220–1228
Pisaruk A, Koshel N, Mekhova L, Zabuga O, Ivanov S (2020) Influense of curcumin on lifespan if it is applied at the larval stage of Drosophila melanogaster. Ageing and Longevity 1(2):89–96
Pivari F, Mingione A, Piazzini G, Ceccarani C, Ottaviano E, Brasacchio C et al (2022) Curcumin supplementation (Meriva®) modulates inflammation, lipid peroxidation and gut microbiota composition in chronic kidney disease. Nutrients 14(1):231
Pourjafarian J, Kandi M (2023) The effect of endurance training with nano-curcumin supplementation on p53-p21 axis and TERF2 gene expression in muscle tissue of male wistar rats: an experimental study. J Rafsanjan Univ Med Sci 21(11):1103–1114
Pyo IS, Yun S, Yoon YE, Choi JW, Lee SJ (2020) Mechanisms of aging and the preventive effects of resveratrol on age-related diseases. Molecules 25(20):4649
Rattis BA, Piva HL, Duarte A, Gomes FG, Lellis JR, Soave DF et al (2022) Modulation of the mTOR pathway by curcumin in the heart of septic mice. Pharmaceutics 14(11):2277
Salla M, Karaki N, El Kaderi B, Ayoub AJ, Younes S, Abou Chahla MN et al (2024) Enhancing the bioavailability of resveratrol: combine it, derivatize it, or encapsulate it? Pharmaceutics 16(4): 569
Santos MA, Franco FN, Caldeira CA, de Araújo GR, Vieira A, Chaves MM (2023) Resveratrol has its antioxidant and anti-inflammatory protective mechanisms decreased in aging. Arch Gerontol Geriatr 107:104895
Serati-Nouri H, Mahmoudnezhad A, Bayrami M, Sanajou D, Tozihi M, Roshangar L et al (2021) Sustained delivery efficiency of curcumin through ZSM-5 nanozeolites/electrospun nanofibers for counteracting senescence of human adipose-derived stem cells. J Drug Deliv Sci Technol 66: 102902
Shep D, Khanwelkar C, Gade P, Karad S (2019) Safety and efficacy of curcumin versus diclofenac in knee osteoarthritis: a randomized open-label parallel-arm study. Trials 20:1–11
Shoba G, Joy D, Joseph T, Majeed M, Rajendran R, Srinivas PSSR (1998) Influence of piperine on the pharmacokinetics of curcumin in animals and human volunteers. Planta Med 64(04): 353–356
Singh A, Yadawa AK, Rizvi SI (2024) Curcumin protects against aging-related stress and dysfunction through autophagy activation in rat brain. Mol Biol Rep 51(1):694
Singh A, Soni U, Varadwaj PK, Misra K, Rizvi SI (2025) Anti-inflammatory effect of curcumin in an accelerated senescence model of Wistar rat: an in vivo and in-silico study. J Biomol Struct Dyn 43(3):1459–1470
Sokolik VV, Berchenko OG (2023) The cumulative effect of the combined action of miR-101 and curcumin in a liposome on a model of Alzheimer's disease in mononuclear cells. Front Cell Neurosci 17:1169980
Song J, Liu L, Hao K, Mao S, Tang Y, Tong X, Dai F (2021) Resveratrol elongates the lifespan and improves antioxidant activity in the silkworm Bombyx mori. J Pharma Anal 11(3):374–382
Timmers S, Konings E, Bilet L, Houtkooper RH, van de Weijer T, Goossens GH et al (2011) Calorie restriction-like effects of 30 days of resveratrol supplementation on energy metabolism and metabolic profile in obese humans. Cell Metab 14(5):612–622
Torregrosa-Muñumer R, Vara E, Fernández-Tresguerres JÁ, Gredilla R (2021) Resveratrol supplementation at old age reverts changes associated with aging in inflammatory, oxidative and apoptotic markers in rat heart. Eur J Nutr:1–11
Tresserra-Rimbau A, Medina-Remón A, Pérez-Jiménez J, Martínez-González MA, Covas MI, Corella D et al (2013) Dietary intake and major food sources of polyphenols in a Spanish population at high cardiovascular risk: the PREDIMED study. Nutr Metab Cardiovasc Dis 23(10):953–959

Valizadeh H, Abdolmohammadi-Vahid S, Danshina S, Gencer MZ, Ammari A, Sadeghi A et al (2020) Nano-curcumin therapy, a promising method in modulating inflammatory cytokines in COVID-19 patients. Int Immunopharmacol 89:107088

Vaughn AR, Branum A, Sivamani RK (2016) Effects of turmeric (Curcuma longa) on skin health: a systematic review of the clinical evidence. Phytother Res 30(8):1243–1264

Wilson N, Kataura T, Korsgen ME, Sun C, Sarkar S, Korolchuk VI (2023) The autophagy–NAD axis in longevity and disease. Trends Cell Biol 33(9):788–802

Xu Y, Fang M, Li X, Wang D, Yu L, Ma F et al (2024) Contributions of common foods to resveratrol intake in the Chinese diet. Foods 13(8):1267

Yang ZJ, Huang SY, Zhou DD, Xiong RG, Zhao CN, Fang AP et al (2022) Effects and mechanisms of curcumin for the prevention and management of cancers: An updated review. Antioxidants 11(8):1481

Yıldırım TD (2024) Resveratrol therapeutic potential in age-related diseases. Anti-Aging Eastern Europe 3(4):194–201

Zamora-Ros R, Andres-Lacueva C, Lamuela-Raventós RM, Berenguer T, Jakszyn P, Martínez C et al (2008) Concentrations of resveratrol and derivatives in foods and estimation of dietary intake in a Spanish population: European prospective investigation into cancer and nutrition (EPIC)-Spain cohort. Br J Nutr 100(1):188–196Zhou

Zamora-Ros R et al (2011) Comparison of 24-h volume andcreatinine-corrected total urinary polyphenol as a biomarkerof total dietary polyphenols in the Invecchiare InCHIANTIstudy. Anal Chim Acta 704:110–115

Zheng B, Zanders F, Engwerda AHJ, Hemmelgarn L (2024) Curcuminoids. In: Handbook on natural pigments in food and beverages. Woodhead Publishing, pp 227–247

Zhou L, Liu J, Bu LL, Liao DF, Cheng SW, Zheng XL (2021a) Curcumin acetylsalicylate extends the lifespan of Caenorhabditis elegans. Molecules 26(21):6609

Zhou DD, Luo M, Huang SY, Saimaiti A, Shang A, Gan RY, Li HB (2021b) Effects and mechanisms of resveratrol on aging and age-related diseases. Oxid Med Cell Longev 2021(1): 9932218

Zia A, Farkhondeh T, Pourbagher-Shahri AM, Samarghandian S (2021) The role of curcumin in aging and senescence: molecular mechanisms. Biomed Pharmacother 134:111119

Chapter 23
Hormesis and Hormetins as Drugs for Health and Longevity

Suresh I. S. Rattan

Abstract Hormesis is the universally observed biphasic dose response of physical, chemical and biological stressors which at low levels generally have health beneficial effects. Moderate and repeated physical exercise as a stress-inducing hormetin is the paradigm for this phenomenon of physiological hormesis. Repeated mild stress activates several early and later adaptive and protective pathways in cells and organisms, that can have numerous health-beneficial, ageing-modulatory, and lifespan-extending effects. The essential requirement for the action of hormetins is the initiation of a cascade of stress response pathways, followed by the activation of other metabolic pathways of maintenance and repair. Thus, potential hormetins present in the food or synthesized chemically, combined with other physical and mental hormetins, can be used both as supplements and drugs for maintaining, recovering and enhancing health and longevity.

Keywords Health · Stress · Homeostasis · Homeodynamics · Hormetins · Stress response

23.1 Introduction

Biogerontological understanding of the causes, consequences and mechanisms of ageing shows that ageing is an emergent, epigenetic and a complex meta-phenomenon (Rattan 2018, 2024). Ageing is progressively manifested as the failure of maintenance by the complex and interactive processes of health and homeodynamics (Rattan 2020a). Such understanding implies that instead of the prevalent single target-oriented "anti-ageing" approaches, ageing interventions should aim for the maintenance, enhancement and recovery of overall health (Rattan 2020a, b). Hormesis offers such an holistic approach based in the principle that cellular responses to single or multiple rounds of mild stress generally elicit health

S. I. S. Rattan (✉)
Department of Molecular Biology and Genetics, Aarhus University, Aarhus, Denmark
e-mail: rattan@mbg.au.dk

S. I. S. Rattan, C. Gezer (eds.), *Supplements and Drugs for Healthy Longevity*, Healthy Ageing and Longevity 22, https://doi.org/10.1007/978-3-032-11608-6_23

beneficial effects (Rattan and Kyriazis 2019; Epel 2020; Panossian et al. 2021; Agathokleous et al. 2020).

Hormesis is the universally observed biphasic dose response of physical, chemical and biological stressors which at low levels can have health beneficial effects (Calabrese et al. 2007; Calabrese 2007; Rattan and Kyriazis 2019). The term hormesis is further specified as conditioning hormesis, post-exposure conditioning, and physiological hormesis dealing with specific aspects of the stress responses (Calabrese et al. 2007). Previously, the phenomenon of hormesis has also been known as U-shaped curve, inverse-U-shaped curve, J-shaped curve, and adaptive response. Furthermore, the term "hormetin" was proposed to refer to any condition that can induce health-beneficial physiological hormesis (Rattan 2017). Moderate and repeated physical exercise as a stress-inducing hormetin is the paradigm for physiological hormesis (Radak and Rattan 2024).

The aim of this chapter is to provide a brief overview of the phenomenon and the process of hormesis brought about by hormetin, which can be considered as drugs FOR health, in contrast to the usual expectation of drugs AGAINST this or that disease or a problem.

23.2 Molecular Basis of Hormesis in Biogerontology

Stress response (SR) is a foundational characteristic of the survival ability and homeodynamic space of all living systems (Rattan 2020a). A disruption of homeostasis, leading to the activation of one or more SR pathways to counteract the disruption, is the fundamental molecular basis of hormesis. Starting from the exposure to a stressor to the potential occurrence of physiological hormesis, SR processes are characterised in the following conceptual steps: (i) disturbance in the homeodynamic equilibrium; (ii) activation of signalling pathways to initiate SR pathways; (iii) activation of SR pathway-mediated effector responses; and (iv) restoration of homeodynamics, and eventual enhanced adaptive ability (Bhattacharya and Rattan 2019; Rabouw et al. 2019; Rossnerova et al. 2020; Calabrese and Kozumbo 2021). A successful SR not only results in the re-establishment of homeodynamics, but also strengthens the homeodynamic space (Rattan 2020a).

The molecular mechanisms of SR pathways are well described, along with the identification of initiating stressors within the cells, immediate or early activational responses, and the biochemical sensors, mediators and effectors of the responses (Table 23.1).

It should be noted that the activation of the AMP kinase (AMPK) signaling pathway is the earliest mediator of overall and integrated SR, which, depending on the nature of the induced stress, become more specific later on (Salminen et al. 2016; Labbadia et al. 2017; Rossnerova et al. 2020; Agathokleous et al. 2020). A crucial aspect of SR is that the immediate and primary SR initiated by a stressor is followed by the activation or suppression of one or more other SR pathways, thus creating a

Table 23.1 Main molecular level cellular stress response pathways as the basis of hormesis

Stress response *(arranged in alphabetical order)*	Some common stressors	Molecular sensors and effectors
Autophagic response (AR)	Nutritional limitation, hypoxia, damaged organelles	Autophagosomes, lysosomes
DNA damage response (DDR)	Radiation, reactive oxygen species	DNA damage sensors (ATM, ATR), p53, DNA repair proteins
Energy-deficiency response (EDR)	Energy depletion, metabolic imbalance	Sirtuins
Heat shock response (HSR)	Heat, exercise, hypergravity, heavy metals, natural and synthetic small molecules, antibiotics	Heat shock transcription factors (HSF), heat shock proteins (HSP), chaperones, proteasome
Inflammatory response (IR)	Pathogens, allergens, damaged macromolecules	Nuclear factor-κB transcription factors, cytokines, nitric oxide synthase
Oxidative stress response (OSR)	Oxidants, free radicals, reactive oxygen species	Transcription factors (Nrf2, FOXO), hemeoxygenase, antioxidative enzymes (SOD, catalase, glutathione)
Unfolded protein response (UPR)	Unfolded and misfolded proteins in endoplasmic reticulum (ER) and mitochondria, cytokines	ER-chaperones, mitochondrial-chaperonins, proteasome

Modified from Bhattacharya and Rattan (2019)

full SR profile (SRP) (Rattan et al. 2018; Demirovic and Rattan 2013). Immediate or early induction of one or more SR is an essential condition for the potential health-beneficial hormetic effects of a hormetin that initially causes some kind of molecular damage (Rossnerova et al. 2020).

23.3 Hormetins in Health Maintenance and Promotion

Physical exercise is the most studied and well-documented stress-inducing hormetin for its whole-body level health beneficial effects (Broskey et al. 2019; Thirupathi et al. 2021; Radak and Rattan 2024). These holistic effects include improved immune system, improved cognition, improved neuro-muscular coordination, increased muscle strength and mass, increased life expectancy and improved overall quality of life (Ramsey et al. 2021; Sies 2021; Campos et al. 2023; Farmand et al. 2025). Thus, physical exercise is considered as a paradigm for hormesis and as a drug for health, often performing better than specific target-oriented drugs, such as metformin (Glossmann and Lutz 2019).

Some examples of health-promoting hormetins are listed below, and are organised in three principal categories of hormetins:

- *Physical hormetins:* exercise, radiation, hot and cold temperature shock, hypoxia, gravity, and electromagnetism (David et al. 2022; Le Bourg 2021; Boulares et al. 2025; Bennett et al. 2024; Almiron et al. 2024; Oliveira et al. 2018; Zhu et al. 2018; Romeo et al. 2016; Guisle et al. 2022).
- *Nutritional/chemical hormetins:* calorie restriction, fasting, dietary components, minerals, trace elements and nanoparticles (Franco et al. 2019; Ingram and Roth 2020; Kazmierczak-Baranska et al. 2020; Lee et al. 2021; Madeo et al. 2019; Murakami 2022; Luo et al. 2022; Calabrese 2021; Mytych et al. 2016; Duregon et al. 2021)
- *Biological hormetins:* microbes and mental exercises, including meditation (Bana and Cabreiro 2019; Wargo 2020; Epel 2020; Zhang et al. 2021; Nord 2023).

Most of the research on discovering and developing novel ageing interventions, such as prebiotics, probiotics, postbiotics, nutraceuticals, functional foods, and various cognitive methods can be put into one or the other hormetin category (Wang et al. 2017; Calabrese et al. 2021; Davinelli et al. 2020; Sanchez-Morate et al. 2020; Annunziata et al. 2021; Pinchuk et al. 2021; Shannon et al. 2021). Furthermore, a majority of the more than 250 so-called geroprotectors listed in the curated database (http://geroprotectors.org/) are potential hormetins owing to their mode of action involving the induction of SR in cells and organisms.

Since the activation of one or more pathways of SR is the essential condition for the health benefits of hormesis, the following nomenclature for hormetins has been proposed (Table 23.2), which can facilitate the discovery and applications of novel hormetins.

Table 23.2 A proposed nomenclature for hormetins based on their effect in inducing an early and primary stress response

Hormetin	Primary Stress Response	Examples
Hormetin—A	Autophagy	Fasting, calorie restriction (CR) and CR-mimetics,
Hormetin—D	DNA damage/repair response	Radiation, and chemical and biological mutagens
Hormetin—E	Energy deficiency response	sirtuin activators, flavonoids, polyphenols
Hormetin—H	Heat shock stress response	Physical and mental exercise, celasterol, temperature shock, hypoxia
Hormetin—I	Inflammatory stress response	Microbes, prebiotics, probiotics
Hormetin—O	Oxidative stress response	Pro-oxidants, spices, polyphenols

Modified from Rattan (2022)

This nomenclature does not pick and choose any particular compound(s) or conditions as the "active molecules", but points towards the primary SR which can and do initiate a series of physiological processes.

23.4 Conclusions and Challenges

The health significance of mild stress is often difficult to appreciate because the initial and immediate effects are generally small when measured after one exposure (Calabrese et al. 2018). However, after a period of repeated exposures to mild stress, the overall biological outcomes, such as enhanced robustness, resilience, functional performance, and possibly increased survival and longevity, are amplified, cumulative, synergistic and pleiotropic (Rackova et al. 2021). Physical exercise is the best example of the importance and significance of repeated exposure to mild stress resulting in a wide range of physiological and psychological health benefits. This re-asserts the fact that hormetins-induced hormesis results in the biological amplification and cascade of adaptive responses leading to the improvement in physiological functions and performance in a holistic manner.

Although at a conceptual level, hormetins and hormesis do "work" in human beings, there are several issues and challenges yet to be resolved for the extensive application of the principles of hormesis in maintaining, recovering and enhancing health. Two of these critical issues are:

1. What should be the optimal stress exposure regimens in terms of the intensity and frequency at different biological levels, while considering factors such as age and the overall health and disease status of an individual? This is crucial for distinguishing between "mild" and "severe" levels of stress, which can have very different consequences from being health beneficial to being fatal, respectively.
2. What are the interactive and pleiotropic effects of multiple stresses after a single or multiple exposure? This is a crucial aspect of using hormetins because in real life situations, unlike the traditional experimental protocols in the laboratory, there is never a single or an isolated stress that one gets exposed to. Therefore, novel experimental approaches, in which multiple combinations of hormetins can be tested, and their individual and combined effects can be fully elucidated, need to be developed.

Thus, physical, nutritional and biological hormetins can be the basis of adopting a hormetic way of life through "stress of choice", especially for human beings. What is important is that the use of single or multiple hormetins must induce one or more SR pathways over and above their basal levels present normally in daily life activities and diet (Stage et al. 2025). Thus, hormesis and hormetins can be both a kind of supplement and a drug for promoting health and longevity, and strengthening the homeodynamic space, robustness and resilience.

References

Agathokleous E, Kitao M, Calabrese EJ (2020) Hormesis: highly generalizable and beyond laboratory. Trends Plant Sci 25:1076–1086

Almiron M, Gomez FH, Sambucetti P, Norry FM (2024) Heat-induced hormesis in longevity is linked to heat-stress sensitivity across laboratory populations from diverse altitude of origin in Drosophila buzzatii. Biogerontology 25:183–190

Annunziata G, Sureda A, Orhan IE, Battino M, Arnone A, Jimenez-Garcia M, Capo X, Cabot J, Sanadgol N, Giampieri F, Tenore GC, Kashani HRK, Silva AS, Habtemariam S, Nabavi SF, Nabavi SM (2021) The neuroprotective effects of polyphenols, their role in innate immunity and the interplay with the microbiota. Neurosci Biobehav Rev 128:437–453

Bana B, Cabreiro F (2019) The microbiome and aging. Annu Rev Genet 53:239–261

Bennett S, Tiollier E, Owens DJ, Brocherie F, Louis JB (2024) Implications of heat stress-induced metabolic alterations for endurance training. Int J Sports Med 45(6):422–435

Bhattacharya S, Rattan SIS (2019) Primary stress response pathways for pre-conditioning and physiological hormesis. In: Rattan SIS, Kyriazis M (eds) The science of hormesis in health and longevity. Academic Press

Boulares A, Jdidi H, Douzi W (2025) Cold and longevity: can cold exposure counteract aging? Life Sci 364:123431

Broskey NT, Marlatt KL, Most J, Erickson ML, Irving BA, Redman LM (2019) The panacea of human aging: calorie restriction versus exercise. Exerc Sport Sci Rev 47:169–175

Calabrese E (2007) Threshold - dose - response model - RIP: 1911 to 2006. Bioessays 29:686–688

Calabrese EJ (2021) Hormesis mediates acquired resilience: using plant-derived chemicals to enhance health. Annu Rev Food Sci Technol 12:355–381

Calabrese EJ, Kozumbo WJ (2021) The hormetic dose-response mechanism: Nrf2 activation. Pharmacol Res 167:105526

Calabrese EJ, Bachmann KA, Bailer AJ et al (2007) Biological stress response terminology: integrating the concepts of adaptive response and preconditioning stress within a hormetic dose-response framework. Toxicol Appl Pharmacol 222:122–128

Calabrese EJ, Agathokleous E, Kozumbo WJ, Stanek EJ 3rd, Leonard D (2018) Estimating the range of the maximum hormetic stimulatory response. Environ Res 170:337–343

Calabrese EJ, Agathokleous E, Kapoor R, Dhawan G, Kozumbo WJ, Calabrese V (2021) Metformin-enhances resilience via hormesis. Ageing Res Rev 71:101418

Campos JC, Marchesi Bozi LH, Krum B et al (2023) Exercise preserves physical fitness during aging through AMPK and mitochondrial dynamics. Proc Natl Acad Sci U S A 120: e2204750120

David E, Bitan R, Atlas S, Wolfson M, Fraifeld VE (2022) Correlative links between natural radiation and life expectancy in the US population. Biogerontology 23:425–430

Davinelli S, De Stefani D, De Vivo I, Scapagnini G (2020) Polyphenols as caloric restriction mimetics regulating mitochondrial biogenesis and Mitophagy. Trends Endocrinol Metab 31: 536–550

Demirovic D, Rattan SIS (2013) Establishing cellular stress response profiles as biomarkers of homeodynamics, health and hormesis. Exp Gerontol 48:94–98

Duregon E, Pomatto-Watson L, Bernier M, Price NL, De Cabo R (2021) Intermittent fasting: from calories to time restriction. Geroscience 43:1083–1092

Epel ES (2020) The geroscience agenda: toxic stress, hormetic stress, and the rate of aging. Ageing Res Rev 63:101167

Farmand S, Du Preez A, Kim C, De Lucia C, Ruepp MD, Stubbs B, Thuret S (2025) Cognition on the move: examining the role of physical exercise and neurogenesis in counteracting cognitive aging. Ageing Res Rev 107:102725

Franco R, Navarro G, Martinez-Pinilla E (2019) Hormetic and mitochondria-related mechanisms of antioxidant action of phytochemicals. Antioxidants (Basel) 8:373

Glossmann HH, Lutz OMD (2019) Metformin and aging: a review. Gerontology 65:1–10

Guisle I, Canet G, Petry S, Fereydouni-Forouzandeh P, Morin F, Kerauden R, Whittington RA, Calon F, Hebert SS, Planel E (2022) Sauna-like conditions or menthol treatment reduce tau phosphorylation through mild hyperthermia. Neurobiol Aging 113:118–130

Ingram DK, Roth GS (2020) Glycolytic inhibition: an effective strategy for developing calorie restriction mimetics. Geroscience 43(3):1159–1169

Kazmierczak-Baranska J, Boguszewska K, Karwowski BT (2020) Nutrition can help DNA repair in the case of aging. Nutrients 12:3364

Labbadia J, Brielmann RM, Neto MF, Lin YF, Haynes CM, Morimoto RI (2017) Mitochondrial stress restores the heat shock response and prevents proteostasis collapse during aging. Cell Rep 21:1481–1494

Le Bourg E (2021) A mild heat stress increases resistance to heat of dFOXO Drosophila melanogaster mutants but less in wild-type flies. Biogerontology 22:237–251

Lee MB, Hill CM, Bitto A, Kaeberlein M (2021) Antiaging diets: separating fact from fiction. Science 374:eabe7365

Luo M, Mai M, Song W, Yuan Q, Feng X, Xia E, Guo H (2022) The antiaging activities of phytochemicals in dark-colored plant foods: involvement of the autophagy- and apoptosis-associated pathways. Int J Mol Sci 23:11038

Madeo F, Carmona-Gutierrez D, Hofer SJ, Kroemer G (2019) Caloric restriction Mimetics against age-associated disease: targets, mechanisms, and therapeutic potential. Cell Metab 29:592–610

Murakami A (2022) Novel mechanisms underlying bioactivities of polyphenols via hormesis. Curr Opin Toxicol 30:100337

Mytych J, Wnuk M, Rattan SI (2016) Low doses of nanodiamonds and silica nanoparticles have beneficial hormetic effects in normal human skin fibroblasts in culture. Chemosphere 148:307–315

Nord C (2023) The balanced brain. Penguin Random House

Oliveira MF, Geihs MA, Franca TFA, Moreira DC, Hermes-Lima M (2018) Is "preparation for oxidative stress" a case of physiological conditioning Hormesis? Front Physiol 9:945

Panossian AG, Efferth T, Shikov AN, Pozharitskaya ON, Kuchta K, Mukherjee PK, Banerjee S, Heinrich M, Wu W, Guo DA, Wagner H (2021) Evolution of the adaptogenic concept from traditional use to medical systems: pharmacology of stress- and aging-related diseases. Med Res Rev 41:630–703

Pinchuk I, Kohen R, Stuetz W, Weber D, Franceschi C et al (2021) Do low molecular weight antioxidants contribute to the protection against oxidative damage? The interrelation between oxidative stress and low molecular weight antioxidants based on data from the MARK-AGE study. Arch Biochem Biophys 713:109061

Rabouw HH, Langereis MA, Anand AA, Visser LJ, De Groot RJ, Walter P, Van Kuppeveld FJM (2019) Small molecule ISRIB suppresses the integrated stress response within a defined window of activation. Proc Natl Acad Sci U S A 116(6):2097–2102

Rackova L, Mach M, Brnoliakova Z (2021) An update in toxicology of ageing. Environ Toxicol Pharmacol 84:103611

Radak Z, Rattan SIS (2024) Exercise, hormesis and ageing: a new section in biogerontology. Biogerontology 26:26

Ramsey KA, Rojer AGM, D'Andrea L, Otten RHJ et al (2021) The association of objectively measured physical activity and sedentary behavior with skeletal muscle strength and muscle power in older adults: a systematic review and meta-analysis. Ageing Res Rev 67:101266

Rattan SIS (2017) Hormetins as drugs for healthy aging. In: Vaiserman M (ed) Anti-aging drugs: from basic research to clinical practice. The Royal Society of Chemistry

Rattan SIS (2018) Biogerontology: research status, challenges and opportunities. Acta Biomed 89: 291–301

Rattan SIS (2020a) Homeostasis, homeodynamics and aging. In: Encyclopedia of biomedical gerontology, vol 2, pp 238–241

Rattan SIS (2020b) Naive extrapolations, overhyped claims and empty promises in ageing research and interventions need avoidance. Biogerontology 21:415–421

Rattan SIS (2022) Physiological hormesis and hormetins in biogerontology. Curr Opin Toxicol 29: 19–24
Rattan SIS (2024) Seven knowledge gaps in modern biogerontology. Biogerontology 25:1–8
Rattan SIS, Kyriazis M (eds) (2019) The science of Hormesis in health and longevity. Academic Press
Rattan SIS, Demirovic D, Nizard C (2018) A preliminary attempt to establish multiple stress response profiles of human skin fibroblasts exposed to mild or severe stress during ageing in vitro. Mech Ageing Dev 170:92–97
Romeo S, Sannino A, Scarfi MR, Massa R, D'Angelo R, Zeni O (2016) Lack of effects on key cellular parameters of MRC-5 human lung fibroblasts exposed to 370 mT static magnetic field. Sci Rep 6:19398
Rossnerova A, Izzotti A, Pulliero A, Bast A, Rattan SIS, Rossner P (2020) The molecular mechanisms of adaptive response related to environmental stress. Int J Mol Sci 21:7053
Salminen A, Kaarniranta K, Kauppinen A (2016) AMPK and HIF signaling pathways regulate both longevity and cancer growth: the good news and the bad news about survival mechanisms. Biogerontology 17:655–680
Sanchez-Morate E, Gimeno-Mallench L, Stromsnes K, Sanz-Ros J, Roman-Dominguez A, Parejo-Pedrajas S, Ingles M, Olaso G, Gambini J, Mas-Bargues C (2020) Relationship between diet, microbiota, and healthy aging. Biomedicines 8:287
Shannon OM, Ashor AW, Scialo F, Saretzki G, Martin-Ruiz C, Lara J, Matu J, Griffiths A, Robinson N, Lilla L, Stevenson E, Stephan BCM, Minihane AM, Siervo M, Mathers JC (2021) Mediterranean diet and the hallmarks of ageing. Eur J Clin Nutr 75(8):1176–1192
Sies H (2021) Oxidative eustress: on constant alert for redox homeostasis. Redox Biol 41:101867
Stage A, Wibaek R, Ronn PF, Bjornsbo KS, Brons C, Allesoe K, Holtermann A, Vaag AA, Linneberg A, Aadahl M (2025) The physical activity health paradox in type 2 diabetes. Am J Prev Med 68:545–554
Thirupathi A, Wang M, Lin JK, Fekete G, Istvan B, Baker JS, Gu Y (2021) Effect of different exercise modalities on oxidative stress: a systematic review. Biomed Res Int 2021:1947928
Wang D, Calabrese EJ, Lian B, Lin Z, Calabrese V (2017) Hormesis as a mechanistic approach to understanding herbal treatments in traditional Chinese medicine. Pharmacol Ther 184:42–50
Wargo JA (2020) Modulating gut microbes. Science 369:1302–1303
Zhang X, Zong B, Zhao W, Li L (2021) Effects of mind-body exercise on brain structure and function: a systematic review on MRI studies. Brain Sci 11
Zhu B, Dacso CC, O'Malley BW (2018) Unveiling "Musica Universalis" of the cell: a brief history of biological 12-hour rhythms. J Endocr Soc 2:727–752

Part IV
Conclusion

Chapter 24
Supplements and Drugs for Health and Longevity: Recommendations, Conclusions and Challenges

Ceren Gezer and Suresh I. S. Rattan

Abstract In this concluding chapter, the strength of evidence), and possible recommendations, for supplements and drugs are summarized, particularly focusing on the human usage. An analysis of all the chapters included in this book shows that although there is a lot of important scientific information available from studies performed on cell culture- and animal-based model systems, SOE from human studies is generally weak due to the limitations of the study design, sample size, limited number of studies, and inconsistent results among independent studies. Therefore, in order to recommend and promote the use of single or multiple supplements and drugs by human beings, much more reliable scientific evidence is still needed, especially with respect to the biomarkers of ageing, health and lifespan. An unconditional and excessive use of the so-called supplements and drugs for achieving a longer and healthier lifespan is presently nothing more than wishful thinking and a marketing tool.

Keywords Supplement · Drug · Ageing · Longevity · Health · Evidence · Recommendations

24.1 Introduction

The aim of compiling this book was to review, evaluate and discuss the plethora of marketed supplements and drugs in the context of the changing paradigm from anti-ageing to healthy longevity. The two categories, supplements and drugs, represent two distinct approaches towards ageing interventions. While supplements are mainly the natural compounds and conditions which are usually a regular part of the

---

C. Gezer
Department of Nutrition and Dietetics, Faculty of Health Sciences, Eastern Mediterranean University, Famagusta, Cyprus
e-mail: ceren.gezer@emu.edu.tr

S. I. S. Rattan (✉)
Department of Molecular Biology and Genetics, Aarhus University, Aarhus, Denmark
e-mail: rattan@mbg.au.dk

S. I. S. Rattan, C. Gezer (eds.), *Supplements and Drugs for Healthy Longevity*, Healthy Ageing and Longevity 22, https://doi.org/10.1007/978-3-032-11608-6_24

lifestyle, drugs are generally the target-oriented formulations against a problem or a pathological situation. For example, nutrients, and micronutrients, including vitamins, that are normally obtained from food and diet, can be taken in extra (supplementary) quantities, both to compensate for any deficiency and to attain any additional benefits. Drugs, on the other hand, are defined as natural or synthetic substances designed and used for the diagnosis, treatment, or prevention of diseases or other maladies, and are primarily used in clinical practice (Visioli 2022).

So, the critical question is whether the use of supplements and drugs could be recommended as a strategy FOR promoting health and longevity without specifically targeting AGAINST any diseases. For this purpose, we have used a position paper, published in 2018, by the Academy of Nutrition and Dietetics (USA) (https://www.eatright.org/), providing its members the "Evidence-Based Nutrition Practice Guidelines (EBNPGs)" for systematic reviews, position papers, and practice papers. These resources were designed to develop a strength of evidence (SOE) grading system depending on the "Evidence Analysis Library Conclusion Grading System" (Handu et al. 2018).

The SOE grading system includes 5 items, such as the quality, consistency, the extent of the scientific research, and the clinical impact and generalizability. According to this grading system for determining SOE, Grade I refers to Good, Grade II refers to Fair, Grade III refers to Limited, Grade IV refers to Expert opinion only, and Grade V refers to grade not assignable. Further description of the grading system includes that the "Grade I-Good" means that there is a solid, reproducible and trustworthy scientific evidence; "Grade II-Fair" includes studies with strong design and methodology with minor concerns and inconsistencies, studies by independent researchers with minor concerns of statistical and clinical significance, and minor concerns for generalizability. Similarly, the "Grade III-Limited" means including studies with weak design, limited sample size or limited number of studies and inconsistent results of independent studies along with small size effect of statistical and clinical significance, and major concerns for generalizability. Grade IV means not including any studies mainly due to inconsistent and unsubstantiated subjective data (Handu et al. 2018).

We have evaluated all chapters included in this book with the above scope, and have graded them for two subtypes of research strategies: cell culture-/animal-based studies, and human-based studies (Table 24.1). This analysis performed by us complements those provided by the authors themselves in their respective chapters, where they summarize the current status of different supplements and drugs, and make recommendations, particularly focusing on the human usage, appropriate dosage, and generalizability.

Table 24.1 Evaluating the strength of evidence (SOE) and possible recommendations for human usage of supplements and drugs discussed in this book

		SOE Grade		Recommendation
Chapter number/title	Compounds/Compound sources	Cell culture/ animal model-based studies	Human-based studies	
Chapter 1 Vita-amines, 'molecules for life': do they live up to their name?	Vitamins A, B1, B2, B3, B5, B6, B7, B9, B12, C, D, E, K	II	III	Natural food sources should be preferred; supplements recommended only in case of deficiency
Chapter 2 Micronutrients and nutraceuticals for health and longevity	Pyrroloquinoline quinone, Ergothioneine, Selenium, Zinc, Magnesium	II	III	Natural food sources should be preferred; supplements recommended only in case of deficiency
Chapter 3 Pro-, post- and pre-biotics for health and longevity	Prebiotics: Xylooligosaccharides, Inulin, Oligofructose, Arabinoxylan, Mannan, Arabinogalactan, Glucomannan, Anthocyanins Gerobiotics: *Lactobacillus casei* *Streptococcus thermophilus* *Bifidobacterium longum* *Bifidobacterium animalis* *Lactobacillus plantarum* *Bifidobacterium breve*	III	III	Natural food sources should be preferred; more studies needed for the identification of probiotic strains and optimal doses
Chapter 4 Hormonal supplements for health and longevity	Melatonin, Phytomelatonin, Dehydroepiandrosterone, Phytoestrogen, Insulin-like growth factor-I	II	III	Not recommended for general, non-clinical usage; several issues yet to be resolved, for example off label use, gender, pathophysiological disorders, and ethnic variations
Chapter 5 NAD^+ and its precursors for health and longevity	Nicotinamide riboside (NR), Nicotinamide mononucleotide (NMN)	II	III	Not recommended unconditionally; large variations in product concentrations; lack of authorization from any governmental authorities

(continued)

Table 24.1 (continued)

Chapter number/title	Compounds/Compound sources	SOE Grade		Recommendation
		Cell culture/ animal model-based studies	Human-based studies	
Chapter 6 CoQ10 supplementation for healthy longevity	CoQ10	III	IV	Supplementation of CoQ_{10} in old persons is recommended under medical advice
Chapter 7 Iodine for health and longevity	Iodine	III	III	Casual supplementation is not recommended, but iodine fortification programmes are recommended for public health
Chapter 8 *Tinospora cordifolia* for health and longevity	*Tinospora cordifolia* extracts/formulations	III	III	Not recommended; several knowledge gaps remain to be filled
Chapter 9 Neuronutrition: evaluating the efficacy of natural compounds in neurodiversity for brain health	Boswellia *Coriolus versicolor* *Hericium erinaceus*	III	IV	Not recommended; lack of sufficient standardization, pharmacokinetic data, and human trials to support broader use
Chapter 10 Ayurvedic supplements for healthy brain and longevity	Chyawanprash, Triphala rasayan, Brahmi ghrita, Ashwagandharishta, Saraswatarishta, Amalaki rasayana, Medhya rasayana	III	III	Although used in the Indian folk tradition, these cannot be recommended universally unconditionally
Chapter 11 Natural healthy-aging agents: wisdom from traditional Chinese medicine	Asian ginseng, Astragalus, Cistanche, Angelica sinensis, Goji berry, Dendrobium, Rosa rugosa, Poria, Dried tangerine peel, Panax notoginseng	II	III	Although used in the Chinese folk tradition, these cannot be recommended universally unconditionally
Chapter 12 African-medicinal formulations for health and longevity	More than 20 plants listed	III	IV	Although used in the African folk tradition, these cannot be recommended universally unconditionally

(continued)

Table 24.1 (continued)

Chapter number/title	Compounds/Compound sources	SOE Grade		Recommendation
		Cell culture/ animal model-based studies	Human-based studies	
Chapter 13 Greek medicinal phytochemicals with anti-ageing properties	Oleuropein, Silymarin, Chlorogenic acid Crocin, Secoisolariciresinol diglucoside A-Terpineol, Linolenic acid, Pinocarveol	II III III	III III IV	Not recommended for general use since several knowledge gaps hinder full clinical translation
Chapter 14 Mediterranean-diet based supplements for healthy longevity	Gallic Acid, Ellagic Acid, Chlorogenic Acid, Ferulic acid, Kaempherol, Quercetin, Myrcetin, Naringenin, Naringin, Hesperetin, Hesperidin, Apigenin, Luteolin, Delphinidin, Cyanidin, Pelargonidin, Malvidin, Peonidin, Petunidin, Tyrosol, Secoisolariciresinol, Hydroxy Tyrosol, Oleoropein	II	III	Natural food sources should be preferred; supplementation lacks sufficient standardization, pharmacokinetic data, and human trials to support broader use
Chapter 15 Geroprotective effects of plant-extracts	Green tea extract, Honeysuckle, Bilberry, Aronia extract, Maral root	II	III	Not recommended as general supplements since several knowledge gaps hinder full clinical translation
Chapter 16 Metformin as a senotherapeutic agent	Metformin	II	III	Planned clinical trials awaited integrating circadian biomarkers, drug interaction profiles, and principles of geriatric pharmacology for personalized use in aging populations
Chapter 17 Effects of procaine, Gerovital H3, on some chronic conditions often encountered in older people	Procaine, Gerovital H3	II	II	Recommended conditionally and mainly as an adjunct to classical therapies

(continued)

Table 24.1 (continued)

Chapter number/title	Compounds/Compound sources	SOE Grade		Recommendation
		Cell culture/animal model-based studies	Human-based studies	
Chapter 18 Medicinal peptide drugs and Khavinson Peptides	Khavinson peptides®	II	II	Recommended for conditional therapeutic use
Chapter 19 Calorie-restriction mimetics as drugs for healthy longevity	Resveratrol, Metformin, Spermidine, Berberine	II	III	Not recommended since lack of sufficient standardization, pharmacokinetic data, and human trials to support broader use.
Chapter 20 Dasatinib and quercetin synolytics for health and longevity	Dasatinib+Quercetin	II	III	Not recommended due to very limited information about optimal doses, frequency, duration of treatment, and interactions with other medications
Chapter 21 Carotenoids for healthy longevity	Carotenoids	II	III	Natural food sources are recommended; however, lutein and zeaxanthin supplementation may be recommended for the treatment of age-related macular degeneration.
Chapter 22 Curcumin and resveratrol for healthy longevity	Curcumin, Resveratrol	II	III	Natural food sources are recommended; presently lack of sufficient standardization, pharmacokinetic data, and human trials to support broader use.
Chapter 23 Hormesis and hormetins as drugs for health and longevity	Life style-based physical, mental and nutritional hormetins	III	IV	A general advice with principles of hormetic way of life by "stress of choice" to support life mechanisms

SOE Strength of Evidence, *Grade I* Good, *Grade II* Fair, *Grade III* Limited, *Grade IV* Expert opinion only, *Grade V* Grade not assignable

24.2 Conclusion and Challenges

It seems that the overcrowded market selling hundreds of the so-called supplements and drugs for health, anti-ageing, and lifespan extension is, at present, mostly a wishful and naïve extrapolation of the scientific data gathered from the cell culture- and animal-based studies. Whatever the scientific strengths and significance of such experimental studies, extrapolating these to human beings is one of the major knowledge gaps and challenges in biogerontology (Rattan 2024). Our analysis (Table 24.1) shows that the SOE from human studies is generally of the grade III type, meaning that these studies are with weak design, limited sample size, limited number of studies, and inconsistent results of independent studies.

We do not mean to imply that the well-curated database of the so-called geroprotectors, which lists almost 260 compounds (https://geroprotectors.org/) has no scientific reliability and no potential human applications. Each one of the tested compounds may have various biological effects and some health- and longevity-benefits within the scope and limitations of the experimental model system used in research (Gao et al. 2023; Leri et al. 2020; Liu et al. 2024; Park et al. 2024). Furthermore, a lot is known about the basic molecular mechanisms through which such compounds seem to work, such as either directly acting on the molecular target (s), or indirectly acting as hormetins by inducing physiological hormesis, that in turn activates one or several stress response-mediated repair and defense processes (Rattan 2017). A majority of the plant-derived supplements and drugs containing polyphenols, flavonoids and other active molecules are hormetins.

In conclusion, the information, knowledge and wisdom collated from the 23 chapters included in this book is summarized in the following 7 points:

1. In the absence of any deficiency detected by clinical testing, all nutrients and micronutrients can be adequately obtained from the food and diet, without any need for supplementation (Fekete et al. 2023; Martel et al. 2019; Tessier et al. 2025).
2. Taking higher doses of one or more supplements may, at best, give only some psychological satisfaction, and at worst, can be harmful in becoming toxic and mutagenic, especially over long-term usage (Kahraman et al. 2023; Murakami 2020, 2022; Rackova et al. 2021; Rattan 2022).
3. The antagonistic and/or synergistic effects of combining various supplements and drugs are poorly understood at present, and serious precautions should be carefully considered for their use (Duan et al. 2022; Rackova et al. 2021).
4. Supplements and drugs for health are no replacement for balanced food. Our traditional dietary habits are mostly geo-culturally evolved, and there is neither a perfectly healthy food and dietary practice nor is there an absolutely harmful one (Rattan and Kaur 2022) (except for the recent emergence of ultra-processed foods (Dai et al. 2024; Lane et al. 2024; Vitale et al. 2024)).
5. Food matrix, bioavailability, nutrigenetics, and chronobiology are among the factors which determine the ultimate effects of supplements and drugs (Bertelli et al. 2021; Truzzi et al. 2021).

6. The gut microbiota, as a footprint of the health of an individual, is affected positively or negatively by supplements and drugs, along with other factors such as the normal diet, genetic profile, and the physical and mental health status (Chen et al. 2024; Todorova et al. 2024).
7. Hormesis and a hormetic way of life, incorporating principles of pleasure, moderation and variety, can be the basis for using any supplements and drugs for health and longevity (Rattan 2022).

Finally, we think that at present using one or more nutritional and other supplements and drugs for maintaining, improving and recovering health is mostly an act of faith and hope, rather than on strong scientific evidence-based data. Generally, at low doses, such compounds, extracts and formulations, even if only tested in cell culture- or animal-based model systems, may not have any harmful toxic effects on the human body. However, in order to be sure of their wider dose-responses, and negative and positive effects over a longer period of usage, much more scientific research is necessary. Such research will require testing multiple combinations of compounds in multiple situations, and under strict agreed norms of scientific investigations. In addition, ensuring product quality, accurate labelling, and pre-market evaluation for safety and efficacy is extremely important. Furthermore, interdisciplinary collaboration among biogerontologists, clinical researchers, nutrition scientists, and regulatory bodies must be encouraged. Moving on from hype, to hope to reality is a big challenge yet to be realized systematically.

References

Bertelli A, Biagi M, Corsini M, Baini G, Cappellucci G, Miraldi E (2021) Polyphenols: from theory to practice. Foods. https://doi.org/10.3390/foods10112595

Chen JC, Wang R, Wei CC (2024) Anti-aging effects of dietary phytochemicals: from Caenorhabditis elegans, Drosophila melanogaster, rodents to clinical studies. Crit Rev Food Sci Nutr. https://doi.org/10.1080/10408398.2022.2160961

Dai S, Wellens J, Yang N, Li D, Wang J, Wang L, Yuan S, He Y, Song P, Munger R, Kent MP, MacFarlane AJ, Mullie P, Duthie S, Little J, Theodoratou E, Li X (2024) Ultra-processed foods and human health: an umbrella review and updated meta-analyses of observational evidence. Clin Nutr 43:1386–1394. https://doi.org/10.1016/j.clnu.2024.04.016

Duan H, Pan J, Guo M, Li J, Yu L, Fan L (2022) Dietary strategies with anti-aging potential: dietary patterns and supplements. Food Res Int. https://doi.org/10.1016/j.foodres.2022.111501

Fekete M, Lehoczki A, Tarantini S, Fazekas-Pongor V, Csípő T, Csizmadia Z, Varga JT (2023) Improving cognitive function with nutritional supplements in aging: a comprehensive narrative review of clinical studies investigating the effects of vitamins, minerals, antioxidants, and other dietary supplements. Nutrients. https://doi.org/10.3390/nu15245116

Gao L, Liu X, Luo X, Lou X, Li P, Li X, Liu X (2023) Antiaging effects of dietary supplements and natural products. Front Pharmacol. https://doi.org/10.3389/fphar.2023.1192714

Handu D, Moloney L, Rozga MR, Cheng F, Wickstrom D, Acosta A (2018) Evolving the academy position paper process: commitment to evidence-based practice. J Acad Nutr Diet 118. https://doi.org/10.1016/j.jand.2018.07.004

Kahraman C, Kaya Bilecenoglu D, Sabuncuoglu S, Cankaya IT (2023) Toxicology of pharmaceutical and nutritional longevity compounds. Expert Rev Mol Med. https://doi.org/10.1017/erm.2023.18

Lane MM, Gamage E, Du S, Ashtree DN, McGuinness AJ, Gauci S, Baker P, Lawrence M, Rebholz CM, Srour B, Touvier M, Jacka FN, O'Neil A, Segasby T, Marx W (2024) Ultra-processed food exposure and adverse health outcomes: umbrella review of epidemiological meta-analyses. BMJ. https://doi.org/10.1136/bmj-2023-077310

Leri M, Scuto M, Ontario ML, Calabrese V, Calabrese EJ, Bucciantini M, Stefani M (2020) Healthy effects of plant polyphenols: molecular mechanisms. Int J Mol Sci. https://doi.org/10.3390/ijms21041250

Liu Y, Fang M, Tu X, Mo X, Zhang L, Yang B, Wang F, Kim Y-B, Huang C, Chen L, Fan S (2024) Dietary polyphenols as anti-aging agents: targeting the hallmarks of aging. Nutrients 16. https://doi.org/10.3390/nu16193305

Martel J, Ojcius DM, Ko YF, Ke PY, Wu CY, Peng HH, Young JD (2019) Hormetic effects of phytochemicals on health and longevity. Trends Endocrinol Metab. https://doi.org/10.1016/j.tem.2019.04.001

Murakami A (2020) Hormesis-mediated mechanisms underlying bioactivities of phytochemicals. Curr Pharmacol Rep. https://doi.org/10.1007/s40495-020-00235-4

Murakami A (2022) Novel mechanisms underlying bioactivities of polyphenols via hormesis. Curr Opin Toxicol. https://doi.org/10.1016/j.cotox.2022.02.010

Park SH, Lee DH, Lee DH, Jung CH (2024) Scientific evidence of foods that improve the lifespan and healthspan of different organisms. Nutr Res Rev 37:169–178. https://doi.org/10.1017/S0954422423000136

Rackova L, Mach M, Brnoliakova Z (2021) An update in toxicology of ageing. Environ Toxicol Pharmacol 84. https://doi.org/10.1016/j.etap.2021.103611

Rattan SIS (2017) Hormetins as drugs for healthy aging. RSC Drug Discov Ser. https://doi.org/10.1039/9781782626602-00170

Rattan SIS (2022) Physiological hormesis and hormetins in biogerontology. Curr Opin Toxicol. https://doi.org/10.1016/j.cotox.2022.01.001

Rattan SIS (2024) Seven knowledge gaps in modern biogerontology. Biogerontology. https://doi.org/10.1007/s10522-023-10089-0

Rattan SIS, Kaur G (2022) Nutrition, food and diet in health and longevity: we eat what we are. Nutrients. https://doi.org/10.3390/nu14245376

Tessier AJ, Wang F, Korat AA, Eliassen AH, Chavarro J, Grodstein F, Li J, Liang L, Willett WC, Sun Q, Stampfer MJ, Hu FB, Guasch-Ferré M (2025) Optimal dietary patterns for healthy aging. Nat Med 31:1644–1652. https://doi.org/10.1038/s41591-025-03570-5

Todorova MN, Savova MS, Mihaylova LV, Georgiev MI (2024) Nurturing longevity through natural compounds: where do we stand, and where do we go? Food Front. https://doi.org/10.1002/fft2.350

Truzzi F, Tibaldi C, Zhang Y, Dinelli G, D′Amen E (2021) An overview on dietary polyphenols and their biopharmaceutical classification system (Bcs). Int J Mol Sci. https://doi.org/10.3390/ijms22115514

Visioli F (2022) Science and claims of the arena of food bioactives: comparison of drugs, nutrients, supplements, and nutraceuticals. Food Funct. https://doi.org/10.1039/d2fo02593k

Vitale M, Costabile G, Testa R, D'Abbronzo G, Nettore IC, Macchia PE, Giacco R (2024) Ultra-processed foods and human health: a systematic review and meta-analysis of prospective cohort studies. Adv Nutr. https://doi.org/10.1016/j.advnut.2023.09.009

Index

S. I. S. Rattan, C. Gezer (eds.), *Supplements and Drugs for Healthy Longevity*, Healthy Ageing and Longevity 22, https://doi.org/10.1007/978-3-032-11608-6

M

N

O

P

R

S

T

V

X

GPSR Compliance
The European Union's (EU) General Product Safety Regulation (GPSR) is a set of rules that requires consumer products to be safe and our obligations to ensure this.

If you have any concerns about our products, you can contact us on

ProductSafety@springernature.com

In case Publisher is established outside the EU, the EU authorized representative is:

Springer Nature Customer Service Center GmbH
Europaplatz 3
69115 Heidelberg, Germany

www.ingramcontent.com/pod-product-compliance
Lightning Source LLC
Chambersburg PA
CBHW061939160926
51622CB00003B/30
* 9 7 8 3 0 3 2 1 1 6 0 7 9 *